Jia-Su Wang and Su-Yu Wang

# High Temperature Superconducting Magnetic Levitation

DE GRUYTER

北京大學出版社
PEKING UNIVERSITY PRESS

**Mathematics Subject Classification 2010**
**Physics and Astronomy Classification Scheme 2010**
Primary: 82D55; Secondary: 82D44

**Authors**
Jia-Su Wang
Applied Superconductivity Laboratory (ASCLab) of Southwest Jiaotong University
Lan Fengyuan Court, No. 816 Campus Road, Xipu Town, PI Du District, Chengdu, Sichuan 611731, China.
asclaban@gmail.com

Su-Yu Wang
Applied Superconductivity Laboratory (ASCLab) of Southwest Jiaotong University
Lan Fengyuan Court, No. 816 Campus Road, Xipu Town, PI Du District, Chengdu, Sichuan 611731, China.
jsywang@swjtu.edu.cn

ISBN 978-3-11-053818-2
e-ISBN (PDF) 978-3-11-053843-4
e-ISBN (EPUB) 978-3-11-053836-6
Set-ISBN 978-3-11-053844-1

**Library of Congress Cataloging-in-Publication Data**
A CIP catalog record for this book has been applied for at the Library of Congress.

**Bibliographic information published by the Deutsche Nationalbibliothek**
The Deutsche Nationalbibliothek lists this publication in the Deutsche Nationalbibliografie; detailed bibliographic data are available on the Internet at http://dnb.dnb.de.


Typesetting: Compuscript Ltd., Shannon, Ireland
Printing and binding: CPI books GmbH, Leck
Cover image: Nikada / E+ / Getty Images
♾ Printed on acid-free paper
Printed in Germany

www.degruyter.com

# Contents

# Preface

Levitation or suspension is a phenomenon that an object occupies a fixed position in a gravitational field without any direct physical contact. Scientists have found and suggested a number of suspension methods, for example aerodynamic, acoustic, optical, electric, magnetic, and radio-frequency (RF) suspension. The levitation phenomenon attracted not only wide attention of the public, but also the interest of scientists and engineers. Magnetic levitation (Maglev) is a promising one in various suspension phenomena, and it is implemented using the principle of independent suspension with magnetic field. Maglev technologies have a series of potential advantages including high or ultra high speed, low energy consumption, environment friendly, low maintenance, moderate footprint, low operating costs, etc. Maglev is currently attracting worldwide interest for both high-speed intercity vehicles and low-speed urban transit vehicles, especially high- or ultra-high-speed intercity traffic. After nearly a century long development, Maglev finally began to penetrate ground transportation markets and presented the possibility of rapid development. Traditional and superconducting Maglev have different characteristics. The high-temperature superconducting (HTS) bulk Maglev can realize stable levitation without any active control. The physical characteristics of self-stabilization of HTS bulk Maglev is a unique phenomenon in nature. This is especially attractive for practical applications of Maglev transportation. Experts worldwide are trying to complete the HTS Maglev vehicle test. This book is devoted to HTS Maglev. Fundamentals of superconductivity are introduced in Chapter 1. Superconductivity is a deep and complex subject. We can only provide a brief overview about basic superconductivity in this book, which is required to understand how HTS Maglev works. In Chapter 2, we discuss the superconducting materials. After a brief introduction of superconducting materials, we will focus on the HTS bulk and its thermal, mechanical, and trapped flux properties, which are all essential in HTS Maglev applications. Magnetic levitation will be discussed in Chapter 3. To help the reader understand Maglev transportation, we start with the value of the ground track-based traffic system and the limitations of the traditional railway transport, then move on to levitation phenomenon, magnetic levitation concepts, and non-superconducting magnetic levitation applications like the permanent magnet Maglev and normal conductive Maglev. In addition to the advantages of permanent magnet and normal conductive Maglev, superconducting Maglev gains further improvement including more energy-saving, environmental protection, high speed, and most important, the possibility of ultra high speed. The superconducting Maglev, LTS and HTS Maglev, will be discussed in Chapter 4. Various types of Maglev have different performance. In the constant pursuit of the "perfect" Maglev transportation, the new HTS Maglev train stands out to be one of the best candidates. With the previous Maglev testing method, the YBCO bulk is placed in a simple liquid nitrogen container which is below the permanent magnet. Other than

https://doi.org/10.1515/9783110538434-202

being highly inaccurate, this simple method is not even close to real conditions of Maglev applications. To to investigate the levitation forces, guidance forces, as well as the cross stiffness, dynamic rigidity, relaxation time, etc. of the HTS Maglev over a PM guideway (PMG), we developed several Maglev measurement systems from 1999 to 2014. The measurement systems are different from the earlier system that the measured HTS bulk is above the PMG. As a result, a cryogenic vessel with a super-thin bottom is needed. To investigate HTS Maglev bearing and flywheel energy storage technology, the HTS Maglev bearing measurement system was designed and successfully developed in 2014. The principles, methods, structure, functions, and specifications of the several HTS Maglev measurement systems are presented in detail in Chapter 5. Several HTS Maglev measurement systems play an important role in the research and development of the HTS Maglev. The subsequent chapters will present these research results and their applications in developing the prototype of HTS Maglev. The earlier study was carried out in the HTS Maglev measurement system SCML-01, and according to these massive research results, the first manned HTS Maglev test vehicle in the world, "Century", was successfully developed on December 31, 2000. In Chapter 6, the research results and properties of the entire Maglev test vehicle are presented in detail. After the success of the first manned HTS Maglev vehicle, the theoretical and experimental researches of HTS Maglev are carried out by the authors' team unremittingly. The research results of 2D and 3D numerical simulations of HTS Maglev in ASCLab (Applied Superconductivity Laboratory) are presented in Chapter 7. In Chapter 8, new research progress about HTS Maglev is presented. These include the new research results on HTS Maglev properties between HTS bulks and PMG and the ETT HTS Maglev vehicle developed using the new results. Research results and the experimental prototype of HTS Maglev bearing and 5-kW h flywheel energy storage system are presented in Chapter 9. HTS Maglev launch technology prototype system and HTS linear synchronous motors are presented in Chapter 10. The book presents the research findings from our team of the HTS Maglev R&D, and I wish to thank all members of our team for their hard work and selfless dedication. We especially thank our previous graduate students Dr. Xiao-Rong Wang of Lawrence Berkeley National Laboratory, Dr. Hong-Hai Song of Michigan State University, Dr. Zhong-You Ren of Siemens Shenzhen Magnetic Resonance Ltd., research scientist He Jiang of Aerospace Science Research Institute, Dr. Min Zhu of Institute of Electronics of Chinese Academy of Sciences, Dr. Wei Wang of Cambridge University, Dr. Rong Zeng, Dr. Long-Cai Zhang, Dr. Lu Liu, Dr. Si-Ting Pan, and others for offering unpublished papers or summaries of their original research work on HTS Maglev. The birth of "Century" 16 years ago is the reason why we wrote this book. Therefore, we would like have to thank Zhong-You Ren, He Jiang, Min Zhu, Xiao-Rong Wang, Chang-Yan Deng, You-Wen Zeng, Qi-Xue Tang, Hi-Yu Huang, Xu-Ming Shen, and others for their important and substantive contributions to the development of "Century". We are very grateful to Hong-Tao Ren and his group, the Beijing General Research Institute for Nonferrous Metals, for providing high-quality HTS bulk materials, and

Li-Ming Pan and his group, the Yuxin Machinery Co., Ltd., for providing help for the rectangle-shape thin-bottom liquid nitrogen vessel. This book is based on the research work of the ASCLab group that is composed of the above-mentioned researchers and ACSLab alumni including Dr. Min-Xian Liu, Hua Jing, Qing-Yong He, Yu-Jie Qing, Dong-Hui Jiang, Chang-Qing Ye, and Xing-Zhi Wang, whose work we are also thankful for. We also thank my colleagues and friends outside our team for discussions and exchanges about the HTS Maglev during their visits to our laboratory: Prof. Y. Iwasa of the Francis Bitter Magnet Laboratory of MIT in USA, Dr. Frank N. Werfel of the Adelwitz Technologiezentrum GmbH (ATZ), Dr. E. H. Brandt of the Institut für Physik, and Dr. Wolfgang Gawalek of the IPHT-Jena in Germany, Prof. M. Izumi of the Tokyo University of Marine Science and Technology, Prof. Masato Murakami of the Shibaura Institute of Technology, Dr. Hiroyuki Fujimoto of the Railway Technical Research Institute in Japan, Prof. Richard M. Stephan of the Federal University of Rio de Janeiro in Brazil, and Prof. Du-Xing Chen of the Universitat Autonoma Barcelona in Spain. During his visit, Prof. Iwasa presented draft chapters of his book *Case Studies in Superconducting Magnets, Second Edition* in three lectures. Not long after his visit to ASCLab, Dr. Brandt got terminally ill. He gave his entire research archive to us before he passed away. We will always remember the important contribution he made to superconductor research and his devotion to scientific research. The continuous collaboration between Werfel's group and our team has been highly productive. As part of the project of science and technology cooperation between the government of China and Germany in 2004–2005, extensive and in-depth exchanges between our team and Dr. Ludwig Schultz's group of the IFW was conducted, and this cooperation promoted the HTS Maglev development. Prof. Archie Campbell of Cambridge University in the UK sent the details of his research findings on Maglev in 2001. We also acknowledge the National High-tech R&D Program (National 863 Program) and the National Natural Science Foundation in China for their long-term research grants for the HTS Maglev project from 1990 to 2010. Last but not least, we thank Wei Wang at Cambridge University and Xin-Ju Wang at Intel for proofreading Chapters 1 to 6. We highly appreciate Prof. D. G. Naugle at Texas A&M University for reviewing this book rigorously.

Jia-Su Wang
Su-Yu Wang
Chengdu, P. R. China
January 5, 2016

Jia-Su Wang and Su-Yu Wang

# 1 Fundamentals of superconductivity

## 1.1 Introduction

The electrical resistance of a metallic conductor decreases gradually as temperature drops. Even near absolute zero, metallic conductor still has residual resistance. For a superconductor, the resistance drops abruptly to zero below its critical temperature $T_c$.

Conventional superconductors are those that can be explained by the Bardeen-CooperSchrieffer (BCS) theory or its derivatives. Superconductors that cannot be explained by the BCS theory are defined as the unconventional superconductors.

Superconductivity was discovered first in mercury by Onnes in 1911, [1] and the perfect diamagnetic phenomenon was discovered in 1933. Various superconducting theories appeared in succession. The characteristic parameters of superconductors can be well described by phenomenological macroscopic theories. These are the well-known London theory [2] and the Ginsburg-Landau theory. [3] The latter has been extended by Abrikosov [4] and Gorkov [5] and is therefore often called the GLAG theory. In 1957, Bardeen et al. [6] proposed the BCS superconducting microscopic theories. This theory assumed that the charge carriers of the superconducting currents are the bound electron pairs (Cooper pairs) and that there exists an energy gap between the normal and superconductive states.

High-$T_c$ superconductor (HTSC) is a family of superconducting ceramic materials. "High" temperature in this context refers to the critical temperature which is higher than the boiling point of liquid nitrogen (77 K or –196°C). The construction and operational cost of the cryogenic system for holding the liquid nitrogen are much cheaper than that for liquid helium. The HTSC offers the highest transition temperatures of all superconductors. Thus, the HTSC is dominant for commercial applications since liquid nitrogen is much easier to maintain. For these reasons, the HTSC is particularly attractive for the practical engineering applications.

The London and GLAG theories form a complete set of superconducting macroscopic theories and are always used to explain the magnetic properties of type II HTSC. However, these macroscopic theories are approximations of the BCS microscopic theory. Conventional superconductors can be properly explained by the BCS theory; however, to date, there is no widely accepted theory to explain the superconductivity of HTSC materials as well as other unconventional superconductors.

Superconductors can be classified into types I and II superconductors. Type I superconductors have a single critical field $H_c$, above which their superconductivity disappears. Type II superconductors have two critical fields, lower critical field $H_{c1}$ and

https://doi.org/10.1515/9783110538434-001

upper critical field $H_{c2}$, between which they allow partial penetration of the magnetic field (mixed state).

In the past few decades, superconductivity has been described in detail in many works. This chapter only briefly introduces the essential features that directly relate to the theme of this book, high-temperature superconducting magnetic levitation (Maglev), in order to provide the readers with some necessary background knowledge for later chapters. For further details about superconductivity, related bibliographies are listed at the end of this chapter.

## 1.2 Discovery of superconductivity

In 1908, Heike Kamerlingh Onnes found the way to liquefy helium and to reach temperatures as low as 4 K. Superconductivity was discovered first in mercury by Onnes in 1911, [1] in the University of Leiden, Netherlands. The experiment was performed on October 26, 1911, with the historic plot showing that the resistance dropped abruptly to zero when the material was cooled below its critical temperature at 4.2 K. [7] A type I superconductor is a material that exhibits two features, namely zero electrical resistance and perfect diamagnetism, when it is cooled below critical temperature $T_c$. The feature of the zero electrical resistance is that the electrical resistance of the material suddenly drops to zero as the temperature decreases. The ratio of the resistances between normal state and the superconducting state was measured to exceed $10^{14}$. In 1933, Meissner and Ochsenfeld [8] found that when a superconducting sphere was cooled below its transition temperature in an applied magnetic field, it excluded the magnetic flux. This perfect diamagnetic phenomenon is called the Meissner-Ochsenfeld effect.

In 1913, the superconductivity of lead was found at 7 K. In 1937, at Kharkov, Shubnikov et al. [9] experimentally discovered the phenomenon of type II superconductivity in single-crystal, single-phase superconducting alloys. In 1941, the superconductivity of niobium nitride was found at 16 K. However, even though thousands of superconductors were discovered for many years, the highest critical temperature remained below 20 K, generally defined as low-temperature superconductor (LTSC). Among LTSCs, only Nb-Ti and $Nb_3Sn$ are deemed to have promising engineering applications, but rather low operating temperatures (4.2 K, liquid helium) are required. After the discovery of $Nb_3Ge$ ($T_c$ = 23.2 K) in 1973, the critical temperature of LTSC did not increase for more than 10 years. Nearly 30 years later, in 2001, magnesium diboride ($MgB_2$) was discovered. which has a transition temperature of 39 K. The $MgB_2$ has the highest critical temperature among conventional superconductors.

In 1986, Bednorz and Müller, [10] of IBM Research Laboratory near Zurich, Switzerland, discovered superconductivity in a lanthanum-based cuprate perovskite material ($La_{2-x}Ba_xCuO_4$) with a transition temperature of 35 K.

In January of 1987, Chu, [11] from the University of Houston, and his associates Wu and Ashburn, from the University of Alabama at Huntsville, discovered the ceramic yttrium barium copper oxide ($YBa_2Cu_3O$) superconductor with a critical temperature above the temperature of liquid nitrogen (77 K). Meanwhile, Zhao et al., [12] from the Institute of Physics of the Chinese Academy of Sciences, synthesized the same YBaCuO HTSCs in February of 1987.

For a normal conductor, the electric current can be visualized as a fluid of electrons moving through the ionic lattice. The electrons constantly collide with the ions in the lattice, and during each collision, some of the dynamic energy carried by the electrons is absorbed by the lattice and converted into heat. This is the phenomenon of electrical resistance in the normal conductors (non-superconductivity). The resistance in normal metals decreases continuously with the decrease in temperature and reaches a constant value, known as the residual resistance that arises due to the presence of impurities. However, the DC electrical resistance in a superconductor drops abruptly to an immeasurably small value, i.e. almost zero below the critical temperature, $T_c$.

For a conventional superconductor, the superconducting carriers consist of bound pairs of electrons known as Cooper pairs. This pair is constrained by an attractive force between electrons through the exchange of phonons. The Cooper pair fluid is a superfluid, which means it can flow without energy dissipation.

This zero resistance phenomenon is completely different from the steadily decreasing resistance of normal metals. This is an important characteristic parameter of superconductors, i.e. the critical temperature $T_c$. Although the theoretical transition range $\Delta T$ from the normal to the superconducting state is very sharp, it sometimes occurs gradually. For a pure and homogeneous superconductor, the transition range $\Delta T$ can be in the order of millikelvin.

The conventional superconducting theory, for example, the BCS theory, cannot fully explain the phenomenon of HTS. The best available theory model of high-temperature superconductivity is still somehow incomplete.

Superconductivity describes a thermodynamic state below certain critical temperature $T_c$. The several parameters that characterize the superconductivity are critical temperature $T_c$, energy gap $E_g$, coherence length $\zeta$, penetration depth $\lambda_L$, Ginzburg-Landau (GL) parameter $\kappa$, critical current density $J_c$, thermodynamic critical field $H_c$, lower critical field $H_{c1}$, upper critical field $H_{c2}$, and so on. Generally, after knowing or measuring a few parameters, one can derive the other related parameters.

## 1.3 Meissner-Ochsenfeld effects

The Meissner-Ochsenfeld effect [8] (often abbreviated as the Meissner effect) is one important signature of superconductivity. Perfect diamagnetism is observed due to expulsion of the magnetic flux from inside the superconducting material for fields below a critical field when the temperature is below the critical temperature $T_c$. This important property allows us to distinguish a superconducting material from a perfect conductor.

According to Lenz's law, when a changing magnetic field is applied to a conductor, it will induce an electric current in the conductor that creates an opposing magnetic field. However, the Meissner-Ochsenfeld effect is the spontaneous expulsion which occurs during the transition to the superconducting state. Suppose there is a superconductor in the normal state placed in a constant external magnetic field. When it is cooled below the critical temperature $T_c$, the abrupt expulsion of the internal magnetic field would occur, which is not expected based on Lenz's law.

The dimensionless magnetic susceptibility is defined as $\chi = M/H$, and the magnetic flux density can be expressed as

$$B = \mu_0 H(1 + \chi) = \mu_0(H + M), \tag{1.1}$$

while $\mu_0$ is the permeability of free space. The property of perfect diamagnetism means that the susceptibility $\chi = -1$, i.e. the magnetization $M = -H$, so there can be no $B$ field inside a superconductor.

If a superconductor in the normal state is cooled below $T_c$ by zero field-cooled (ZFC) method, the normal state to the superconducting state transition happens without the presence of any magnetic field. Thereafter, the cooled superconductor is placed in an external magnetic field, and the magnetic field will be expelled from the superconductor. If this superconductor in normal state is placed in an external magnetic field, the field will penetrate and have the same value inside and outside. When a superconductor in the normal state is cooled below its $T_c$ by field-cooled (FC) method, the normal state to the superconducting state transition happens with the presence of a magnetic field; thereafter, the field will be expelled from the superconductor. This phenomenon of superconductors is named as the Meissner-Ochsenfeld effect, and a similar phenomenon cannot be found in conventional conductors and demonstrates that superconductivity is a new thermodynamic phase.

The superconductors that totally exclude applied magnetic flux are known as type I superconductors. The Meissner-Ochsenfeld effect exists in type I superconductors below the critical field $H_c$, but the superconductivity will be destroyed above $H_c$. In type I superconductors whose thicknesses are much greater than the penetration depth $\lambda$, internal magnetic fields and transport currents are merely able to exist only in a surface layer of thickness $\lambda$.

The magnetic properties of type II superconductor are more complex. type II superconductors are similar to type I superconductors but the presence of a Meissner-Ochsenfeld effect only happens when the magnetic applied field is smaller than a lower critical field $H_{c1}$. When the applied magnetic field is higher than $H_{c1}$ but lower than its upper critical field $H_{c2}$, magnetic vortices begin to penetrate into the superconductor and the material enters the mixed state. The magnetic field can partially penetrate into the superconductor, and the field is only partially excluded, i.e. diamagnetism is not perfect.

Movement of vortices in a type II superconductor can cause heat dissipation, i.e. a finite resistivity, called flux flow resistivity, the superconductor is no longer in the superconducting state of type I superconductor. If the current is sufficiently small, the vortices are stationary, and the resistivity is zero. The resistance below the critical current density $J_c$ is negligible compared with that of normal conductors. Indeed, there is no complete Meissner-Ochsenfeld effect in type II superconductors above $H_{c1}$, since in the mixed state, the magnetic field penetrates into the superconductor in the shape of the vortices.

Interaction between the magnetic field and type I superconductor in Meissner-Ochsenfeld state produces a magnetic levitation force, but this force is much smaller than that of a type II superconductor in the mixed state. This is a very important matter, but also confusing. Maglev in type II superconductor looks similar to the Meissner-Ochsenfeld effect, but there is a critical difference. The Meissner-Ochsenfeld effect completely excludes the magnetic fields below $H_{c1}$ to the field penetrating depth $\lambda$ on the superconductor surface, and it gives rise to a small levitation forces. In the mixed state of a type II superconductors, the pinning forces hold the superconductor in a place, thereby allowing it to levitate stably. The levitation forces of type II superconductors in the mixed state are much larger than the levitation force in a pure Meissner-Ochsenfeld state in type I superconductors.

Svedlindh et al. [13] reported the observation of a paramagnetic Meissner-Ochsenfeld effect (PME) [14] in a BiSrCaCuO superconductor. The Meissner-Ochsenfeld effect is characterized by a diamagnetic response of the superconducting material when a magnetic field is applied, but in several cases, this magnetic response in FC experiments can be paramagnetic, challenging the conventional Meissner-Ochsenfeld effect.

## 1.4 Superconducting thermodynamics [15]

The phase transition from the normal state to the superconducting state can be dealt with the general principles of thermodynamics. Considering the magnetization effect for a magnetic medium, the Gibbs free energy $G$ can be expressed as

$$G = U - TS + pV - \mu_0 HM. \tag{1.2}$$

Here, $U$ is the internal energy, $T$ is the temperature, $S$ is the entropy, $p$ is pressure, $V$ is the volume, $\mu_0$ is the permeability of free space, and $M$ is the magnetization. At constant temperature and constant pressure conditions,

$$\mathrm{d}G = -\mu_0 M \mathrm{d}H. \tag{1.3}$$

The Gibbs free energy can be obtained by the integral

$$G(T,p,H) - G(T,p,0) = -\mu_0 \int_0^H M\mathrm{d}H. \tag{1.4}$$

A magnetic field which is applied to the superconductor can produce negative magnetization, i.e. the magnetization $M = -H$. The magnetic flux of the negative magnetization exactly offsets that caused by the external magnetic field – this is the Meissner-Ochsenfeld effect that describes the superconductor, i.e. $B = 0$. Therefore, the Gibbs free energy, $g$ per unit volume

$$g_s(T,p,H) = g_s(T,p,0) + \frac{1}{2}\mu_0 H^2, \tag{1.5}$$

where the subscripts indicates the superconducting state, and the magnetic energy density, $\mu_0 H^2/2$, is independent of temperature. Thus, the superconducting state free energy in an external magnetic field will be increased due to the negative magnetization of superconductors.

In the normal state, $M = \chi H$, magnetic susceptibility $\chi$ is very small (about $10^{-5}$), it can be described as

$$g_n(T,p,H) \approx g_n(T,p,0), \tag{1.6}$$

where the subscript n indicates the normal state. The Gibbs free energy of the superconductor in the normal state is unchanged before and after applying the magnetic field. According to the balance conditions of phase transitions between superconducting state and normal state in the critical magnetic field $H_c$,

$$g_n(T,p,H_c) = g_s(T,p,H_c), \tag{1.7}$$

then Eq. (1.5) becomes

$$g_n(T,p,0) - g_s(T,p,0) = \frac{1}{2}\mu_0 H_c^2. \tag{1.8}$$

The Gibbs free energy in the superconducting state depends on the value of the critical field at that temperature. This confirms that there is a close relationship between superconductivity and magnetism. It shows that the free energy density of the superconducting state is lower than that of the normal state. Usually, this energy density is called as the condensation energy of the superconducting state. The thermodynamic critical field $H_c$ of type I superconductor can be derived from Eq. (1.8).

The above discussion deals with the phase transition between the normal state and the superconducting state at $T < T_c$ in a magnetic field. It is a first-order transition and which requires a latent heat for the phase change. At $T = T_c$ the superconducting phase transition has no latent heat, but it has a specific heat jump. This shows that the superconducting phase transition at $T = T_c$ is a second-order phase transition.

## 1.5 London equations [16]

London and London [2] assumed that the electrons move in a frictionless state and derived a phenomenological macroscopic theory of superconductivity.

In the normal conducting state, the current density $\boldsymbol{J}_n$ and the electric field $\boldsymbol{E}$ are connected by Ohm's law $\boldsymbol{J}_n = \sigma_n \boldsymbol{E}$, where $\sigma_n$ is the normal conductivity. In the normal phase, the current density in the steady state is given by

$$\boldsymbol{J}_n = \sigma_n \boldsymbol{E} = \frac{ne^2\tau}{m}\boldsymbol{E}, \tag{1.9}$$

where $\tau$ is the relaxation time and $\sigma_n$ is the conductivity in the normal state.

According to the two-fluid model, the total density $n$ of electrons is a sum of normal electrons, $n_n$, and superconducting electrons, $n_s$. The superconducting electrons $n_s$ are not scattered by either impurities or lattice vibrations, i.e., phonons do not contribute to the resistivity. These electrons are freely accelerated by an electric field. The equation of the superconducting electrons motion is

$$m\frac{\partial \boldsymbol{v}_s}{\partial t} = e\boldsymbol{E}, \tag{1.10}$$

where $\boldsymbol{v}_s$ is superconducting electrons velocity.

From Maxwell's equations, since the displacement current $\boldsymbol{D}$ inside the superconductor vanishes, one obtains

$$\nabla \times \boldsymbol{E} = -\frac{\partial \boldsymbol{B}}{\partial t}, \tag{1.11}$$

$$\nabla \times \boldsymbol{B} = \mu_0 \boldsymbol{J}. \tag{1.12}$$

The superconducting current density $\boldsymbol{J}_s = n_s e \boldsymbol{v}_s$ obeys the following equation (the first London equation)

$$\frac{\partial}{\partial t}\boldsymbol{J}_s = \frac{n_s e^2}{m}\boldsymbol{E}. \tag{1.13}$$

Adding curl to both sides, we have

$$\frac{\partial}{\partial t}\nabla \times \boldsymbol{J}_s = \frac{n_s e^2}{m}\nabla \times \boldsymbol{E}, \tag{1.14}$$

where ∇ is the curl operator, $\boldsymbol{E}$ is the electric field, and $\boldsymbol{B}$ is the magnetic field. We have

$$\frac{\partial}{\partial t}\nabla \times \nabla \times \boldsymbol{B} = \frac{\mu_0 n_s e^2}{m}\nabla \times \boldsymbol{E} = -\frac{\mu_0 n_s e^2}{m} \cdot \frac{\partial \boldsymbol{B}}{\partial t}. \tag{1.15}$$

Because

$$\nabla \times \nabla \times \frac{\partial \boldsymbol{B}}{\partial t} = \nabla\left(\nabla \cdot \frac{\partial \boldsymbol{B}}{\partial t}\right) - \nabla^2 \frac{\partial \boldsymbol{B}}{\partial t}, \tag{1.16}$$

and $\nabla \cdot \boldsymbol{B} = 0$, one obtains

$$\frac{\mu_0 n_s e^2}{m}\frac{\partial \boldsymbol{B}}{\partial t} = \nabla^2 \frac{\partial \boldsymbol{B}}{\partial t}. \tag{1.17}$$

Thus, $B$ must satisfy this differential equation in superconductors, leading to a characteristic length scale known as the London penetration depth, $\lambda_L$,

$$\lambda_L = \left(\frac{m}{\mu_0 n_s e^2}\right)^{1/2}. \tag{1.18}$$

One obtain

$$\nabla^2 \boldsymbol{B} = \frac{1}{\lambda_L^2}\boldsymbol{B}, \tag{1.19}$$

or

$$\nabla \times \boldsymbol{J}_s = \frac{n_s e^2}{m}\boldsymbol{B} = \frac{1}{\mu_0 \lambda_L^2}\boldsymbol{B}. \tag{1.20}$$

This is the second London equation, and the first London equation can be written as

$$\frac{\partial}{\partial t}\boldsymbol{J}_s = \frac{n_s e^2}{m}\boldsymbol{E} = \frac{1}{\mu_0 \lambda_L^2}\boldsymbol{E}. \tag{1.21}$$

The second London equation leads to the Meissner-Ochsenfeld effect, and can also be derived from the BCS theory by assuming that the BCS state is spatially homogeneous. The London equation predicts that the magnetic field in a superconductor decays exponentially from the surface into the superconducting material.

## 1.6 Penetration depth

The London penetration depths $\lambda_L$ describes the distance to which the magnetic flux penetrates into the superconductor from the surface between air and metal.

For the special case of a semi-infinite superconductor with a flat boundary within free space, if the magnetic field outside the superconductor is constant and parallel to the superconducting boundary plane, the differential equation (1.19) becomes one-dimensional with a solution

$$B(x) = B(0) \exp\left[-x\Big/\left(\frac{m}{\mu_0 n_s e^2}\right)^{1/2}\right] = B(0)\exp(-x/\lambda_L), \tag{1.22}$$

where $B(0)$ is the magnetic field at the surface of the superconductor. Here we see that when a magnetic field is present, superconducting currents are induced to shield the magnetic field in the interior of a superconductor. The physical meaning of the London penetration depth $\lambda_L$ can easily be understood.

Thus, the London equations imply a characteristic length scale $\lambda_L$ over which external magnetic field is exponentially suppressed.

$$\lambda_L(0) = \sqrt{\frac{m}{\mu_0 e^2 n_s}}, \tag{1.23}$$

where $\lambda_L(0)$ is the penetration depth at the absolute zero and is one of the inherent characteristic parameters of a superconductor. According to the temperature dependence of the density of superconducting electrons in the two-fluid model, the penetration depth depends on temperature (BCS calculation also gives a similar relationship) as shown below,

$$\lambda_L(T) = \sqrt{\frac{m}{\mu_0 e^2 n_s(T)}} = \frac{\lambda_L(0)}{[1-t^4]^{1/2}}, \tag{1.24}$$

where $t = T/T_c$. Close to $T_c$, $\lambda_L$ tends to infinity, and at a temperature very close to $T_c$, Eq. (1.24) can be expressed as

$$\lambda_L(T) = \frac{1}{2}\lambda_L(0)(1-t)^{-\frac{1}{2}}. \tag{1.25}$$

It should be noted that for superconductors in the superconducting state, the field can only penetrate into the superconductor by a penetration depth $\lambda_L$ for both DC and AC magnetic fields. For superconductors in the normal state, there is no penetration depth, namely the magnetic field cannot be shielded.

## 1.7 Coherence length

The coherence length $\xi$ is one of the most important parameters describing superconductivity. It can be intuitively understood as the mutual correlation length between which the two electrons that constitute the Cooper pair. It refers to the space scale of the electronic wave function. The BCS theory gives the coherence length as

$$\xi_0 = \frac{\hbar v_F}{\pi \Delta_0} = 0.18 \frac{\hbar v_F}{k_B T_c}, \tag{1.26}$$

where $\hbar$ is the Planck constant, $h$, divided by $2\pi$, $v_F$ is the Fermi velocity, and $2\Delta_0 = 3.528 k_B T_c$ is the value of the superconducting energy gap $E_g$ at zero temperature in the superconducting state, $k_B$ is the Boltzmann constant, and $\xi_0$ is the coherence length at zero temperature of a pure (without impurities) material.

When there is only a small size difference between the electron mean free path $l$ and the coherence length $\xi_0$, Pippard [17] gives an empirical formula for the effective coherence length, $\xi_l$, for an impure conductor,

$$\frac{1}{\xi_l} = \frac{1}{\xi_0} + \frac{1}{\alpha l}, \tag{1.27}$$

where $\alpha$ is a constant and is about 0.8. In this case, the effective penetration depth [18] $\lambda_e$ is

$$\lambda_e = \lambda_L \left(1 + \frac{\xi_0}{l}\right)^{1/2}. \tag{1.28}$$

The HTSC materials differ remarkably from conventional superconductors in which they have much smaller coherence lengths. In the LTSC materials, $\xi$ is the order of a few thousand angstroms, but in the HTSC materials, it is in the order of 1 to 10 Å. The small size of $\xi$ affects the HTSC thermodynamic and electromagnetic properties.

The coherence length $\xi_0$ is related to the GL coherence length, $\xi_{GL}$, through the expression

$$\xi_{GL}(T) = \frac{\alpha \xi_0}{\left[1 - \left(\frac{T}{T_c}\right)^2\right]}, \tag{1.29}$$

where $\alpha$ is a constant.

The ratio of the penetration depths $\lambda_L$ and the coherence lengths $\xi$ is called as the GL parameter $\kappa$:

$$\kappa = \lambda_L / \xi. \tag{1.30}$$

$\kappa$ is an important parameter that characterizes the superconducting material and distinguishes type I from type II superconductors.

The coherence lengths $\xi$ of the HTSC materials have strong anisotropy. The conductivity along the chains ($b$-axis) is more than twice that in the $a$-axis direction which is perpendicular to the chains. [19] This results in large anisotropy of the coherence lengths and the values of the superconducting gap at the same critical temperature.

## 1.8 Critical current densities

The critical current density $J_c$ (A/m$^2$) of a superconductor is the transport current which destroys the Cooper pairs and the superconducting state. However, in real applications of type II superconductors, the critical current densities $J_c$ refers to the value starting the flux flow, which is several orders of magnitude lower than the value which destroys the Cooper pairs.

The critical current density $J_c$ which destroys the Cooper pairs, is determined by the magnetic field $H$ which is produced by the critical current density. If a current $I$ flows in the axial direction of a superconducting cylinder wire of radius $R$, the wire causes a circular magnetic field $H_0$ around it. This field decreases with radius $r$ which from the wire outside surface into the center, and could be expressed as

$$H_0 = \frac{I}{2\pi r}. \tag{1.31}$$

If the current $I$ produces a magnetic field $H_0$ which is higher than $H_c$, the superconducting state will be destroyed. Thus, $I_c$ is intuitively defined as the value of $H_0 = H_c$ in the type I superconductors, i.e.

$$I_c = 2\pi R H_c(T). \tag{1.32}$$

In fact, since the magnetic flux is expelled from the inside of a superconducting wire, the current density $J$ cannot be uniform. The current $I$ only flows in the surface layer at the penetration depth $\lambda_L$. The total current $I$ over the cross section of the superconducting wire is expressed as

$$I = 2\pi R \lambda J, \tag{1.33}$$

where $J$ is the current density and the quantity $2\pi R\lambda$ is the effective cross section area of the surface layer. Thus, the highest intrinsic current, which destroys Cooper pairs and superconductivity of a superconductor, can be calculated using the thermodynamic critical field $H_c$, i.e.

$$H_c = \lambda J. \tag{1.34}$$

If the values of $H_c$ and $\lambda$ of a type II superconductor are 0.1 T and 100 nm, the critical current density $J_c$ is $10^8$ A/cm$^2$.

The GL critical current density is expressed as [20]

$$J_c = \frac{\Phi_0}{3\sqrt{3}\pi\mu_0\lambda^2\xi}. \tag{1.35}$$

According to Eqs. (1.24) and (1.42),

$$J_c(T) = J_c(0)\left[1-\left(\frac{T}{T_c}\right)^2\right]\left[1-\left(\frac{T}{T_c}\right)^4\right]^{1/2}. \tag{1.36}$$

Since the current of a type I superconductor is concentrated within the penetration depth of the surface, the critical current of a type I superconductor wire is directly proportional to the outer circumference of superconducting wire. For a type II superconductor which works in a mixed state between $H_{c1}$ and $H_{c2}$ for practical applications, the distribution of the superconducting current is more complicated.

For HTSC with planar structure, the critical current densities are smaller when the external field is applied perpendicular to the *c*-axis than when applied parallel to the axis. [20]

## 1.9 Critical magnetic fields

A superconductor in a sufficiently strong magnetic field will return from the superconducting state to the normal state. This field is named as the critical magnetic field $H_c$. For type II superconductors, the upper critical magnetic field $H_{c2}$ depends on the vortex structure. For type I superconductors, the superconducting state can be destroyed by the thermodynamic critical magnetic field $H_c$. The critical magnetic field increases as temperature is lowered and reaches the maximum value at $T = 0$ K. Generally, the thermodynamic critical field $H_c$ is lower than $H_{c2}$ for type II.

The superconducting state can also be destroyed by the self-field from a large transport current through the material. This maximum transport current is the so-called critical current density $J_c$ (see section 1.8). It depends on the properties and the geometry of the superconductor. Both the critical currents and the critical magnetic fields are directly related to the temperature. With lower temperature, the better property is achieved. This critical magnetic field is also a very important parameter for industrial applications.

Type I superconductors have a single critical field $H_c$, above which superconductivity vanishes. Type II superconductors have two critical fields, the lower critical magnetic field $H_{c1}$, and the upper critical magnetic field $H_{c2}$, between which they allow partial penetration of the magnetic field in the form of magnetic vortices.

The splitting energy level of the electron ground state in a magnetic field is called as the Zeeman splitting energy level. When the Zeeman splitting energy level $\Delta E$ of electrons becomes comparable to the energy gap $2\Delta$, Cooper pairs are broken and superconductivity vanishes. This magnetic field is named as the Pauli-limiting field and is the theoretical upper critical field $H_{c2}$ of superconductors.

Type I superconductors have only one critical field $H_c$, and Eq. (1.8) has shown the difference $G_n - G_s$ in the Gibbs free energy between the normal and the superconducting states, which is proportional to the critical magnetic field $H_c^2$,

$$G_n - G_s = \frac{1}{2}\mu_0 H_c^2. \tag{1.37}$$

Since this is a thermodynamic expression, $H_c$ is named as the thermodynamic critical field. Both type I and type II superconductors have thermodynamic critical fields. In addition, a type II superconductor has both a lower and an upper critical fields, $H_{c1}$ and $H_{c2}$, respectively. The lower critical field $H_{c1}$ is given by [20]

$$H_{c1} = \frac{\Phi_0 \ln \kappa}{4\pi\mu_0\lambda^2}. \tag{1.38}$$

The upper critical field $H_{c2}$ occurs when the flux density is so dense that the cores of vortices overlap. It can also be described by the GL coherence length $\xi_{GL}$ and the quantum of magnetic flux $\Phi_0$,

$$H_{c2} = \frac{\Phi_0}{2\pi\mu_0\xi_{GL}^2}, \tag{1.39}$$

where $\mu_0$ is the permeability of free space and $\Phi_0$ is the quantized flux which is expressed as $\Phi_0 = hc/2e = (2.07 \times 10^{-15}\text{Wb})$.

Combining Eqs. (1.30), (1.38) and (1.39), we have

$$H_{c2}/H_{c1} = \frac{2\lambda^2}{\xi_{GL}^2 \ln \kappa} = 2\kappa^2/\ln \kappa. \tag{1.40}$$

The value of $H_c$ at $T < T_c$ varies with temperature; however, the critical magnetic field at absolute zero $H_c(0)$ is constant. The properties and functions depend on the material itself. The relation between the critical magnetic field $H_c$ and temperature $T$ is [15]

$$H_c = H_c(0)\left[1 - \left(\frac{T}{T_c}\right)^2\right]. \tag{1.41}$$

If the applied magnetic field is larger than $H_c(0)$ at 0 K, then even at absolute zero, the superconducting state will be destroyed. The magnetic field which destroys superconductivity is a very important parameter, as it determines the maximum current density (which generates a strong magnetic field) in practice applications.

The upper critical field $H_{c2}(0)$ of a type II superconductor is very high so that direct measurement is not realistic; however, it can be estimated using the slope of $H_{c2}$ with increasing temperature, the approximate formula is [16]

$$H_{c2} = 0.69T_c \left|\frac{dH_{c2}}{dT}\right|_{T=T_c} . \quad (1.42)$$

For HTSC with planar structure, the critical magnetic fields are larger when the external field applied is perpendicular to the *c*-axis than when it is applied parallel to this axis.

## 1.10 Type I superconductors

Superconductors can be categorized into two classes, types I and II. The ratio (1.30) of the penetration depth $\lambda$ to the superconducting coherence length $\xi$ is called as the GL parameter $\kappa$, which determines whether a material is type I or type II superconductor. For type I superconductors, this ratio is

$$0 < \lambda/\xi < 1/\sqrt{2}, \quad (1.43)$$

and for type II superconductors,

$$\lambda/\xi > 1\sqrt{2}. \quad (1.44)$$

Most pure elemental superconductors, except niobium and carbon nanotubes, are type I superconductors, while almost all impure and compound superconductors are type II superconductors. Type 1.5 superconductor refers to a multicomponent superconductor which is characterized by two or more coherence lengths, and $\xi_1 < \sqrt{2}\lambda < \xi_2$.

Superconducting materials that completely expel magnetic flux until they become completely normal are called as type I superconductors. The superconductor is in the Meissner-Ochsenfeld state at small magnetic field. The Meissner-Ochsenfeld state breaks down when the applied magnetic field has a certain larger value.

## 1.11 Type II superconductors

Type II superconductors [4, 9, 21] have two critical fields, the lower critical magnetic field $H_{c1}$ and the upper critical magnetic field $H_{c2}$, between which they allow partial penetration of the magnetic field.

In type II superconductors, [22, 23] increasing the applied field more than the lower critical field $H_{c1}$ forces the superconductors to enter the mixed state (namely the vortex state). In the mixed state, the normal state exists in the cores of the vortices and the superconducting state coexists outside the cores. With an increasing amount of magnetic flux that penetrates into the material, there remains no resistance to the flow of superconducting current if the current is not too large. The mixed state is actually caused by the vortices in the electronic superfluid, sometimes called fluxons since the amount of flux carried by these vortices is quantized. The vortices play a very special role in applications of type II superconductors.

In the mixed state, the magnetic fluxes parallel to the external magnetic field penetrates within a cylinder. The cylinder is called the normal core. The normal vortex cores in the mixed state are surrounded by superconducting regions. Vortex shielding currents are formed around the normal core, and the magnetic field is gradually shielded. The circulating vortex shielding current generates and maintains the flux line in the normal core. The magnetic flux is the magnetic flux quantum.

The vortices have a core radius equal to the coherence length $\xi$ and a surrounding outer region with radius equal to the penetration depth $\lambda$, where superconducting screening currents flow around the core in the thin layer $\lambda$. Its behavior is analogous to the exponential decay [see Eq. (1.22)] in type I superconductors. In applied fields $H(H_{c1} < H < H_{c2})$, the density of vortices is approximately proportional to the applied field.

When the superconductor carries a current $\boldsymbol{I}$, the flux lines experience a Lorentz force:

$$\boldsymbol{F}_{\mathrm{L}} = \boldsymbol{I} \times \boldsymbol{B}, \tag{1.45}$$

where $\boldsymbol{B} = \mu_0 \boldsymbol{H}$. Under this Lorentz force $F_{\mathrm{L}}$, the flux lines start to move through the lattice if it overcomes the pinning force $F_{\mathrm{P}}$ of the lattice. If the pinning force $F_{\mathrm{P}}$ is stronger than the Lorentz force $F_{\mathrm{L}}$, i.e. at temperatures far below $T_{\mathrm{c}}$, the flux line lattice is pinned by the pinning center. However, thermal fluctuations will cause some vortices to escape from the pinning center. This phenomenon is named as flux creep. However, if the Lorentz force $F_{\mathrm{L}}$ is larger than the pinning force $F_{\mathrm{P}}$, the flux line lattice will be driven by the Lorentz force $F_{\mathrm{L}}$ and move. This phenomenon is named as flux flow. The motion of vortices generate energy dissipation because the vortex cores are in the normal state.

In an ideal type II superconductor, there is nothing to hinder the motion of flux lines. Since the magnetic gradient cannot be formed due to the absence of pinning centers, this leads to a vanishing critical current density $J_{\mathrm{c}}$. Such superconductors have little value for applications.

In a real superconductor, there are always lattice defects or impurities and small particles which serve as pinning centers for the flux line lattice. The pinning centers play an important role to prevent the flux line from motion. The better the flux-pinning properties, the higher the critical current density. When the magnetic field is sufficiently strong, the flux becomes free from the pinning and can move freely. This magnetic field is called as the irreversible field $H_{irr}$, and it is one of the important parameters for superconducting applications. At the irreversible field $H_{irr}$, the critical current density reduces to zero. The irreversible field $H_{irr}$ is determined by the mechanism of flux creep, and can be derived from the $E$-$J$ characteristics caused by the flux creep. [24] As the critical temperature is approached, the normal cores are more closely packed and eventually overlap so that the superconducting state is lost. At this upper critical field $H_{c2}$, superconductivity is destroyed.

To avoid the motion of vortices and thus ensure zero resistance of a superconductor, various defects such as granular structure, lattice defects, or artificial defects are introduced into the superconducting material. Each defect attracts vortices, and these vortices produce a series of pinning centers in the superconductor. Below the critical current density $J_c$, a hard type II superconductor can have zero resistance up to very high magnetic fields. The high magnetic field is not $H_{c2}$, but is the irreversible field $H_{c3}$ and $H_{c1} < H_{c3} < H_{c2}$.

HTSCs are limited by the irreversibility field $H_{c3}$, above which magnetic vortices melt or decouple. Even though BSCCO has a higher upper critical field $H_{c2}$ than YBCO, it has a much lower $H_{c3}$ (typically smaller by a factor of 100), thus limiting its use for making high-field magnets. It is for this reason that conductors of YBCO are preferred to BSCCO.

## 1.12 Specific characteristics of HTSC

From the London theory, GL theory to BCS theory, the theory of superconductivity has experienced a long history. Although the BCS theory can explain the characteristic properties of conventional superconductors, the HTSC materials still have no widely accepted theory to explain their properties. The HTSCs with their layered structures have strong anisotropy and giant flux creep effects. It is complicated and difficult to study HTS. Although the properties of HTSCs are in many respects similar to those of conventional superconductors, the HTSCs have a series of special characteristics compared with LTSCs. The critical temperatures $T_c$ are higher by almost one order of magnitude, and the superconducting energy gaps $2\Delta$ are also larger by one order of magnitude. In addition, the penetration depths $\lambda$ are also larger for one order of magnitude, and the superconducting coherence lengths $\xi$ are much shorter, thus, the GL parameters $\kappa$ are larger (on the order of 100). The superconducting wave function has d-wave symmetry in HTSCs, but that of LTSCs shows s-wave symmetry. A series of the differences and problems between HTSCs and LTSCs have posed serious challenges for theoretical explanation of the physical mechanisms of HTSCs.

Axial symmetry is assumed for the case of in-plane $a$ and $b$ directions, so that the screening currents in $a$ and $b$ directions are equivalent to each other. Such an assumption is a good approximation for orthorhombic HTSCs. Thus, there are two different penetration depths $\lambda_{ab}$, $\lambda_c$, and coherence lengths $\xi_{ab}$, $\xi_c$. The penetration depth $\lambda_{ab}$ and the coherence length $\xi_{ab}$ are in the $a$, $b$ plane, and $\lambda_c$ and $\xi_c$ are along the $c$ direction. Two different GL parameters $\kappa_{ab}$ and $\kappa_c$ as for the applied field in the $a$-$b$ plane and along the $c$ direction, respectively, [25, 26]

$$\kappa_{ab} = \left[\frac{\lambda_{ab}\lambda_c}{\xi_{ab}\xi_c}\right]^{1/2}, \tag{1.46}$$

$$\kappa_c = \frac{\lambda_{ab}}{\xi_{ab}}. \tag{1.47}$$

The thermodynamic critical field $H_c$ is given by

$$H_c = \frac{\Phi_0}{2\sqrt{2}\pi\mu_0\lambda_{ab}\xi_{ab}}. \tag{1.48}$$

There are expressions for the lower and upper critical fields. For the particular case of axial symmetry, the critical fields in the $a$-$b$ plane and along the $c$ direction [24] are

$$H_{c1,ab} = \frac{\Phi_0 \ln\kappa}{4\pi\mu_0\lambda_{ab}\lambda_c}, \tag{1.49}$$

$$H_{c1,c} = \frac{\Phi_0 \ln\kappa}{4\pi\mu_0\lambda_{ab}^2}, \tag{1.50}$$

$$H_{c2,ab} = \frac{\Phi_0}{2\pi\mu_0\xi_{ab}\xi_c}, \tag{1.51}$$

$$H_{c2,c} = \frac{\Phi_0}{2\pi\mu_0\xi_{ab}^2}, \tag{1.52}$$

where $\mu_0$ is the permeability of free space and $\Phi_0$ is the quantized flux expressed by $\Phi_0 = hc/2e$.

The following relations hold generally for the characteristic length scales in HTSC:

$$\xi_c < \xi_{ab} \ll \lambda_{ab} < \lambda_c,$$

therefore,

$$H_{c1\perp c} < H_{c1\parallel c} \ll H_{c2\parallel c} < H_{c2\perp c}.$$

## References

[1] Onnes HK. Disappearance of the electrical resistance of mercury at helium temperatures. Commun Phys Lab Univ Leiden. 1911;12–120.
[2] London F, London H. The electromagnetic equations of the supraconductor. Proc R Soc. 1935;A149–71.
[3] Ginzburg VL, Landau LD. Theorii sverkhrovodimosti. Zh. Eksp. (English translation: On the theory of superconductivity). Zh Eksp Teor Fiz K. 1950;29–748.
[4] Abrikosov AA. Type II superconductors and vortices from the 1950s to the 1990s. Sov Phys JETP. 1957;5–1174.
[5] Gorkov LP. Critical supercooling field in superconductivity theory. [BCS theory]. Sov Phys JETP. 1960;10:998.
[6] Bardeen J, Cooper L, Schrieffer J. Theory of superconductivity. Phys Rev. 1957;108–1175.
[7] Haag PKD. Report in Superconductivity Centennial Conference. Hague, Netherlands, September 21, 2011.
[8] Meissner W, Ochsenfeld R. Short initial announcements. Naturwiss. 1933;33–787.
[9] Shubnikov LV, Khotkevich VI, Shepelev Yu D, Ryabinin Yu N. Magnetic properties of type-II superconductors. 1937;7–221.
[10] Bednorz JG, Müller KA. Possible high $T_c$ superconductivity in the Ba–La–Cu–O system. Z Phys. 1986;B6:189.
[11] Wu MK, Ashburn JR, Torng CJ, Hor PH, Meng RL, Gao L, et al. High-pressure study of the new Yb–Ba–Cu–O superconducting compound system. Phys Rev Lett. 1987;58–908.
[12] Zhao ZX, Chen LQ, Cui CG, Huang YZ, Liu JX, Chen GH, et al. The high critical temperature superconductivity of Sr(Ba)–La–Cu oxide. Chin Sci Bull 1987; 32–177 (in Chinese).
[13] Svedlindh P, Niskanen K, Norling P, Nordblad P, Lundgren L, Lönberg B, et al. Anti-Meissner effect in the BiSrCaCuO-system. Physica C. 1989;162:164–1365.
[14] Roa JJ, Dias FT, Segarra M. Magnetical response and mechanical properties of high temperature superconductors, $YBaCu_3O_{7-x}$ materials. In: Grigorashvili Y, editor. Croatia: InTech. 2012;193–218.
[15] Zhang LY, Zhang JL, Cui GJ. Superconductor physics. Beijing: Electronics Industry Press. 1995 (in Chinese).
[16] Wang JS, Wang SY. Application of superconductivity. Chengdu, University of Science and Technology Chengdu Press, 1995 (in Chinese).
[17] Pippard AB. An experimental and theoretical study of the relation between magnetic field and current in a superconductor. London: Proc R Soc A. 1953;216–547.
[18] M Tinkham. Introduction to superconductivity McGraw-Hill[J]. Physics Today. 1996;49(8):1996.
[19] Friedmann TA, Rabin MW, Giapintzakis J, Rice JP, Ginsberg DM. Direct measurement of the anisotropy of the resistivity in the *a-b* plane of twin-free, single-crystal, superconducting $YBa_2Cu_3O_{7-\delta}$. Phys Rev B Condens Matter. 1990;42:6217–6221.
[20] Poole CP Jr, Farach HA, Creswick RJ, Prozorov R. Superconductivity. Second edition. New york: Academic Press, 2007.
[21] Abrikosov AA. Type II superconductors and vortex lattice. Rev Mod Phys. 2004;76:975–979.
[22] Saint-James D, Sarma G, Thomas EJ. Type II superconductivity. Oxford: Pergamon Press, 1969.
[23] Shepelev AG. The discovery of type II superconductors (Shubnikov phase). In: Luiz AM, editor. Croatia: InTech 2010:17–46.
[24] Matsushita T. Flux pinning in superconductors. Berlin, Heidelberg: Springer-Verlag, 2007.
[25] Chudnovsky EM. Pinning by oxygen vacancies in high-$T_c$ superconductors. Phys Rev Lett. 1990;65:3060–3062.

## Further bibliography

[1] Chu CW. Superconductivity. High-temperature encyclopedia of applied physics. Volume 20. Weinbeim: VCH Publishers, 1997:213–247.
[2] Herrmann PF. Current leads. In: Seeber B, editor. Handbook of applied superconductivity. Florida: Institute of Physics Publishing, 1998:chapter D10.
[3] Pavese F. Handbook of applied superconductivity. Florida: Institute of Physics Publishing, 1998.
[4] Van Duzer T, Charles C, Turner W. Principles of superconductive devices and circuits. New Jersey: Prentice Hall PTR, 1999.
[5] Poole C, Farach HA, Creswick RJ. Handbook of superconductivity. New york: Academic Press, 1999.
[6] Ketterson JB, Song SN. Superconductivity. Oxford: Cambridge University Press, 1999.
[7] Lehndorff BR. High-$T_c$ Superconductors for magnet and energy technology: fundamental aspects. Berlin, Heidelberg: Springer-Verlag, 2001.
[8] Cardwell DA, Ginley DS. Handbook of superconducting materials. Volume I: Superconductivity, materials and processes. Florida: Institute of Physics Publishing, 2003.
[9] Cardwell DA, Ginley DS. Handbook of superconducting materials. Volume II: Characterization, applications and cryogenics. Florida: Institute of Physics Publishing, 2003.
[10] Annett JF. Superconductivity, superfluids and condensates. New York: Oxford University Press, 2004.
[11] Tinkham M. Introduction to superconductivity. Second edition. New York: Dover Publications, 2004.
[12] Ginzburg VL, Andryushin EA. Superconductivity. Singapore: World Scientific Publishing, 2004.
[13] Krabbes G, Fuchs G, Canders WR, May H, Palka R. High temperature superconductor bulk materials. Weinheim: Wiley-VCH Verlag, 2006.
[14] Schrieffer JR, Brooks JS. Handbook of high-temperature superconductivity. Berlin, Heidelberg: Springer-Verlag, 2007.
[15] Poole CP Jr, Farach HA, Creswick RJ, Prozorov R. Superconductivity. Second edition. New York: Academic Press, 2007.
[16] Matsushita T. Flux pinning in superconductors. Heidelberg: Springer-Verlag, 2007.
[17] Kakani SL, Kakani S. Superconductivity. Haicheng: Anshan Ltd, 2009.
[18] Blundell SJ. Superconductivity: a very short introduction. New York: Oxford University Press, 2009.
[19] Iwasa Y. Case studies in superconducting magnets: design and operational issues. Berlin Heidelberg: Springer, 2009.
[20] Kalsi S. Applications of high temperature superconductors to electric power equipment. New York: John Wiley and Sons, 2011.
[21] Ireson G. Discovering superconductivity: an investigative approach. New York John Wiley & Sons, 2012.
[22] Edi MM. Superconductivity: recent development and new production technologies. New York: Nova Science Publishers, 2012.
[23] Han RS. Superconductivity centennial. BeiJing: Peking University Press, 2013.

Jia-Su Wang and Su-Yu Wang

# 2 Superconducting materials

Superconductors are divided into conventional and unconventional superconductors and can also be divided into high- and low-temperature superconductors (HTSC and LTSC, respectively). The HTSCs can be cooled with liquid nitrogen (77 K) instead of liquid helium (4.2 K). This is commercially important because liquid nitrogen is much cheaper, which makes HTSCs particularly attractive for practical applications.

The superconducting material can be produced in the forms of single crystals, films, wires, and tapes. The most common LTSCs in applications are the Nb-Ti (niobium-titanium) and the $Nb_3Sn$ (niobium-tin) wires and tapes. Superconducting films are mainly used for electronic devices. In principle, for the HTSC materials, there are bulk and wire materials for the electrical engineering applications. The HTSC bulks are mainly used for the HTS Maglev train, Maglev bearings, Maglev flywheel energy storage devices, and so on.

## 2.1 LTS materials

The development of superconducting materials has experienced a long tortuous path from initial discovery to HTSC (Fig. 2.1). A LTSC material typically refers to the Nb-based alloy (most commonly Nb-Ti) and A15 ($Nb_3Sn$ and $Nb_3Al$ etc.) superconductors with lower critical temperature $T_c$. The most commonly conventional superconductors used in applications are Nb-Ti (niobium-titanium) alloy and $Nb_3Sn$ (niobium-tin) compounds [1]. The well-established Nb-Ti and $Nb_3Sn$ LTSCs have occupied a large market share of commercial market of superconducting wires.

Long and flexible wires are required to build a high-field magnet. Although $Nb_3Sn$ was found before Nb-Ti, it is a brittle material and hard to fabricate into wires. Despite a lower transition temperature of 10 K and a lower upper critical field $H_{c2}$ of 14.5 T, the Nb-Ti wire offers the significant advantage of flexibility and formability. Therefore, magnets using Nb-Ti wires were the first commercially available superconducting one in the 1960s. In 1962, the first commercial Nb-Ti alloy superconducting wire was developed [3]. Consequently, the first practical superconducting magnet was manufactured using the Nb-Ti wire. It opened a new epoch for the application of superconducting magnets. Up to now, the Nb-Ti has achieved a critical current density in excess of $10^6$ A/cm$^2$ at 4.2 K and 5 T [4, 5].

In 1961, Kunzler et al. [6] reported that the compound $Nb_3Sn$ was capable of sustaining enormous current densities without resistance in very high magnetic fields. Since $Nb_3Sn$ has higher critical temperature $T_c$, higher critical current density $J_c$, and higher critical field $H_{c2}$, it is used for higher field applications. The maximum critical fields

https://doi.org/10.1515/9783110538434-002

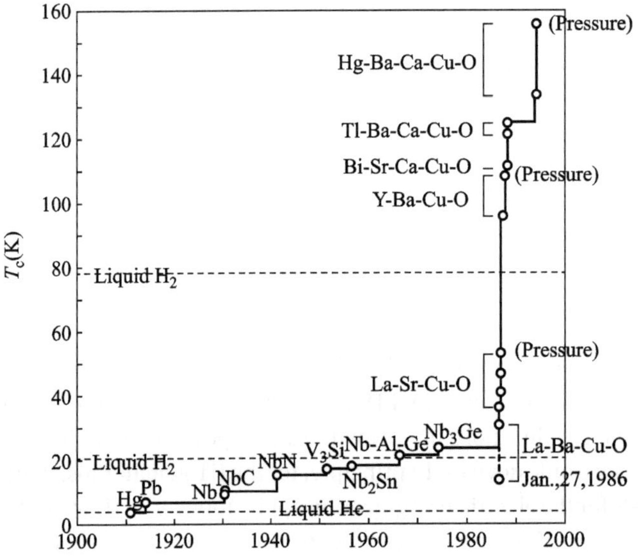


**Fig. 2.1:** Critical temperatures of superconductors from LTSC to HTSC [2].

$H_{c2}$ are 14.5 T (Nb-Ti) and 30 T ($Nb_3Sn$), respectively [7, 8]. These superconducting wires are widely used in magnetic resonance imaging (MRI) and nuclear magnetic resonance (NMR), particle accelerators, magnetic separation, transformers, energy storage devices, electric power cables, fault current limiters, electric motors or linear motors, Maglev trains, etc. Since LTS devices which were cooled with liquid helium turned out to be uneconomic, the magnet technology [9, 10], especially magnets for particle accelerators and magnetic resonance imaging (MRI), are the only commercial applications for those superconductors at present.

Magnet applications mainly depend on the critical current density at a certain temperature and field conditions. It can be seen from Fig. 2.2 that the critical current density [12] of Nb-Ti is lower than that of $Nb_3Sn$. The performance of $Nb_3Sn$ has been improved significantly over that of Nb-Ti. Although Nb-Ti is the most commonly used superconducting material, its upper critical field is lower than $Nb_3Sn$ (8 T at 4.2 K), i.e. $Nb_3Sn$ can generate a magnetic field higher than 12 T (4.2 K). However, due to stress and strain, especially under transverse compression, the degradation in $J_c$ restricts the use of $Nb_3Sn$ in large-scale applications. In practical engineering, the selection of superconducting materials is mainly based on the price. So far, the price of Nb-Ti wires is cheaper than that of $Nb_3Sn$.

The $Nb_3Al$ superconductor has outstanding features of high critical field and excellent strain tolerance in critical current performance. The test results demonstrate that the $Nb_3Al$ conductor is suitable for applications in high field magnets. So far, the critical current density $J_c$ of $Nb_3Al$ superconducting wire has achieved $9 \times 10^4$ A/cm². (20% higher than that of conventional wires) at 4.2 K and 15 T [13].

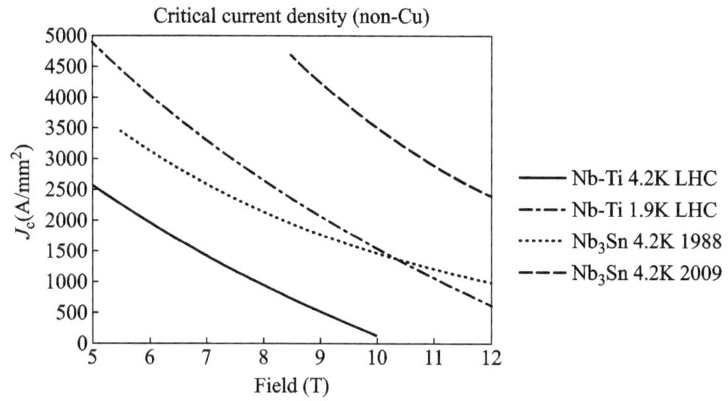


**Fig. 2.2:** Critical current density of Nb-Ti, at 4.2 and 1.9 K, compared with that of $Nb_3Sn$ at 4.2 K [11].

This demonstrates that $Nb_3Al$ superconducting wires can be used in accelerator magnets.

In 2001, Nagamatsu et al. [14] discovered superconductivity in a magnesium diboride ($MgB_2$), which had a transition temperature of 39 K. This is the highest critical temperature so far achieved in a conventional superconductor. The material has attracted considerable interest because of its potential for low cost, higher stability, and relatively simple deposition techniques, which enable $MgB_2$ to be comparative with Nb-Ti in the future. The operating temperature of the $MgB_2$ can be kept around 20 K by a cryocooler, but Nb-Ti requires 4.2 K and even lower. However, in comparison to standard Nb-Ti, $MgB_2$ has some disadvantages in the current density, mechanics, and uniformity.

The properties of major LTSC materials are shown in Tab. 2.1.

HTSC wires of Bi-2212 or YBCO are interesting for magnetic fields above 16 T, since this field is barely achievable with $Nb_3Sn$ at 1.9 K. The limits of the upper critical field $H_{c2}$ of $Nb_3Sn$ drive the development of HTSC materials with higher $H_{c2}$ values and greater critical current density ($J_c$) for ultra-high field applications.

**Tab. 2.1:** Characteristic parameters of LTSC materials [15–18].

| Materials | $T_c$(K) | $\mu_0 H_{c1}$(T) | $\mu_0 H_c$(T) | $\mu_0 H_{c2}$(T) | $\lambda_L$(Å) | $\xi$(Å) |
|---|---|---|---|---|---|---|
| Nb | 9.3 | 0.181 | 0.199 | 2 | 850 | 400 |
| Nb-Ti | 9.5 | | 0.253 | 15 | 3000 | 40 |
| $Nb_3Sn$ | 18 | 0.035 | 0.53 | 32 | 650 | 30 |
| $Nb_3Al$ | 18.7 | | | 34 | | |
| $Nb_3Ge$ | 23.2 | | | 38 | 900 | 30 |
| $MgB_2$ | 39 | | 0.66 | 74 | 1850 | 50 |

## 2.2 LTS bulk materials

There are no practical applications of LTS bulk superconductors, mostly due to their thermal instability [19]. Local heat generated inside the LHS bulk materials easily leads to a giant magnetic flux jump, which quenches the superconducting state. However, the specific heat of YBCO is different from these conventional BCS superconductors. Thus, HTS bulks are thermally stable even in large sample sizes due to their relatively large specific heat in the superconducting state.

The $MgB_2$ in LTS materials is a special case, since it has a higher specific heat. Flux jump issues of $MgB_2$ arise at very low temperatures, but tend to disappear at temperatures higher than 10 K [20]. Thus, $MgB_2$ can be used to make a large bulk material that produces more trapped flux. Thus, $MgB_2$ bulk has a promising potential as a superconducting permanent magnet (SCPM). The HTS REBaCuO bulk magnet can produce tesla-order SCPMs. However, it is difficult to fabricate a large single-domain REBaCuO bulk over 60 mm in diameter. In comparison to the REBaCuO bulk magnet, the $MgB_2$ bulk magnet has several attractive properties, for instance, low-cost, easy manufacturing of large samples, high mechanical strength, and homogeneous trapped field distribution. Furthermore, the problem of weak links at grain boundaries can be ignored in the $MgB_2$ polycrystalline bulk due to their long coherence length $\zeta$ [21]. These characteristics enable us to produce better and larger polycrystalline $MgB_2$ bulk magnets for use below their transition temperature $T_c = 39$ K. Bulk $MgB_2$ PMs may have the potential to be used in power systems and magnetic levitated trains using liquid $H_2$ or a cryocooler operation at 20 K.

Several groups have already reported the trapped fields in the $MgB_2$ bulk by the field-cooled (FC) magnetization and obtained a trapped field over 1.5 T at low temperatures [20–25]. Naito et al. [26] have reported that the maximum of the trapped field at the top surface is 1.43 T at 13.4 K for $MgB_2$ bulk 20 mm in diameter and is 1.50 T at 16.4 K for one that is 30 mm in diameter.

The achievable maximum flux density in a bulk material is the key factor in the development of high-power-density superconducting electrical machines. The engineering issues relevant to the development of rotors of synchronous machines using magnetized bulk $MgB_2$ have been experimentally investigated by Marignetti et al., [27] and they have verified the possibility of using bulk $MgB_2$ magnets for the field excitation of synchronous machines. Both the experiments and the analysis indicate a maximum trapped field limit of 1.2 T at 15 K.

## 2.3 HTS materials

HTSCs constitute a large family of superconducting ceramic materials. They offer the highest transition temperatures of all superconductors. The ability to use relatively cheap and easily handled liquid nitrogen as a cryogen has increased the range of practical applications of superconductivity.

The families of HTS materials include YBaCuO, BiSrCaCuO, TlBaCaCuO, HgBaCuO, HgBaCaCuO, etc. Among them, only the bismuth strontium calcium copper oxide (BSCCO) and the yttrium barium copper oxide (YBCO) are commercially available. In general, HTSC can be fabricated in the shape of single crystal, bulk, thin film, and wires or tapes. The forms used for electrical engineering applications include bulk and wire materials. Wires and tapes have the potential to be used in magnet, power device, and transportation as a more efficient and more compact replacement for copper technology. Thin films have the potential for use in electronics. Bulk materials have the potential to generate large trapped magnetic fluxes that are much greater than those achievable in conventional PMs. HTS bulk is emphatically introduced in the book of HTS Maglev. Melt-textured rare-earth Ba-Cu-O (REBCO, RE=Y, Nd, Sm, Eu, Gd etc) bulk has high critical current density and high critical magnetic field, which can produce strong and stable levitation. However, the critical current density of the bulk is two orders of magnitude lower than that of thin films. It is necessary to improve further the properties of HTS bulk.

In 1986, Bednorz and Müller [28] discovered superconductivity in a lanthanum-based cuprate perovskite material ($La_{2-x}Ba_xCuO_4$). Soon after, in January of 1987, Chu, Wu, et al. [29] discovered ceramic yttrium barium copper oxide ($YBa_2$ $Cu_3O$) superconductors with a critical temperature above the boiling temperature of liquid nitrogen. Meanwhile, Zhao, Chen, et al. [30] prepared $YBa_2Cu_3O$ in February of 1987. Thereafter, the research on HTS materials and their applications surged rapidly all over the world.

BiSrCaCuO (BSCCO) is a new class of superconductors. Michel et al. [31] first discovered superconductivity in Bi-2201 in 1987; shortly afterwards, the superconductivity of Bi-2212 [32] and Bi-2223 [33, 34] were discovered. The BSCCO system has three phases, Bi-2201, Bi-2212, and Bi-2223, with transition temperatures of 20, 85, and 110 K, respectively. The BSCCO family is analogous to the thallium family (TBCCO) and the mercury family (HBCCO), which are all HTSCs.

BSCCO was the first HTSC material to be used to make practical superconducting wires. Therefore, it is referred to as the first-generation (1G) HTS wires. Bi-2212 [35] wire has many features, such as isotropic electromagnetic performance, filament structure, compatibility with conventional cabling and braided strand insulation methods. The upper critical field $H_{c2}$ in Bi-2212 polycrystalline samples at 4.2 K has been measured as 200 ± 25 T [36]. Bi-2212 is one of the most promising materials for superconducting magnets which can generate a magnetic field above 25 T [37]. The irreversible field $H_{irr}$ (id. $H_3$) is an important parameter for applications, since in practical applications, the HTSCs are limited by the irreversible field $H_{irr}$, above which the magnetic vortices melt or decouple. To use Bi-2223 tapes in magnetic fields, the irreversible region lies below 2 T (at below 50 K); for Bi-2212 in the same field, it lies below 30 K. In order to produce higher magnetic fields, the magnets have to be operated at very low temperatures (e.g. 20 K). The Bi-2212 wires have better transport properties than Bi-2223 at 4.2 K. First-generation HTS wire was commercially available since the late 1990s. The commercial long-length wires have reached 1000 km in 2003. Moreover, several power system devices [38] have been developed. The commercial manufacturing technology for HTS BSCCO tapes is listed in Tab. 2.2. [39]. Table 2.2 shows that the HTS BSCCO tapes made in China have achieved a higher technical level, long length of 100–1000 m, critical current ($I_c$) of 90 A at 77 K and 0 T, and engineering critical current density ($J_e$) > 9 kA/cm$^2$ at 77 K and 0 T. However, the high cost of Ag-sheathed Bi-2223 tapes, their lower irreversible field $H_{irr}$, and their high AC losses are limiting factors for power applications.

Therefore, even though the BSCCO has a higher upper critical field than YBCO, its properties limit its use in making high-field magnets. Since HTS BSCCO wires exhibit a poor pinning performance, they cannot be used for power devices even with liquid nitrogen cooling. It is for this reason that coated conductors of YBCO are preferred to BSCCO. The corresponding YBCO wires are referred to as second-generation (2G) HTS wires. 2G HTS wire is fabricated by depositing rare earth, barium-copper-oxide (REBCO) HTS materials layered onto flexible thin metal substrate tapes with buffering metal oxides. The main advantages of 2G HTS wire [40–42] are high critical current density, large-scale production, lower manufacturing cost, better mechanical properties, and better performance under high magnetic fields. The 2G HTS REBCO wires are very attractive for various low- and high-field applications.

YBCO (yttrium barium copper oxide, $YBa_2Cu_3O_{7-x}$) contains the proportions of the 3 different metals in the mole ratio of 1 to 2 to 3 for yttrium to barium to copper, respectively. YBCO research focuses on optimizing properties in relation to applications of HTS bulk (to see next section) and tape materials. The powder-in-tube process (PIT) is successful for manufacturing LTS $Nb_3Sn$, $MgB_2$ and HTS Bi-family superconductors. However it cannot be used for YBCO conductors. Thin films of YBCO have demonstrated better superconducting properties for use at liquid nitrogen temperatures with higher critical current density $J_c$. The manufacturing process of the 2G HTS REBCO tapes is mainly concentrated on substrates, buffer layers, the superconducting film over layer, and their complex interactions. Preparation of YBCO

**Tab. 2.2:** Technology level of the HTS BSCCO wires [39].

| **Nations** | **Developer companies** | **Wires lengths (m)** | **Critical currents current $I_c$ (A) at 77 K, 0 T** | **Engineering critical densities $J_e$ (kA/cm$^2$) at 77 K, 0 T** |
|---|---|---|---|---|
| U. S. A. | AMSC | 100–400 | 115 | >12 |
| China | Innova | 100–1000 | 90 | >9 |
| Japan | Sumitomo | 100–400 | 100 | >10 |
| Germany | EAS | 100–1000 | 70 | >7 |
| Germany | Trithor | 100–1000 | 60 | >7 |

tapes is achieved by coated film conductor processing technology. REBCO-coated conductors are one of the most promising candidates for widespread applications because of their higher $J_c$ values in external magnetic fields. The $J_c$ of 2G tapes strongly depends on the direction of the applied field, and it requires a near-perfect texture to achieve a high $J_c$. The 2G tapes have a magnetic anisotropy. The upper critical magnetic field $H_{c2\perp}$ is about 120 T for fields perpendicular and $H_{c2\parallel}$ about 250 T for fields parallel to the $CuO_2$ planes [43]. Li et al. [44] have investigated the influence of deposition parameters on the orientation, in-plane texture, surface morphology, and roughness of films at a tape speed of 20–50 m/h. Li et al., in Shanghai Jiao Tong University, have achieved 2G YBCO tapes with 1 cm width and 100 meters length, and a critical current $I_c > 500$ A. [45] 2G HTS YBCO wires [46, 47] are now available in long lengths of high quality suitable for practice applications. A HTS magnet using 2G HTS wire was developed by Song et al. [48] in SuperPower Inc.

The irreversible field $H_{irr}$ for the BSCCO superconductors is much lower than for YBCO. One restriction on YBCO is the critical temperature $T_c$ of only 92 K, which could be an issue for applications. TlBaCaCuO (Tl-1223) or HgBaCaCuO (Hg-123) offer better in field performance with a higher critical temperature $T_c$ and irreversible field $H_{irr}$.

The TlBaCuO (thallium barium copper oxide) HTSC, critical temperature $T_c$ of 90 K, was discovered by Sheng et al. [49] in 1988. Shortly later, this group succeeded in preparing TlBaCaCuO superconductors with $T_c = 120$ K [50]. The thallium-based superconductor family is large and includes Tl-1223, Tl-2223, Tl-1212, Tl-2212, and Tl-2223 compounds. It holds the record for the highest critical temperature $T_c$ of 125 K. These compounds are very interesting for applications. Tl-2223 not only has a high critical temperature $T_c$, but also has a high irreversibility field (at $T_c$ of 120 K). In addition, both Tl-1212 and Tl-2212 have low microwave surface resistance. Therefore, they are used widely in microwave passive devices.

Superconductivity in mercury barium copper oxide (HgBaCuO) was discovered by Putilin et al. in 1993. [51]. Shortly thereafter, the mercury barium calcium copper oxide (HgBaCaCuO) superconductor with a $T_c$ of 135 K ($T_c$ of 150 K under high pressures, around 30 GPa) was discovered by Schilling et al. [52] and Chu et al. [53] The HTS Hg-1223 not only exhibits the highest $T_c$ and high irreversibility fields $H_{irr}$, but also shows remarkable critical current densities above 77 K. This is more important for power system applications.

In addition to the above HTSCs, a new HTS LaFeAsO family, iron-based superconductors [54, 55], was discovered in 2006 in Tokyo Institute of Technology. This new type of superconductor is based on conducting layers of iron and pnictide. It is a chemical compound which contains iron layers and the copper elements. In 2008, superconductivity of the analogous material LaFFeAsO was found at 43 K [56, 57]. Soon after, the discovery, the superconducting transition temperature $T_c$ was raised up to 55 K by replacing La with other rare earth (RE) elements in the Institute of Physics, China Academy of Sciences [58, 59]. The following researches from other groups suggest that, by replacing the lanthanum in $LaO_{1-x}F_xFeAs$ with other rare earth elements such as cerium, samarium, neodymium, and praseodymium, superconductivity can also be found. In addition to higher critical temperature, $T_c$, the iron-based superconductors also have very low anisotropy, very high critical magnetic field, $H_{c2}$, and large critical current density, $J_c$. It is very suitable for high magnetic field application. For the engineering application of the 20-K temperature region, the iron-based superconductors may be the better choice than Nb-Ti and $Nb_3Sn$. Recent progress about iron-based superconductors can be found elsewhere [60].

The properties of main HTSC materials are shown in Tab. 2.3.

## 2.4 HTS bulk materials

HTS bulk materials have a series of unique features in terms of processing techniques, characteristic properties, and areas of applications. HTS bulk materials are especially attractive for practical applications because of their large current transport capacity in the presence of strong magnetic fields and their ability to trap very high fields in a compact space, particularly essential features for self-stable levitation without active control systems. Although the critical current density of bulk HTSC is about two orders of magnitude lower than that of thin films, the bulks offer a large effective critical current density ($J_c$) since metal substrates are not required. Thus, HTS bulks are widely used in flywheel energy storage system, transportation vehicles, current leads, fault current limiters, wind power generators, ship motors, compact NMR/MRI, etc [62]. In order to realize these applications, large size and excellent HTS bulks are required. It is necessary to improve the bulk properties further. Specific details about HTS bulk materials have been dissussed elsewhere [63]. This section only directly relates to the basic knowledge of HTS bulks for magnetic levitation.

### 2.4.1 Sintering HTS bulk materials

A single crystal is important for characterizing materials properties. However, single crystals do not have large transport current because the effective flux-pinning properties are only provided by defects and impurities; thus, they have no value for engineering applications.

In the early stage of HTS bulk material development, HTS bulks were synthesized by a sintering technique, which is a common ceramic processing route and generally easy to apply. The sintering techniques have a severe weak-link problem associated with grain boundaries; as a result, the techniques are most often employed in the search for and development of new compounds. The sintering bulk process includes the precursor powder preparation, pressing and shaping, melt casting, and recrystallization. HTS bulk materials can be achieved by a more conventional sintering approach. In June 1987, scientists in Hefei Research Institute of Cryogenics and

**Tab. 2.3:** Characteristic parameters of HTSC materials [61].

| Materials | $T_c$ (K) | $\mu_0 H_{c1ab}$ (T) | $\mu_0 H_{c1c}$ (T) | $\mu_0 H_{cab}$ (T) | $\mu_0 H_{c2ab}$ (T) | $\mu_0 H_{c2c}$ (T) | $\lambda_{Lab}$ (Å) | $\lambda_{Lc}$ | $\xi_{ab}$ (Å) | $\xi_c$ (Å) |
|---|---|---|---|---|---|---|---|---|---|---|
| 1. Y-123 | 92 | 520 | 53 | | 650 | 140 | 1250–4500 | 260–1300 Å | 270–310 | 40–60 |
| 2. Bi-2212 | 85 | 0.0001 | 0.019–0.022 | 0.022 | 3403 | 57–59 | 1800–2500 | 100 μm | 19–27 | 0.45 |
| 3. Bi-2223 | 110 | 0.0001 | 0.0135 | | 1210 | 39 | 2450 | 7.6 μm | 29 | 0.93 |
| 4. Tl-2223 | 125 | | | | | | 2050 | 4800 Å | 13 | |
| 5. Tl-1223 | 120 | | | | | | 2000 | | 18 | |
| 6. Tl-2212 | 110 | | | | | | 2150 | | 22 | 5 |
| 7. Hg-1223 | 134 | | | | 158–166 | 166 | | | 14.1–14.4 | |
| 8. $LaO_{1-x}F_xFeAs$ | 55 | | | | >100 | | | | 28 | |

Electronics Technology successfully prepared large bulk HTS samples using the sintering technique. The critical temperature of a sintered YBCO bulk sample with diameter 43 mm and thickness 2 mm was 86 K [64]. At the same time, the microwave surface resistance of this large YBCO bulk was measured. The results were not published, because the measured results were poor. Thereafter, the other group published results are also poor, because the sintered bulks have no good surface of the conductive properties and have severe weak link.

To compare with the conventional copper leads, HTS current leads can greatly reduce the heat leakage to the superconducting magnets operating at low temperatures. This is because the thermal conductivity of the HTS is two orders of magnitude smaller than that of copper. HTS bulks composed entirely of superconducting materials do not have metallic sheaths of high thermal conductivity as in HTS wires. Thus, the heat leakage of bulk HTS current leads is expected to be low. According to different application needs, the HTS bulk-sintered materials can be processed into various shapes (bars, tubes, plate, etc.). Both bulk-sintered BSCCO [65] and YBCO [66] can be made for current leads. The sintering technique is an important technology to produce BSCCO bars that can be used for current leads to supply power for superconducting magnets. A *c*-axis-oriented sintered BSCCO bulk can achieved a $J_c$ of $2.3 \times 10^4$ A/cm$^2$ at 20 K, 1 kOe) [67]. The heat leakage of the commercial Bi-based HTS current leads is a tenth that of copper current leads. In comparison to conventional copper leads, the thermal load and consumption of liquid helium were greatly reduced. Sintering techniques were also employed to fabricate tubes to shield external fields for the medical diagnosis of the human brain or heart [68]. Nishikubo et al. [69] have developed a RF shield using a HTS bulk. Rabbers et al. [70] have demonstrated the shielding of a DC magnetic field at 4.2 K, in fields up to 2 T, with a thick LHT $MgB_2$ bulk cylinder, 70 mm long and 18 mm bore.

A further improvement of HTS current leads is possible by replacing the Bi-based conductor with a Y-based one, since the HTS YBCO bulk, especially the melt texture growth (MTG) YBCO, can increase the current capacity and the reliability in magnetic fields. Thus, the MTG bulk HTS YBCO can be used for current leads. Large single-domain materials can shield higher fields. For instance, the shielding factor of a tube with 5-mm single-domain YBCO exceeds the value for sintered YBCO by one order of magnitude in the low-frequency range (1–200 Hz) [71]. Endoh et al. [66] has prepared a 500-A class HTS current lead package using the YBCO rod whose size is $\phi 3 \times 30$ mm, with the heat leakage under typical practical conditions, from 80 K hot end to 20 K cold end, being 163 mW. Ohsemochi et al. [72] has successfully achieved a HTS bulk current lead system with a rated current of 3000 A.

### 2.4.2 Melt process HTS bulk materials

The energy density of magnetized HTS bulks at 77 K is greater than that with HTS wires and tapes, since HTS bulks are capable of supporting a large persistent eddy current. However, for sintered YBCO HTS bulk, the $J_c$ is low due to the existence of weak links at the large numbers of grain boundaries. The sintered materials cannot be used in HTS magnetic levitation engineering; thus, a new preparation technique for bulk YBCO is required. Fabrication techniques for bulk (RE)BCO materials in the practical applications must overcome the weak-link problem associated with grain boundaries.

To overcome the $J_c$ limitations, due to the weak-link properties of the grain boundaries in sintered YBCO ceramics, Jin et al. [73] developed, in 1988, a melt process called MTG. The melt textured YBCO bulk can support a $J_c$ at least three orders of magnitude higher than that of sintered samples. The value of the transport current density $J_c$ of the MTG samples obtained by Jin et al. [74] had exceeded $10^4$ A/cm$^2$ in zero magnetic field at liquid nitrogen temperature of 77 K. This result indicates that grain alignment may have extensively reduced the weak links and lead to a great improvement of the material. In order to improve the $J_c$ values in a magnetic field, effective pinning centers must be introduced into the materials. In the following years, this process technique was modified and improved by several researchers. The first improvement was achieved by Salama et al. [75] by introducing the liquid-phase processing technique. Murakami et al. [76, 77] developed the melt powder melt growth (MPMG) method. REBCO prepared by MPMG has a high $J_c$ at 77 K in high magnetic fields. MPMG can easily increase domain size to a cubic centimeter. Microstructural observations reveal that the size of the 211 phase is much finer and its distribution is much more uniform than that of the classical MTG samples. MPMG also improves the $J_c$ values, especially under a magnetic field of 1 T. The $J_c(B)$ variations which were calculated from the magnetization measurements on MPMG processed samples exhibit $5–6 \times 10^4$ A/cm$^2$ in the self field and over $2 \times 10^4$ A/cm$^2$ in fields under 1 T.

In order to synthesize products for engineering applications, numerous approaches have been tried using a variety of synthesis technologies, such as MTG, melt powder melt growth (MPMG), top-seeded melt-textured growth (TSMG), and seeded infiltration and growth (SIG) [78] processes, etc. Up to now, the melt processes of the HTS (RE) BCO bulk generally use both MPMG and SIG process methods. Recently, Li et al. [79] and Chen et al. [80] give detailed reviews of materials processing of single-grain REBCO HTS bulks. The TSMG process for bulk (RE)BCO materials has been established as an effective method to fabricate large- and single-domain bulk REBCO with high performance. The TSMG is conductive to multi-block processing at the same time and controlled multi-domain bulk growth. However, the TSMG process technique has drawbacks, such as easy coarsening, large shrinkage, and liquid outflow. Macroscopic cracks and pores are also easy to form in the melt growth. The chemical composition of the final product is inhomogeneous in general. In comparison to the TSMG process, the major advantage of the SIG process is the ability to provide fine-sized spherical RE-211 precipitates in the REBCO matrix, even without addition of compounds such as Pt and $CeO_2$. However, the SIG process has the disadvantage that the uniformity of microstructure and $J_c$ is uncertain [81].

### 2.4.3 Developments of HTS bulk materials

The rare-earth Ba-Cu-O (REBCO, RE=Y, La, Nd, Sm, Eu, Gd, Dy, Ho, Er, Tm and Yb etc.) HTS bulks have high critical current density $J_c$ and high trapped magnetic flux. The maximum trapped magnetic field is approximately an order of magnitude higher than that for the presently used conventional PMs. Early studies of HTS bulk focused on YBCO. The critical current of HTS YBCO bulk materials had increased one thousand times in a few years from 1987 to 1991 [82]. Since then, HTS bulk materials and their applications have been considerable [79]. With a modified SIG process, the value of $J_c$ is up to $2.3 \times 10^5$ A/cm$^2$ under zero field and $10^4$ A/cm$^2$ up to 7 T at 77 K in YBCO [83]. Muralidhar et al. [84, 85] in the Railway Technical Research Institute (RTRI) reported achievement in the (Nd, Eu, Gd) BaCuO (NEG-123) system of superior $J_c$ values ($9.25 \times 10^5$ A/cm$^2$ at 65 K, $6.40 \times 10^5$ A/cm$^2$ at 77 K, and even $1 \times 10^5$ A/cm$^2$ at 90 K). A trapped field $B_T$ near 10 T may be possible at 77 K if proper chemical pinning centers are created [86].

A breakthrough in the inexpensive growth of large amounts of high-quality YBCO samples was made by the TSMG method [87]. In this TSMG technique, RE123 (Sm123 or Nd123) crystals with a higher melting temperature than Y123 were put on top of the precursor pellet. The processing conditions for the fabrication of NdBCO and SmBCO are considerably more complex than those for YBCO. A large number of single domain REBCO samples of various shapes were prepared with this method [88–93].

The melt process for the RE (Nd, Sm, or Eu) BCO systems under controlled oxygen partial pressure ($p_{O_2}$) is known as the oxygen-controlled melt growth (OCMG) process [94]. The OCMG-processed REBCO superconductors exhibit a larger $J_c$ in high magnetic fields and a considerably improved irreversible field $H_{irr}$ at 77 K, implying that stronger flux pinning can be realized in a commercially feasible way. The GdBCO sample exhibited the highest zero-field critical current density of $8.8 \times 10^4$ A/cm$^2$ at 77 K for $H_{\perp c}$, while the highest secondary peak $J_c$ of $6.5 \times 10^4$ A/cm$^2$ was achieved in NdBCO at 1 T and 77 K for $H_{\parallel c}$ [95].

A critical current density $J_c$ value above $10^5$A/cm$^2$ for melt processed YBCO superconductors has been obtained especially under high magnetic fields. For an external field parallel to both *a-b* planes and the direction of the current flow, critical current densities of $1.6 \times 10^4$ A/cm$^2$ in self field and $1.1 \times 10^4$ A/cm$^2$ under 1 T have been obtained [96].

In the following years, many efforts have been made to increase the textured domain size and trapped flux. A single bulk GdBCO sample 50 mm in diameter exhibited a trapped field of 2.6 T at 77 K, with the Hall sensor in contact with the surface [97].

A single bulk GdBCO sample 65 mm in diameter exhibited a trapped field of 3.05 T at 77 K. One study measured the trapped field between two GdBCO bulks in order to minimize the demagnetizing effect and found that it reached 4.3 T at 77 K [98]. Enlargement of the single-grain bulk REBCO materials is effective in improving their

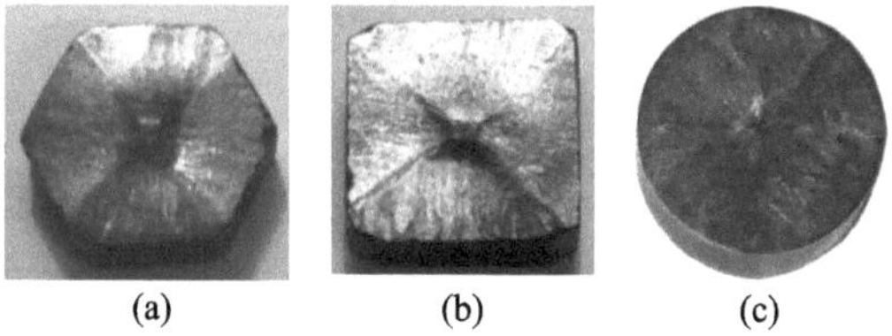


**Fig. 2.3:** Three shapes of HTS YBCO bulks in the top view, (a) regular hexagon, (b) square and (c) cylinder (Beijing General Research Institute for Nonferrous Metals, China).

ability to trap magnetic flux. However, it is difficult to obtain a single-grain bulk larger than about 80 mm because of undesirable nucleation at the positions away from the seed crystal [99].

In order to develop manned Maglev trains, high-quality MTG YBCO bulks [100] (Fig. 2.3a–c) were prepared by the Beijing General Research Institute for Nonferrous Metals in 2000. The superconductors used in the first manned HTS Maglev test vehicle in the world were the YBCO bulk 30-mm-diameter samples (Fig. 2.3c). Single- and multi-seeded melt-textured HTS YBCO bulks [101, 102] have been developed by the Adelwitz Technologiezentrum GmbH (ATZ), Germany. Some basic material parameters of ATZ melt-textured YBCO are summarized in Tab. 2.4. The three-seed YBCO bulks have a size of $67 \times 35 \times 15$ mm (Fig. 2.4) and were used in the Maglev vehicle at the University of Rio de Janeiro in 2011 and the Applied Superconductivity Laboratory (ASCLab), Southwest Jiaotong University, China, in 2013.

A microgravity experiment for growing large GdBCO bulks, 127 mm in diameter and 20 mm in thickness, was successfully performed on the spacecraft in 2003 [103]. Colleagues in the field of superconductor research are looking forward for more data from this large GdBCO bulk.

**Tab. 2.4:** YBCO bulk material parameters [101].

| Parameters | Value |
|---|---|
| Transition temperature $T_c$ | 92 K (−181°C) |
| Specific resistance $\rho_{300}$ | 0.6–0.8 mΩ·cm |
| Density (theoretical) | 6.38 g/cm$^3$ |
| Density (experimental) | 5.95 g/cm$^2$ (93%) |
| Specific heat capacity $c_{300}$ | 0.18 W·s·g$^{-1}$· K$^{-1}$ |
| Calorimeter meas. $c_{RT}$ | 0.25 W·s·K$^{-1}$·g$^{-1}$ |
| Coefficient heat transfer $\lambda$ | 4 W·m$^{-1}$·K$^{-1}$ $\parallel c$ |
| | 9 W·m$^{-1}$·K$^{-1}$ $\parallel a, b$ |
| | 6–7 W·m$^{-1}$·K$^{-1}$ poly |
| Lin. expansion coefficient | $5 - 8 \times 10^{-6}$ K$^{-1}$ |
| Tensile strength | 25–30 MPa |
| Critical current density $J_c$ | $10^4$–$10^5$ A·cm$^{-2}$ (77 K, 0 T) |

**Fig. 2.4:** Single- and multi-seeded melt-textured HTS YBCO bulks (Adelwitz Technologiezentrum GmbH (ATZ), Germany) [101].

In order to obtain a large single-grain bulk, Morita et al. [104, 105] in Nippon Steel Corporation (NSC) succeeded in developing the RE compositional gradient technique and prepared large-grained YBCO bulk superconductors with high $J_c$ values. Since then, the research and development of HTS bulks has made steady progress [106]. Nippon Steel Corporation reported fabrication of large single-grained REBCO bulk superconductors 150 mm in diameter using the RE compositional gradient technique. Figure 2.5 shows a single-grain, 150 mm in diameter GdBCO bulk with a compositional gradient of Dy. The insert in Fig. 2.5 is the trapped field distribution at 87 K, indicating that the sample is a single-grain bulk with no serious weak links.

Another important superconducting characteristic of the melt textured YBCO samples for the application is trapping flux. The magnitude of the trapped flux of REBCO HTS bulks is proportional to its critical current density $J_c$ and the volume. Krabbes et al. [107] prepared a cylindrical YBCO bulk with a modified melt crystallization process (MMCP) [108]. A maximum field of 16 T was trapped at 24 K in the gap of a

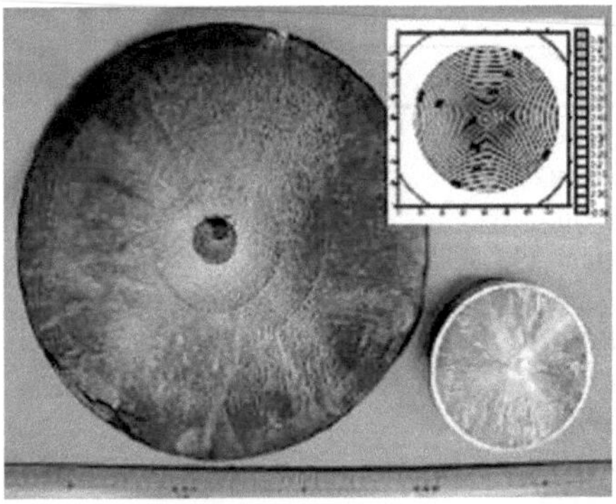

**Fig. 2.5:** REBCO large single-grain bulk 150 mm in diameter fabricated (Nippon Steel Corporation) [106].

mini-magnet made of two YBCO samples. $B_0$ on top of a single sample was 12.5 T at 20 K and 9 T at 40 K.

A bulk YBCO superconductor as small as 2.4 cm in diameter covered with carbon fiber fabric can trap an extremely high static field of 13.55 T at 34 K, and a bulk SmBCO superconductor as small as 2.4 cm in diameter can trap an extremely high static field of 13.69 T at 47 K [109]. Tomita and Murakami [110] prepared a YBCO bulk sample with high flux trapping after improving the mechanical stability and the thermal conductivity of the YBCO bulk. Their YBCO bulk was 2.65 cm diameter and trapped a magnetic field of 17.24 T at 29 K. This is the highest trapped magnetic flux so far.

Using NdBCO/YBCO/MgO film seeds and the cold-seeding method in TSMG, Xu et al. [111] successfully reprocessed failed bulks, and demonstrated a novel, convenient, and effective process for recycling the failed REBCO bulks. Peng et al. [112] reported that a large size SmBCO, 32 mm in diameter, was successfully grown in air by the cold-seeding method. In the conventional TSIG process, three types of powders, such as $Gd_2BaCuO_5$, $GdBa_2Cu_3O_{7-x}$, and $Ba_3Cu_5O_8$, must be prepared.

Yang et al. [113] has a new modified TSIG process technique, where only $BaCuO_2$ powders are required during the fabrication of the single-domain GdBCO bulk superconductors. Yang et al. [114] have observed in real time the growth process of single-domain YBCO bulk superconductors using an in situ high-temperature video camera, and obtained the growth rate of single domain YBCO bulk. The technique is helpful to find solutions for modifying preparation processes of YBCO bulk as well as for improve batch product quality.

Yang et al. [115] has reported a way to optimize the quality of the YBCO crystal by TSMG fabrication, with NdBCO thin film seeds. The optimal growth conditions of TSMG processing with NdBCO thin film seeds have been determined, and the optimal growth method is helpful for engineering fabrication. Wu et al. [116] reported a new approach for growing large single domains of YBCO by a TSMG process. Large YBCO single domains of 53 mm in diameter have been successfully produced using this method, and they predicted that samples with 75 mm in diameter can also be grown.

Shi et al. [117] have reported a successful multi-seeding technique for the fabrication of fully aligned, artificial (0° misalignment) grain boundaries within large-grain YBCO bulk superconductors using bridge-shaped seeds. Plechacek et al. [118] have developed a process capable of simultaneously fabricating up to 64 pieces of HTS bulks. Zhou et al. [119, 120] have reported that the SIG of GdBCO single grains using a YBCO-pressed pellet as the liquid source can successfully settle the liquid source leakage problem, even at $T_{max} \sim 1100°C$. The present method resulted in a significant enhancement of the trapped flux density of the GdBCO grains.

In order to improve the processing of single domains with such large dimensions, Noudem et al. [121, 122] proposed drilling artificial holes in the sintered powder before the crystal growth. The single domain of YBCO bulk multiple holes can improve mechanical properties, thermal stability, and the process of oxidation, and increase interfacial flux pinning if the pores can be made sufficiently small. More efficient heat transfer, faster oxygenation and less microcracking, possibility of reinforcement and of interlocking connections, etc. may result from the development of processing methods of the perforated and textured Y123 with a high performance which can open new pathways towards practical applications. Lousberg et al. [123] have reported the effects of filling the holes of drilled HTS samples with a soft ferromagnetic powder. The magnetic properties of the trapped field magnet were measured, and the experiments demonstrated an increase in flux trapping ability.

Sawh et al. [124] recently reported the results of studies on 53 melt-textured YBCO trapped field magnets (TFMs), 2 cm in diameter. The average trapped field on the seed-side surface was 2.04 T at 77 K.

## 2.5 Thermal properties of HTS bulk

It is well known that the critical temperature $T_c$, critical current density $J_c$, and upper critical field $H_{c2}$ of HTS bulks are the most important parameters for high-field applications of bulk superconductors, which have been discussed above. HTS bulks have a series of unique properties, which are very important for the applications. These problems should be mainly for the physicists and materials scientists to investigate. In addition to the three critical parameters $T_c$, $J_c$, and $H_{c2}$, some thermal and mechanical properties of HTS bulk materials associated with the applications will be briefly discussed in the section. Details of the HTS bulk properties can be found elsewhere [125].

### 2.5.1 Specific heat

The specific heat capacity, often simply called specific heat, is the heat capacity per unit mass of a material. An object's heat capacity $C$ is defined as the ratio of the amount of heat energy transferred to an object and the resulting increase in temperature of the object,

$$Q = Cm\Delta T, \tag{2.1}$$

where $Q$ is the amount of heat, $C$ is the specific heat capacity, $m$ is the mass, and $\Delta T$ is the temperature change after absorption or release of heat. In the International System of Units, heat capacity has the unit J/(kg·K).

The specific heat $C$ in a normal conductor consists of two contributions: from electrons in the conduction band $C_{el}$ and from the lattice or phonons $C_{ph}$. The electronic specific heat $C_{el}$ is defined as the ratio of that portion of the heat used by the electrons to the rise in temperature of the system. The free electron contribution to the specific heat is typically less than 1% of the phonon specific heat at room temperature.

When a small amount of heat is put into a superconductor, some of the heat is used to increase the lattice vibrations, and the rest is used to increase the energy of the conduction electrons. The specific heat of the electrons in a superconductor varies with the temperature in the normal and the superconducting state. The electronic specific heat in the superconducting state $C_{els}$ is smaller than in the normal state $C_{eln}$ at enough low temperatures, and the precise measurements indicate that at temperatures considerably below the transition temperature, the logarithm of the electronic specific heat is inversely proportional to the temperature. However, $C_{els}$ becomes much larger than $C_{eln}$ as the transition temperature $T_c$ is approached. The transition to the superconducting state is accompanied by quite drastic thermodynamic changes in the superconductor. The specific heat at the transition from the normal to the superconducting state in zero magnetic field appears as a jump at the critical temperature $T_c$.

Specific heat experiments on bulk superconductors are insensitive to the phase of the order parameter. However, they can provide valuable information on the density of states near the Fermi level because the electronic specific heat $C_e$ is proportional to the density of states.

Due to the low specific heat of LTS materials in superconducting state, practical LTS wires are produced as multifilamentary composites in order to prevent quenching. In contrast to LHS materials, there are some peculiar features in the specific heat of YBCO compared with those of conventional BCS superconductors, namely bulk HTS are thermally stable even in large sample sizes due to their relatively large specific heat in the superconducting state [126]. However, when a HTS bulk magnet is activated, another new thermal instability is brought in by the flux motion. For thermal stability, the cooling power of the system must be larger than the local heat generation.

Myers et al. [126] reported the specific heats of four superconducting materials at temperatures from 0 to 300 K and in magnetic fields from 0 to 14 T (Fig. 2.6). The specific heat of the Bi2212 samples was relatively independent of applied field, the zero-field specific heat of a two-dimensional random oriented single stack sample increased from 0.155 J/(kg·K) at 4 K to 254 J/(kg·K) at 250 K. At 2 K, the specific heat of a $Nb_3Sn$ rod-in-tube strand increased from 0.0257 J/(kg·K) at 0 T to 0.0716 J/(kg·K) at 14 T. In the zero field, the specific heat of a $MgB_2$ sample increased from 0.623 J/(kg·K) at 4 K to 382 J/(kg·K) at 250 K.

Naitoa et al. [127] have measured the temperature dependence of specific heat $C(T)$ for a Ag (10-20 μm thick) deposited YBCO coated conductor (YCC) film (about 1.5 μm thick), YCC reinforced by a thin Cu tape (300 μm thick), and a Hastelloy substrate with a buffer layer. Figure 2.7 shows the temperature dependence of the specific heat of YCC-Ag20 (20 μm), YCC-Ag10 (10 μm)+Cu300(300 μm), and Hastelloy+buffer. The $C(T)$ of all the samples decreases monotonically with decreasing temperature. Absolute values of $C(T)$ at 300 K are about 370, 390, and 400 J/(kg·K) for YCC-Ag20, YCC-Ag10Cu300, and Hastelloy+buffer, respectively.

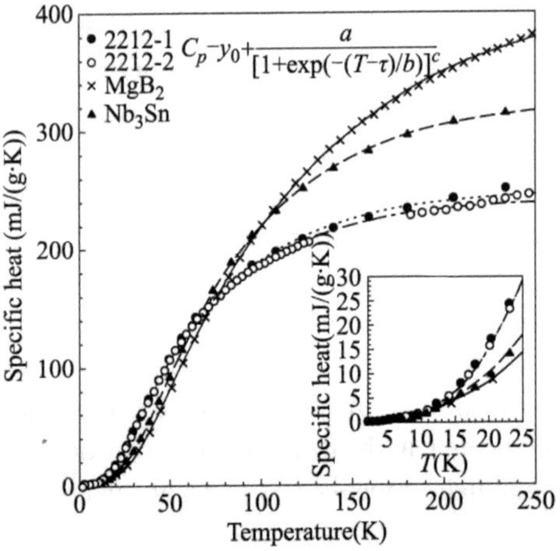


**Fig. 2.6:** Temperature dependence of the specific heat $C(T)$ of the Bi2212, $Nb_3Sn$, and $MgB_2$ samples at 0 T [126].

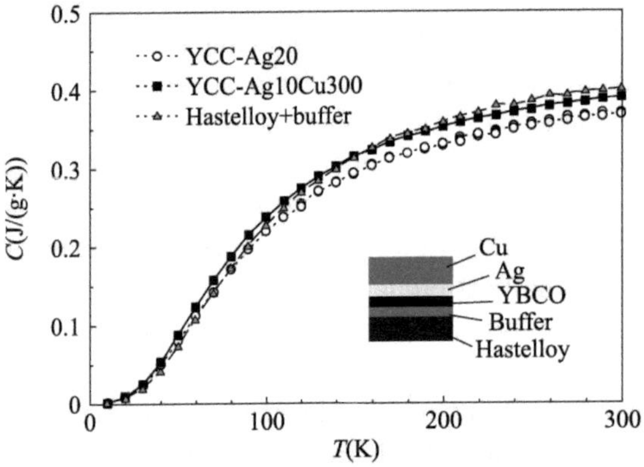


**Fig. 2.7:** Temperature dependence of the specific heat $C(T)$ of a YBCO-coated conductor YCC-Ag20, YCC-Ag10Cu300, and Hastelloy+buffer tapes [127].

Gahtori et al. [128] reported the temperature dependence of the specific heat $C(T)$ of $GdBa_2(Cu_{1-x}Mn_x)_3O_{7-\delta}$ for different $x$ (see Fig. 2.8) and gave the theoretical calculation results [129]. The model based on separate electron and phonon contributions was used to interpret the specific heat data. The substitution of Mn in the system has been found to effectively suppress the specific heat jump observed in the pristine compound. In the lower right corner of Fig. 2.6, it can be seen clearly that the specific heat of HTS Bi2212 is higher than LTS $Nb_3Sn$ and $MgB_2$ in the same temperature range.

Knowledge of specific heat and thermal conductivity of HTS is essential to understand the response of the superconductor to heat released due to variations of the applied magnetic field, for instance, the heat response of the superconductor during

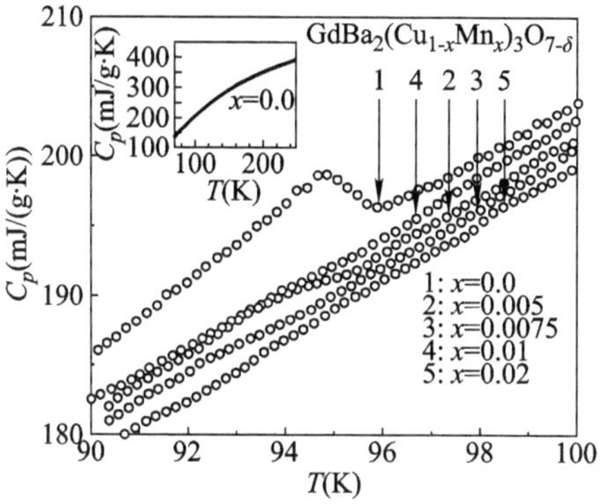


**Fig. 2.8:** Variation of specific heat $C(T)$ of $GdBa_2(Cu_{1-x}Mn_x)_3O_{7-\delta}$ for $x \leqslant 0.02$ in the temperature range of 90–100 K [128].

local heating when HTS bulks are activated by pulsed fields and the phenomenon of flux jumps at low temperature.

### 2.5.2 Thermal conductivity

According to Fourier's law, the temperature gradient $\Delta T$ imposed across an isotropic sample of a cross-sectional area $A$ results in a heat flow $Q$ given by

$$Q = -kA\Delta T, \tag{2.2}$$

With the minus sign indicating that heat always flows from a warmer to a colder region of a substance. The units of thermal conductivity are W/(m·K). The thermal conductivity $k$ consists of both the thermal conductivity of the electrons $k_{el}$ and the lattice or phonon $k_{ph}$.

In pure metals, the electronic component accounts for nearly all the heat conducted, while the lattice component is almost negligible. In superconductors at temperatures well below $T_c$, electrons condense into Cooper pairs that cannot transport entropy. Thus, they do not contribute to the thermal conductivity. The phenomenon can also be understood in terms of the superconducting state being a perfectly ordered state, i.e. one of zero entropy, thus vanishing thermal conductivity due to the electrons. At sufficiently low temperatures, the thermal conductivity is attributable entirely to lattice waves and is similar to the form of the thermal conductivity of an insulating material. The thermal conductivity in the normal state ($k_n$) approaches that in the superconducting state ($k_s$) as the temperature approaches the transition temperature for all materials, either pure or impure. In normal metals like copper, large electrical conductivity is accompanied by large thermal conductivity. The ratio between the two is approximately constant. The thermal conductivity in the superconducting state is smaller than that in the normal state and almost vanishes at very low temperatures. Even in the superconducting state, the existence of normal electrons cannot be completely avoided, due to the so-called quasiparticles in the superconducting state. The amounts of these quasiparticles in YBCO will affect the thermal conductivity, and their effect has been observed to decrease rapidly with temperature. It is now widely accepted that the rapid increase of $k$ below $T_c$ and the peak of $k$ are mostly due to the contribution of quaispartides located in the $CuO_2$ planes. However, further study is needed to clarify the effects of phonons and quasiparticles to the specific contribution.

Marchal et al. [130] reported the dependence of thermal conductivity, thermoelectric power, and electrical resistivity on temperature for a bulk, large-grain melt-processed YBCO HTS containing two grains separated by a well-defined grain boundary (see Fig. 2.9). Transport measurements at temperatures between 10 and 300 K were carried out both within a single grain (intra-granular properties) and across the grain boundary (inter-granular properties). The influence of an applied external magnetic field of up to 8 T on the measured sample properties was also investigated. The presence of the grain boundary is found to strongly affect the electrical resistivity of the melt-processed bulk samples; however, it has almost no effect on its thermoelectric power and thermal conductivity, within experimental error. The results of this study provide direct evidence that the heat flow in multi-granular melt-processed YBCO bulk samples should be virtually unaffected by the presence of grain boundaries in the material.

The coefficient of thermal conductivity of HTS bulk is about 2-10 W/mK at 77 K and very small below 10 K. Generally, there is a peak at around 50 K (see Fig. 2.9a–c). The planar structure of HTSCs makes the thermal conductivity anisotropic. The thermal conductivity of bulk YBCO is anisotropic: 3.5 W/mK along the *c*-axis and 14 W/mK along the *a*-*b* plane [110]. Other anisotropic results are 4 and 20 W/mK along the *c*-axis and the *a*-*b* plane, respectively [131].

### 2.5.3 Thermal expansion

The thermal expansion coefficients, $\alpha$, are defined as the change of length or volume per degree of temperature under constant pressure. The linear thermal expansion coefficient is defined by $\alpha_l = \mathrm{d}l/(l\mathrm{d}T)$, $\mathrm{d}l/l$ is the change rate of length, $\mathrm{d}T$ is temperature change. The unit of linear thermal expansion coefficient is $K.^{-1}$ Thermal expansion of superconducting materials, the change of the superconductor dimension between room and low operating temperature, is important for both trapped flux and engineering applications. The stresses resulting from the differential thermal expansion should be as small as possible. This is especially important for highly brittle HTS bulk materials, where these stresses may actually damage the superconducting bulks. The stresses also directly affect the maximum trapped flux in YBCO bulks. The

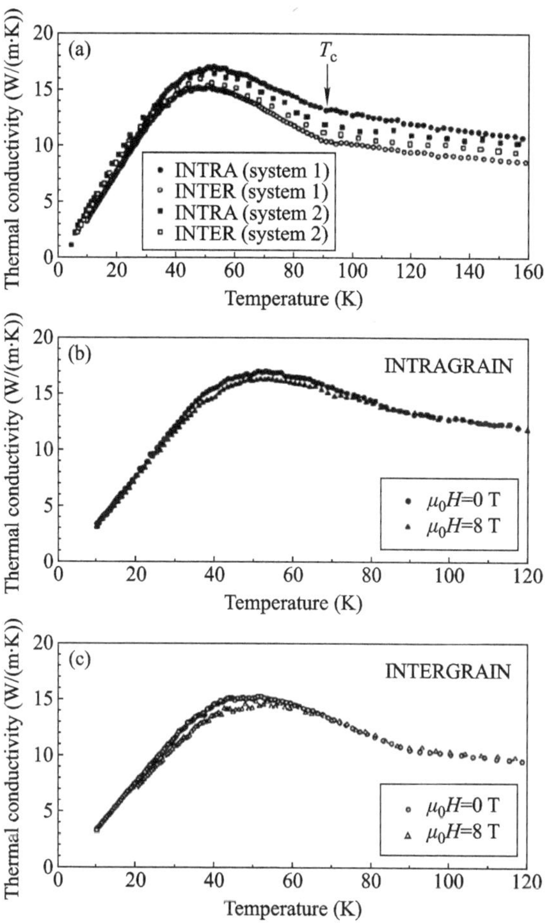


**Fig. 2.9:** (a) Comparison of the thermal conductivity measured within a YBCO grain and across a grain boundary in zero applied magnetic field, measured using: a physical property measurement system with the thermal transport option (TTO) (system 1) and a home-made high-sensitivity system with three thermal radiation screens (system 2). (b) Intra- and (c) inter-granular thermal conductivities at 0 and 8 T [130].

thermal expansion is closely related to the specific heat. Thermal expansion is also very useful for predicting how solid state properties respond to volume changes in the limit of zero applied pressure.

The thermal expansion of sintered HTS bulk is similar to that of non-superconducting ceramics. The HTS single crystals are necessary for study of structural anisotropy. Since HTS bulk has a layered orthorhombic structure, one needs to measure the linear coefficients of thermal expansion along all three crystallographic axes, i.e. the *a*- and *b*-axes in *a*-*b* plane and in the *c*-direction perpendicular to the $CuO_2$ planes. The linear thermal expansion coefficient of HTS bulk is in the range

$1 \times 10^{-6}$ to $1 \times 10^{-5}$ K.$^{-1}$ in temperatures of 40–300 K in *a-b* plane. The coefficient is in the range $3 \times 10^{-6}$ to $1.5 \times 10^{-5}$ K$^{-1}$ in temperatures of 40–300 K along the *c*-axis and rises approximately linearly up to 150 K, then more slowly rising [132]. Some measurement methods and data of thermal expansion can be found elsewhere [133]. The thermal expansion associated with trapped flux will be discussed in Section 2.7.

## 2.6 Mechanical properties of HTS bulk

In addition to the thermal properties above, the mechanical properties of the HTS bulks are also extremely important for industrial applications. HTS bulks are brittle materials with poor mechanical properties. This is because a number of voids are produced by oxygen formation or gas trapped in the MTG YBCO process. In addition, material fractures are caused by the existence and propagation of microcracks in the superconductor. In particular, thermal stress and fatigue are produced with repeated and rapid changes between room and low operation temperatures and hence cause failure of YBCO superconductors. When the highest trapped fields are generated by the external magnetic fields, large stress/strains will be induced and cause more serious rupture, which can be very dangerous. A number of researches have been carried out to improve the mechanical properties of melt-textured YBCO. The Vickers hardness test is one of the convenient methods to estimate the mechanical properties of materials. The mechanical properties of HTS bulk associated with the application are briefly discussed in this section. Detailed data of the mechanical properties for YBCO bulks can be found elsewhere [63, 134].

The mechanical properties of YBCO samples can be determined using the nanoindentation technique. Stress-strain curves ($\sigma$-$\varepsilon$) and mechanical properties such as hardness ($H$), elastic modulus ($E$), fracture toughness ($K_{IC}$), fracture strength ($\sigma_f$), and yield stress ($\sigma_{ys}$) can be obtained from the applied load ($P$) vs. indentation depth ($h$) data and with the corresponding tip indenter.

The density of the HTS materials depends strongly on the preparation and processing technique. The density of the HTS bulk materials prepared by MTG is larger than that by the sintering process. The density of HTS films and single crystals is the largest. For either MTG or the sintering process, some porosity problems in the interior of bulk materials are inevitable. The density of the HTS bulk ceramic materials $\rho$ depends strongly on their porosity. The magnitude of the elastic modulus of ceramic samples depends strongly on their density $\rho$, which is less than the theoretical maximum density $\rho_{max} = 6.383\ \mathrm{g/cm^3}$. The ratio $\rho/\rho_{max}$ denotes the relative density.

The hardness ($H$) correlates with yield strength and Young's modulus of the material. Vickers hardness is the preferred method since the indentation traces are small, typically 10–100 μm.

The elastic modulus $E$ of an object is defined as the slope of its stress-strain curve in the elastic deformation region. The elastic modulus (Young's or elastic modulus)

$E$ is defined by Hooke's law, $\sigma = E\varepsilon$, where $\sigma$ and $\varepsilon$ are tensile stress and strain, respectively.

The shear modulus $G$ in a certain deformation range is defined by the relation $\tau = G\gamma$, where $\tau$ is the shear stress and $\gamma$ is the shear strain.

Fracture toughness $K_{IC}$ is described as the ability of a material containing a crack to resist fracture. Fracture toughness reveals the resistance against crack propagation.

The normal stress at the beginning of fracture at the elastic limit is defined as the fracture strength $\sigma_f$. It is calculated from the load at the beginning of fracture during a tension test and the original cross-sectional area of the test sample, i.e. ratio of the maximum stress and fracture cross-sectional area at material breakage. HTS RE123 bulks are very brittle ceramics. Their fracture strength is not an invariant quantity, but is rather controlled by the microcracks within the bulk material. These microcracks set a limit for the fracture strength of the material because they start to propagate when the external tensile stress reaches the fracture toughness of the material. The fracture strength of YBCO-123 was reported to range from 40 MPa to 200 MPa, depending on the sample quality.

The tensile strength ($R_m$) of bulk YBCO along the *a-b* plane is relatively low, ranging from 10 to 30 MPa [135]. The high-trapped fields of HTS bulk samples are limited primarily by the tensile strength, because there is a relatively large Lorentz force in the HTS bulk.

The yield stress of the material $\sigma_{ys}$ can be obtained using the following expression $p_m = 1.1\sigma_{ys}$, where the mean contact pressure $p_m$ between the indenter and the material is the applied load divided by the contact area.

The mechanical properties for YBCO bulks are listed in Tab. 2.5. The mechanical properties of YBCO bulks have been investigated and measured using various methods. It is difficult to get standard data. This table lists only the present representative values for the reader's reference. The data for the Young's modulus and shear modulus show a weak increase slightly with decreasing temperature.

Fujimoto [143] reported the mechanical properties of densified bulk single-domain GdBCO. The average flexural stress of the specimens in this experiment was 72 MPa for densified bulks and 58 MPa for standard bulks. The Young's modulus was 145 GPa for the densified bulk and 120 GPa for the standard bulk. The measured fracture toughness $K_{IC}$ was 1.5 for the standard bulk and 1.7 for the densified bulk.

Fujimoto and Murakami [144] made a thorough study on the mechanical properties of a low-void-density Gd123 bulk and a conventional Gd123 bulk with voids evaluated at 77 K. The fracture toughness $K_{ICave}$ of the low-void-density bulk was higher than those of the porous bulk. The $K_{ICave}$ of the porous bulk was 1.36 MPa·m$^{1/2}$ for as-grown and 2.27MPa·m$^{1/2}$ for annealed. $K_{ICave}$ of the densified bulk was 1.43 MPa·m$^{1/2}$ for as-grown and 2.44 MPa·m$^{1/2}$ for annealed. Obviously, the fracture toughness of annealed bulks was higher than that of as-grown bulks. The increase in the Young's modulus and flexural strength at 77 K, compared to those at room temperature, was ascribed to the decrease in inter-atomic distance by cooling.

Tab. 2.5: The mechanical properties for REBCO bulk materials.

| Parameters | Materials | Units | Values | Ref. |
|---|---|---|---|---|
| Density ($\rho$) | YBCO-MTG | g/cm$^3$ | 6.383 | Theoretical max. value |
| Vickers hardness ($H$) | YBCO-TSMG (T) | GPa | $10.5 \pm 0.1$ | Roa et al. [136] |
| | YBCO-TSMG (O) | GPa | $7.8 \pm 0.1$ | Roa et al. [136] |
| Elastic modulus ($E$) | YBCO-TSMG pure | GPa | $177 \pm 10$ | Foerster et al. [137] |
| | YBCO-TSMG +10% $Ag_2O$ | GPa | $149 \pm 4$ | Foerster et al. [137] |
| | YBCO-TSMG (O) at 300 K | GPa | $193 \pm 4$ | Roa et al. [136] |
| | YBCO-TSMG (T) at 300 K | GPa | $192 \pm 3$ | Roa et al. [136] |
| | YBCO-TSMG (T)/(O) | GPa | $120 \pm 5$ | Bartolomé [138] |
| | GdBCO-MTG | GPa | 145 | Fujimoto et al. [139] |
| | DyBCO-TSMG | GPa | 120–160 | Murakami et al. [145] |
| Shear modulus ($G$) | YBCO-MTG | GPa | 37–40 | Krabbes et al. [63] |
| Fracture toughness ($K_{IC}$) | YBCO-TSMG pure | MPa·m$^{1/2}$ | $1.43 \pm 0.02$ | Foerster et al. [137] |
| | YBCO-TSMG +10% $Ag_2O$ | MPa·m$^{1/2}$ | $1.6 \pm 0.1$ | Foerster et al. [137] |
| | GdBCO-MTG | MPa·m$^{1/2}$ | 2.2 | Fujimoto et al. [139] |
| | DyBCO-TSMG | MPa·m$^{1/2}$ | 0.8–1.6 | Murakami et al. [145] |
| Fracture strength($\sigma_f$) | YBCO-MTG | MPa | 38–87 | Koblischka-Veneva et al. [140] |
| | REBCO-TSMG | MPa | ~70 | Murakami et al. [145] |
| | GdBCO | MPa | 72 | Fujimoto et al. [143] |
| Tensile strength ($R_m$) | YBCO | MPa | 10–30 | Sakai et al. [135] |
| | YBCO | MPa | 18.4–77.4 | Tomita [141] |
| Yield stress ($\sigma_{ys}$) | YBCO-TSMG | GPa | $3.58 \pm 0.10$ | Roa et al. [142] |

YBCO-TSMG (T) and (O) are the tetragonal of *a-b* plane and the orthorhombic phases along the *c*-axis, respectively.

Murakami et al. [145, 146] reported the mechanical properties of the low-porosity DyBCO bulks with different Dy211 contents and found that the bending strength and the Young's modulus of the low-porosity bulks were higher than those of porous Dy123 bulks.

The evaluation on the mechanical properties of an REBCO large single grain bulk of diameter 150 mm at liquid nitrogen temperature (77 K) was carried out by Murakami et al. [147] in 2013. Fracture strength data at 77 K were obtained through bending tests for planar specimens. The average fracture strength ($\sigma_{fave}$) was about 70 MPa. The $\sigma_{fave}$ of the large single-grain bulk at 77 K was slightly higher than that at room temperature.

Another important problem in the engineering applications of HTS bulks is the rapid and irreversible degradation due to exposure to ambient atmospheric conditions. In addition, other environments, such as fabrication, handling, transport, and storage, would also cause similar irreversible degradation. In particular, the degradation is very serious when HTS REBCO bulks are exposed to humid atmosphere environments. In order to prevent the performance degradation for REBCO bulk materials exposed to harmful environments, it is very necessary to investigate protection methods for REBCO bulk materials.

Rod et al. [148] studied the degradation of textured YBCO under operation in high humidity conditions and experimentally proved that the YBCO samples after degradation contain a high amount of Y-211 phases in the degraded region due to the reaction of YBCO with water. The results show that the initial formation of barium hydroxide and the non-superconducting green phase, $Y_2BaCuO_5$, induces surface degradation and degradation of mechanical properties.

In the research and development of the first man-loading HTS Maglev test vehicle [149] in the world in the last century, thin copper box sealing technology was used. YBCO bulks were placed in a copper box, then sealed by welding, vacuum pumped, injected with dry nitrogen with a pressure slightly higher than atmosphere, and then permanently sealed. Dry nitrogen improves heat conduction. The eight HTS Maglev equipments are composed of 344 blocks of YBCO. The magnetic levitation forces of each were measured every 6 months. The results showed that there was almost no change in the magnetic levitation forces after more than 10 years. After opening the copper boxes enclosing the HTSC, the YBCO bulks were almost unchanged. For permanent applications, this may be a very effective method (see Sections 6.24 and 6.29). There is another method that uses the epoxy and cryogenic adhesive to seal the YBCO bulks; however, long-term protection is not as effective as the above-mentioned methods.

## 2.7 Trapped fluxes in HTS bulk

Rare-earth PMs with high energy density and high intrinsic coercive force are widely used, bringing great technical and economic benefits. Superconductors can trap a high magnetic flux and thereby can become a PM. The superconducting PMs have much higher energy density than the rare-earth PMs, and their potential industrial applications are very promising.

Trapped fields in LTS bulk $Nb_3Sn$ were investigated at the temperature of 4.2 K (liquid helium) to 1.5 K in 1965. A $Nb_3Sn$ hollow cylinder, 18.6 mm long, 10.2 mm outer diameter, and 2.7 mm inside diameter, was synthesized by combines niobium-tin powder and sintering technique. The sintered bulk $Nb_3Sn$ hollow cylinder had demonstrated that a magnetic field of 0.2, 0.4, and 2.7 T can be trapped at 4.2, 2.2, and 1.5 K, respectively [150].

In 1977, a hollow $Nb_3Sn$ superconducting cylinder of 2.45 cm in diameter was used to trap a magnetic field of 2.24 T at 4.2 K. [151]. The cylinder of 7.62 cm was constructed by helically wrapping 44 layers of superconducting $Nb_3Sn$ ribbon 2.54 cm wide by 75 μm thick around a mandrel.

However, the trapped field in LTS bulk superconductors at low temperatures was found to be strongly limited by thermomagnetic instabilities, resulting from flux jumps as the external field exceeds certain values of the applied magnetic field. Due to the low specific heat of LTS materials in superconducting state, the thermomagnetic instabilities are very pronounced at low temperatures (see Section 2.5.1).

In contrast to LTS materials, there are some peculiar features in the specific heat of YBCO compared with those of conventional BCS superconductors, namely HTS bulks are thermally stable even in large sample sizes due to their relatively high specific heat in the superconducting state. Due to the larger specific heat of HTSC, the heat within the superconductor is released to the surrounding cryogen. Therefore, thermal instabilities have no influence on the trapped field in HTSC, at least at temperatures above 30 K. Very high trapped fields can be achieved by the HTS bulk material. The superconducting PM and its applications have become a reality because of HTSC material.

In engineering applications, high-texture and *c*-axis-oriented single domains are required. Large-sized, high-performance REBCO single domains are now commercially available. The trapped flux density ($B_{trap}$) due to flux pinning and the associated superconducting currents flowing persistently in a REBCO can be expressed in a simple model, as [152]

$$B_{trap} = A\mu_0 J_c r, \tag{2.3}$$

where $A$ is a geometrical constant, $\mu_0$ is the permeability of the vacuum, and $r$ is the radius of the grain. In order to increase the trapped flux, both the critical current density, the dimension and the homogeneity of HTSC bulk must be enhanced. Increase of the critical current density is accomplished by the improvement of flux pinning properties.

$$J_c = A\frac{V_f}{d}, \tag{2.4}$$

where $A$ is constant, $V_f$ is the volume fraction, and $d$ is the average diameter of Y211 particles. Thus, under constant volume of the second phase particles, $J_c$ is inversely proportional to the size of the Y211 particles [153].

In order to avoid the influence of a temperature rise on thermal stability, the cooling power of the system must be larger than local heat generation. Since the thermal conductivity of HTS bulk is small (see Section 2.5.2, about 2–10 W/mK at 77 K), cooling by cryogen at the bulk's surface is not sufficient to cool the interior region, leading to a local temperature rise. The temperature rise will reduce the flux pinning capability and induced flux jumps. Once the flux jumps appear in a HTS bulk, it is no

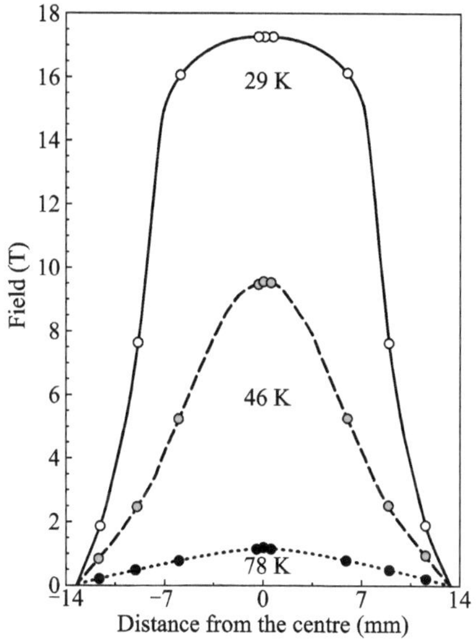


**Fig. 2.10:** The effect of temperature on trapped-field distribution. The field was trapped between two 26.5-mm-diameter YBCO disks for 29, 46, and 78 K [110].

longer stable. When the flux jump is serious enough, the HTS bulk will be destroyed due to the crack of formation from the large electromagnetic forces. Thus, for the superconducting PMs, it is very important to maintain the thermal stability.

These are serious challenges for superconducting materials experts. It is necessary to prevent the generation of microcracks and micro-pores during the melt processing. There are some methods to improve the HTS bulk performance, for example, addition of silver, sample reinforcement with metal rings, etc. Tomita and Murakami have developed a post-fabrication treatment, which improves the mechanical properties as well as the thermal conductivity of a YBCO bulk magnet, thereby increasing its field-trapping capacity. These methods use resin impregnation and wrap the materials in carbon fiber and also include a small hole drilled into the center of the magnet to allow the impregnation of BiPbSnCd alloy into the superconductor. These methods greatly enhance the thermal stability and internal mechanical strength. As a result, 17.24 T could be trapped, without fracturing, in a bulk Y-Ba-Cu-O sample of 2.65 cm diameter at 29 K (Fig. 2.10) [110].

There are three typical ways to magnetize a HTS bulk: field cooling (FC) magnetization, zero-field cooling (ZFC) magnetization, and pulse field magnetization (PFM). When the HTS bulk is magnetized in a magnetic field, the HTS bulk can trap flux. After removing the field, a superconducting permanent magnet can be obtained. In comparison, the ZFC magnetization is more difficult, because a very high external magnetic field is generally needed. The FC magnetization is convenient. The applied magnetic field needs to reach at least somewhat above the level of the residual magnetic field.

The magnetization process may lead to a local temperature rise in superconductors. Oka et al. [154] reported this warming effect. The temperature rise of a SmBCO single-domain bulk superconductor was precisely measured during the ZFC process. The highest temperature rise was reported as 7.5 K when the sample was magnetized at a sweeping rate of 5.06 mT/s at 50 K.

The pulsed field magnetization (PFM) is considered to be the most practical method because it is inexpensive, small-volume, and mobile experimental set-up to apply a magnetic field. However, the field trapped by PFM was smaller than that attained by FCM below 77 K, because of the large temperature rise caused by the magnetic flux motion in the bulk. In order to improve the magnetization performance of HTS PM using PFM method, much work has been done. Several approaches were tried and succeeded in enhancing trapped flux around 77 K, including such as iteratively magnetizing pulsed field operation with reducing amplitudes (IMRA) [155] and multi-pulse technique with stepwise cooling (MPSC) [156]. A single-pulse magnetization gives a poor result compared to the ZFC, IMRA and MPSC magnetization. The trapped flux of the latter can be made to be almost as high as for the ideal FC case.

Fujishiro et al. [157] succeeded in establishing a record of high-field trapping, $B_{trap} = 5.20$ T on the $\phi$45 mm GdBCO bulk, using a 6.7-T applied field at 29 K. A maximum $B_{trap}$ of 3.6 T was attained by a single-pulse application.

Trapped field profiles for a 26-mm-diameter GdBCO bulk magnetized at 77 K in FCM and PFM were investigated [158]. FCM trapped a higher peak field (0.90 T) than PFM (0.63 T), but using IMRA allows pulsed magnetization to give approximately 95% of the FC trapped flux. The real effectiveness of PFM came when using both the IMRA and MPSC together.

In general, the pulsed magnetic field is provided by the solenoid in PFM. The pulsed field of the solenoid is not uniform and imposes inhomogeneous pulsed-field distribution upon the magnetization of HTS bulk samples. Ida et al. [159] studied the magnetization of GdBCO which was inserted between two disks as pulsed-current vortex-type Cu coils. As a result, the pulse current was smaller and the magnetic field distribution was more uniform than for a solenoid tape.

Deng [160] presented the experimental results for different penetration patterns of magnetic flux when employing vortex-type coils in PFM, that is, the magnetic flux will primarily penetrate inside the bulk from the upper and lower surfaces using vortex-type coils. In general, the size and current of vortex-type coils is smaller than that of a solenoid; thus, so a reduced heat generation and temperature rise can be obtained by vortex coils, which is very important for PFM applications.

A new compact NMR magnet using HTS bulk magnets has been developed. However, it is difficult to trap a uniform magnetic field above 4.7 T at liquid nitrogen temperature. Kim et al. [161] presented a remagnetization method in order to improve the field homogeneity of a HTS bulk NMR relaxometry device, and obtained the 1.58 T and 9 ppm/cm$^3$ using the single HTS bulk with 20 mm inner diameter, 60 mm outer diameter, and 80 mm height by the remagnetization process.

Durrell et al. [162] reported a trapped field of 17.6 T in a stack of two GdBCO HTS bulks from a 17.8-T magnetizing field at 26 K. This is the highest field trapped in a bulk superconductor reported to date. The silver-doped GdBCO bulks were 25 mm in diameter and fabricated by TSMG and reinforced with shrink-fit stainless steel.

LTS bulks show the large and sharp magnetic flux jumps that prevent their use. HTS bulks show higher thermal stability and can be used as superconducting PMs. The $MgB_2$ is a new type of LTSCs. The $MgB_2$ bulk has also a promising potential as superconducting PM. It has several attractive features for bulk magnets, such as low cost, light weight, and homogeneous trapped field distribution. In addition, flux jumps in $MgB_2$ arise at low temperature (4.2 K), but tend to disappear at temperatures larger than 10 K. Perimi et al. [20] have studied the magnetic field trapping capability of $MgB_2$ discs of different shapes, at 10-K temperature. The SIMS (superconductive inserts in metallic substrates) devices showed a higher stability of the trapped fields with respect to the bulk discs. Typical trapped fields, measured at 1 mm from the surface of the device, are of the order of 1 T.

Fujishino et al. [24] performed PFM on a large $MgB_2$ bulk 50 mm in diameter, in which a trapped field $B_{trap} = 0.47$ T was achieved at 23 K. Thereafter, they reported a maximum trapped field $B_{trap} = 0.71$ T, which was realized at the center of the bulk surface at 14 K after a magnetic pulse application of $B_{ex} = 1.55$ T [24]. They found that the flux dynamics and heat generation in a $MgB_2$ bulk during PFM were clearly in contrast to those for REBCO bulks, because of the small specific heat, large thermal conductivity, and narrow temperature margin against $T_c$ in the $MgB_2$ bulk.

In order to reduce the cost and meet the needs of more and more applications, batch production of bulk HTSC been improved. Plechacek et al. [118] have reported development and successful tests of a melt-powder-melt-growth process capable of simultaneous fabrication of up to 64 YBCO levitation disks, a necessary step towards the planned production capacity of several thousand pieces a year. Diameters of the HTS bulks varied from 14 to 56 mm. The trapped magnetic field measured at 77 K on the bulks arbitrarily chosen from different batches was in all cases higher than 0.5 T. Levitation force at 77 K with a levitation gap of 1 mm reached the same value of 80 N in all samples. The self-field critical current density $J_c$ at 77 K exhibited values between 50 and 90 kA/cm$^2$ at the bulk center and 70–95 kA/cm$^2$ at the bulk edge.

A production run of 60 melt-textured YBCO trapped field magnets, 2 cm in diameter, was reported by Sawh et al. [163] The HTS PMs have pinning centers of Y211 and $Pt_{0.4}U_{0.6}/YBa_3O_6$ deposits and also damage tracks due to ions from uranium fission. The resulting average trapped field at the center of the seed-side surface is 2.04 T at 77 K.

Trapped flux can be measured by magnetic sensor scanning, which will be introduced in Chapter 5.

HTS PM can be widely used in construction of practically frictionless bearings, non-contact spinners, Maglev flywheel energy storage devices, Maglev trains, motors, wind or tidal power generators, NMR/MRI, magnetic drag delivery, magnetic separation, magnetron sputtering, and contact-less liquid pumps, etc.

## References

[1] Matthias BT, Geballe TH, Geller S, Corenzwit superconductivity of $Nb_3Sn$. E. Phys Rev. 1954;95:1435.

[2] Chu CW. Strong electron–phonon interactions in $HgBa_2Ca_3Cu_4O_{10+\delta}$ superconductor. In: Encyclopedia of applied physics, Vol. 20, Weinheim: VCH Publishers, Inc.; 1997. pp. 213–47.

[3] Berlincourt TG, Hake RR. Upper critical fields of transition metal alloy superconductors. Phys Soc II. 1962; 7:293–95.

[4] Kadyrov E, Gurevich A, Larbalestier DC. High critical current densities in Nb47%Ti multilayers with a planar copper flux pinning nanostructure. Appl Phys Lett. 1996;68:1567–69.

[5] Wilson MN. NbTi superconductors with low ac loss: A review. Cryogenics. 2008;48:381–95.

[6] Kunzler JE, Buehler E, Hsu FS, Wernick JH. Superconductivity in $Nb_3Sn$ at high current density in a magnetic field of 88 kgauss. Phys Rev Lett. 1961;7:215.

[7] Godeke A, Jewell MC, Fischer CM, Squitieri AA, Lee PJ, Larbalestier DC. The upper critical field of filamentary $Nb_3Sn$ conductors. J Appl Phys. 2005;97:093909.

[8] Godeke A, Cheng D, Dietderich DR, Ferracin P, Prestemon SO, Sabbi G, Scanlan RM. Limits of NbTi and $Nb_3Sn$, and development of W&R Bi-2212 high field accelerator magnets. IEEE Trans on Appl Supercond. 2007;17(2):1149–52.

[9] Wilson MN. Superconducting magnets. New York: Oxford University Press; 1987.

[10] Iwasa Y. Case studies in superconducting magnets: design and operational issues. Heidelberg: Springer; 2009.

[11] Rossi L, Bottura L. Superconducting magnets for particle accelerators. vol. 5. Singapore: World Scientific Publishing Company; 2012. p. 51–89.

[12] Bottura L, Godeke A. Superconducting materials and conductors, fabrication, and limiting parameters. vol. 5. Singapore: World Scientific Publishing Company; 2012. p. 25–50.

[13] Nakagawa K. Development of $Nb_3Al$ Superconducting Wires Fabricated by the RHQT Process. Teion Kogaku, 2010;33(9):623–628.

[14] Nagamatsu J, Nakagawa N, Muranaka T, Zenitani Y, Akimitsu J. Superconductivity at 39 K in magnesium diboride. Nature. 2001;410:63–4.

[15] Roberts BW. Survey of superconductive materials and critical evaluation of selected properties. J. Phys Chem. 1976;5:581–822.

[16] Orlando TP, Delin KA. Foundations of applied superconductivity. New Jersey: Addison-Wesley; 1991.

[17] Cardwell DA, Ginley DS. Handbook of superconducting materials, Vol I: Superconductivity, materials and processes. Florida: Institute of Physics Publishing; 2003. 1:2009

[18] Matsushita T. Flux pinning in superconductors. Heidelberg: Springer-Verlag; 2007.

[19] Tachikawa K, Togano K. Potential methods for the fabrication of high-$T_c$ superconductors for wires and cables. Proc IEEE. 1989;77:1124–1131.

[20] Perini E, Giunchi G, Saglietti L, Albisetti A, Matrone A, Cavaliere V. Magnetic field trapping in MgB bulks and inserts. IEEE Trans on Appl Supercond. 2011;21:2690–2693.

[21] Kambara M, Babu NH, Sadki ES, Cooper JR, Minami H, Cardwell DA, Campbell AM, Inoue IH. Rapid communication: High intergranular critical currents in metallic $MgB_2$ superconductor. Supercond Sci Technol. 2001;14:L5.

[22] Yamamoto A, Yumoto H, Shimoyama J, Kishio K, Ishihara A, Tomita M. $MgB_2$ bulk superconducting magnet. Physics. 2012;8:1845.

[23] Viznichenko RV, Kordyuk AA, Fuchs G, Nenkov K, Mnuller KH, Prikhna TA, Gawalek W. Temperature dependence of trapped magnetic field in $MgB_2$ bulk superconductor. Appl Phys Lett. 2003;83:4360–4362.

[24] Fujishiro H, Naito T, Oyama M, Arayashiki T, Tamura T, Sasaki T, Giunchi G. Numerical Simulation of trapped field and temperature rise in MgB bulks magnetized by pulsed field. IEEE Trans on Appl Supercond. 2013;23:6800804.

[25] Muralidhar M, Ishihara A, Suzuki K, Fukumoto Y, Yamamoto Y, Tomita M. Optimization of the fabrication process for high trapped field $MgB_2$ bulks. Physica C. 2013;494:85–88.

[26] Naito T, Sasaki T, Fujishiro H. Trapped magnetic field and vortex pinning properties of $MgB_2$ superconducting bulk fabricated by a capsule method. Supercond Sci Technol. 2012;25:095012.

[27] Marignetti F, Cavaliere V, Giunchi G, Messina G, Celentano G, Matrone A. Use of MgB Superconductors for excitation field in synchronous machines—Part I: bulk magnets. IEEE Trans on Appl Supercond. 2013;23:8002506.

[28] Bednorz JG, Müler KA. Possible high-$T_c$ superconductivity in Ba–La–Cu–O system. Physik Z. 1986;B64(1):189–193.

[29] Wu MK, Ashburn JR, Torng CJ, Hor PH, Meng RL, Gao L, Huang ZJ, Wang YQ, Chu CW. High-pressure study of the new Yb–Ba–Cu–O superconducting compound system. Phys Rev Lett. 1987;58:908–912.

[30] Zhao ZX, Chen LQ, Cui CG, Huang YZ, Liu JX, Chen GH, Li SL, Guo SQ, He YY. High $T_c$ superconductivity of Sr (Ba)–La–Cu. Chinese science bulletin. 1987;32:177–179. (In Chinese).

[31] Michel C, Hervieu M, Borel MM, Grandin A, Deslandes F, Provost J, Raveau B. Superconductivity in the Bi–Sr–Cu–O system. Z. Phys B. 1987;68:421–423.

[32] Tarascon JM, LePage Y, Barboux P, Bagley BG, Greene LH, McKinnon WR, Hull GW, Giroud M, Hwang DM. Crystal substructure and physical properties of the superconducting phase $Bi_4(Sr, Cr)_6Cu_4O_{16+x}$. Phys Rev B. 1988;37:9382.

[33] Maeda H, Tanaka Y, Fukutumi M, Asano T. The strength of rubbing worked on polyimide films for aligning nematic and chiral smectic liquid crystals: Controlling pretilt angles and some electrooptic performances of LCDs. Jpn J Appl Phys. 1988;27:827–829.

[34] Zandbergen HW, Huang YK, Menken MJV, Li JN, Kadowaki K, Menovsky AA, vanTendeloo G, Amelinckx S. Electron microscopy on the $T_c$ = 110 K (midpoint) phase in the system $Bi_2O_3$–SrO–CaO–CuO. Nature. 1988;332:620–623.

[35] Huang Y, Miao H, Hong S, Parrell JA. Recent advances in Bi-2212 round wire performance for high field applications. IEEE Trans on Appl Supercond. 2014;24:6400205.

[36] Golovashkin AI, Shabanova NP. Direct determination of electronic density of states N*(EF) for $Nb_3Sn$ from upper critical field $H_{C2}$ and resistivity $\sigma$. Physica C: Supercond. 1991;185–189:2709–2710.

[37] Weijers HW, Trociewitz UP, Marken KR, Meinesz M, Miao H, Schwartz J. The generation of 25.05 T using a 5.11 T $Bi_2Sr_2CaCu_2Ox$ superconducting insert magnet. Supercond Sci Technol. 2004;17:636–636.

[38] Kalsi S. Applications of high temperature superconductors to electric power equipment. New York: John Wiley and Sons; 2011.

[39] Yi HP, Zhang JS, Liu Q, Han ZH. The chinese journal of nonferrous metals. 2004;14S1:341–346. (In Chinese).

[40] Rupich MW, Li X, Thieme C, Sathyamurthy S, Fleshler S, Tucker D, Thompson E, Schreiber J, Lynch J, Buczek D, Demoranville K, Inch J, Cedrone P, Slack J. Advances in second generation high temperature superconducting wire manufacturing and R&D at American Superconductor Corporation. Supercond Sci Technol. 2010;23:014015.
[41] Rupich MW, Li X, Sathyamurthy S, Thieme CLH, DeMoranville K, Gannon J, Fleshler S. Second Generation Wire Development at AMSC. IEEE Trans Appl Supercond. 2013;23:6601205.
[42] Hazelton D, Xie Y, Selvamanickam V, Anthony R, Llambes JC, Lehner T. Proc IEEE CITRES; 2010. pp. 42–46.
[43] Larbalestier DC. In: Rogalla H, Kes PH, editors. 100 years of superconductivity. Florida: CRC Press; 2012. 627–642.
[44] Li YJ, Zhao ZC, Liu LF, Ye QH, Zheng H. IEEE Trans on Appl Supercond. 2009;19:3295–3298.
[45] Li YJ, 2G HTS wire in Shanghai Jiao Tong University. Email message (2014).
[46] Selvamanickam V, Chen Y, Kesgin I, Guevara A, Shi T, Yao Y, et al. IEEE Trans on Appl Supercond. 2011;21:3049–3054.
[47] Li X, Rupich MW, Thieme CLH, Teplitsky M, Sathyamurthy S, Thompson E, Siegal E, Buczek D, Schreiber J, DeMoranville K, Hannus D, Lynch J, Inch J, Tucker D, Savoy R, Fleshler S. The development of second generation HTS wire at American Superconductor. IEEE Trans on Appl Supercond. 2009;19:3231–3235.
[48] Song HH, Brownsey P, Zhang YF, Waterman J, Fukushima T, Hazelton D. 2G HTS coil technology development at SuperPower. IEEE Trans on Appl Supercond. 2014;23:4600806.
[49] Sheng ZZ, Hermann AM, Ali AE, Almasan C, Estrada J, Datta T, Matson R. Superconductivity at 90 K in the Tl-Ba-Cu-O System. Phy Rev Lett. 1988;60:937–940.
[50] Sheng ZZ, Hermann AM. Nature. Bulk superconductivity at 120 K in the Tl–Ca–Ba–Cu–O system. 1988;19:138–139.
[51] Putilin SN, Antipov EV, Chmaissem O, Marezio M. Superconductivity at 94 K in $HgBa_2CuO_{4+\delta}$. Nature. 1993;362(6417):226–228.
[52] Schilling A, Cantoni M, Guo JD, Ott HR. Nature. Superconductivity above 130 K in the Hg–Ba–Ca–Cu–O system. Nature. 1993;363(6424):56–58.
[53] Chu CW, Gao L, Chen F, Huang ZH, Mcng RL, Xue YY. Superconductivity above 150 K in $HgBa_2Ca_2Cu_3O_{8+\delta}$ at high pressures. Nature. 1993;365(6444):323–325.
[54] Kamihara Y, Hiramatsu H, Hirano M, Kawamura R, Yanagi H, Kamiya T, Hosono H. Iron-based layered superconductor: LaOFeP. J Am ChemSoc. 2006;128(31):10012.
[55] Kordyuk AA. Iron-based superconductors: Magnetism, superconductivity, and electronic structure (Review Article). Low Temp Phys. 2012;38(9):888–899.
[56] Kamihara Y, Watanabe T, Hirano M, Hosono H. Iron-based layered superconductor $La[O_{1-x}F_x]FeAs$ (x = 0.05–0.12) with $T_c = 26$ K. J Am Chem Soc. 2008;130(11):3296–7.
[57] Takahashi H, Igawa K, Arii K, Kamihara Y, Hirano M, Hosono H. Superconductivity at 43 K in an iron-based layered compound $LaO_{1-x}F_xFeAs$. Nature. 2008;453(7193):376–8.
[58] Ren ZA, Lu W, Yang J, Yi W, Shen XL, Li ZC, Che GC, Dong XL, Sun LL, Zhou F, Zhao ZX. Superconductivity in the iron-based F-doped layered quaternary compound $Nd[O_{1-x}F_x]FeAs$. Chin Phys Lett. 2008;82(6):2215–2216.
[59] Ren ZA, Zhao ZX. Research and Prospects of Iron-Based Superconductors. Adv Mater. 2009;21(6):4584.
[60] Ma YW. Progress in wire fabrication of iron-based superconductors. Supercond Sci Technol. 2012;25(11):912–914.
[61] Cardwell DA, Ginley DS. Handbook of superconducting materials, Vol I: superconductivity, materials and processes. Bristol: Institute of Physics Publishing; 2003. p. 927, 982, 1022, 1041.

[62] Hull JR, Murakami M. Applications of bulk high-temperature superconductors. Proc IEEE. 2004;92:1705–1718.

[63] Krabbes G, Fuchs G, Canders WR, May H, Palka R. High temperature superconciluctor bulk materials. Weinheim: Wiley-Vch Verlag GmbH & Co. KGaA; 2006.

[64] Wang SY, Wang JS, Wang JY, Wei XQ, Zhang HB, Jia J, Yang SH, Wang C, Tang B. Superconductivity of Bulk Y–Ba–Cu–O Oxide. Superconductivity and cryogenics. 1987;15(3):47–48.

[65] Tamura H, Mito T, Yamada Y, Watanabe M, Ohkubo J, Heller R. Bi2212 HTS tubular bulk with conical shape for current lead. Journal of Physics: conference series. 2006;43:1035–1038.

[66] Endoh R, Kato H, Izumi T, Shiohara Y. Evaluation of the invasion heat for the HTS current lead using YBCO. Physica C. 2003;392–396:1167–1170.

[67] Obata K, Shimoyama J, Yamamoto A, Ogino H, Kishio K, Kobayashi S, Hayashi K. Critical current properties of c-axis oriented Bi (Pb) 2223 bulks sintered under high gas pressures. Physics procedia, 2012;36:665–668.

[68] Fagnard JF, Vanderheyden B, Vanderbemden P. Magnetic shielding with bulk high temperature superconductors: factors influencing the magnetic field penetration in hollow cylinders. Miryala M, ed. Superconductivity: recent development and new production technologies. New York: Nova Science Publishers; 2012. pp. 127–150.

[69] Nishikubo T, Endo H, Itoh M. Orientation characteristics in RF magnetic shielding effects of BPSCCO Plates. IEEE Trans on Appl Supercond. 2009;19(3):2245–2248.

[70] Rabbers JJ, POomen M, Bassani E, Ripamonti G, Giunchi G. Magnetic shielding capability of $MgB_2$ cylinders. Supercond Sci Technol. 2010;23(12):125003.

[71] Fang H, Claycomb JR, Zhou XY, Putman PT, Padmanhabhan S, Miller JH, Ravi-Chandar K, Salama K. Melt-textured YBCO superconducting tube for magnetic shielding. IEEE Trans on Appl Supercond. 2003;13(2):3103–3105.

[72] Ohsemochi K, Ono Mi, Nomura S, Kuriyama T, Kasahara H, Akita S, Koso S. Development of 3kA conduction cooled HTS current lead system . Cryogenics. 2003;43(10–11):643–648.

[73] Jin S, Tiefel TH, Sherwood RC, Van Dover RB, Davis ME, Kammlott GW, Fastnacht RA. Melt-textured growth of polycrystalline $YBa_2Cu_3O_{7-\delta}$, with high transport $J_c$, at 77 K. Phys Rev. 1988;37(13):7850–7853.

[74] Jin S, Tiefel TH, Sherwood RC, Van Dover RB, Davis ME, Kammlott GW, Fastnacht RA, Kuth HD. High critical currents in Y–Ba–Cu–O superconductors. Appl Phys Lett. 1988;52(24): 2074–2076.

[75] Salama K, Selvamanicham V, Gao L, Sun K. High current density in bulk $YBa_2Cu_3O_x$ superconductor. Appl Phys Lett. 1989;54(23):2352–2354.

[76] Murakami M, Morita M, Doi K, Miyamoto K. A new process with the promise of high $J_c$ in Oxide Superconductors. Jpn J Appl Phys. 1989;28(7):1189–1194.

[77] Murakami M, Gotoh S, Fujimoto H, Yamagushi K, Koshizuka N, Tanaka S. Flux pinning and critical currents in melt processed YBaCuO superconductors. Supercond Sci Technol. 1991;4(4):S43.

[78] Chen YL, Chan HM, Harmer MP, Todt VR, Sengupta S, Shi D. A new method for net-shape forming of large, single-domain $YBa_2Cu_3O_{6+x}$. Physica C. 1994;234(3–4):232–236.

[79] Li BZ, Zhou DF, Xu K, Hara S, Tsuzuki K, Miki M, Felder B, Deng ZG, Izumi M. Materials process and applications of single grain (RE)–Ba–Cu–O bulk high-temperature superconductors. Physica C. 2012;482(11):50–57.

[80] Chen PW, Chen SY, Chen IG, Wu MK. Melt processing and flux pinning of bulk RE–Ba–Cu–O high temperature superconductors. Miryala M, (Ed.) Superconductivity: recent development and new production technologies. New York: Nova Science Publishers; 2012. pp. 105–125.

[81] Iida K, Babu NH, Shi Y, Cardwell DA. Seeded infiltration and growth of large, single domain Y–Ba–Cu–O bulk superconductors with very high critical current densities. Supercond Sci Technol. 2005;18(11):1421–1427.
[82] Ogawa N, Hirabayashi I, Tanaka S. Preparation of a high-$J_c$, YBCO bulk superconductor by the platinum doped melt growth method. Physica C. 1991;177(1–3):101–105.
[83] Kumar ND, Rajasekharan T, Muraleedharan K, et al. Unprecedented current density to high fields in $YBa_2Cu_3O_{7-\delta}$ superconductor through nano-defects generated by preform optimization in infiltration growth process. Supercond Sci Technol. 2010;23(10):105020.
[84] Muralidhar M, Sakai N, Jirsa M, Murakami M, Hirabayashi I. Highly enhanced flux pinning in melt-textured NEG-123 doped Nb and Mo nano particles. Appl Phys Lett. 2008;92:162512.
[85] Muralidhar M, Sakai N, Jirsa M, Tomita M. LRE-$Ba_2Cu_3OY$ superconductors on the way to applications. In: Miryala M, editor. Superconductivity: recent development and new production technologies, Nova Science Publishers; 2012. pp. 11–29.
[86] Weinstein R, Parks D, Sawh RP, Davey K. Measurements of flux pumping activation of trapped field magnets. Supercond Sci Technol. 2010;23(11):115015–115021(7).
[87] Lee DF, Partsinevelos CS, Presswood RG, Jr., Salama K. Melt texturing of preferentially aligned $YBa_2Cu_3O_x$ superconductor by a seeded directional solidification method. J Appl Phys.1994;76:603–605.
[88] Fuchs G, Stoye P, Staiger T, Krabbes G, Schäzle P, Gawalek W, et al. Melt textured YBCO samples for trapped field magnets and levitating bearings. IEEE Trans on Appl Supercond. 1997;7(2):1949–1952.
[89] Kaiser AW, Koch HJBR. Processing technique for fabrication of advanced YBCO bulk materials for industrial applications. Inst Phys Conf Ser. 1997;158:837–840.
[90] Cardwell DA, Lo W, Thorpe HDE, Roberts A. A controllable temperature gradient furnace for the fabrication of large grain YBCO ceramics. J Mater Sci Lett.1995;14(20):1444–1447.
[91] Yoo SI, Sakai N, Takaichi H, Higuchi T, Murakami M. Melt processing for obtaining $NdBa_2Cu_3O_y$ superconductors with high $T_c$ and large $J_c$. Appl Phys Lett. 1994;65(5):633–635.
[92] Gawalek W, Habisreuther T, Zeisberger M, Litzkendorf D, Surzhenko O, Kracunovska S, et al. Batch-processed melt-textured YBCO with improved quality for motor and bearing applications. Supercond Sci Technol. 2004;17(10):1185.
[93] Murakami M. Materials developments and applications of bulk Re–Ba–Cu–O superconductors. In: Narlikar A, editor. Frontiers in superconducting materials, Springer Berlin Heidelberg; 2005. pp. 869–884.
[94] Murakami M, Yoo SI, Higuchi T, Skai N, Weltz J, Koshizuka N, et al. Flux Pinning in Melt-Grown $NdBa_2Cu_3O_y$ and $SmBa_2Cu_3O_y$ Superconductors. Jpn J Appl Phys. 1994;33(5):L715–L717.
[95] Muralidhar M, Murakami M. Synthesis and characterization of $LRE_{1+x}Ba_{2-x}Cu_3O_y$ (LRE: Nd, Eu, Gd, NEG) superconductors: a low oxygen partial pressure (0.05% $O_2$+Ar). Supercond Sci Technol. 2002;15(5):683.
[96] Lees MR, Bourgault D, Braithwaite D, de Rango P, Lejay P, Sulpice A, et al. Transport properties of magnetically textured $YBa_2Cu_3O_{7-\delta}$. Physica C. 1992;191(3–4):414–418.
[97] Nariki S, Sakai N, Murakami M. Processing of high-performance Gd–Ba–Cu–O bulk superconductor with Ag addition. Supercond Sci Technol. 2002;15(15):648.
[98] Nariki S, Sakai N, Murakami M. Superconducting properties of (Gd, Y)–Ba–Cu–O bulk materials. The 4$^{th}$ International PASREG workshop; June30–July 2, Jena Germany; 2003.
[99] Teshima H, Morita M, Okajima N, Terao Y, Sekino M, Ohsaki H, et al. Magnetization characteristics of multiple bulk superconductors for higher field applications. In: Proc. ICEC. 24–ICMC; May 14–18; Fukuoka, Japan; 2012. 879–882.
[100] Ren H, Xiao L, Jiao Y, Zheng M. Processing and characterization of YBCO superconductors by top-seeded melt growth method in batch process. Physica C. 2004;412(412):597–601.

[101] Werfel FN, Floegel-Delor U, Rothfeld R, Riedel T, Goebel B, Wippich D, et al. Superconductor bearings, flywheels and transportation. Supercond Sci Technol. 2011;25(1):14007–14002(16).

[102] Werfel FN, Floegel-Delor U, Riedel T, Goebel B, Rothfeld R, Schirrmeister P, et al. Large-scale HTS bulks for magnetic application. Physica C. 2013;484(1):6–11.

[103] Sakai N, Inoue K, Shisa A, Hirata H, Murakami M. Fabrication of large-grain bulk superconductors in microgravity environment. Acta Astronautica. 2003;53(1):35–43.

[104] Morita M, Miyamoto K, Doi K, Murakami M, Sawano K, Matsuda S. Processing and superconducting properties of high-$J_c$ bulk YBaCuO prepared by melt process. Physica C. 1990;172(3–4):383–387.

[105] Morita M, Sawamura M, Takebayashi S, Kimura K, Teshima H, Tanaka M, et al. Processing and properties of QMG materials. Physica C. 1994;235–240(6):209–212.

[106] Teshima H, Morita M. Physics recent progress in HTS bulk technology and performance at NSC. Procedia. 2012;36:572–575.

[107] Krabbes G, Fuchs G, Verges P, Diko P, Stöer G, Gruss S. 16 T trapped fields in modified YBaCuO: materials aspects. Physica C. 2002;378–381(15):636–640.

[108] Schäzle P, Krabbes G, Gruβ S, Fuchs G. YBCO/Ag bulk material by melt crystalization for cryomagnetic applications. IEEE Trans on Appl Supercond. 1999;9(2):2022–2025.

[109] Tomita M, Murakami M. Improvement of the trapping capacity of bulk superconductor with resin impregnation. IEEE Trans on Appl Supercond. 2003;13(2):3095–3098.

[110] Tomita M, Murakami M. High-temperature superconductor bulk magnets that can trap magnetic fields of over 17 tesla at 29 K. Nature. 2003;421(6922):517–20.

[111] Xu HH, Chen YY, Cheng L, Yan SB, Yu DJ, Guo LS, et al. YBCO-Buffered NdBCO Film with higher thermal stability in seeding REBCO growth and recycling failed bulk YBCO superconductors. J Supercond Nov Magn. 2013;26(4):919–922.

[112] Peng BN, Cheng L, Zhuang YF, Xu HH, Yao X. Large size and high $T_c$, $J_c$ SmBCO bulk superconductor with addition of Sm242 particles grown in air. Physica C. 2014;496(496):11–13.

[113] Yang WM, Zhi X, Chen SL, Wang M, Li JW, Ma J, et al. Fabrication of single domain GdBCO bulk superconductors by a new modified TSIG technique. Physica C. 2014;496(1):1–4.

[114] Yang WM, Guo XD, Wan F, Li GZ. Real-time observation and analysis of single-domain YBCO bulk superconductor by TSIG process. Crystal growth & design. 2011;11(7):3056–3059.

[115] Yang CM, Wang SY, Huang YC, Chen PW, Chen IG, Wu MK. The optimal growth of single grain bulk Y–Ba–Cu–O superconductors with Nd–Ba–Cu–O thin film seed. IEEE Trans on Appl Supercond.2014;23(3):6800204–6800204.

[116] Wu XD, Xu KX, Fang H, Jiao YL, Xiao L, Zheng MH. A new seeding approach to the melt texture growth of a large YBCO single domain with diameter above 53 mm. Supercond Sci Technol. 2009;22(12):125003.

[117] Shi YH, Durrell JH, Dennis AR, Babu NH, Mancini CE, Cardwell DA. Properties of grain boundaries in bulk, melt processed Y–Ba–Cu–O fabricated using bridge-shaped seeds. Supercond Sci Technol. 2012;25(4):45006–45012(7).

[118] Plechacek V, Jirsa M, Rames M, Muralidhar M. Batch production of YBCO disks for levitation applications. Physics procedia. 2012;36:538–543.

[119] Zhou DF, Hara S, Li B, Xu KZ, Noudem J, Izumi M. Significant improvement of trapped flux in bulk Gd–Ba–Cu–O grains fabricated by a modified top-seeded melt growth process. Supercond Sci Technol. 2013;26(1):015003.

[120] Zhou DF. Doctoral Dissertation, Tokyo University of Marine Science and Technology, 2013;26(1):015003.

[121] Noudem JG, Meslin S, Harnois C, Chateigner D, Chaud X. Melt textured $YBa_2Cu_3O_y$ bulks with artificially patterned holes: a new way of processing c-axis fault current limiter meanders. Supercond Sci Technol. 2004;17(7):931–936.

[122] Noudem JG. From porous $Y_2O_3/Y_2BaCuO_5$ to single domain $YBa_2Cu_3O_7$ Superconductors. In: Miryala M, editor. Superconductivity: recent development and new production technologies, Nova Science Publishers; 2012. pp. 65–79.

[123] Lousberg GP, Fagnard JF, Chaud X, Ausloos M, Vanderbemden P, Vanderheyden B. Magnetic properties of drilled bulk high-temperature superconductors filled with a ferromagnetic powder. Supercond Sci Technol. 2011;24(3):035008.

[124] Sawh RP, Weinstein R, Carpenter K, Parks D, Davey K. Production run of 2 cm diameter YBCO trapped field magnets with surface field of 2 T at 77 K. Supercond Sci Technol. 2013;26(10):105014.

[125] Koblischka-Veneva A, Sakai N, Tajima S, Murakami M. Handbook of superconducting materials. In: Cardwell DA, Ginley DS (Eds.). Vol I: superconductivity, materials and processes, Institute of Physics Publishing; 2003. 893–945.

[126] Myers CS, Susner MA, Motowidlo L, Distin J, Sumption MD, Collings EW. Transport critical current densities and \$n\$-values of multifilamentary MgB wires at various temperatures and magnetic fields. IEEE Trans on Appl Supercond. 2014;24(3):1–5.

[127] Naitoa T, Fujishiroa H, Yamamurab Y, Saitob K, Okamotoc H, Hayashic H, et al. Specific heat and thermal diffusivity of YBCO coated conductors. Physics procedia. 2012;36:1609–1613.

[128] Gahtori B, Lal R, Agarwal SK, Rao A, Lin YK, Sivakumar KM, et al. Thermal transport in $(Y,Gd)Ba_2(Cu_{1-x}Mn_x)_3O_{7-\delta}$ for $x \leqslant 0.02$. J Phys Condens. 2007;19(25):256212.

[129] Gahtori B, Agarwal SK, Chakraborty T, Rao A, Kuo YK. Specific heat and correlation between resistivity and thermoelectric power of $GdBa_2(Cu_{1-x}Mn_x)_3O_{7-\delta}$ HTSC system for $x \leqslant 0.02$. Physica C. 2009;469:27–29.

[130] Marchal C, Fagnard JF, Shi YH, Cardwell DA, Mucha J, Misiorek H, et al. The influence of a grain boundary on the thermal transport properties of bulk, melt-processed Y–Ba–Cu–O. Supercond Sci Technol. 2013;26(1):88–93.

[131] Ainslie MD, Fujishiro H, Ujiie T, Zou J, Dennis AR, Shi YH, et al. Modelling and comparison of trapped fields in (RE)BCO bulk superconductors for activation using pulsed field magnetization. Supercond Sci Technol. 2014;27(6):065008.

[132] Meingast C, Blank B, Bükle H, Obst B, Wolf T, Wül H. Anisotropic pressure dependence of $T_c$ in single-crystal $YBa_2Cu_3O_7$ via thermal expansion. Phys Rev B. 1990;41(16):11299.

[133] Meingast C. Thermal properties: thermal expansion. In: Cardwell DA, Ginley DS, editor. Handbook of superconducting materials Vol II: superconductivity, materials and processes, Institute of Physics Publishing; 2003. pp. 1513–1525.

[134] Koblischka-Veneva A, Sakai N, Tajima S, Murakami M. In: Cardwell DA, Ginley DS, editor. Handbook of superconducting materials Vol II: characterization, applications and cryogenic, Institute of Physics Publishing; 2003. pp. 1527–1550.

[135] Sakai N, Seo SJ, Inoue K, Miyamoto T, Murakami M. In: Koshizuka N, Tajima S, editor. Advances in superconductivity XI, Springer; 1999. pp. 685–688.

[136] Roa JJ, Dias FT, Martinez M, Padilla JA, Segarra M. Oxygenation kinetics of YBCO-TSMG samples using the nanoindentation technique. Journal of the European Ceramic Society. 2012;32(2):425–431.

[137] Foester CE, Lima E, Rodriguez P, Serbena FC, Lepienski CM, Cantao MP, et al. Brazilian mechanical properties of Ag-doped top-seeded melt-grown YBCO pellets. J Phys. 2008;38(3a):341–345.

[138] Bartolomé E, Roa JJ, Bozzo B, Segarra M, Granados X. Effective silver-assisted welding of YBCO blocks: mechanical versus electrical properties. Supercond Sci Technol. 2010;23(4):045013.
[139] Fujimoto H, Teshima H, Morita M. Improvement of microstructure and mechanical properties of large single domain GdBaCuO superconductor with low void density. Physica C. 2010;470(22):S23–S25.
[140] Koblischka-Veneva A, Sakai N, Tajima S, Murakami M. In: Cardwell DA, Ginley DS, editor. Handbook of superconducting materials Vol I: Superconductivity, materials and processes, Institute of Physics Publishing; 2003. p. 911.
[141] Tomita M, Murakami M, Katagiri K. Reliability of mechanical properties for bulk superconductors with resin impregnation. Physica C. 2002;378–381(5):783–787.
[142] Roa JJ, Morales M, Martinez M, Capdevila XG, Segarra M. Curva tensión-deformación mediante indentación esférica para materiales cerámicos superconductores de YBCO-TSMG. Anales de Mecanica de la Fractura. 2009;26(2):495.
[143] Fujimoto H. Preparation and mechanical properties of large single domain GdBaCuO superconductor with low void density. IEEE Trans Appl Supercond. 2009;19(3):2933–2936.
[144] Fujimoto H, Murakami A. Fracture toughness properties of Gd123 superconducting bulks. Physics procedia. 2012;36:458–462.
[145] Murakami A, Miyata H, Hashimoto R, Otaka K, Katagiri K, Iwamoto A. Mechanical properties of low porosity Dy123 bulks with different Dy211 content. Physica C.2009;469(15):1207–1210.
[146] Murakamia A, Otakab K, Miurab T, Iwamoto A. Effects of Pt content on mechanical properties of Dy123 bulks physics procedia. 2012;36:551–556.
[147] Murakami A, Fujimoto H, Iwamoto A. Low temperature mechanical properties of RE–Ba–Cu–O large single-grain Bulk 150 mm in diameter. IEEE Trans on Appl Supercond. 2013;23(3):6800505–6800505.
[148] Roa JJ, Jiméez-Piqué E, Díz J, Morales M, Calleja A, Segarra M. Corrosion induced degradation of textured YBCO under operation in high humidity conditions. Surface & Coatings Technology. 2012;206(19–20):4256–4261.
[149] Wang JS, Wang SY, Zeng YW, Huang H, Luo F. The first man-loading high temperature superconducting Maglev test vehicle in the world. Physica C. 2002; s378381 (1):809–814.
[150] Lange F. The effect of rates of change of magnetic field and of temperature on flux jumps in $Nb_3Sn$ hollow cylinders. Cryogenics. 1965;5(3):143–145.
[151] Rabinowitz M, Arrowsmith HW, Dahlgren SD. Dependence of maximum trappable field on superconducting $Nb_3Sn$ cylinder wall thickness. Appl Phys Lett. 1977;30(11):607–609.
[152] Takizawa T, Murakami M (Eds). Critical current in superconductors. Tokyo. Nihon University College of Humanities and Sciences. Fuzambo International; 2005.
[153] Murakami M. Materials developments and applications of bulk RE–Ba–Cu–O superconductors.Berlin. In: Narlikar A, editor. Frontiers in superconducting materials, Springer Verlag; 2005. 569–884;81(2):481–488.
[154] Oka T, Yokoyama K, Fujishiro H, Noto K. Thermal and magnetic behaviors of a melt-textured superconducting bulk magnet in the zero-field-cooling magnetizing process. Supercond Sci Technol. 2009;22(6):065014
[155] Yanagi Y, Itoh Y, Yoshikawa M, Oka T, Hosokawa T, Ishihara H, et al. Trapped field distribution on Sm–Ba–Cu–O bulk superconductor by pulsed-field magnetization. Berlin. Springer Verlag. 2000;470–472.
[156] Sander M, Sutter U, Koch R, Klaser M. Pulsed magnetization of HTS bulk parts at T < 77 K. Supercond Sci Technol. 2000;13(6):841.
[157] Fujishiro H, Tateiwa T, Fujiwara A, Oka T, Hayashi H. Higher trapped field over 5 Ton HTSC bulk by modified pulse field magnetizing. Physica C Supercond. 2006;s445–448:334–338.

[158] Patel A, Giunchi G, Albisetti AF, Shi Y, Hopkins SC, Palka R, et al. High force magnetic levitation using magnetized superconducting bulks as a field source for bearing applications. Physics procedia. 2012;937–942.

[159] Ida T, Matsuzaki H, Akita Y, Izumi M, Sugimoto H, Hondou Y. Magnetization properties for Gd–Ba–Cu–O bulk superconductors with a couple of pulsed-field vortex-type coils. Physica C. 2004;s412–414:638–645.

[160] Deng Z, Shinohara N, Miki M, Felder B, Tsuzuki K, Watasaki M, et al. Pulsed-field magnetization properties of bulk superconductors by employment of vortex-type coils. Physics procedia. 2012;36(3):958–962.

[161] Kim SB, Hayashi N, Kitamura H, Eritate I. Study on the remagnetization process to improve the field homogeneity of the stacked HTS bulk magnets. IEEE Trans on Appl Supercond. 2013;24(3):1–5.4301405.

[162] Durrell JH, Dennis AR, Jaroszynski J, Ainslie MD, Palmer KGB, Shi YH, et al. A trapped field of 17.6 T in melt-processed, bulk Gd–Ba–Cu–O reinforced with shrink-fit steel. Supercond Sci Technol. 2014;27:082001.

[163] Sawh RP, Weinstein R, Carpenter K, Parks D, Davey K. Production run of 2 cm diameter YBCO trapped field magnets with surface field of 2 T at 77 K. Supercond Sci Technol. 2013;26(10):105014.

Jia-Su Wang and Su-Yu Wang

# 3 Magnetic levitation

Levitation and suspension have been for centuries associated with psychic and strange phenomena in which an object occupies a fixed position in a gravitational field without any direct physical contact. The interaction between objects in this suspended state will not bring friction, energy loss, and heat generation at the interface. Magnetic levitation refers to the suspension phenomenon associated with a magnetic field. The characteristic is a suspension phenomenon without direct contact between the object and the magnetic sources. Magnetic levitation is abbreviated as Maglev. Maglev is a general term, but in some cases, it is often thought to refer to the Maglev train. This is perhaps due to the influence of these applications in the field of traffic. In this book, we use the following definition, as not to cause unnecessary confusion. The term "Maglev" maintains its original meaning. It can be used as an attributive to define other specialized magnetic levitation, for example, Maglev train, Maglev bearing, etc. In this book, the first letter of Maglev is capitalized.

## 3.1 Introduction

After the development of the high-speed information technology, the human race dreams of a "door-to-door" high-speed transport system just like the modern information technology. The conventional automobile transport system has many disadvantages, such as traffic accidents, energy consumption, pollution, traffic jams, etc. Therefore, ground transportation with tracks is considered as one of the mainstream transport methods of the 21st century. Traditional railway transport has not solved the future problems of ground transportation at high speed. High-speed railway (HSR) emerges as the times require. Overall, airline, road traffic, and HSR modes compete actively with each other in the distance range of 300 to 1000 km, and the result is that the HSR occupies obvious advantages in the market. The speed of 250–350 km/h with HSR cannot meet the demand for higher speed or ultra-high speed. In a 1000-km journey, people readily choose airline traffic. However, Maglev has a higher speed and better safety, so that the Maglev technology becomes one of the best choices for the future ultra-high-speed transportation. The Maglev system can achieve speeds of 500 km/h or even thousands of km/h, have lower energy consumption and life cycle costs, attract more passengers, and produce less noise and vibration than HSR. In order to facilitate the reader's understanding of Maglev transportation. This chapter will first introduce a general synopsis of the traditional railway and then discuss the levitation phenomenon, magnetic levitation concepts, permanent magnet (PM) levitation, and the normal conductive Maglev. The superconducting Maglev, LTS and HTS Maglev, will be discussed in the next chapter. Various

https://doi.org/10.1515/9783110538434-003

Maglev versions have different unique performances. In the constant pursuit of the "perfect" Maglev transportation, the new HTS Maglev train may be one of the best candidates.

More books and review articles about the Maglev vehicles can be found in the references (Maglev books and reviews at the end of this chapter).

## 3.2 Rail traffic – the mainstream of the 21st-century transportation

Comparing the rail traffic with other traffic modes, we find that rail traffic is undoubtedly the most reliable traffic mode. Its reliability results not only from a lower number of traffic accidents, but also less loss incurred from these accidents. Car and airplane accidents were 8.6 and 7.1 times, respectively, more frequent than railway accidents in Japan in 1985. Compared with the automobile and aircraft accident rate, rail traffic accidents are nearly negligible. The mortality rate of the Japanese Shinkansen was zero during 34 years of operation. The mortality rates (people per 10 billion person-kilometers [pkm]) for cars, planes, and trains were 501, 5.31, and 0.78, respectively [1]. The total number of people killed worldwide in road traffic accidents is more than 22 million [2], which is close to the 26 million soldiers killed in World War II.

In addition, railway traffic has a series of advantages [3]. Compared to automobile or air transport, railway transport produces the lowest amount of emissions. Railway traffic is an efficient and environment-friendly transport system in most cases, where large volumes of goods can be transported for long distances quickly and with a minor impact on environment. Transport's share of $CO_2$ emissions from fuel combustion is about 30% of all emissions. $CO_2$ emissions for railway, bus, aviation, and automobile are 19, 51, 109, and 147 g/pkm, respectively [4]. From the total costs of reduction of negative impact on environment in the transport industry, only 8% comes from railway transport, whereas up to 90% comes from road transport, even though its traffic performance is lower by 50% than the traffic performance of railway transport. Transportation taxes are based on the environmental (pollution, noise, etc.) cost; thus, it is more competitive to increase railway than highway and aviation transport.

Another important argument placing railway transport above road transport is a lower occupation of agricultural land. While road transport occupies almost 74% of such land, railway transport occupies only 26%, even though its traffic performance is almost twice as big as that of road traffic. In general, the total area for highways is three to four times that required for the railway.

In the national economy, railway traffic has its specific and irreplaceable position. It enables and facilitates transportation of goods at much cheaper cost. In comparison with road carriers, railway transport also has lower restrictions. Road carriers must respect several regulations, such as limitations in utilizing highways and primary roads, limitations in peak traffic, or obligations for driver safety breaks. Road traffic is a disordered traffic tool. Traffic "jams" often block roads, causing a great waste of time and economic loss. In 1991, in 37 American cities each with a population over 1 million, 91% of the cars experienced traffic jams with a total economic loss near $100 billion.

Investigations of consumer preferences for air and HSR services in Japan were conducted by Fu [5]. Overall, airline, road traffic, and HSR modes compete actively with each other in the distance range of 300 to 1000 km. The results show that the rail market share was 53%, 69%, and 36% for 300–500-, 500–700-, and 700–1000-km distance ranges, respectively. Airline's market share was 5%, 19%, and 56%, respectively, for the same distance ranges in intra-Japan markets. Road traffics' market shares are 42%, 12%, and 8%, respectively, for the same distance ranges in intra-Japan markets. Thus, it can be seen that the HSR for 250–300-km/h speed has obvious market advantages for the 1000-km journey.

For those reasons, more than 200 countries will construct urban railway transit systems, and the vast majority are funded by government. Rail transportation is the mainstream of the 21st-century transportation. All countries in the European Union have express trains running between their major cities with several countries also operating high-speed trains. The HSR is also being developed very rapidly in China.

## 3.3 Developments in railway traffic

The traditional railway has an almost 300-year history (since 1825). Railway traffic plays an important role in human progress. Now, traditional railway traffic will be quickly replaced by HSR. The limitations on HSR traffic will be discussed in this section.

The advent of HSR in 1964, with its subsequent improvements in operating speed, brought a new economic era to Japan. The Japanese technology has significantly improved since then, and nowadays, the Shinkansen is running at 250–300 km/h. Before the Chinese HSR, the Japanese HSR operation routes were longest in the world.

Developers of the German Intercity Express (ICE) set the goal that HSR should offer an average traveling speed twice that of the car and half as high as that of air travel. The German ICE test train achieved 406 km/h in 1988, and the German ICE system achieved an average speed of 350 km/h.

The French Train Grande Vitesse (TGV) achieved a new speed record of 417 km/h on the long line between Paris and Lyon in 1981. The French TGV system achieved an average speed of 317 km/h. This was followed by another leap in the speed record in 1991, on an experimental run, the French TGV train established the record speed for rail systems of 515 km/h [6]. It is worth mentioning that a new world rail speed record was set on April 3, 2007, by the French SNCF with a speed of 574.8 km/h on the new East European high-speed line [7].

By the end of 2012, Chinese HSR (250–350 km/h) completed the traffic line – the "Four North-South and Four East-West Network". China's HSR total operating mileage reached 19,000 km in 2015. At present, more than a dozen countries in the world have operational HSR. China has the largest HSR operation mileage in the world. The total HSR operating mileage is half that of the whole world [8]. According to the long-term extension plan, the HSR line will have 30,000 km in China by 2020.

The rapid development of HSR fully demonstrated its superiority compared with road and airline traffic. Without further development and progress of science and technology, HSR will not be able to satisfy the need for ultra-high speed, safety, comfort, and other requirements.

Although HSR has reached the speeds of more than 500 km/h, its operating speed is generally at 200–350 km/h. When the velocity of ground railway transportation is higher than 350 km/h, the train consumes large amounts of energy and produces huge noise. Therefore, in March 1998, the International Railway Association of Innovation proposed that the maximum speed of ground transportation should be 350 km/h.

Such an operating speed is far less than the speed of the Japanese and the German Maglev. With further increased speed, wheel-rail HSR will bring a series of problems, for example, increased power consumption, vibration, noise, rail deterioration. Wheel-rail high-speed trains cannot solve the ground traffic problem of ultra-high speed, but Maglev can be the best choice. Table 3.1 shows the ground rail transit speed.

There have been many discussions about the advantages and disadvantages of Maglev [6, 9–11]. Indisputably, Maglev is a new ground transportation technology that is a fundamentally new innovation since the invention of the railway. The Maglev train offers numerous advantages over the conventional wheel-rail system: (1) Maglev vehicles are capable of traveling at speed up to 500 and 600 km/h [12], or even theoretically feasible over 1000 km/h [13, 14], and elimination of wheel and track wear provides a reduction in maintenance costs. (2) The distributed weight-load reduces the construction costs of the guideway. (3) Maglev train has no danger of derailment due to the strong electromagnetic force. (4) Maglev has no noise and vibration caused by the wheels and railway. (5) Maglev has a better climbing ability. (6) Guideway curves require a smaller radius. (7) Maintenance costs are low because no mechanical moving parts. (8) They are less affected by climate. (9) Swift acceleration and deceleration prevent sliding during operation. Table 3.2 shows the comparison of Maglev and wheel-rail systems.

An ultra-high-speed transport system can be developed by the Maglev. A Maglev train has no direct mechanical contact between the vehicle and its guideway, and the vertical support, lateral guidance, and longitudinal propulsion at levitation can be achieved simultaneously.

**Tab. 3.1:** Ground rail transit speed.

| Dates | Categories | Types | Speeds (km/h) | Countries |
|---|---|---|---|---|
| 2015, 04, 12 | LTS Maglev | MLX-01 | 603 | Japan |
| 2003, 12, 02 | LTS Maglev | MLX-01 | 581.7 | Japan |
| 2007, 04, 03 | HSR | TGV | 574.8 | France |
| 1999, 04, 14 | LTS Maglev | MLX-01 | 552 | Japan |
| 1997, 12, 12 | LTS Maglev | MLX-01 | 531 | Japan |
| 1979, 12, 21 | LTS Maglev | MLL-500 | 517 | Japan |
| 1990, 05, 18 | HSR | TGV | 515.3 | France |
| 2003, 11, 12 | NC Maglev | TR07 | 501 | China |
| 2010, 09, 28 | HSR | CRH380 | 486.1 | China |
| 1989, 12, 05 | HSR | TGV | 482.4 | France |
| 1996 | HSR Shinkansen | Bullet Train | 443 | Japan |
| 1988, 01, 22 | NC Maglev | TR06 | 412.6 | Germany |
| 1988, 05, 01 | HSR | ICE | 406.9 | Germany |
| 1976, 02, 19 | NC Maglev | TR04 | 401.3 | Germany |
| 1981, 02, 26 | HSR | TGV | 380 | France |
| 1955, 03, 29 | HSR | BB 9004 | 331 | France |
| 1972, 12, 08 | HSR | TGV Turbo | 318 | France |
| 1938, 07, 03 | Railway | Mallard | 202.6 | Britain |

**Tab. 3.2:** Comparison of Maglev and wheel-rail systems.

| | Maglev system | Wheel-rail system |
|---|---|---|
| Speed | >500 km/h | <350 km/h |
| Vibrations | No mechanical contact, weak | Contact between wheels and rails, relatively strong |
| Noise | 60–65 dB | 75–80 dB |
| Safety | No possibility of derailment | Derails from a minor defect |
| Guideway | Light vehicle & distributed load → light-weight | Heavy & concentrated load → hardy structure |
| Maintenance | Very little | Periodic replacement of wheels, gears, rails, etc. |
| Grade | About 80–100 | About 30–50 |
| Curve | In 30 m in radius | In 150 m in radius |

One of the ultra-high-speed transport systems is the LTS Maglev train in Japan. On December 21, 1979, the prototype ML-500 test train reached an unmanned speed of 517 km/h on a 7-km track, which proved the tremendous potential of Maglev trains for high speed. Also, when two trains going in opposite directions passed at the same time, a relative speed of 1026.3 km/h was recorded on November 16, 2004. The maximum speed attained was 603 km/h on April 21, 2015, which verified that this system had enough capability in high-speed operation.

Powell [14] compared the energy efficiency per passenger mile by Maglev, automobiles, and airplanes. The Maglev energy consumed per passenger-kilometer is lesser by about a factor of 10 than that of automobiles and airplanes. If all automobile, air, and rail passenger traffic in the USA States was accomplished using Maglev at an average speed of 322 km/h, the total electrical energy usage would be only about 3% of all current traffic electric energy usage. In 1998, the USA announced the Transportation Equity Act TEA-21 for the 21st century to provide the legal form for the development of Maglev transportation technology in the USA, and a Maglev line of 969 km was planned and reviewed.

China, Japan, Germany, USA, and other countries have begun to put Maglev trains into commercial operation, commissioning and project construction. We believe that in the near future, Maglev train will bring a new era of traffic.

## 3.4 Levitation phenomenon

Levitation phenomenon not only attracted wide attention of the public, but also attracted the interest of scientists and engineers. In fact, levitation and suspension can be achieved by some physical methods.

Maglev technologies have a series of advantages (see Tab. 3.2). Maglev is currently attracting worldwide interest for both high-speed intercity vehicles and low-speed urban transit vehicles, especially high- or ultra-high-speed intercity traffic.

In 1842, Earnshaw [15] theoretically studied the classical inverse-square law forces for gravity, electrostatics, and magnetism field and proved mathematically that stable levitation is impossible for an object placed in a repulsive or attractive static force field. Earnshaw wrote it before Maxwell's work and the discovery of the electron. According to Earnshaw's theorem, magnetic levitation in many common situations is not possible. However, this is not generally true. The theorem applies only to individual particles. Braunbeck [16] went further to show that the instability is due to the relative permeability of the materials involved being greater than unity.

In the 19th century, buoyancy was considered and studied to support objects without contact to the ground. Scientists have proposed several suspension principles, for example, levitation using water pressure, levitation based on the aircushion principle, levitation and propulsion with a blower, etc. A model train levitated by water pressure based on an idea of Girard was demonstrated at the Paris Exposition of 1889 [17].

On March 19, 1912, Bachelet [18] was awarded a patent for an electromagnetic suspension (EMS) system, the "levitating transmitting apparatus". His invention was first intended to be applied to smaller mail-carrying systems, but its potential application to larger train-like vehicles is certainly clear. Bachelet's invention opened the door to human Maglev application.

In 1934, Kemper [19] was awarded a patent for a “monorail vehicle with no wheels attached”. These inventions eventually led to the development of the Transrapid Maglev train program begun in 1969 in Germany. This magnetic levitation was called the electromagnetic suspension (EMS). EMS was achieved by the attraction forces between an electromagnet and the iron rail. Stable levitation was produced by electromagnets using an active feedback loop.

Powell and Danby [20, 21] are the inventors of the superconducting Maglev transportation systems. They filed a patent application (No. 684775) for an electromagnetic inductive suspension and stabilization system for a ground vehicle on November 21, 1967, and the patent was granted (No. 3470828) on October 7, 1969. This magnetic levitation is called the electrodynamic suspension (EDS). The time-varying magnetic fields induce eddy currents in the conductors which create a mutually repulsive magnetic field. The time-varying magnetic fields can be produced by motion of a magnet. Their inventions included inductive levitation and stable guidance, null flux geometry, and the linear synchronous motor for vehicle propulsion. This invention of the EDS using LTS magnets is the basis for the 500-km/h LTS Maglev train route from Tokyo to Osaka now under construction.

In 1933, Meissner and Ochsenfeld [22] found that when a sphere is cooled below its transition temperature in a magnetic field, it expels the magnetic flux. In 1945, the stable levitation of a permanent magnet over a superconductor was first discovered for a magnet above a concave lead (low $T_c$ superconductor) disk by Arkadiev [23, 24], at Moscow State University, USSR.

For both diamagnetic materials [16, 25] and high-$T_c$ superconductors [26, 27], stable levitation is possible in static magnetic fields. The levitation of high-$T_c$ superconductors is a most conspicuous phenomenon. It is really static (requires no energy input), quiet, and exceedingly stable. It is called the unique self-stable levitation. A high-$T_c$ superconductor can be levitated above or below a PM.

HTS bulk could realize very stable levitation without any active control. The physical characteristics of self-stabilizing of a HTS bulk Maglev is a unique phenomenon in nature. A lot of progresses have been made in the theory and application of HTS Maglev, especially in the fields of energy storage and transportation. In 2001, Jia-Su Wang and Su-Yu Wang were awarded a patent for “a superconducting Maglev system” [28], for a manned HTS Maglev vehicle.

Other types of stable levitation such as aerodynamic, acoustic, optical, electric, and radio-frequency (RF) levitation, etc., can be found elsewhere [29].

The invitation keynote speeches from Japanese, Chinese, German, American, and Korean experts at the 2011 Conference (21st International Conference on Magnetically Levitated Systems and Linear Drives, October 10–13, 2011, Daejeon, Korea) reported several main Maglev systems using EMS, EDS, HTS, etc.

In order to compare and understand the unique advantage of HTS Maglev, traditional magnetic levitation will be reviewed first.

## 3.5 Magnetic levitation

Magnetic levitation refers to the suspension phenomenon associated with the magnetic field. The Maglev train has no direct mechanical contact between the vehicle and its guideway, which achieves vertical support, lateral guidance, and propulsion in the levitation state simultaneously.

The conditions necessary for magnetic levitation of a magnetizable electrically conducting sphere depend on its conductivity, permeability, radius, and the spatial dependence of magnetic field [30].

Earnshaw [15] has shown that no system of permanent magnets can be utilized to achieve a stable magnetic suspension.

Braunbeck [16] has shown that the instability was due to the values of permeability $\mu$ for the materials used being greater than their free-space values, i.e. in static magnetic fields paramagnets ($\mu > \mu_0$) and ferromagnets ($\mu \gg \mu_0$) cannot levitate steadily. Therefore, there are exceptions to Earnshaw's theorem in respect to diamagnetic materials and superconductors. Stable suspensions have been achieved since using diamagnetic materials ($\mu < \mu_0$) and superconductors ($\mu = \mu_0$). Truly stable levitation without consumption of energy is possible only in magnetic fields. Energy is stored in the magnetic field. The energy density $U$ per unit volume contained in a magnetic field $B$ in MKS is given by

$$U = \frac{1}{2\mu_0} B^2, \tag{3.1}$$

where $\mu_0 = 4\pi \times 10^{-7}$ H/m is the permeability of free space.

Although levitation of diamagnetic materials [16, 25] seems to be of no more than academic interest, levitation of HTS materials [26, 27] is the most spectacular and promising.

According to attractive and repulsive levitation forces between the vehicle and guideway, Maglev systems include permanent magnet suspension (PMS) [31], EMS [32–38], EDS [35–42], and HTS Maglev [42–48]. The PMS includes Maglev between on-board PMs and DC electromagnet guideways or PMG, as well as Maglev between an onboard combination of PMs and DC electromagnets and steel rails, etc. The EMS is a Maglev between on-board DC electromagnets and steel rails. The EDS is a Maglev using eddy currents induced by an electromagnet moving over a conducting plate. The HTS Maglev is between on-board HTSCs and PMs and has several modes of stable levitation: repulsive levitation at Meissner state in a type I superconductor (small levitation forces), repulsive levitation based on both partial flux exclusion (levitation forces) and flux pinning (guidance forces) in type II superconductors, and suspension levitation based on flux pinning forces in a type II superconductor.

EMS and EDS are two basic types of magnetic suspensions developed prior to HTS Maglev. The technology is basically mature and has already begun trial operation development.

EMS is an attractive suspension with the on-board conventional electromagnets suspended below a steel rail, thus providing an attractive force between the steel rail and the vehicle. Since the suspension method is inherently unstable, the position between the magnets and relation to the rail is monitored on a continuous basis and active control is employed to ensure stability. The suspension achieves its maximum lift in the stationary state, and at higher speeds, the lift force is degraded due to induced eddy currents in the rail. EMS results in a small levitation clearance owing to the relatively low field strength of the on-board electromagnet. In addition, it requires feedback for stability and an on-board power supply system which gives extra volume and weight, and also requires a high precision guideway due to small clearance.

EDS is the repulsive levitation in which superconducting magnets or PMs are used on board. It employs the high magnetic field strengths of superconducting magnets to achieve large levitation clearance and is self-stabilizing without feedback. If an on-board PM is used, levitation clearance is smaller because the magnetic field strength of the PM is far less than that of the superconducting magnet. These on-board magnets with forward speed can produce eddy currents in the conducting guideway, and thus produce levitation by repulsion. Since there is no lift while stationary, the vehicle is supported by wheels at low speeds, and it will be like an aircraft takeoff as the speed reaches a certain value. An updated EDS system, Maglev 2000 [14], adopts a high-speed switch and quadrupole magnet technology, in order for the Maglev to be driven at high speed along a single guideway and it may not slow down at the stopover station. The track line only uses a double guideway in the station. This method is possible to promote the EDS system engineering application at a lower building cost of the train route.

The EDS and EMS systems require relative speed or active controls, Respectively, for stable levitation. The mixed-$\mu$ (relative permeability) system is a combination of different $\mu$ materials, namely ferromagnetic material (iron), diamagnetic material (superconducting screen), and air (electric coil space). Therefore, the levitation device is called mixed-$\mu$ system [49, 50]. An iron block is set inside the racetrack-shaped exciting coils. A vertical-direction-restoring force is generated, but a horizontal direction is not stable due to the attractive magnetic force between each other. Once the superconducting screens are installed between the iron block and coil, the attractive force decreases because the strength of magnetic field is weakened near the screen and the iron block can be stabilized horizontally. This system can keep a large air gap between the iron block and the superconducting screens.

In addition to the above magnetic levitation, a strange magnetic suspension phenomenon under the action of magnetic field also attracts the interest of scientists. For example, frogs, grasshoppers, strawberries, and other diamagnetic organisms may be suspended in a strong magnetic field environment [51].

In the early 1970s, the developments of Maglev transportation technology began in Germany, the USA, Japan, Canada, and Great Britain. At that time, the USA, Japan, and Canada often adopted a superconducting EDS scheme, while Great Britain used the normal conductive EDS scheme. In the initial development, Germany adopted two schemes, EDS and EMS, simultaneously, later abandoning the first, but continuing to develop EMS. At the beginning of development, Japan also adopted two schemes simultaneously, high-speed EDS with superconducting magnets and low-speed EMS with normal conductive magnets, and adhered to the two project development thereafter.

Magnetic levitation has many types, for example, permanent magnet (PM), normal conductive coil magnet, LTS coil magnet, HTS coil magnet, and HTS bulk material Maglev, etc. These Maglev types can be differentiated into EMS and EDS Maglev. Thus far, the normal conductive coil Maglev only has an EMS version, because the magnetic field of a normal conductive magnet is much lower.

The development and application of Maglev in bearing and ground transportation (sometimes called a linear bearing) areas have made good progress [38, 52]. and it is likely to become the most promising means of transport of the 21st century [14, 53]. The flywheel energy storage device with a HTS Maglev bearing is probably the most competitive energy storage system for the 21st century.

## 3.6 Permanent magnet Maglev

The levitation technology based on attractive forces of magnets is a PM suspension (PMS) which has favorable effects in respect to energy consumption in comparison to the electromagnetic suspension (EMS). PMS and EMS both have a lift-to-weight ratio of about 7:1. At maximum speed, the PMS control requires only 50 W/Mg with a gap of 20 mm as compared to more than 1 kW/Mg for EMS with a gap of 10 mm [54], i.e. the power consumption for suspension unit weight in the PMS with a gap of 20 mm is 1/20th of that in the EMS with a gap of 10 mm under high-speed condition. Many schemes for levitating vehicles with PMs, usually combined with propulsion by a linear electric motor, have been proposed in the past few years. Ferrite, alnico, and rare-earth PMs can be used in magnetically supported and propelled Maglev transport systems [55, 56].

Although the suspension vehicle using PM were started in the early 1960s, substantial researches were conducted after the advent of Rare-earth permanent magnets (RePM) in the 1970s and especially with neodymium-iron-boron (NdFeB) magnets in the mid-1980s [57]. In the middle of 1970s, tests with the PM Maglev system were commenced [58]. A dozen years ago, the magnetic energy product from NdFeB PM has reached 55 MGOe [59]. Thus far, the magnetic energy product of batch produced NdFeB PM (N52) in China is about 400 kJ/m$^3$ (50.3 MGOe), clearly suitable for Maglev application [31]. In contrast, the thermal and chemical stability of cobalt-samarium (Sm-Co) is better than NdFeB. However, NdFeB is cheaper. PMs in the vehicle undercarriage serve a dual function, as part of the levitation and of the propulsion systems.

According to the Earnshaw's theorem, PMs cannot suspend each other in stable configuration. PMs cannot be entirely supported by other PMs alone; thus, at least one degree of freedom must be eliminated by a mechanical constraint. In order to realize the stable suspension, it is necessary to adopt special technology and ingenious design. Although the levitation between two PMs itself is difficult, a combination of PMs and electromagnets or superconductors can achieve stable magnetic levitation. PM levitation can achieved by various design schemes: repulsive levitation between both PMs with the same poles facing each other, PM levitated above a DC electromagnet or superconductor and vice versa, attractive suspension between PMs as moving components and DC electromagnets as fixed components, attractive suspension between a combination of PMs and DC electromagnets and steel rails, etc.

The PMs levitated above a DC electromagnet are unstable in the lateral direction. Wang and Ilene [60] have successfully demonstrated the stable repulsive levitation method using PMs and air-core electromagnets. The mobile carrier PMs are levitated based on the repulsive force above the oblong-shaped electromagnets, which constitute one part of the guideways. The stability of the levitated PMs is regulated by another part of the guideways. A state feedback controller with integral compensator is designed for the stability control. The PM levitation approach is confirmed successfully by a silicon wafer transport in a microelectronic fabrication line. The levitation gap of wafer carrier is 0.5 mm above the guideway surface, and the maximum velocity is 22.2 cm/s in a traveling range of 1.0 m.

Chen et al. [61] have demonstrated a larger moving range dual-axis repulsive system using four active guideways. The feasibility and effectiveness of the Maglev system have been clearly demonstrated experimentally. The features of the two-axis Maglev guideway system include repulsive levitation, hybrid magnets, a passive carrier and active guideway, an oblong coil, and a four-guideway design. This PM Maglev system is similar to the system above, but its structure is more complex because the moving range functions are at the dual axis.

In PM Maglev, generally, the weight of the vehicle is levitated by attractive forces of PMs installed in the undercarriages of the vehicle. To control the attractive forces of PMs, the levitation air gap between PMs and the guideway should be controlled by active feedback. Tsunashima and Abe [62] have described the design and performance of a mechanical air gap controller for a PM Maglev transport vehicle without active feedback control. The mechanical air gap controller, consisting of a control lever and small guide wheels which work as a fulcrum, can control the air gap within required limits to levitate the vehicle. The effect of a dead zone in the mechanical air gap controller influences the levitation performance. Field test results show that the mechanical air gap controller can achieve 95% static levitation when the load variation is less than ±20% of the vehicle total weight, and stable levitation is achieved by a control lever ratio that is larger than 3 [62].

Wei [63], in Shanghai Normal University, invented and developed a manned PMS Maglev vehicle prototype and exhibited it in the 6th Shanghai International Industry

Fair in 2004. Specifications of the Maglev vehicle are as follows [64]: track length of 11 m and width of 1.7 m, vehicle length of 2.63 m and width of 1.25 m, maximum levitation force of 20 kN (can be manned by 6).

Magplane Technology proposed the new concept using a Halbach array [65] of PMs instead of the original LTS magnets [66]. This reduces the vehicle operational complexity and the number of potential failure modes, but results in some compromises: (1) to limit the weight of the PM systems, the operating gap has been reduced from 15 to 10 cm; (2) the weight of the PMs in the baseline case is 40% greater than the weight of the previous superconducting system. A past cruise speed of 500 km/h has been modified to 160 km/h.

The on-board levitation and propulsion PMs use NdFeB. The Halbach array magnets result in the maximum magnetic field at the guideway for a given weight of magnets and at the same time result in the minimum magnetic field in the passenger cabin. The maximum lift for a given weight of magnet is achieved when the magnets are concentrated over relatively limited areas, but with significant array height. A lift pad magnet weight of 5.9 tonnes, and sufficient lift can be achieved with four pads (Fig. 3.1). The lift-to-magnet weight ratio is 5.4. The surface field at the pad and at 10 cm from the guideway surface is 0.85 and 0.45 T, respectively [66].

Compared to other countries, Maglev systems in the USA will be important not only as carriers of passengers but also as carriers of truck-type freight. There are four USA urban Maglev systems under development at American Maglev Technology (AMT), General Atomics, LaunchPoint Technologies, and MagneMotion. Both AMT and MagneMotion use the EMS levitation technology. General Atomics uses the Inductrack levitation technology, which is essentially an EDS system with PM Halbach arrays instead of superconducting magnets. Unlike the other three, the General Atomics Maglev system theoretically has no stability issue, and control is only needed for a satisfactory ride quality. Therefore, in terms of reliability, the General Atomics Maglev system is better than the other three systems. LaunchPoint Technologies also uses PM Halbach arrays for levitation, but it is different from the PM Halbach arrays in the Inductrack systems.

MagneMotion and Old Dominion University (ODU) in the USA undertook a cooperative agreement to demonstrate the MagneMotion M3 urban Maglev technology [67, 68] on an existing elevated guideway on the ODU campus. The suspension and guidance forces are provided by the PMs rather than the electromagnets. The onboard suspension power using PMs is smaller than that with electromagnets. The use of PMs on the guideway above and below in the Maglev M3 system allows doubling the Maglev gap. Coils winding around the PMs are excited so as to stabilize the suspension and control the magnetic gap. The key parameters for the MagneMotion M3 Maglev system are 8.8-m vehicle length, 2.55-m vehicle width, 5.5-Mg vehicle weight, 9.0-Mg maximum load, 1.40-Mg/m guideway girder mass, and 0.28-Mg/m levitation mass [54]. The long stator linear synchronous motor (LSM) for propulsion allows all propulsion power and control systems to be on the guideway and the LSM

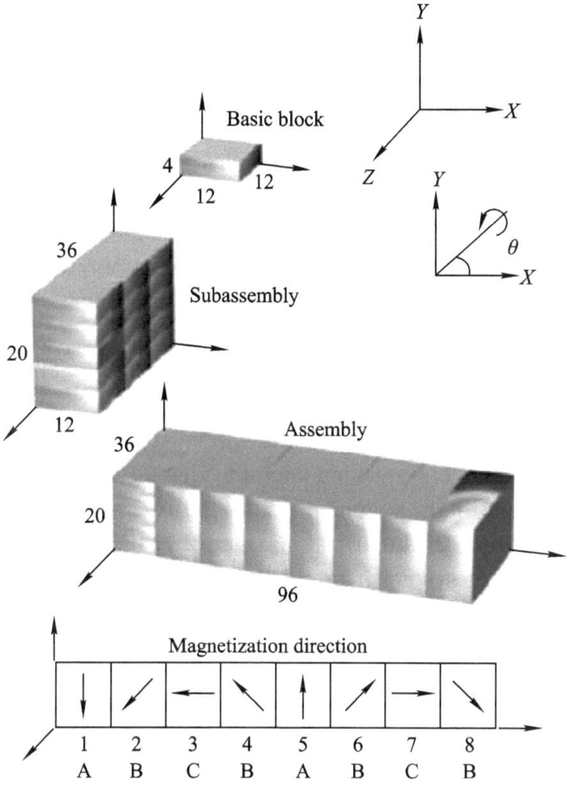


Fig. 3.1: Lift pad magnetic structure built from magnetized unit blocks in the Magplane Technology [66].

can be less expensive and more efficient than linear induction motor (LIM) propulsion for Maglev at any speed. There is no need to transfer propulsion power or safety-critical information to the vehicle or to depend on mechanical forces for acceleration and braking [54].

The AMT scheme is similar to MagneMotion, but the differences in performance and specification are bigger. Coils around the PMs adjust the magnetic forces 10,000 times per second to maintain this frictionless air gap while simultaneously controlling guidance and vertical stability. The level of energy usage for AMT is significantly less than the conventional wheels-on-rail technologies. In contrast to the average conventional transportation energy usage rate of 6 kWh/km, this technology uses approximately 1.8 kWh/km, or 70% net less energy [69]. The AMT's passenger capacity is 100 seats, about three times that of MagneMotion. However, AMT's operating speed is lower.

The PM-EDS has several advantages, simple structure, small size, low cost, etc. The PM-EDS is, however, used for small systems only because of the absence of high-powered PMs. The field concentration and the change of the magnetic field distribution for the PM Halbach arrays [65] used in Maglev may overcome this problem. The PM Maglev systems of General Atomics and LaunchPoint Technologies use the PM Halbach array technology.

In contrast to AMT and MagneMotion, LaunchPoint Technologies proposed the concept of Personal Rapid Transit (PRT), which would employ a low-cost guideway with many small, low-speed vehicles to provide high passenger capacity [70]. LaunchPoint Technologies proposed a variety of guideway design ideas, depending on the installation requirements. Guideways can be evacuated tubes ("Magtubes") or tubes open to the air. The Magtube system is flexible for both passengers and freight.

Stable levitation forces and vertical forces are provided by stabilized permanent magnet (SPM) technology [70]. For LaunchPoint Technologies, repulsors assembled from PM Halbach arrays are used for levitation. Due to the special configuration of repulsors, the system is stable vertically, but unstable laterally. Therefore, feedback control is necessary to monitor and control the alignment of the repulsors with respect to the rails. The vehicle magnets, the EM lateral stabilization, and the guideway magnets together form the SPM levitation system. In fact, SPM is the control current in the coil winding at the on-board Halbach arrays, thus, the stability is achieved. Lateral stabilization keeps the vehicle centered with minimal power requirement (~ 100 W/ton of vehicle weight). High-strength NdFeB PMs arranged in opposing Halbach arrays (see Fig. 3.2) provide levitation with a gap size of 30–80 mm at all speeds (no power requirement for levitation). In order to reduce the cost of track, the magnetic field of the track is weak. The magnetic field of on-board Halbach arrays is strong on one side of the track, but weak on the other side, in order to ensure human safety. A LSM propels the vehicle along the track. On-board motor coils interact with a permanent magnet track to move the vehicle forward. The design speed of this system is 800 km/h.

General Atomics reports an urban PM Maglev program [71]. Concepts developed during the research on passive magnetic bearing systems at the Lawrence Livermore National Laboratory gave a new approach to magnetic levitation, the Inductrack [72]. This Maglev system is unique in that operates with a large air gap of ~ 30 mm with no feedback control required for stable levitation for the PM Maglev. Furthermore, the packaging of the levitation magnets allows the vehicle to achieve a very tight turn radius of 18.3 m [73]. The General Atomics Urban Maglev system can offer many

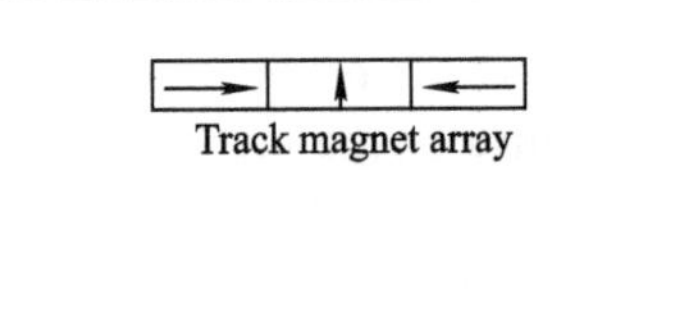


**Fig. 3.2:** Halbach arrays of LaunchPoint [70].

benefits, including low-noise operation, the ability to operate in challenging terrain with steep grades and tight turns, all-weather operation, low maintenance, rapid acceleration, and the potential for high speed.

The levitation system uses a vehicle mounted on PM of double Halbach arrays. The orientations of the magnetization of the magnets in the Halbach arrays are arranged to concentrate the field lines below the array while nearly canceling the field above the array. This results in a system which requires no active magnetic shielding of the passenger compartment. In a double Halbach array, the strong sides of two Halbach arrays oppose each other with the track between the double Halbach array. The track and vehicle chassis cross-section, as well as the magnetic configuration of the double Halbach array are shown in Fig. 3.3. The dotted lines show the position of the surface of the levitation track position. The guidance force is provided passively by the propulsion magnets (on the vehicle) interacting with the laminated iron core of the LSM winding (on the track). The Inductrack systems of the PM-EDS require the vehicle to travel at a certain speed of roughly around 100 km/h, to produce strong enough repulsive magnetic forces for levitation.

The test track is full scale, 120 m in length, with a 50-m radius curve. The track was completed in November 2004. A passive secondary suspension system provides damping and improves the passenger ride quality. Testing with chassis weight up to 10,000 kg, a speed of 10 m/s, air gaps up to ~ 30 mm, and acceleration up to 2.8 m/s$^2$ has been achieved [71].

Hoburg [74] studied the distribution of the magnetic field in the coach of a Maglev vehicle based on Halbach arrays. The results showed that the magnitude of the magnetic field due to superimposed fields from the propulsion, upper levitation, and lower levitation magnets were about 1.5 gauss at floor level, 0.5 gauss at seat level, and 0.2 gauss at head level over the area of the passenger compartment.

Cho et al. [75] analyzed the characteristics of an electrodynamic suspension PM-EDS device with a PM Halbach array system. In order to validate the characteristic analysis scheme for PM-EDS devices, the dynamic performance was tested using a high-speed rotary-type dynamic test facility with linear peripheral speed up to 250 km/h. A comparison was made between the analysis and the experimental results to demonstrate the design considerations of PM-EDS device for high-speed Maglev.

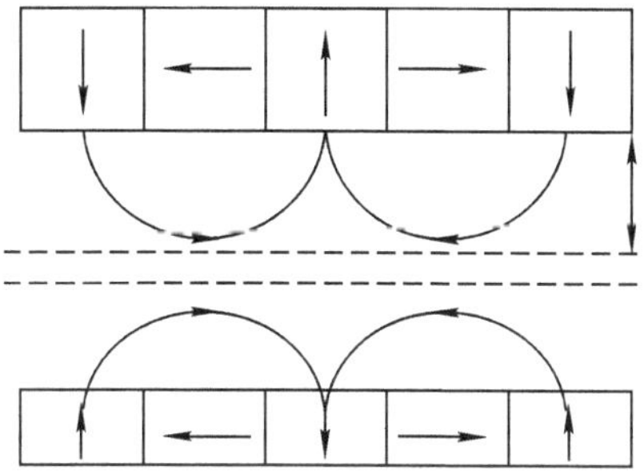

**Fig. 3.3:** Halbach arrays of General Atomics [71]. The dotted lines show the surface of the levitation track position.

It showed that three-dimensional analyses are more effective than two-dimensional ones.

In general, the PMs in Maglev are the traditional ones, for example, NdFeB. HTS PMs have much larger magnetic energy products than the traditional PMs and may be better candidates for future Maglev applications. When the HTS bulk is magnetized to a high magnetic field, part of this field is trapped in the bulk to create the superconducting PM. The trapped magnetic field of the superconducting bulk magnet has been reported to be better than that of a conventional PM (for example, NdFeB). The maximum trapped fields were 16 T [76] (at 24 K) and 11.2 T (at 47 K), and trapped fields of 9 T (at 46 K) and 17.24 T [77] (at 29 K) were recorded. Even a large HTS bulk magnet is believed to trap a very high magnetic field of 5 T at 77 K [78].

Fujimoto et al. [79] describes a superconducting bulk magnet for a Maglev train with liquid nitrogen refrigeration, instead of Nb-Ti superconducting racetrack coils which require liquid helium as a coolant. If this is successful, it will have a great impact on the future transport systems.

## 3.7 Normal-conductive Maglev

In 1934, Hermann Kemper was awarded a patent for a "monorail vehicle with no wheels attached" [19]. This invention eventually led to the development of the Transrapid Maglev train program begun in 1969 in Germany and high-speed surface transport (HSST) Maglev train program in Japan. This magnetic levitation is called the EMS. EMS is achieved by the attraction forces between the electromagnet and the iron rail, and stable levitation is produced by electromagnets with an active feedback loop. EMS systems of the Transrapid in Germany and the HSST in Japan use on-board normal-conductive magnets, and the EMS systems of AMT and MagneMotion in the USA use on-board PMs (see Section 3.6).

In the EMS, vertical suspension, lateral guidance, and propulsion are achieved without direct mechanical contact between the vehicle and guideway. Both vertical suspension and lateral guidance are produced by attractive forces between on-board normal conductive electromagnets and the steel rails. Control of both vertical and lateral forces is achieved by checking the gap change and adjusting the corresponding magnetic fields. This method is similar to that of conventional active control.

The EMS is a basic type of magnetic suspensions. The technology is mature and has already begun trial operation and development. In the early 1970s, the developments of EMS Maglev transportation technology began in Germany, Japan, the USA, and Great Britain.

The first commercial automated EMS Maglev (Fig. 3.4) system in the world [81] was a low-speed vehicle that ran from the airport terminal of Birmingham International Airport to the nearby Birmingham International Railway Station in 1984. The length

**Fig. 3.4:** The first commercial automated EMS Maglev system in UK [80].

of the track was 600 m, the levitation gap was 15 mm, and it was propelled by linear induction motors. This system stopped operating in 1995 due to aging of the electrical system rather than the Maglev technology itself.

The German Transrapid [82–84] Maglev system is an intercity high-speed system and is the fundamental innovation in railroad engineering since the construction of the first railroads. Instead of wheels and rails, the Transrapid uses non-contact electromagnetic levitation, guidance, and propulsion systems. Feedback control electronic systems guarantee that the suspension clearance remains constant (about 10 mm). The suspension system is supplied by on-board batteries which are recharged in the process of operation by linear generators integrated into the support magnets. The synchronous long stator linear motor is used both for propulsion and braking, and the motor primary coils are mounted on the guideway. The braking energy can be fed back into the electrical network.

In the early 1970s, Germany began Maglev development. From the start of Transrapid 02 in 1971 to the successful development of the current Transrapid 09, the Germans underwent development road of more than 30 years. The Transrapid 09 (Fig. 3.5) started test operation in July 2008.

During commissioning in Shanghai in November 2002, the Chinese bullet train repeatedly reached a speed of more than 440 km/h, thereby exceeding its designed revenue service operation speed of 430 km/h. The maiden trip of the Transrapid train on its first commercially operated route worldwide started from Shanghai's Long Yang Road to the Pudong International Airport on December 31, 2002 [85]. This was the first commercial demonstration operation line for an EMS high-speed Maglev system in the world. On November 12, 2003, a Shanghai Transrapid carried its passengers to a new world record for standard-specification ground-transport vehicles: 501 km/h. The maximum speed is not used in daily service due to the relatively short route length of 30 km [86].

The HSST Maglev technology [87] is based on a normal conductive system levitated by the attractive force of magnets and propelled by LIM without wheels. Japan

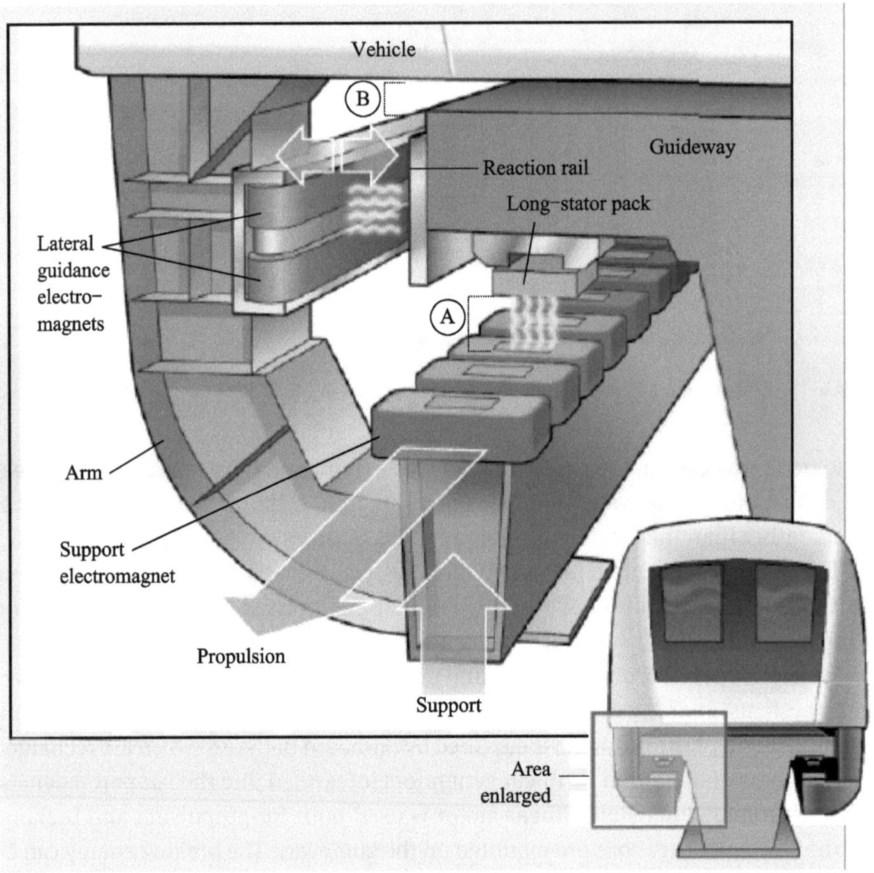


**Fig. 3.5:** Philip Holmer. Faster than a speeding bullet train: China is throttling up a 430-km/h magnetically levitated train to link Shanghai and its airport [96].

and Germany are developing different types of normal-conductive Maglev trains. In comparison with Germany's Transrapid, Japanese HSST is actually an urban transportation system with a relatively low-speed urban Maglev vehicle for the EMS. The two systems are similar in the sense that they both use linear motors for propulsion and electromagnets for levitation. However, the HSST is propelled by LIMs. The HSST primary coils are attached to the carriage body and the secondary aluminum reaction plates are on the track.

Japan Airlines began developing normal conductive Maglev HSST as a transportation system to provide high-speed access to airports in 1972. HSST-03 (50 seats) made a demonstration run at the Tsukuba Science Expo from March to September 1985, and 610,000 passengers were transported. Fig. 3.6 shows a cross-section of the scheme for the Japanese HSST system.

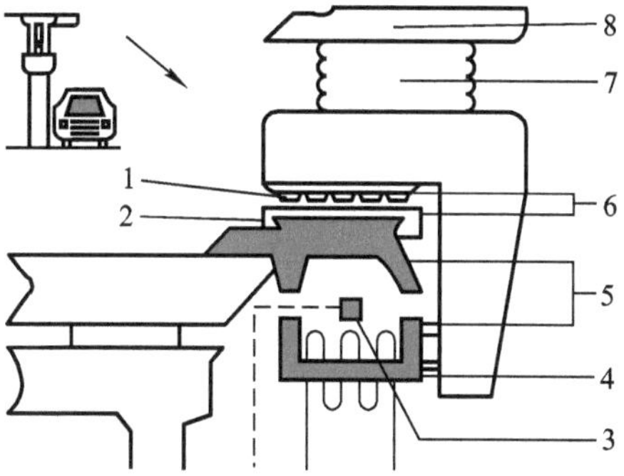


**Fig. 3.6:** Cross section of the scheme of the Japanese HSST system. (1) Windings. (2) Reaction plate. (3) Gap sensor. (4) Levitation magnet. (5) Levitation system. (6) Propulsion system. (7) Air spring. (8) Vehicle [96].

The first commercial train of the HSST Maglev system (HSST-100 type) [88, 89] began revenue service on the Tobu Kyuryo Line in the suburbs of Nagoya City in Japan on March 6, 2005. This route is 9.2 km from the Fujigaoka subway station in Meito Ward, Nagoya (Aichi Prefecture) through Nagakute Town to Yakusa Station on the Aichi Kanjo (loop) line in Yakusacho of Toyota City and has nine stations. There are 3-car trains and the capacity is 3500 passengers/hour. End-to-end trip time is 15 minutes, with 6-minute interval (frequencies) in the peak period and 10-minute interval during the off-peak period. This is the first full-scale commercial application of the medium-/low-speed urban Maglev technology in the world.

In 1989, the R&D project for a low-to-medium speed EMS Maglev system started in Korea. The urban Maglev system engineering program [90, 91] started in December 2006. The program is composed of three core projects: systems engineering, vehicle development, and demonstration line construction. The prototype vehicle of two permanently coupled Maglev vehicles was built in December 2009 and transferred to a 1.3-km test track in KIMM, Daejon, Korea. Jia-Su Wang and Su-Yu Wang took the prototype Maglev vehicle during the Maglev International Conference in 2011. The urban EMS Maglev vehicle system operation is stable and comfortable. An objective of the program is to develop a competitive urban transit Maglev vehicle and to construct a 6.1-km urban Maglev demonstration line at Incheon International Airport. In February 3, 2016, this urban Maglev line was opened to link Incheon International Airport and Yongyu Station and holds an operating speed of 110 km/h. therefore, Korea has the world's second commercially operating unmanned unban Maglev line after Japan's Linimo.

Since the late 1980s, under the support of the Chinese Ministry of Science and Technology, a national program has started to study the key technologies of Maglev. Four research and development groups were organized at the National Railway Academy, the Southwest Jiaotong University (SWJTU) (Fig. 3.7), the National Defence University, and the Institute of Electrical Engineering of the Chinese Academy of Sciences. Significant success has been achieved, and several prototype vehicles were built. The scheme is similar to the HSST in Japan, which uses EMS normal-conductive Maglev, i.e. an electromagnet for levitation and a LIM for propulsion.

**Fig. 3.7:** Maglev test vehicle [92] of the Southwest Jiaotong University in China.

The Beijing Enterprises Holdings Limited together with the National University of Defense Technology (NUDT) built a CMS-03 test vehicle and a 204-m-long test line in 2001 in Changsha [92–94]. Up to now, the vehicle has run over 7000 km with over 20,000-test runs and 40,000 start and stop operations. Its safety and reliability have been proved. The vehicle's main parameters are length of 15.5 m, width of 3 m, height of 3.83 m, empty weight of 21 t, nominal load capacity of 9 t, a maximum speed of 150 km/h, an operational acceleration of 1.1 $m/s^2$, a minimum road radius of 75 m, and a maximum climbing grade of 7%. The construction of the first test line in Beijing of the medium- to low-speed Maglev (Fig. 3.8) was started on February 28, 2011.

On November 4, 1999, the Ministry of Science and Technology and Transrapid International Inc. signed an agreement to conduct a cooperative pre-feasibility study for construction of a test-operation line in China [93]. On June 30, 2000, an agreement of cooperative feasibility study demonstration and line operation for a Shanghai Maglev was signed. The first Maglev vehicle (Fig. 3.9), composed of three sections, accomplished its trial run on a single track on December 31, 2002. Following a 12-month system commissioning, the Shanghai Maglev line was officially put into trial commercial service on May 1, 2004, running 9 hours per day in the initial period

**Fig. 3.8:** Maglev test vehicle of the National University of Defense Technology in China [93].

Fig. 3.9: Shanghai Transrapid Maglev [97] running line in China.

and currently 14 hours [95]. The relative speed between two opposite-direction trains passing each other on neighboring tracks was up to 860 km/h on July 14, 2003. In April 2006, the Shanghai Maglev Line Project passed national acceptance and was officially put into commercial operation. This is the first commercial automated high-speed EMS Maglev system in the world [96].

Numerous studies have been completed in the USA, but no commercial Maglev systems have been deployed [98]. The USA first invented Maglev in 1912 [18] and put forward the concept of a superconducting Maglev in 1966. [20, 21] The American government implemented the National Maglev Initiative (NMI) program in the early 1990. There are some Maglev projects, either low speed or high speed, in the USA [99]. On January 29, 1999, the US Federal Transit Administration announced the Urban Maglev Project (UMP) with a vision to "develop American magnetic levitation technology to improve urban mass transportation."

Since the mid- to low-speed EMS Maglev vehicle operation started in UK in 1984, Japan and Korea also built the EMS Maglev vehicle system with better performance. The HSST Maglev System (HSST-100 type) on a 9.2-km route began revenue service on the Tobu Kyuryo Line in the suburbs of Nagoya in Japan, on March 6, 2005. The urban Maglev system engineering program started in December 2006. This EMS Maglev vehicle will be commercially operated on a 6.1-km line from the downtown area to the Incheon International Airport.

China's first Maglev train railway started construction in Changsha on May 16, 2014. The Maglev railway will be completed in 2016. This will be the first commercial operation in China of a low-speed Maglev train [100]. Changsha Maglev Railway is owned by the Hunan Maglev Transportation Development Limited, and the design and construction is being undertaken by China Railway Construction Corporation Limited (CRCC). The Maglev line is 18.5 km in length from the Changsha Train South Station to the Changsha Huanghua Airport, and the estimated total investment is about 4 billion yuan. The maximum design speed is 120 km/h, so that it will take only about 10 minutes to reach the airport by the Maglev train [100].

## References


[1] http://www.wyzxsx.com/Article/Class4/201108/252441.html.

[2] Zhao JZ. http://wenku.baidu.com/view/d3e98309581b6bd97f19ea01.html?from=rec&pos=2&weight=44&lastweight=42&count=5.

[3] http://www.zscargo.sk/en/customers/why-to-transport-with-us/advantages-of-railway-traflc/.

[4] Morita Y, Shimizu K, Kato H, Shibahara N, Yamasaki T. A study for the measurement of environmental impact by railway construction. Proceedings of the ASME/ASCE/IEEE 2011 Joint Rail Conference JRC2011 March 16–18, 2011, Pueblo, Colorado, USA.

[5] Fu X, Oum T H, Yan J. An analysis of travel demand in Japan's intercity market empirical estimation and policy simulation. Journal of Transport Economics & Policy, 2014, 48(1).

[6] Vuchic VR, Casello JM. An evaluation of Maglev technology and its comparison with high speed rail. Transportation Quarterly, 2002;56(2):33–49.

[7] Anthoine J, Gouriet J, Rambaud P. Reduction of the sonic boom from a high-speed train entering a gallery. Proceedings of the 13th AIAA/CEAS Aeroacoustics Conference (28th AIAA Aeroacoustics Conference), Aeroacoustics Conferences; 2007 May 21–23; Rome, Italy AIAA 2007–3559.

[8] http://news.xinhuanet.com/yzyd/tech/20140306/c_119631631.htm.

[9] Gran RJ. The benefits of Maglev technology. Proceedings of the 23rd Fluid Dynamics, Plasmadynamics, and Lasers Conference, Proceedings of the Fluid Dynamics and Co-located Conferences; 1993 July 6–9; Orlando, FL AIAA 93–2949.

[10] Lee HW, Kim KC, Lee J. Review of Maglev train technologies. IEEE transactions on magnetics, 2006;42(7):1917–1925.

[11] http://lrrpublic.cli.det.nsw.edu.au/lrrSecure/Sites/Web/physics_explorer/physics/lo/superc_08/superc_08_03.htm.

[12] Wang JS, Wang SY, Deng CY, Zeng YW, Song HH, Zheng J, Wang XZ, Huang HY, Li F. Design consideration of a high temperature superconductor Maglev vehicle system. IEEE Trans Appl Supercond. 2005;15(2):2273–2276.

[13] Wang JS, Wang SY, Deng CY, Zeng YW, Song HH, Huang HY. A superhigh speed HTS Maglev vehicle, International Journal of Modern Physics B. 2005;19(01n03):399–401.

[14] Powell JR, Danby G. MAGLEV: The new mode of transport for the 21st century, The Summer Issue of 21st Century Science & Technology, 2003. pp. 43–57.

[15] Earnshaw S. On the nature of the molecular forces which regulate the constitution of the luminiferous ether, Trans. Camb. Phil. Soc, 1842;7:97–112.

[16] Braunbeck W. Magnetic levitation of graphite. Z. Phys. 1939;112:753–763.

[17] Masada E. The First 100 Years of Maglev. Proceedings of the 21st International Maglev Conf., Daejeon, Korea, 2011 Oct 10–13; Daejeon, South Korea.

[18] Bachelet E. Levitating transmitting apparatus. US Patent no. 1020942, On March 19, 1912.

[19] Kemper H. Monorail vehicle with no wheels attached. Reich Patent number 643316, 1934.

[20] Powell JR, Danby G. High speed transport by magnetically suspended trains. Proceedings of the ASME 1966 Winter Annual Meeting; 1966 Nov 27–Dec 1; New York, USA.

[21] Powell JR, Danby G. A 300-mph magnetically suspended train. Mechanical Engineering, 1967;89(11):30–35.

[22] Meissner W, Ochsenfeld R. A new effect in penetration of superconductors. Die Naturwissenschaften. 1933;21:787–788.

[23] Arkadiev V. Hovering of a magnet over a superconductor. J Phys (USSR). 1945;9:148.

[24] Arkadiev V. A floating magnet. Nature. 1947;160(4062):330–330.

[25] Simon MD, Heflinger LO, Geim AK. Diamagnetically stabilized magnet levitation. Am J Phys. 2001;69(6):702–713.

[26] Hellman F, Gyorgy EM, Johnson DW, Jr., O'Bryan HM, Sherwood RC. Erratum: "Levitation of a magnet over a flat type II superconductor". Journal of Applied Physics. 1988;63(10):5196–5196.

[27] Peters PN, Sisk RC, Urban EW, Huang CY, Wu MK. Observation of enhanced properties in samples of silver oxide doped $YBa_2Cu_3O_x$. Applied Physics Letters. 1988;52(24):2066–2067.

[28] Wang JS, Wang SY. A superconducting Maglev system. Chinese patent, No. ZL 01128867.1, 2001.

[29] Brandt E H. Levitation in physics. Science. 1989;243(4889):349–55.

[30] Jones T B. A necessary condition for magnetic levitation. Journal of Applied Physics. 1979;50(7):5057–5058.

[31] Strnat KJ. Modern permanent magnets for applications in electro-technology. Proceedings of the IEEE. 1990;78(6):923–946.

[32] Laithwaite ER. Electromagnetic levitation. Proceedings of the Institution of Electrical Engineers. IET Digital Library, 1965;112(12):2361–2375.

[33] Jayawant BV. Electromagnetic suspension and levitation. Rep Prog Phys. 1981;44(4):411.

[34] Glatzel K, Khurdok G, Rogg D. The development of the magnetically suspended transportation system in the Federal Republic of Germany. IEEE Transactions on Vehicular Technology, 1980;29(1):3–17.

[35] Rogg D. Development of magnetic levitation transport systems in the federal republic of germany: survey, present state, prospects, and reasons. Proceedings of the international conference on Maglev transport '85. 1985, sept; Tokyo, Japan. 1985, IEE Japan, pp. 1–11.

[36] Sinha P K. Electromagnetic suspension dynamics & control. Magnetic Levitation Vehicles, 1987.

[37] Weh H. Magnetic levitation technology and its development potential. Proceedings of the 11th International Conference on Magnetically Levitated Systems and Linear Drives: Maglev '89; 1989 July 7–11; Yokohama, Japan, IEE Japan, pp. 1–9.

[38] Nagaike T, Takatsuka H. Present status and prospect of HSST. Proceedings of the 11th International Conference on Magnetically Levitated Systems and Linear Drives: Maglev '89; 1989 July 7–11; Yokohama, Japan, IEE Japan, pp. 29–35.

[39] Powell JR, Danby GR. High speed transport by magnetically suspended trains. Recent Adv Eng Sci. 1968;5(J):159.

[40] Fujie J. Current Status of EDS System in Japan. Proceedings of the 11th International Conference on Magnetically Levitated Systems and Linear Drives: Maglev '89; 1989 July 7–11; Yokohama, Japan, IEE Japan, pp. 81–83.

[41] Wang JS, Wang SY. Applications of superconductivity. Chengdu: University of Science and Technology Chengdu Press; 1995. pp. 123–163. (In Chinese).

[42] Moon FC. Superconducting levitation: applications to bearings and magnetic transportation. New York:Wiley; 1994.

[43] Lehndorff B. High-$T_c$ superconductors for magnet and energy technology: fundamental aspects. New York: Springer Science & Business Media; 2001.

[44] Wang JS editor. Proceedings of International Workshop on HTS Maglev (IS'2002); 2002 June 25–27; Chengdu, China.

[45] Hull JR. Levitation, Cardwell DA, Ginley DS editors. Handbook of superconducting materials Vol II: characterization, applications and cryogenics. Britain: Institute of Physics Publishing. 2003. pp. 1663–1673.

[46] Ma KB, Postrekhin YV, Chu WK. Superconductor and magnet levitation devices. Rev of Sci Instruments. 2003;74(12):4989–5017.

[47] Murakami M, Norlikar editor. Materials developments and applications of bulk Re–Ba–Cu–O superconductors. Frontiers in superconducting materials. Berlin: Springer Verlag; 2005. pp. 869–884.
[48] Wang JS, Wang SY, Narlihar A editor. Synthesis of bulk superconductors and their properties on permanent magnet guideway. Frontiers in superconducting materials. Berlin: Springer Verlag; 2005. pp. 885–912.
[49] Kashiwagi T, Kubota M, Suzuki E, Matsuda T, Hirakawa M, Nakashima H, Ohsaki H, Mizuno T. Levitation characteristics of the superconducting mixed-$\mu$ system. Physica C Superconductivity & Its Applications. 2003;392(03):654–658.
[50] Asher GM, Williams JT, Paul RJA, Brown J. Current research into mixed-$\mu$ levitator designs for high speed Maglev vehicles. J de Phys. 1984;45(C1):733.
[51] Berry MV, Geim AK. Of flying frogs and levitrons. European Journal of Physics. 1997;18(4):307.
[52] Werfel FN, Floegel-Delor U, Rothfeld R, Riedel T, Goebel B, Wippich D, Schirrmeister P. Superconductor bearings, flywheels and transportation. Supercond Sci Technol. 2011;25:014007.
[53] Hecht L. Magnetrain. The Road to the 21st Century, the Winter 2000–2001 issue of 21st Century Science & Technology, http://www.21stcenturysciencetech.com/articles/magnetrain.html.
[54] Thornton RD. Efficient and affordable Maglev opportunities in the United States. Proceedings of the IEEE. 2009;97(11): pp. 1901–1921.
[55] Weh H. Linear synchronous motor development for urban and rapid transit systems. IEEE Trans on Magnetics. 1979;15(6):1422.
[56] Atherton DL. Maglev using permanent magnets. IEEE Trans on Magnetics. 1980;16(1):146.
[57] Sagawa M, Fujimura S, Togawa N, Yamamoto H, Matsuura Y. Invited paper on NdFeB from the manufacturer of NEOMAX. Journal of Applied Physics. 1984;55(6):2083.
[58] Gutberlet HG. The german magnetic transportation program. IEEE Trans on Magnetics. 1974;10(3):417.
[59] Kaneko Y. Highest performance of Nd–Fe–B magnet over 55 MGOe. IEEE Trans on Magnetics. 2000;36(5):3275.
[60] Wang IYA, Busch-Vishniac I. A new repulsive magnetic levitation approach using permanent magnets and air-core electromagnets. IEEE Trans on Magnetics. 1994;30(4):1422.
[61] Chen MY, Wang MJ, Fu LC. A novel dual-axis repulsive Maglev guiding system with permanent magnet: modeling and controller design. IEEE/ASME Trans on Mechatronics. 2003;8(1):77.
[62] Tsunashima H, Abe M. Static and dynamic performance of permanent magnet suspension for Maglev transport vehicle. Vehicle System Dynamics. 1998;29(2):83.
[63] Wei LH. Train and its levitation, guidance and propulsion systems. CN Patent 1431114 A, 2003.
[64] http://kjc.shnu.edu.cn/Default.aspx?tabid=6985.
[65] Halbach K. Application of permanent magnets in accelerators and electron storage rings (invited). Journal of Applied Physics. 1985;57(8):3605.
[66] Montgomery DB. Overview of the 2004 Magplane Design. Magplane Technology. Maglev 2004 Overview Paper 6.10.04.
[67] Thornton R, Clark T, Perreault B, et al. An M3 Maglev system for old dominion university. 20th International Conference on Maglev, San Diego, CA, December 15–18, 2008, No.78.
[68] Wieler JG, Thornton RD. Urban Maglev – Development Plans and Prospects. 21st International Conference on Maglev, Daejeon, Korea, October 10–13, 2011, No.152.
[69] http://american-Maglev.com/.
[70] Fiske OJ. LaunchPoint Technologies, Inc., http://www.launchpnt.com/Detail_View.67+M58ef53687fc.0.html.

[71] Gurol S, Baldi B, Post R. General atomics urban Maglev program status. 19th Maglev Conference; 2006 Sep 13–15; Dresden, Germany, 2006. p.65–68.
[72] Post RF, Ryutov DD. The inductrack: a simpler approach to magnetic levitation. IEEE Trans Appl Supercond. 2000, 10:901–904. Richard Post. U. S. Patent, Inductrack configuration. US 6629503, Oct 7, 2003.
[73] Gurol S, Baldi B, Post RF. The general atomics low speed urban Maglev technology development program. 82ed Annual Meeting of the Transportation Research Board (TRB); 2003 January 12–16; Washington, D. C.; 2003. 24.
[74] Hoburg JF. Modeling Maglev passenger compartment static magnetic fields from linear Halbach permanent-magnet arrays. IEEE Transactions on Magnetics, 2004;40(1):59–64.
[75] Cho HW, Han HS, Bang JS, et al. Characteristic analysis of electrodynamic suspension device with permanent magnet Halbach array. Journal of Applied Physics. 2009;105(7):08P510.
[76] Gruss S, Fuchs G, Krabbes G, Verges P, Stöver G, Müller K-H, Fink J, Schultz L. Superconducting bulk magnets: very high trapped fields and cracking. Applied Physics Letters. 2001;79:3131–3133.
[77] Tomita M, Murakami M. High-temperature superconductor bulk magnets that can trap magnetic fields of over 17 tesla at 29 K. Nature, 2003;421(6922):517–20.
[78] Yoo SI, Sakai N, Takaichi H, Higuchi T, Murakami M. Melt processing for obtaining $NdBa_2Cu_3O_y$ superconductors with high $T_c$ and large $J_c$. Appl Phys Lett. 1994;65(5):633–635.
[79] Fujimoto H, Kamijo H, Higuchi T, Nakamura Y, Nagashimi K, Murakami M, Yoo S. Preliminary study of a superconducting bulk magnet for the Maglev train. IEEE Trans Appl Supercond. 1999;9(2):301–304.
[80] http://www.airportsinternational.com/2010/12/first-commercial-sold/4449.
[81] Jayawant BV. Proc R Soc. Lond. Review Lecture. Electromagnetic suspension and levitation techniques. Proceedings of the Royal Society of London, 1988;416(1851):245–320.
[82] Heinrich K, Kretzchmar R. Transrapid MagLev System. Darmstadt, Germany: Hestra-Verlag Darmstad, 1989; Liu Hua-Qing, et al. Compile. German transrapid Maglev Train. University of Electronic Science and Technology Press; 1995. (In Chinese).
[83] http://www.transrapid.de/cgi/en/basics.prg?session=da5845895316818e_657093&a_no=29.
[84] Loeser F. Implementation strategy for transrapid Maglev projects. 21st International Conference on Maglev. 2011 Oct 10–13; Daejeon, Korea; 2011. Keynote Speech No. 5.
[85] http://www.smtdc.com/en/.
[86] UK Ultraspeed Factbook. Expanded 2nd Edition, October 2006, http://www.500kmh.com/UKU_Factbook2.pdf.
[87] Sawada K, Murai M, Tanaka M. Magnetic levitation (Maglev) technologies. Japan Railway & Transport Review. 2000:58–76.
[88] Yasuda Y, Fujino M, Tanaka M, Ishimoto S. The first HSST Maglev commercial train in Japan. Proceedings of the 18th International Conference on Maglev; 2004 Oct 26–28; Shanghai, China; 2004. pp. 76–85.
[89] http://faculty.washington.edu/jbs/itrans/hsstpage.htm.
[90] Shin BC, Kim WJ, Park DY, Beak JG, Kang HS. Progress of urban Maglev program in Korea. 9th World Congress on Railway Research; 2011 May 22–26; Lille, France. 2011.
[91] Shin BC, Kim WJ, Park DY, et al. Recent progress of urban Maglev program in Korea. The 21st International Conference on Magnetically Levitated Systems and Linear Drives. 2011. Oct 10–13; Daejeon, Korea; 2011.Keynote Speech No. 4.
[92] Ruihua Z, Luguang Y, Li Yaohua Wang Ke. Calculation and simulation of power parameters for inverter-fed induction motor. Proceeding of International Conference on Electrical Machines and Systems; 2007, Oct. 8–11; Seoul, Korea; 2007:1066–1070.

[93] http://www.nudt.edu.cn/articleshow_eng.asp?id=5986.
[94] Yan L. Development and application of the Maglev transportation system. IEEE Trans Appl Supercond. 2008;18(2):92–99.
[95] Xiangming W. Experience in operation and maintenance of shanghai Maglev demonstration line and further application of Maglev in China. The 19th International Conference on Magnetically Levitated Systems and Linear Drives. 2006 Sep 13–15; Dresden, Germany; 2006.
[96] Holmer P. Faster than a speeding bullet train. IEEE Spectrum, 2003;40(8):30–34.
[97] http://www.smtdc.com/cn/xnlv2.html.
[98] Rose CR, Peterson DE, Leung EM. Implementation of cargo MagLev in the United States. Maglev 2008: 20th International Conference on Maglev Systems and Drives; 2008 Dec 15–18; San Diego, CA; 2008.
[99] Blow LE. Status of Maglev projects in the USA. 21st International Conference on Maglev; 2011 October 10–13; Daejeon, Korea; Paper No. 3. 2011. Keynote Speech.
[100] http://www.gov.cn/xinwen/2014-05/16/content_2680848.htm.

## Maglev books and reviews

[1] Ma B, Postrekhin YV, Chu WK. Superconductor and magnet levitation devices. Rev of Sci Instruments. 2003;74:4989–5017.
[2] Lehndorff BR. High-$T_c$ superconductors for magnet and energy technology. Berlin: Springer; 2001.
[3] Davis C, Wilkie DF. Analysis of motion of magnetic levitation systems: implications for high-speed vehicles. J Appl Phys. 1971;42:4779–4793.
[4] Rote DM, Cai Y. Review of dynamic stability of repulsive-force Maglev suspension systems. IEEE Trans on magnetics. 2002;38:1383–1390.
[5] Rogg D. Development of magnetic levitation transport systems in the Federal Republic of Germany, survey, present state, prospects and reasons. Proceedings of International Conference on Maglev Transport '85; 1985 Sep 17–20; Tokyo, Japan. IEE Japan; 1985. P. 1–11
[6] Rossing D, Hull JR. Magnetic levitation. The Physics Teacher. 1991;29:552–562.
[7] Thornton D. Magnetic levitation and propulsion. IEEE Trans on Magnetics. 1975;11:981–995.
[8] Brandt EH. Levitation in physics. Science. 1989;243:349–355.
[9] Jayawant BV. Electromagnetic suspension and levitation. Rep Prog Phys. 1981;44:411–477.
[10] Laithwaite ER. Electromagnetic levitation. Proc IEE. 1965;112:2361–2375.
[11] Moon FC. Chang PZ. Superconducting Levitation: applications to bearings and magnetic transportation. New York: Wiley-interscience; 1994.
[12] Werfel FN, Floegel-Delor U, Rothfeld R, Riedel T, Goebel B, Wippich D, et al. Superconductor bearings, flywheels and transportation. Supercond Sci Technol. 2012;25:014007.
[13] Werfel FN, Floegel-Delor U, Riedel T, Goebel B, Rothfeld R, Schirrmeister P, et al. Large-scale HTS bulks for magnetic application. Physica C. 2013;484:6–11.
[14] Werfel FN, Floegel-Delor U, Riedel T, Wippich D, Goebel B, Rothfeld R, et al. HTS magnetic bulk superconductors towards large-scale application. New York: Nova Science Publishers; 2012; p. 183–201.
[15] Wilkie DF. Control and ride quality of a magnetically levitated high speed ground vehicle. Transpn Res. 1972;6:343–369.
[16] Rhodes G, Mulhall BE. Magnetic levitation for rail transport. Oxford: Clarendon Press; 1981.
[17] Ma GT, Wang JS, Wang SY. 3-D finite-element modelling of a Maglev system using bulk high-$T_c$ superconductor and its application. In: Luiz AM editor. Overview of possible applications of high-$T_c$ superconductivity. Croatia: InTech; 2011. p. 119–146.

[18] Boerdijk H. Technical aspects of levitation. Philips Res Reports. 1956;11:45–56.
[19] Brandt AH. Science. 1989;243:349.
[20] Weh H. Magnetic levitation technology and its development potential. Proceeding of International Conference on Maglev '89, July 7–11, 1989, Yokohama, Japan, IEE Japan, p 1–9.
[21] Song HH, Wang JS, Wang SY, Ren ZY, Wang XR, de Haas O, et al. Studies of YBCO electromagnetic properties for high-temperature superconductor Maglev technology. In: Matins BP, editor. New topics in superconductivity research. New York: Nova Science Publishers; 2006. p. 107–156.
[22] Lee H, Kim K, Lee J. Review of Maglev train technologies. IEEE Trans on magnetics. 2006;42:1917–1925.
[23] Geary PJ. Magnetic and electric suspensions. Sci Instr Res Assoc. 1964.
[24] Hull JR. Flywheels on a roll. IEEE Spectrum. 1997;34:20–25.
[25] Hull JR. Levitation. In: Cardwell DA, Ginlen DS, editors. Handbook of superconducting materials Vol II: Characterization, applications and cryogenics. London UK: Institute of Physics Publishing; 2003. p. 1663–1673.
[26] Hull JR. Magnetic levitation. In: Webster JG editor. Wiley encyclopedia of electrical and electronics engineering. New York: John Wiley and Sons, 1991.
[27] Hull JR. Superconducting bearings. Supercond Sci Technol. 2000;13:1.
[28] Hull JR. Using high-temperature superconductors for levitation applications. JOM. 1999; 51:13–18.
[29] Powell JR, Danby GR. Dynamically stable cryogenic, magnetic suspensions for vehicles in very high velocity transport systems. Recent Adv Eng Sci. 1970;5:159–182.
[30] Powell JR, Danby GR. Maglev: The new mode of transport for the 21st century. 21st Century Science and Technology. 2003;16:43–57.
[31] Powell JR, Danby GR, Jordan J, Morena J, Wagner T, editor. A national Maglev network for the U.S. – design and capabilities. Proceeding of the 20th International Conference on Magnetic Levitated Systems and Linear Drivers; 2008 Dec 15–18; California, USA.
[32] Wang JS, Wang SY. Applications of superconducting technology. Chengdu: University of Science and Technology Chengdu Press; 1995.
[33] Wang JS. editor. Proceedings of International Workshop on HTS Maglev; 2002 June 25–27; Chengdu: China.
[34] Wang JS, Wang SY. Synthesis of bulk superconductors and their properties on permanent magnet guideway. In: Narlikar A, editor. Frontiers in superconducting materials, New York: Springer; 2005. p. 885–912.
[35] Wang JS, Wang SY. High temperature superconducting Maglev measurement system. In: Kr Sharma M, editor. Advances in measurement systems. InTech; 2010. p. 51–80.
[36] Wang JS. Past, present and future R&D of HTS Maglev in China. In: Proceedings of the 21st International Conference on Magnetically Levitated Systems and Linear Drives ; 2011 October 10–13, Daejeon, Korea.
[37] Wang JS, Wang SY. High temperature superconductive Maglev vehicle. In: Fan JD, editor. Endless Quests. Peking: Peking University Press; 2014.
[38] Glatzel K, Khurdok G, Rogg D. The development of the magnetically suspended transportation system in the Federal Republic of Germany. IEEE Trans on Vehicular Technology. 1980;29:3–17.
[39] Strnat KJ. Modern permanent magnets for applications in electro-technology. Proceedings of the IEEE. 1990;78:923–946.
[40] Sawada K. Outlook of the superconducting Maglev. Proceedings of the IEEE. 2009;97: 1881–1885.
[41] Sinha K. Electromagnetic suspension dynamics & control. London UK: Peter Peregrinus; 1987;53–57.

[42] Sinha K, Jayawant BV. Analytical and design aspects of magnetically suspended vehicles. Automatica. 1979;15:539–552.
[43] Nagurka ML, Wang SK. A superconducting Maglev vehicle/guideway system with preview control: Part I – Vehicle, guideway, and magnet modeling. Journal of Dynamic Systems, Measurement, and Control. 1997;119:638–643.
[44] Murakami M. Materials developments and applications of bulk Re–Ba–Cu–O superconductors. In: Narlikar A, editor. Frontiers in superconducting materials, Berlin: Springer; 2005. p. 869–884.
[45] Murakami M. Melt processed high-temperature superconductors. Singapore: World Scientific; 1992.
[46] Magnetic levitation technology for advanced transit systems. Society of Automotive Engineers. 1989.
[47] Mao BH, Huang R, Jia SP. Potentiality analysis on application of Maglev technologies in China. J Transpn Sys Eng & IT. 2008;8:29–39.
[48] Bunting PM. Magnetic suspension for guided transport vehicles. Transportation Planning Technology. 1972;1:49–74.
[49] Gran RJ. The benefits of Maglev technology. In: Proceeding of the 24th Fluid Dynamics Conference; 1993 July 6–9, Orlando, US. AIAA, 1993–2949.
[50] Lehndorff R. High-$T_c$ superconductors for magnet and energy technology: fundamental aspects. Berlin: Springer; 2001.
[51] Prikhna TA. Modern superconductive materials for electrical machines and devices working on the principle of levitation. Low Temperature Physics. 2006;32:505–517.
[52] Danby T, Powell JR. Design approaches and parameter for magnetically levitated transport systems. In: Kwok HS, Shaw DT, editors. Superconductivity and its applications. UK: Elsevier; 1988. p. 318–342.
[53] Nagaike T, Takatsuka H. Present status and prospect of HSST. In: Proceeding of International Conference on Maglev '89, July 7–11, 1989, Yokohama, Japan. IEE Japan, p. 29–35.
[54] Jayawant V. Electromagnetic levitation and suspension techniques. London UK: Edward Arnold; 1981.
[55] Vuchic VR, Casello JM. An evaluation of Maglev technology and its comparison with high speed rail. Transportation Quarterly. 2002;56:33–50.
[56] Iwasa Y. High speed magnetically levitated and propelled mass ground transportation. In: Fouer S, Schwartz BB, editors. Superconductivity Machines and Devices, Large Systems Applications, Plenum; 1974. p. 347–399.
[57] Yan LG. Development and application of the Maglev transportation system. IEEE Trans on Appl Supercond. 2008;18:92–99.
[58] Yan LG. The linear motor powered transportation development and application in China. Proceedings of the IEEE. 2009;97:1872–1880.
[59] Yan LG. The Maglev development and commercial application in China. Proceeding of International Conference on Electrical Machines and Systems. 2007, Oct. 8–11; Seoul, Korea. New York: IEEE; 2007. p. 1942–1949.
[60] Deng ZG, Wang JS, Zheng J, Lin QX, Wang SY. Quasi-static optimization of HTS Maglev systems with a focus on bulk superconductor part. In: Miryala M, editor. Superconductivity: recent developments and new production technologies, Nova Science Publishers, New York; 2012. p. 215–239.
[61] Xie YY. Symposium of induction levitation theory and its application. Peking: Scientific and Technical Documentation Press; 1984.

Jia-Su Wang and Su-Yu Wang

# 4 Superconducting magnetic levitation

Superconducting technology has many advantages, such as being environment friendly and energy efficient [1]. In the future field of energy, traffic, and information, it is very competitive. In addition to the advantages of PM and normal conductive Maglev, superconducting Maglev has new advantages, for example, more energy savings, environmental protection, higher speed, especially the possibility of ultra-high speed. The superconducting Maglev types, LTS and HTS Maglev, will be discussed in this chapter. Various Maglev versions have different unique performance. In the constant pursuit for the “perfect” Maglev transportation, the new HTS Maglev train may be one of the best candidates.

## 4.1 Introduction

Powell and Danby [2, 3] are the inventors of the superconducting Maglev transportation systems. They filed a patent application (No. 684775) of electromagnetic inductive suspension and stabilization system for a ground vehicle on November 21, 1967, and the patent was granted (No. 3470828) on October 7, 1969. This magnetic levitation is called the LTS Maglev. The time-varying magnetic fields induce eddy currents in the conductors and create a mutually expulsive magnetic field. The time-varying magnetic fields can be produced by moving a constant magnet. Their inventions, including the inductive levitation and stabilization guideway, null flux geometry, and the linear synchronous motor for vehicle propulsion, and this invention of the EDS using LTS magnets is the basis for the 500-km/h LTS Maglev train line from Tokyo to Osaka now under construction. Additional comments on the LTS EDS Maglev vehicle can be found in Ref. [4].

In 1945, the stable levitation of a permanent magnet over a superconductor was demonstrated first for a magnet above a concave lead (low $T_c$ superconductor) disk by Arkadiev [5, 6], at the University of Moscow.

For both diamagnetic materials [7, 8] and high $T_c$ superconductors [9, 10], stable levitation is possible in static magnetic fields. It is really static (requires no energy input), quiet, and extraordinarily stable, and it is called the unique self-stable levitation. A high-$T_c$ superconductor can be levitated either above or below a permanent magnet.

The levitation of a NdFeB PM of volume 0.7 $cm^3$ above a disk of YBCO bulk superconductor of 2.5 cm diameter and 0.6 cm thickness bathed in liquid nitrogen was observed by Hellman et al. [9] Peter et al. [10] observed a very stable suspension of YBCO samples in the magnetic field below the PM. Soon after, Brandt [11] presented that a HTS bulk could realize stable levitation without any active control. The physical

https://doi.org/10.1515/9783110538434-004

characteristics of self-stabilization of HTS bulk Maglev is a unique phenomenon in nature.

HTSCs form a large family of superconducting ceramic materials. They have a higher critical temperature, critical current density and critical field than LTSCs. The ability to use relatively inexpensive and easily handled liquid nitrogen as a coolant has increased the range of practical applications of superconductivity.

HTS bulk materials are especially attractive for practical applications because of their excellent current transport capacity in the presence of strong magnetic fields and extremely high trapped field ability in a compact space, particularly unique features for self-stable levitation without active control systems.

Since the discovery of the HTS Maglev phenomenon, much progress has been made in theory and application of HTS Maglev, especially in the fields of energy storage [12] and transportation [13]. There are some comprehensive review papers on HTS Maglev which can be found elsewhere [14–18] Special comments on the HTS Maglev vehicle can be found in Refs. [19–26].

## 4.2 LTS Maglev

It is generally known that a superconductor is not resistive. Strictly speaking, the ratio of resistance of the normal conductor to the superconductor has been shown to exceed 10 [14]. Not only do superconductors (LTSCs and HTSCs) have high critical current densities and high critical fields, but also they exhibit perfect diamagnetism (Meissner effect). The magnetic levitation between superconducting bulks and PMs is a unique and inherently self-stable levitation system. These special properties of superconductors are especially suitable for applications.

Up to now, the Nb-Ti has achieved a critical current density in excess of $10^6$ A/cm$^2$ at 4.2 K and 5 T [27, 28]. The maximum critical fields $H_{c2}$ are 14.5 T (Nb-Ti) and 30 T ($Nb_3Sn$) [29, 30]. These superconducting wires can be widely used in a variety of needs for strong magnetic field, of course, including the Maglev.

Normal conductive Maglev is usually the EMS described in Section 3.4. The PM Maglev in the section actually belongs to the EMS type, because it is unstable without feedback control (by Earnshaw's theorem). Full-scale EMS Maglev systems of the Transrapid in Germany, the HSST in Japan, and the UTM-02 in Korea have entered into the commercial stage, because active feedback control technology is quite mature.

At present, LTS Maglev is the EDS version. In the EDS Maglev, the changing magnetic field of the moving coil induces eddy currents in the conducting plate. By Lenz's law, the induced eddy currents generate a magnetic field which opposes the field of the coil, thereby providing a levitation force between the coil and the conducting plate. It will have enough levitation force to lift the entire train if the changing magnetic field is produced by a magnet, superconducting coil magnet at present, on the moving train. Both PM and SCPM can also be used as a suspended magnets of the EDS system if their magnetic fields are strong enough.

The EDS Maglev can be of various types. In the normal flux system, a flat excitation magnet system moves over a conducting roadbed consisting of periodically spaced, independent loops or a continuous plate. In the null flux system, there are double-layer coils in each side of the "U"-type track. The one-layer coil is unpowered, closed-circuit inverse pair loops (null flux). Interaction between it and on-board superconducting magnets has a guidance force for the train. The other layer coil plays the role of propulsion.

The USA first invented the EDS Maglev concept using a superconducting magnet in 1966 [2, 3]. The magnetic levitation and guidance forces of the EDS Maglev are obtained through the interaction of the magnetic field of an on-board superconducting magnet with eddy currents induced in a stationary metallic guideway.

The low-frequency forms of Maxwell's equations are used to analyze the EDS Maglev, and the AC losses of superconductors is very low at this time. This problem can be dealt with by Ampère's circuital law and Faraday's law, and the Lorentz force law. Ampère's circuital law means that the magnetic flux density integrated around a closed loop equals the net current passing through the loop. Faraday's law of electromagnetic induction shows the mechanisms by which a changing magnetic flux generates eddy currents. The Lorentz force law states that a magnetic force results if there is a current flow in a magnetic field.

When the guideway has a certain resistance, the magnetic energy in on-board magnets will be dissipated by the resistance of the guideway. The power loss $P$ gives drag forces $F_{\mathrm{D}}$.

$$F_{\mathrm{D}} = \frac{P}{v}. \tag{4.1}$$

For a given decay constant $L_{\mathrm{s}}/R_{\mathrm{s}}$, energy dissipation is smaller with faster speed. The induced current can only flow in the skin depth of the guideway. The skin depth $\delta$ is

$$\delta = \left[\frac{1}{\pi f \mu_0 \sigma}\right]^{\frac{1}{2}}, \tag{4.2}$$

where $f$ is the frequency of the induced current, $\mu_0$ is the permeability, and $\sigma$ is the conductivity. It can also be expressed as

$$\delta = \frac{k_1}{(v\sigma)^{\frac{1}{2}}}, \tag{4.3}$$

where $v$ is vehicle speed and $k_1$ is a constant determined by the system.

$$P = J_s^2 R_s = J_s^2 \frac{1}{\sigma\delta} = \left[\frac{J_s^2}{k_1}\right]\left[\frac{v}{\sigma}\right]^{\frac{1}{2}}, \tag{4.4}$$

where $J_s$ is surface current density and $R_s$ is surface resistance.

Clearly, $P$ proportionally changes with $v^{\frac{1}{2}}$, and from Eq. (4.1), it can be obtained that $F_D$ is changing proportionally with $v^{-\frac{1}{2}}$. The drag force $F_D$ is smaller with a faster speed $v$. At very low frequency, the magnetic flux distribution is uniform in the guideway, and the induced current depends directly on the speed. $P$ changes proportionally with $v^2$, i.e. $F_D \sim v$. If the skin effect is ignored at a very high speed, the ratio of lift force $F_L$ and drag force $F_D$ can be expressed as

$$\frac{F_L}{F_D} = \frac{v}{v_0}, \tag{4.5}$$

where $v_0$ is a parameter depending on frequencies, the guideway dimensions, material parameters, and structure. Clearly, the drag force $F_D$ is smaller for faster speed. Experimental research [31, 32] proved the correctness of Eq. (4.5).

Guderjahn et al. [33] found that at high velocities, the levitation force can be calculated as if the guideway had infinite conductivity, and they also get similar results.

Coffey et al. [34] confirmed that at finite velocities, the resistivity of the guideway reduces the actual lift force $F_L$ and produces a drag force $F_D$, and they have found the important results: (1) the lift force $F_L$ approaches the magnetic image force $F_i$ asymptotically (at speeds of ~80 km/h for magnets of 0.5 m square, $F_L \approx 0.8F_i$); (2) the drag force at high speeds (>80 km/h) decreases as $v^{-1/2}$; (3) the suspension height is adjustable (150 mm seems reasonable); (4) superconducting magnets of modest dimensions are practical and necessary for this application.

Reitz [31] reported the calculation results of the lift and drag forces of induced eddy currents and their associated fields for several magnet geometries. The ratio of lift to drag is found to be independent of coil geometry, but the velocity dependence of the lift is greatly affected by the geometry. The ratio of lift to coil weight can be as high as 2000 for a superconducting coil moving at 483 km/h (300 mph) at 10 cm above a conducting plate.

On the basis of the above analysis, Miericke and Urankar [35] derived the exact analytical expressions for lift $F_L$ and drag $F_D$ forces on moving flat rectangular current-carrying coils above and below an infinite conducting sheet track of arbitrary thickness. Subsequently, they developed approximate expressions [36] for lift $F_L$ and drag $F_D$ forces in these systems. The numerical computations on the basis of approximate formulae were done with an ordinary computer, and the results agreed so well with costly computer calculations on the basis of the exact expressions.

From here, we see that the results only agree with experiment in specific conditions. Hieronymus et al. [37] measured the lift and drag forces exerted by various small-scale current-carrying superconducting coils, placed near a large rotating disk. Most of the experimental results were compared with exact calculations based upon the general theory given by Miericke and Urankar [35, 36] for infinitely wide but arbitrarily thick sheet tracks. The agreement between experiment and theory was generally good. All numerical results based on these approximate formulae also agreed well with the exact expressions (see Fig. 4.1) [35]. Iwasa [38] detailed the various problems of EDS Maglev, including these theories referred to above.

Lift force $F_L$ can also be understood as the magnetic pressure between the magnetic field of the induced current and the superconducting magnet. Any magnetic field has an associated magnetic pressure $P_B$ and is given by

$$P_B = \frac{B^2}{2\mu_0}, \tag{4.6}$$

i.e., the magnetic pressure is proportional to the square of magnetic field.

In addition to the above theoretical study, in the early 1970s, several groups [39–42] in the USA began some experimental research work on the superconducting Maglev. By the mid-1970s, the effort in the USA had ended with no implementation of a Maglev system. Nearly 20 years after, the US Federal Railway Administration

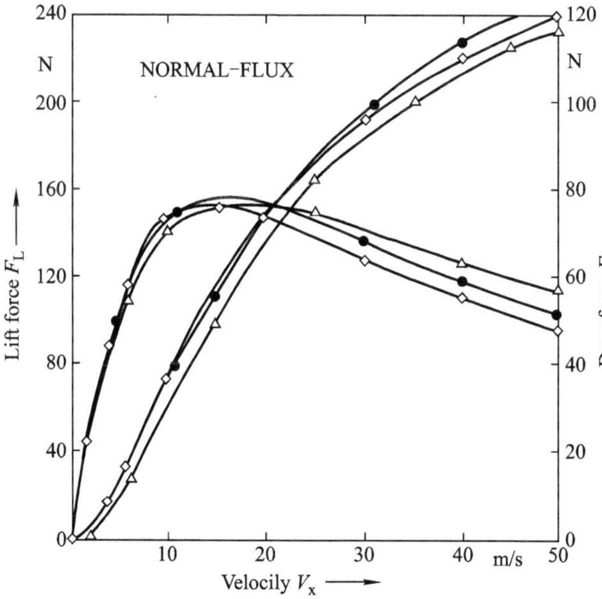


**Fig. 4.1:** Comparison of experimental results of lift forces $F_L$, drag forces $F_D$, and speed $v$ with theoretical values calculated with the exact and approximate equation [36]. The ◇ values were calculated with the exact equation, whereas the ● values were calculated with approximate equation. △ Experimental measurement results.

organized the National Maglev Initiative (NMI) [43, 44] to determine whether it should actively encourage investment in Maglev. The NMI program in the USA was started in 1990. The Government Maglev System Assessment Team operated from 1991 to 1993 as part of the NMI. The NMI's principal tasks were to assess the technical and economic viability of Maglev in the USA and to recommend the most appropriate federal role for its development. It awarded four System Concept Definition (SCD) contracts to teams led by Bechtel Corp., Foster-Miller, Inc., Grumman Aerospace Corp. [45, 46], and Magneplane International, Inc., [47–49]. Based on comparison of the High Speed Surface Transport of TGV in France and Transrapid-07 in Germany, all four US Maglev concepts in the NMI program were LTS Maglev. The Grumman version was EMS Maglev, but the other three schemes were EDS Maglev. The levitation gap of Grumman's EMS Maglev was 40 mm, 4 times as high as in Transrapid-07. The levitation gaps of other three EDS Maglev schemes were greater than 40 mm.

Unlike discrete coil guideways of EDS Maglev in Japan, Magneplane's was a version of the conductive sheet guideway. This conducting sheet for levitation offered potential advantages in construction cost and ride quality over discrete coil guideways, but it had lower electromagnetic lift-to-drag ratios. Sheet guideways could have a larger levitation gap capability. Superconducting magnet systems of comparable weight to PM systems could provide at least three times the levitation gap [47].

The Foster-Miller [43] concept was an EDS similar to the Japanese high-speed superconducting Maglev, but it had some additional features to improve potential performance. The Foster-Miller concept was a vehicle tilting design that would allow it to operate through curves faster than the Japanese system for the same level of passenger comfort. The Bechtel [43] concept was a superconducting EDS system that used a novel configuration of vehicle-mounted, flux-canceling magnets. The vehicle contained six sets of eight superconducting magnets per side and straddled a concrete box-beam guideway. Interaction between the vehicle magnets and a laminated aluminum ladder on each guideway sidewall generated lift.

Numerous studies and development projects [43] have been completed in the USA, but no commercial Maglev systems have been deployed. Powell and Danby [50], the LTS Maglev inventors in the USA, presented a Maglev train for the 21st century. Operating high-speed vehicles from one guideway to another and without having to slow down the trains was to be achieved by an ultra-high-speed electronic switch. The on-board superconducting quadrupole magnets would allow them to smoothly transition, back and forth, between narrow-beam and planar guideways. In this way, the vehicle would ride on the single guideway after leaving the station. Thus, the Maglev system could reduce greatly the construction costs.

Studies and some experimental works for the EDS Maglev system using 10 levitation superconducting magnets were started in Canada [51–53] in the early 1970s. A flat-topped guideway was proposed to minimize ice and snow accumulation, and a new technique was proposed for obtaining lateral guidance using interaction of 50 propulsion magnets with the edges of the levitation strips and with null flux loops overlaying the LSM windings. The Canadian investigation was aimed at developing Maglev vehicles for 480-km/h intercity transit along the Toronto-Ottawa-Montreal corridor in the 1990s. The vehicles were intended to carry 100 passengers, weighing 30 tons and levitated with 15 cm clearance by eight on-board superconducting magnets.

In the initial phase, the research program in the Germany covered the whole variety of possible suspension techniques: air-cushion technique, permanent-magnetic levitation (PMS), electrodynamic levitation (EDS), and electromagnetic levitation (EMS). After careful consideration, they chose normal conductive EMS and superconducting EDS versions [54–56].

The superconducting EDS system in the Germany was a circular test-track facility in Erlangen with 280 m diameter and an inward obliquity angle of 45 degrees. The 17-ton test vehicle was 12 m long and levitated at a gap of 10 cm with a linear induction motor. The test vehicle achieved speeds of 230 km/h in 1976. In 1978, however, they decided to abandon the superconducting EDS development and focus all of its high-speed development efforts on normal conducting EMS systems.

In the late 1980s, under the support of the Ministry of Science and Technology of the P. R. China, a national program was started to study the key technologies of normal conducting EMS Maglev. In the middle of the 1990s, under the support of the Ministry of Science and Technology, a program to establish contacts and academic exchanges with Japanese colleagues about LTS EDS trains was started. A cooperative feasibility study was conducted for a Shanghai-Hangzhou high-speed superconducting EDS Maglev line together with Japanese colleagues. Dr. Zhang from the Ministry of Science and Technology presided over a meeting on the Shanghai-Hangzhou high-speed Maglev train in Hangzhou University in November 26–28, 1996. The topic of the meeting was the feasibility problem of a Shanghai-Hangzhou LTS Maglev train line and to plan a China-Japan cooperative R&D. Experts from both countries and officials in Zhejiang Province attended the meeting. China experts included Wen-Sen Chang, Bing-Heng Shao, Jia-Su Wang, and the engineers from the Fourth Railway Design Institute in China. In the early 1990s, the calculation and design of on-board LTS magnets were done by Jia-Su Wang and Su-Yu Wang. Up to now, China has not developed the LTS Maglev train.

## 4.3 LTS Maglev train

The superconducting EDS Maglev train in Japan is one of the most successful applications of superconducting technology. Similar to the normal conductive EMS Maglev train in Germany, Japan's superconducting Maglev is the model for magnetic suspension successfully applied in the field of transport.

In 1972, Japan National Railway made a test run of the LTS EDL Maglev test vehicle ML-100. It was 7 m long, 2.5 m wide, 2.2 m high and weighed 3.5 tons. It was propelled by a linear induction motor, which had a secondary on-vehicle and

**Fig. 4.2:** The first test LTS Maglev vehicle ML-500 on the Miyazaki Test Track recorded 517 km/h on December 21, 1979 [59].

primaries on-ground. Since the test line was 480 m long, its speed was limited to 65 km/h, and its levitation gap was only 6 cm [57].

In 1977, experiments began in earnest on the Miyazaki Test Track in Japan. On December 21, 1979, the prototype ML-500 test train (Fig. 4.2) reached an unmanned speed of 517 km/h on the 7-km track, proving the tremendous potential for high Maglev speed [58]. Then the test track was remodeled from an inverted-T shape cross section to the more practical U-shape, and the manned vehicle began to run in 1981. Basic tests with four types of test vehicle were conducted over the 20-year period from its opening in 1977 to its closure in 1996.

The construction work for the new 18.4-km Yamanashi Maglev Test line (YMTL) was started in 1988 and was finished in December 1996. The track was modified into a new practical U-shaped guideway. The vehicles (MLX) were of an articulated truck type having a truck at each end (Fig. 4.3). Superconducting magnets were installed on both sides of the truck, containing four vertical superconducting coils and a helium/nitrogen tank with built-in refrigerators, while the levitation, guidance coils, and propulsion coils were installed in the guideway opposite to the superconducting magnet. Electric current fed to the propulsion coils propels and brakes the trains; the system varied the amplitude and frequency to set the velocity and the acceleration/deceleration of the train. On December 12, 1997, a new world record of 531 km/h [58] was set for manned train travel. The main results of the tests were the following [61, 62]: the manned five-car train set registered a record speed of 552 km/h on 14 April 1999; a relative speed of 1003 km/h between two trains passing each other was achieved; substation cross-over tests and passing tests with two trains at the station were demonstrated.

The maximum speed, 581.7 km/h, was attained on December 2, 2003, exceeding the maximum designed speed on the YMTL and the potential maximum speed in

**Fig. 4.3:** The LTS Maglev vehicle MLX01 on the Yamanashi Maglev Test line [60].

commercial service. It is verified that this system has sufficient capability for high-speed operation. A further high-speed passing test of 1026.3 km/h was carried out on November 16, 2004, exceeding the potential passing speed for the commercial service. The stability of both vehicle and ground equipment was confirmed [63]. Through 15 years of tests verifying technology, safety, and reliability at the YMTL, the cumulative travel distance has reached almost 880,000 km with 140,000 test riders having experienced the ride [1].

One of the most important areas of a LTS EDS Maglev train was the development of reliable superconducting magnets [64]. The 16 superconducting magnets were attached on the vehicle's bogie. Within the superconducting magnet cryostat, four superconducting coils made of niobium-titanium (Nb-Ti) wires were installed. The superconducting magnet, capable of generating a maximum magnetic field of 5.5 T, was small (5.4 m long × 1.17 m height) and light (1400 kg), enabling total weight reduction of the vehicle. The superconducting magnet could maintain stability from mechanical vibration or electromagnetic interference applied during high-speed run. Moreover, on-board superconducting magnets required neither cryogen nor power supply except for the refrigerator power source [1].

Existing Superconducting Maglev vehicles in Japan adopt LTS magnets using Nb-Ti wires wound in a racetrack configuration. The LTS coils were cooled by liquid helium in cryogenic vessels and their radiation shield plate was maintained at about 80 K by liquid nitrogen (Figs. 4.4 and 4.5). Evaporated gas helium and nitrogen were completely re-liquefied by on-board refrigerators. The LTS magnets were energized at the train depot and operated in a persistent current mode with the help of a superconducting switch. The decay rate of the magnet current was below 0.1%/day; thus, they did not require any on-board energizing power units [65].

Central Japan Railway Company (JR Central) has built the Tokaido Shinkansen, the world's first high-speed railway. After deliberation, in December of 2007, JR Central announced plans to build a superconducting Maglev train line connecting Tokyo, Nagoya, and Osaka with a maximum speed of 505 km/h [1, 63]. Starting from central Osaka to the heart of Tokyo, the superconducting Maglev train will travel in 67 minutes, 44% faster than the most advanced bullet trains, and traveling by Maglev

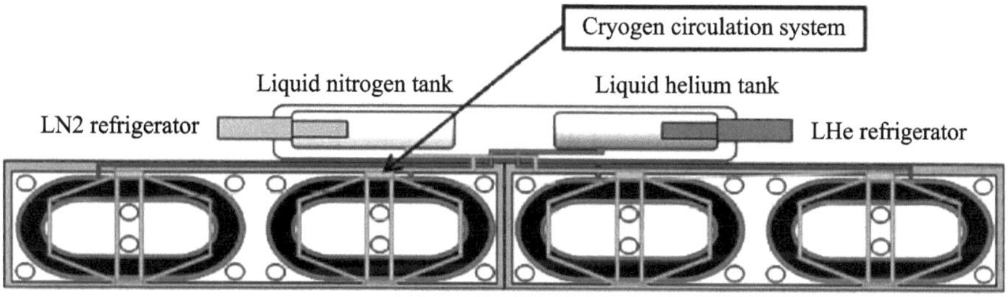


**Fig. 4.4:** Schematic view of LTS magnet using Nb-Ti wires of the EDS Maglev train in Japan [65].

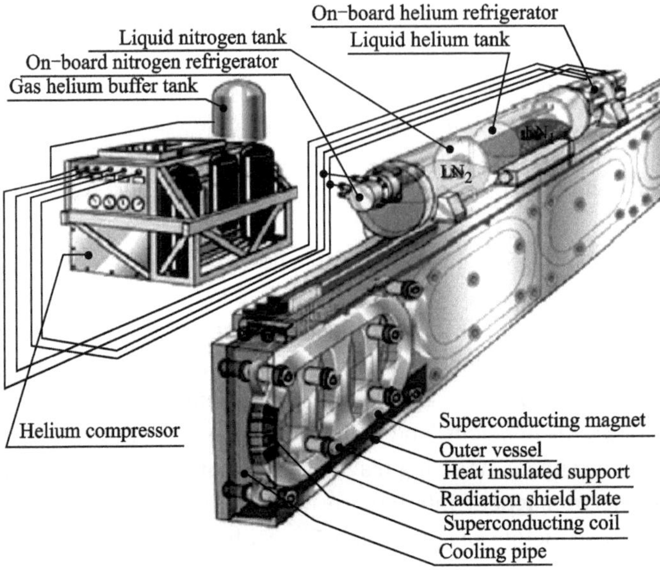


**Fig. 4.5:** Superconducting magnet of the EDS Maglev train in Japan [1].

train will be more convenient than flying because of the travel time to get to the airports. The first part of the Maglev project is planned to be operational in 2027, linking Tokyo with Nagoya, a city halfway between Tokyo and Osaka. The full track between Tokyo and Osaka would be finalized in 2045. Estimated construction cost would be roughly 9 trillion JPY and fully financed by JR Central.

## 4.4 HTS Maglev

BSCCO was the first-found HTS material to be used for making practical superconducting wires. An on-board magnet coil wound by HTS tapes instead of Nb–Ti wires, is very promising. BSCCO HTS magnets have been in development for superconducting Maglev trains since 1999. Because HTS BSCCO wires exhibit poor pinning performance at intermediate temperatures, they cannot be used for levitation even with liquid nitrogen cooling (see Section 2.3). The main advantages of 2G HTS wires are high critical current density, large-scale production, lower manufacturing cost, better mechanical properties, and superior performance under high magnetic fields. The 2G HTS REBCO wires are very attractive for various Maglev applications.

The levitation of a NdFeB PM of volume 0.7 $cm^3$ above a disk of YBCO bulk superconductor 2.5 cm diameter, 0.6 cm thickness bathed in liquid nitrogen was observed by Hellman et al. [9] Very stable suspension of YBCO samples in a magnetic field below the PM was observed by Peter et al. [10] Brandt [11] demonstrated that the HTS bulk magnetic levitation could be stable without any active control.

In 1997, a cooperative project between China and Germany developed a small HTS Maglev model of 20 kg with a levitation gap of 7 mm [66]. International research interest in a manned HTS Maglev vehicle was aroused after the first manned HTS Maglev test vehicle in the world, "Century", was demonstrated in the Applied Superconductivity Laboratory (ASCLab) of Southwest Jiaotong University at 2:26 pm on December 31, 2000 [13]. The second and third manned HTS Maglev vehicles were developed in Germany [67] and Russia [68] in 2004. A full-scale HTS Maglev train as a substitute for an urban light track train is under development in Brazil [69, 70]. In March, 2013, a 45-m-long HTS Maglev ring test line was successfully developed by ASCLab [71]. Moreover, research groups in Japan [72] and Italy [73] have also developed HTS Maglev prototypes.

The HTS Maglev can be widely used in construction of practical Maglev trains, Maglev launch systems, frictionless bearings, non-contact spinners, Maglev flywheel energy storage devices, motors, generators, contact-less liquid pumps, etc [26].

1. HTS Maglev train (see chapters 6 and 8): As mentioned earlier, the principle of the HTS Maglev train is completely different from the normal conductive (non-superconducting) train in Germany and the LTS Maglev train in Japan. The HTS Maglev train is a self-stable Maglev, economical in terms of energy consumption (1/20th of an aircraft), environmentally friendly (no chemical or noise pollution), which can achieve ultra-high speed (thousands of kilometers per hour), and it is a safe and comfortable form of ground transportation.
2. HTS Maglev bearings (see chapter 9): The friction coefficients of a mechanical bearing, an active Maglev bearing, and a HTS Maglev bearing, are $10^{-3}$, $10^{-4}$, and $10^{-7}$, respectively. Thus, the friction coefficient of HTS Maglev bearings is thousands of times lower than that of conventional bearings. HTS Maglev bearings can be used in high-speed or ultra-high-speed rotating systems.
3. HTS Maglev FESS (flywheel energy storage system) (see chapter 9): FESS can store both electrical and mechanical energy. It can directly store mechanical energy without any energy conversion loss, and mechanical energy storage efficiency is 10 times higher than that of a traditional storage system. Energy density has reached 230 Wh/kg. In 5 years, it would reach 2700 Wh/kg, which is three times higher than that of the hydrogen fuel cell.
4. HTS Maglev launch system (inexpensive reusable cold launch system) (see chapter 10): The reusable cold launch system has no ignition or explosion and can be used repeatedly. It has many advantages: does not emit noise or pollution, has all-electric control, mobile, reusable, etc. A HTS Maglev rocket is only 20% of the weight of a conventional propellant rocket; thus, the cost of the space launch can be greatly reduced. With no fuel on the vehicle, super-high speed can be achieved with only 200-kW power. The launch cost can be reduced from $10,000 to $1000

per pound. The HTS Maglev is the right choice that best meets the requirements of reusable cold launch technology compared to other Maglev alternatives.

## 4.5 HTS wire Maglev train

A cryogen-free HTS magnet has been developed in Japan. The magnet coil is wound with HTS tape instead of Nb-Ti wire and it uses a conductive cooling method with direct cooling by refrigeration [65]. Further reduction of operation cost will be expected when the HTS magnet replaces of the LTS magnet.

The R&D project for BSCCO HTS magnets has been in development for the superconducting Maglev trains since 1999. In 2001, JR Central was entrusted with the national project to develop of a trial HTS magnet that contains one "racetrack"-shaped coil. The first trial HTS magnet can be cooled by a cryocooler, using a conduction cooling method. The Ag-sheathed Bi2223 coil is connected with a HTS persistent current switch and cooled by a pulse tube cryocooler below 20 K. The current decay rate is 0.44%/day. This means that the HTS magnet can be operated virtually in the persistent current mode for Maglev use [65]. Schematic view and appearance of HTS magnet using Ag-sheathed Bi2223 coil for the EDS Maglev train in Japan is shown in Figs. 4.6 and 4.7, respectively.

A second HTS magnet consisting of four persistent current HTS coils was developed. The HTS coils were installed in a cryostat and cooled to approximately 15 K by conduction cooling, using two sets of two-stage G-M-type pulse tube cryocoolers. The HTS magnet was operated in a persistent current mode at a rated magneto-motive force of 750 kAt which was the same as that used with LTS magnets. The running tests were executed in the Maglev vehicle "Series L0" on the Yamanashi Maglev Test line, with a top speed of 553 km/h achieved on December 2, 2005 [74, 75]. The running trial

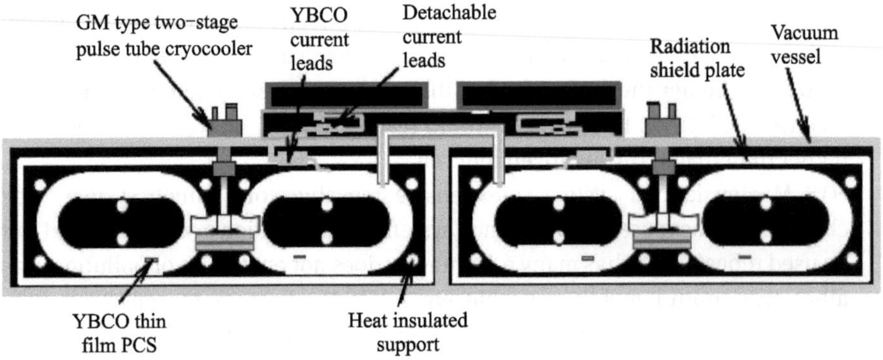


**Fig. 4.6:** Schematic view of HTS magnet using Ag-sheathed Bi2223 coil of the EDS Maglev train in Japan [65].

**Fig. 4.7:** Appearance of HTS magnet using Ag-sheathed Bi2223 coil of the EDS Maglev train in Japan [65].

in the EDS Maglev vehicle proved the practicability of the HTS magnet using G-M type two-stage pulse tube cryocoolers.

In addition to the above EDS Maglev, the practicability of the EMS Maglev using BSCCO HTS wire was studied [76]. The experimental results about the applicability of HTS coils to Maglev systems were reported. EMS Maglev active control for a copper coil magnet has been achieved, but the active control of the HTS coil magnet is complex. This issue should be studied and solved.

The HTS BSCCO wires exhibit a poor pinning performance already at intermediate temperatures and thus cannot be used for levitation even with liquid nitrogen cooling. It is for this reason that coated conductors of YBCO are preferred to BSCCO. The corresponding BSCCO, YBCO superconducting wires are referred to as second-generation (2G) HTS wires. The main advantages of 2G HTS wires are: high critical current density, large-scale production, lower manufacturing cost, better mechanical properties, and superior performance under high magnetic fields. The 2G HTS REBCO wires are very attractive to various Maglev applications (see Section 2.3).

Using 2G wires is especially effective in stability and reliability of Maglev magnets and in simplification of magnet structure. Figure 4.8 shows the merits of applying HTS 2G wires to a superconducting magnet for the Maglev train. The original figure is a schematic view of the LTS magnet which used Nb-Ti coils, and the symbol "≫" shows merits of applying HTS 2G wires. In order to make use of the advantages of 2G wires, the possibility of Maglev applications of the 2G wire with a high critical current density in a high magnetic field was examined. Clearly, the weight and the energy consumption of the on-board HTS 2G magnets and the cryocooler for the Maglev train will decrease. The optimized operation temperature, 40 to 50 K for commercial 2G wires, used to minimize the magnet weight has been reported [77].

The 2G wire enables conduction cooling of the magnet and reduction the total weight of Maglev vehicles. A HTS 2G wire magnet was used to demonstrate [78] that a magnetic flux density of 5 T can be achieved at 45 K and 6 T can be achieved at 40 K. The current density of the 2G coil is 130 A/mm$^2$, and it is similar to that of the existing LTS Nb-Ti coil on-board magnet for the Maglev. The results of these studies have proved that 2G wire can be applied to on-board magnets of the Maglev. The single-stage cryocooler and radiation shieldless cryogenic structure can meet the requirements of

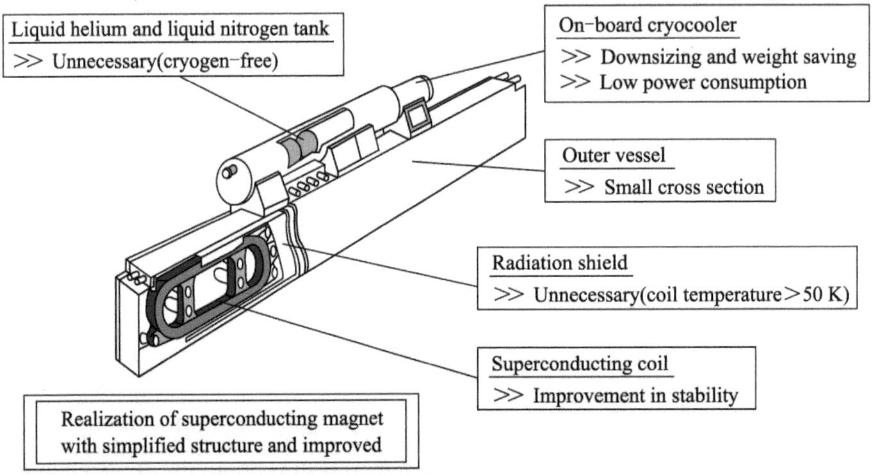


**Fig. 4.8:** Merits of applying HTS 2G wires to a superconducting magnet for the Maglev train. The original figure is a schematic view of the LTS magnet which installed Nb-Ti coils, and symbol ">>" shows merits of applying HTS 2G wires [77].

HTS over 40 K. The distance between the superconducting coil and the outer surface of the cryostat is only 20 mm, which increases interlinkage magnetic flux to the ground coil and improves efficiency of Maglev applications. The eddy current in the ground conductor is proportional to flux linkage, and the levitation force is directly proportional to the eddy current value.

With further improvement of the performance of 2G wire, HTS 2G coil magnets can replace LTS magnets and greatly improve the performance of the EDS Maglev system.

The levitation characteristics of the HTS EDS system were investigated by the relative motion between the HTS levitation magnet of 279.4 mH and the aluminum ground conductor in a rotating state [79]. The experimental results showed that the levitation force saturated after 20 m/s. The measured levitation forces with 20 and 30 A at relative speed of 30 m/s were about 265 and 618 N, respectively. These results can be used in the design of HTS levitation magnets mounted on the very-high-speed Maglev.

There was an interesting experiment using 2G wires as a passive levitator for Maglev vehicles [80]. For that purpose, a superconducting 2G block was built by stacking more than 500 pieces of coated conductors. The 2G block showed good resistance to humidity, which made their storage easier. Also, the 2G wire can be stacked in many different ways, enabling the design of a HTS sample optimized for the magnetic field profile. The levitation and lateral forces between the 2G block and a PM guideway were measured and compared with the results of an YBCO bulk with the approximately same dimensions. The 2G block was conceived to replace a YBCO bulk in the Maglev prototype, but the force between the 2G block and a PM was lower than that for ordinary YBCO bulks. The experiment results may herald the 2G block's potential to be applied in a Maglev bearing and vehicle application.

## 4.6 HTS bulk Maglev

In 1933, Meissner and Ochsenfeld [81] found that when a sphere is cooled below its transition temperature in a magnetic field, it excludes the magnetic flux (see Section 1.3). In 1945, Russian physicist Arkadiev [5], at Maxwell Laboratory in Moscow State University, USSR, published an article, titled "Hovering of a magnet over a superconductor". The stable levitation phenomenon of a magnet over a superconductor was first observed. The photograph [6] in Fig. 4.9 shows a magnet of $4 \times 4 \times 10$ mm in dimensions floating above a concave lead disk (low $T_c$ superconductor) 40 mm. The pure lead is a type I superconductor. The interaction between the magnetic field and a type I superconductor in Meissner-Ochsenfeld state produces magnetic levitation force. The Meissner-Ochsenfeld effect completely expels the magnetic fields (below $H_c$) to the field penetration depth $\lambda$ of the superconductor surface. Thus, the levitation forces are very small. The investigation of bearings using LTSCs was reported in 1953 [82], and the study of a three-phase motor with superconductive bearings was reported in 1965 [83].

There were no practical applications of LTS bulk superconductors before the discovery of HTS, due to their thermal instability [84]. Local heat generation within the LTS bulk materials leads easily to a large magnetic flux jump, namely to a quench of the superconducting state. In contrast to LTS bulk materials, there are some peculiar features in the specific heat of YBCO compared with those of conventional BCS superconductors, namely bulk HTS are thermally stable even in large sample sizes due to their relatively large specific heat (see Section 2.5.1) in the superconducting state.

The LTS $MgB_2$ material is a special case, because it has higher specific heat. Flux jumps in $MgB_2$ arise at very low temperatures, but tend to disappear at temperatures larger than 10 K [85]. Thus, $MgB_2$ can be made as large bulk material and produce a larger trapped flux. $MgB_2$ bulk is also promising as a superconducting permanent magnet.

The type II superconductors are similar to type I superconductors and present a Meissner-Ochsenfeld effect (see Section 1.3) only when the magnetic applied field

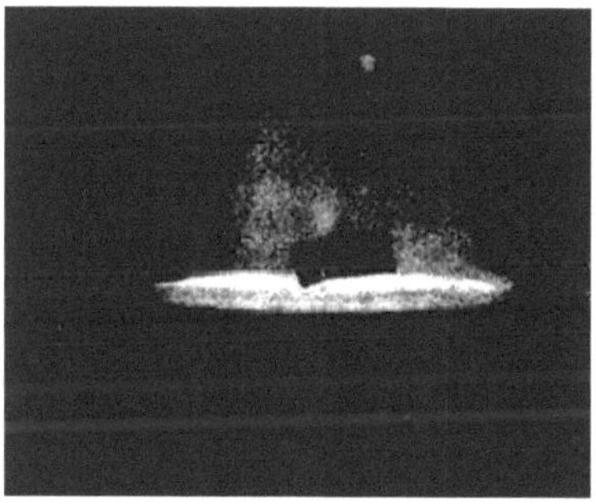

**Fig. 4.9:** The photograph of a magnet of $4 \times 4 \times 10$ mm floating above a concave lead disk 40 mm [6].

is smaller than lower critical field $H_{c1}$. Interaction between the magnetic field and the type II superconductor in the Meissner-Ochsenfeld state produces a magnetic levitation force below $H_{c1}$, but this force is smaller than that of the type II superconductor in the mixed state. It is a very important matter and also is easily confused.

If a superconductor in the normal state is cooled below $T_c$ by the zero field cooled (ZFC) method, and the normal state transitions to the superconducting state without any magnetic field present, when the applied magnetic field is gradually increased with the superconductor in the superconducting state, there will be two different situations: one is for type I superconductors in the superconducting state which completely repel the magnetic field below critical fields $H_c$, the other is the type II superconductors in the superconducting state which also completely repel the magnetic field below lower critical fields $H_{c1}$. When the applied magnetic field reaches $H_c$, the superconducting state of type I superconductors is destroyed. When the applied magnetic field reaches the lower critical field $H_{c1}$, the superconducting state of type II superconductors is not destroyed. Instead, it enters the mixed state. When the applied magnetic field reaches $H_{c2}$, the superconducting state of type II superconductors is destroyed. In practical application, HTS are limited by the irreversibility field $H_{c3}$, above which magnetic vortices melt or decouple. The irreversibility field $H_{c3}$ is between $H_{c1}$ and $H_{c2}$ ($H_{c1} < H_{c3} < H_{c2}$), and $H_{c3}$ is much lower than $H_{c2}$ (see Section 1.11). The mixed state of the type II superconductors at $H_{c1} < H < H_{c3}$ is a necessary condition to produce large levitation forces.

As the applied magnet field is increased to $H_{c1}$, the flux lines of the magnetic field are gradually compressed. When the applied magnetic field increases to a certain value, a part of the magnetic flux penetrates the superconductor edge through the pinning centers (vortex), forms a lossless superconducting shield current. The magnetic field direction of the shield current is the reverse of the external magnetic field; thus, there is a levitation force in the HTS Maglev system. This levitation force between the shield current in the HTS bulk and the applied magnetic field is much larger than that in the type I superconductors. The physical nature of the process is due to flux pinning in type II superconductors; otherwise, it is impossible for the shielding current to continually exist in the HTS bulk.

When a superconductor in the normal state is cooled below $T_c$ by the field cooled (FC) method, the normal-state transitions to the superconducting state with a magnetic field present. Thereafter, the field will be expelled from the superconductor at applied magnetic field below $H_{c1}$ for type II superconductors. When the applied magnetic field is $H_{c1} < H < H_{c3}$, the magnetic flux penetrates the HTS bulk through the pinning centers (vortex) and forms trapped flux. The trapped flux provides the force of lateral stability, and the force is called the guidance force.

The total Maglev energy between the HTS bulks and the PMs is restricted by the sum of magnetic energy of the HTS bulks and the PMs. The magnitudes of the levitation forces and the guidance forces are decided by calculation and experiment. Although the magnetic energy product the HTS bulks is high, it also is certainly limited. In order to obtain the maximum levitation and guidance forces, we must complete the optimum design according to the specific application.

The HTS Maglev phenomenon can be very easily demonstrated and its physical principle is not difficult. However, development of adequate physical models is difficult because the HTS bulk is a special type II superconductor. It is a quasi-crystal containing particles and micro-holes and has a strongly irreversible magnetic behavior, i.e., the magnetization depends on the prehistory of the magnetic field. In addition, HTSCs are often in a very complex magnetic field. Two- and three-dimensional modeling and simulation will be introduced in Chapter 7.

The HTS bulk Maglev is discussed above. The HTS bulk Maglev can be widely used in the construction of practical Maglev trains, Maglev launch systems, frictionless bearings, Maglev flywheel energy storage devices, etc. In addition to this, the magnetic levitation of HTS thin films also has been studied.

Epitaxial YBCO films on single crystalline substrates are mainly used for the superconducting microwave devices. The levitation forces between a permanent magnet and a superconducting thin film have been investigated [86, 87] experimentally, because of the high current density of YBCO film. It has been found that thin films produced very high values of the vertical force $F_z$ and magnetic stiffness $\kappa_z = \delta F_z/\delta z$ per unit volume of superconducting material. The levitation force of a three epitaxial YBCO film stack per unit volume is extremely high [86], yielding 100 N/cm$^3$, whereas the corresponding levitation force for MTG bulk samples is about 1 N/cm [3]. Experimental studies and theoretical modeling of the levitation force between a permanent magnet and a superconducting thin film have been reported [88]. After measurements of the force $F_z$ and magnetic stiffness $\kappa_z$ as functions of distances $z$ between the magnet and superconductor, the explicit formulas for both $F_z$ and $\kappa_z$ are derived. An important result is that this has been established as a basis for using $Fz(z)$ data to infer the critical current density and its field dependence.

## 4.7 HTS bulk Maglev train

### 4.7.1 Unique advantages of HTS bulk Maglev

1. Magnetic levitation of the HTS bulks is a self-stable levitation system, which does not require active control.
2. Energy efficiency: Both levitation and guidance do not require active control or an on-board power supply system. The total operational cost is almost negligible due to the affordability of liquid nitrogen (77 K, 78% in air and easy to liquefy). The price of liquid nitrogen is about 1/100th of that liquid helium (4.2 K). The energy

consumption of an ultra-high-speed ETT HTS Maglev train is only 1/20th of an airplane.

3. Environmental effects: The HTS Maglev train in the ETT has no noise or electromagnetic or chemical pollution.
4. Safety: The HTS Maglev train is an inherently self-stable levitation system. The levitation force will increase exponentially when the levitation height drops. Consequently, it is impossible to contact the permanent magnet guideway (PMG) in all operation circumstances. Safe operation of the guidance system in the horizontal direction can be achieved without controls.
5. Comfort: HTS Maglev train has very high comfort level because of its inherent self-stability.
6. Ultra-high speed: Operation speed can be greater than 1000 km/h in ETT.
7. Low cost: Compared with the urban light rail vehicle (LRV) for the same number of passengers, the HTS Maglev train has lighter vehicle, lighter supporting structure, smaller drive power, and cheaper infrastructure. The HTS Maglev train does not need an elevated reinforced concrete structure, but the LVR must have this structure. Therefore, the total weight of a HTS Maglev train line is lighter than LVR, and the total construction and operation costs are lower than the equivalent LRV system.

### 4.7.2 Moderate-/low-speed HTS Maglev trains

The project of the manned HTS Maglev vehicle was ratified by the National High-tech R&D Program (National 863 Program) of China in 1997. Thereafter, both experimental and theoretical studies of the HTS Maglev were made in detail, and a series of important research results have been obtained [89, 90]. Based on a series of parameters, levitation forces, guidance forces, levitation stiffness, etc. of single- and multi-YBCO bulks have been measured by the HTS Maglev measurement system SCML-01, and the first manned HTS Maglev test vehicle in the world, "Century", was demonstrated in the Applied Superconductivity Laboratory (ASCLab) of Southwest Jiaotong University at 2:26 pm on December 31, 2000 [13], and a patent for "a superconducting Maglev train system" was awarded [91] (see Fig. 4.10).

Figure 4.10a shows that the net levitation gap of the HTS Maglev vehicle was more than 20 mm with five people standing on the vehicle. Figure 4.10b shows the first manned HTS Maglev test vehicle in the world, "Century". "Century" is is 3.5 m long × 1.1 m wide × 0.9 m high with four passenger seats. Figure 4.10c shows the "Century" in Review of 15 Years for National High-tech R&D Program (863 Program) of China in Beijing, 2001. Table 4.1 lists the essential parameters of "Century". The PMG track of the HTS Maglev vehicle has a length of 15.5 m, and they have a concentrating field of 1.2 and 0.4 T at its center surface and 20 mm above the surface, respectively. The

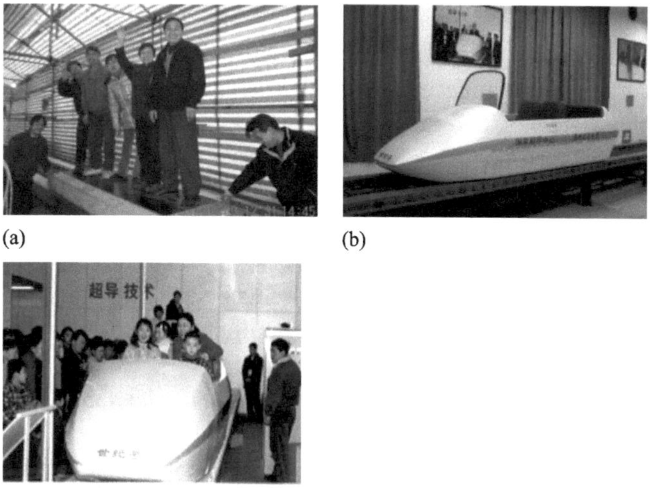

(a) (b) (c)

**Fig. 4.10:** The first manned HTS Maglev test vehicle in the world, "Century". (a) The net levitation gap of the HTS Maglev vehicle was more than 20 mm when five people stood on the vehicle. (b) The first manned HTS Maglev vehicle in the world–"Century" in the ASCLab [92]. (c) "Century" in Review of 15 Years for National High-tech R&D Program (863 Program) of China in Beijing, 2001.

**Tab. 4.1:** Essential parameters of the first manned HTS Maglev test vehicle in the world, "Century".

| Specifications | Units | Value |
|---|---|---|
| Passengers | seats | 4 |
| Length of guideway | m | 15.5 |
| Vehicle body size | mm | $2268 \times 1038 \times 120$ |
| Outside size of vehicle | m | $3.5 \times 1.2 \times 0.8$ |
| Total levitation force | N | 6350 at a gap of 20 mm |
| Levitation gap | mm | 20 |
| Total guidance force | N | 1980 at a gap of 20 mm |
| Type of drive | | linear induction motor |
| acceleration | $m/s^2$ | 1 |

on-board HTS Maglev equipment outside size is $150 \times 516 \times 168$ mm and is composed of a rectangle-shape liquid nitrogen vessel and 43 pieces of YBaCuO bulks. The vessel bottom has a thickness of only 3 mm. The linear motor acceleration is 1 $m/s^2$. Here are the key observations from this HTS Maglev system:

1. In order to minimize the distance between PMG and HTS bulks and increase HTS Maglev efficiency, we developed the innovative rectangular liquid nitrogen cryogenic vessel with a 3-mm-thin bottom wall, which made HTS Maglev vehicle feasible.

2. High-temperature superconductors were protected by dry nitrogen gas in permanently sealed container with internal pressure slightly higher than 1 atm.
3. The total levitation force and guidance force of the whole vehicle at the net levitation gap of 20 mm were 6350 and 1980 N (reference value), respectively.
4. The levitation force of single on-board HTS Maglev equipment assembly over the PMG was 1202 and 1724 N at the levitation gap of 15 and 8 mm, respectively. The total levitation force of eight on-board Maglev equipment assemblies was 10,431 N at a levitation gap of 10 mm.
5. In order to simulate the operation state of the entire vehicle running system, the levitation forces between two on-board HTS Maglev equipment assemblies and two corresponding PMGs were measured.
6. During the 10-month period from July 2001 to May 2002, the levitation forces only decreased by about 5.0% at a levitation gap of 20 mm.
7. By September 2009, more than 40,000 passengers had taken a ride on the Maglev vehicle, and the total recorded mileage was about 400 km.
8. Experiment results verify that the manned running performances of the single bulk YBCO samples above the PMG barely changed after about 9 years.

International interest in the research of manned HTS Maglev vehicle was aroused after "Century" [13] was born. Manned HTS Maglev vehicles were developed in 2004 in Germany [67] and Russia [68]. A full-size HTS Maglev train as a substitute for a light track system for city use is under development in Brazil [69, 70]. Moreover, Japan [72], Italy [73], and other countries had also developed HTS Maglev car models.

Trapped magnetic fields in HTS bulk have achieved 17.24 T at 29 K [93] and 17.6 T at 26 K [94]. This implies that not only do HTS Maglev vehicles have bright prospects, but their practical application process is beyond expectations.

The Brazil group presented a study on comparing the construction costs of a HTS Maglev line of 1.0 km inside the campus of the Federal University of Rio de Janeiro (UFRJ) with those of an LRV. [95]. Preliminary calculations have already shown that this particular Maglev line can be cheaper than a LRV. HTS Maglev cars are lighter than LVR cars for the same number of passengers, the global efforts are approximately 75% lower. The supporting structure weight of a HTS Maglev is 50% of that for the LVR. Figure 4.11 shows a real-scale HTS Maglev vehicle prototype in Brazil [96, 97]. Figure 4.12 shows the on-board assembled HTS Maglev cryostat [98–100]. which was made in ATZ, Germany. Table 4.2 lists specifications of the HTS Maglev cryostat.

The German IFW group has developed the large-scale HTS Maglev vehicle SupraTrans II [101] (Fig. 4.13). On July 1, 2009, the HTS Maglev vehicle project SupraTrans II was approved by the German government after the HTS Maglev vehicle SupraTrans I [67] in IFW, Germany in 2002–2004. The specifications of HTS Maglev vehicle SupraTrans II are as follows: oval guideway length, 80.84 m; ndFeB magnets, 4.85 t; superconductor, YBCO bulk; levitation force, 8.9 kN at 8–10 mm; vehicle mass, about

Fig. 4.11: A real-scale HTS Maglev vehicle prototype in Brazil. The lower right corner is the on-board HTS Maglev cryostat [102] (with permission).

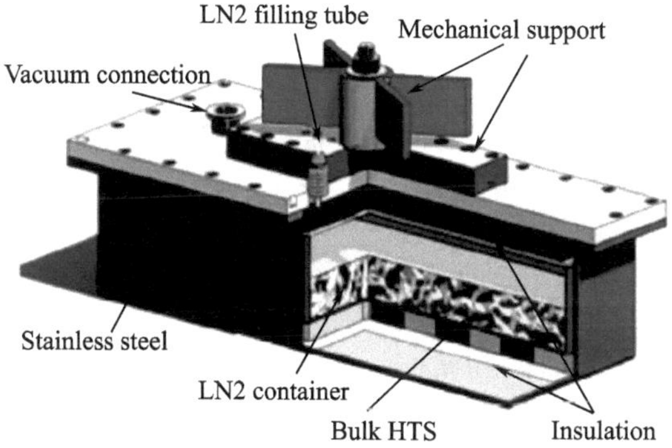


Fig. 4.12: On-board HTS Maglev cryostat in ATZ, Germany [99] (with permission).

400 kg; maximum speed, 20 km/h; levitation height, 10 mm; propulsion, synchronous short stator linear motor.

After the breakthrough of the first manned HTS Maglev test vehicle in December, 2000, authors' group (ASCLab at SWJTU in Chengdu, China) began to conduct test line design for the next research step. In January 2004, Jia-Su Wang presented a HTS Maglev train project, which was approved at a meeting. In May 2008, ASCLab completed a HTS Maglev test line system design. The design calculation validated that the total construction and operation cost of the HTS Maglev line system would be lower than the equivalent LRV system. The weight of the support structure of this HTS Maglev guideway was about 50% that of a LRV. The design specifications of the HTS Maglev vehicle are listed in Tab. 4.3. The test line length was 2 km, and the maximum test speed was 160 km/h.

In March, 2013, a 45-m-long HTS Maglev ring test line was successfully developed by ASCLab under with financial support of State Key Laboratory of Traction Power,

**Tab. 4.2:** Specifications of the Maglev cryostat in ATZ, Germany [99] (with permission).

| | |
|---|---|
| Geometry | 440 × 180 × 120 mm |
| YBCO bulks | 2 × 12 pcs, size(64 × 32 × 13) |
| HTS area | 492 $cm^2$ |
| Cooling | LN2, storage capacity 2.5 l |
| Operation time | 24–30 h, static |
| Cryostat weight | ~ 17 kg |
| Magnetic distance | 2 mm, distance bottom-YBCO |
| Levitation | ~ 2500 N *at* 5 mm; fc30 |
| | ~ 4000 N *at* 5 mm Halbach PM |
| Force density | 5 N/$cm^2$, 8 N/$cm^2$(Halbach) |

**Fig. 4.13:** SupraTrans II large-scale HTS Maglev vehicle in IFW, Germany [102] (Ref. 77, with permission).

**Tab. 4.3:** Design specifications of the HTS Maglev vehicle in the ASCLab.

| Specifications | Units | Value |
|---|---|---|
| Passengers/m | p/m | 10 |
| Passengers weight/m | kgf/m | 750 |
| Vehicle tare/m | Kgf/m | 400 |
| Total weight/m | kgf/m | 1150 |
| Vehicle length | m | 12 |
| Total vehicle weight | t | 13.8 |
| Running speed | km/h | 100 |
| Acceleration | m/$s^2$ | 0.8 |

SWJTU (Fig. 4.14). Deng presents the pictures of the whole scene of the ring line and a close-up of the second-generation HTS Maglev vehicle [71].

The new Maglev vehicle (2.2 m in length, 1.1 m in width) was designed for one passenger with a levitation height of 15–20 mm; the PMG (45 m in length, 0.7 m of track gauge) has a racetrack shape with a curve radius of 6 m, and driving was accomplished

**Fig. 4.14:** A 45 m-long HTS Maglev ring test line developed by ASCLab of SWJTU, China, in March 2013. The whole scene of the ring line (left) and a close-up of the second-generation HTS Maglev vehicle (right) [71] (with permission).

by an induction linear motor with a maximum running speed of 50 km/h. This second-generation HTS Maglev vehicle system was highlighted by the cost-performance and multi-parameter on-board monitoring function. The maximum load capacity was over 1000 kg at a levitation gap of 10 mm, but the cross-sectional area of the PMG was 3000 $mm^2$ and the total length of the linear motor was 3 m composed of four motors installed at one straight section of the track. The parameters of the levitation weight, levitation height, running speed, acceleration, lateral offset, and total running distance of the vehicle are in monitored in real time and displayed on the on-board tablet computer [71].

### 4.7.3 Ultra-high-speed HTS Maglev trains

When the speed of the normal ground traffic transportation is higher than 350 km/h, not only the running noise of the ground traffic vehicle is higher, but also 90% of the driving power is dissipated in the aerodynamic resistance. However, people hope to achieve higher speed, for example, over 3000 km/h. The high speed may be realized when a HTS Maglev vehicle runs in the evacuated tube transportation (ETT).

The long-term stability in both the vertical and the lateral directions of HTS Maglev vehicle is a guarantee for an ultra-high-speed train in the ETT without a control system. Oster [103] was convinced that "automated, silent ETT will work by removing resistance. Ultra lightweight, pressurized cabins travel in tubes on thin wheels, or Maglev. No air is in the tube to cause resistance. Energy is recovered when slowing. Propulsion fuel is not carried on board. High capacity at low cost is achieved using frequent vehicles instead of huge vehicles." Okano et al. [104] proposed a HTS Maglev vehicle in a vacuum passage. The combination of the ETT and the HTS Maglev will bring both advantages into full play, and it may be develop into a new kind of high-speed or ultra-high-speed ground transportation.

On January 12, 2004, Jia-Su Wang and Su-Yu Wang proposed an ultra-high-speed ETT HTS Maglev train project with a maximum speed of more than 600 km/h. The research program and design specifications were similar to those in Tab. 4.3, the

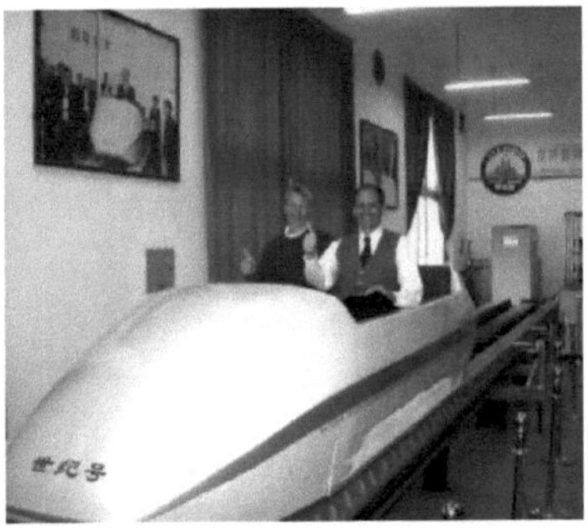

Fig. 4.15: Oster and his wife happily rode on a ake HTS Maglev vehicle (left). He collaborated twice with Jia-Su Wang and Su-Yu Wang in ETT-HTS Maglev development in 2003–2004 (with permission).

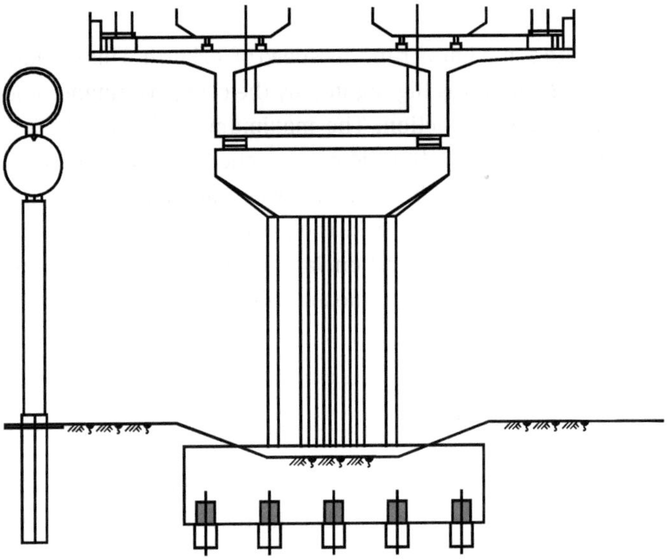

Fig. 4.16: Comparison between of the typical cross section between HTS ETT Maglev train and urban LRV (right) [103] (from Ref. 79, with permission).

only difference is the unmanned operation at ultra-high speeds. Unfortunately, this ultra-high-speed [105–107] HTS Maglev train project has not been approved.

Figure 4.15 is a picture of Daryl Oster and his wife, who happily took the HTS Maglev vehicle. Oster collaborated with author twice to develop the ETT THS Maglev vehicle in 2003 and 2004, after visiting the ASCLab in December 2002. Figure 4.16 shows a comparison between the typical cross section between HTS ETT Maglev train and an urban LRV [103].

There are important differences between the HTS Maglev train and other Maglev trains. The levitation and guidance of the HTS Maglev train is inherently stable, i.e. it does not need any control device. The control technology of the HTS Maglev train system is used only in system operation. Therefore, the HTS Maglev train is especially suitable for ETT transportation. The ETT HTS Maglev train is a new transportation system, and it has different characteristics, namely small vehicle, small guideway, and high frequency dispatch.

Based on the experimental results from the HTS Maglev and the theoretical calculations of the levitation forces of a single on-board HTS Maglev device of high-quality HTS bulk on a new guideway, the design and realization of an ultra-high-speed HTS Maglev is feasible [105].

Figure 4.17 illustrates the main components of an ultra-high-speed HTS Maglev vehicle in a low-pressure tube (LPT) or ETT. The vehicle consists of tube, vehicle body, bogie, low-temperature part, PMG, and linear motor. The bogie is used to transfer the force from the low-temperature part to the vehicle body.

The low-temperature rectangular liquid nitrogen vessel on board has high mechanical strength at 77 K. The continuous operating period is 12 hours. The outside size of the low-temperature vessel is 150 × 516 mm, with inside size 102 × 470 mm and height of 170 mm. The bottom thickness of the vessel is 5 mm in order to increase its mechanical strength.

The total load of the Maglev vehicle is 500 kg including two seats and 150 kg weight of two passengers. The vehicle body is quasi-ellipsoidal in shape, and the outside size of the vehicle with shell is about 2.3 m length, 0.95 m height, and 1.2 m width. The final design results of the Maglev vehicle body are 5000 N of levitation force and 1000 N of guidance force at a 15-mm net levitation gap. The guideway length is 1000 m. The maximum design speed of the vehicle will be over 600 km/h if unmanned, and the manned speed is 100 km/h.

The vehicle would be driven by a linear motor and controlled by a control system on the ground. Drive maximum acceleration can reach 5 g (about 49 m/s$^2$). The vehicle running plan includes the acceleration for 300 m, uniform speed for 400 m, and brake section for 300 m.

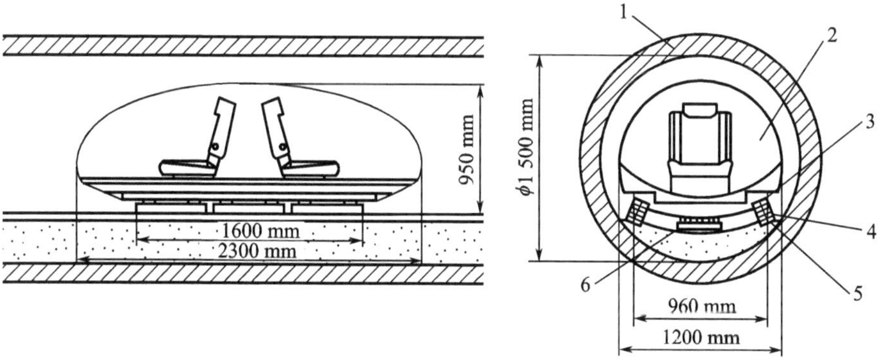


**Fig. 4.17:** The schematic structure for of the ultra-high-speed HTS Maglev in the LPT or ETT [105]. (1) Tube. (2) Vehicle body. (3) Bogie. (4) Low-temperature part (with superconductor inside). (5) Permanent magnet guideway. (6) Linear motor.

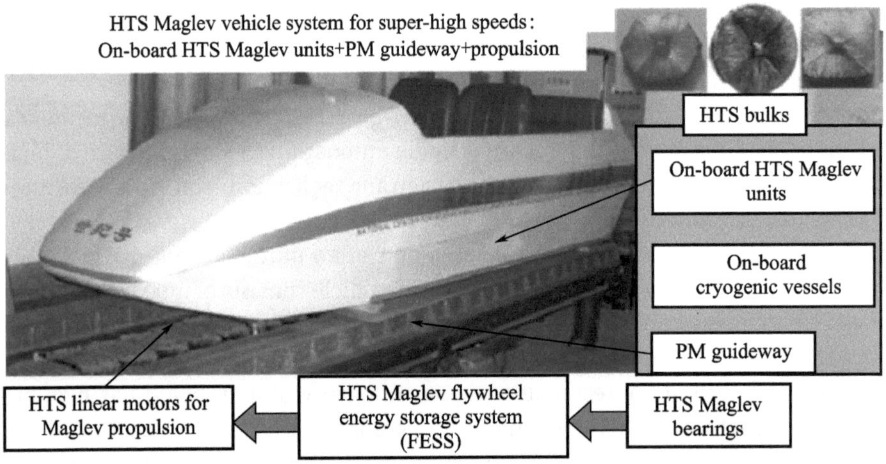


**Fig. 4.18:** Design diagram of the super-high-speed HTS Maglev train.

Figure 4.18 shows the design diagram of the quasi-all-superconducting HTS Maglev train with ultra-high speed (no EET in Fig. 4.18). This project includes HTS Maglev vehicle [13], HTS Maglev bearing [108], HTS Maglev FESS [109, 110], and HTS linear motors [111] for vehicle propulsion. The quasi-all-superconducting HTS Maglev system [112] is fit not only for ultra-high-speed Maglev train but also for a HTS Maglev launch system. The speed of the quasi-all-superconducting HTS Maglev launch system is designed for 3000 km/h, or 1 km/s. Future HTS Maglev trains will be combined with HTS linear motor [113] and the HTS cable [114, 115] to give a full HTS system.

## References

[1] Nishijima S, Eckroad S, Marian A, Choi K, Kim W, Terai M, et al. Superconductivity and the environment: a Roadmap. Supercond Sci Technol. 2013;26:113001.
[2] Powell JR, Danby G. Magnetic suspension system. Paper 66-WA/RR-5, ASME Winter Annual Meeting, 1966. Nov 17–22; Atlanta Georgia.
[3] Powell JR, Danby G. Maglev vehicles-raising transportation advances of the ground. Mechanical Engineering. 1966;89:30.
[4] Wang JS, Wang SY. Application of superconductivity. Press of University of Science and Technology of Chengdu, Chengdu, China; 1995. pp. 123–163. (In Chinese).
[5] Arkadiev V. Hovering of a magnet over a superconductor. J Phys USSR. 1945;9(2):148.
[6] Arkadiev V. A floating magnet. Nature. 1947;160:330.
[7] Braunbeck WZ. Freischwebende Körper im elektrischen und magnetischen Feld. Physics. 1939;112:753.
[8] Simon MD, Heflinger LO, Geim AK. Diamagnetically stabilized magnet levitation. Am J Phys. 2001;69:702.

[9] Hellman F, Gyorgy EM, Johnson DW, Jr., O'Bryan HM, Sherwood RC. Erratum: "Levitation of a magnet over a flat type II superconductor". J Appl Phys. 1988;63:447.

[10] Peters PN, Sisk RC, Urban EW, Huang CY, Wu MK. Observation of enhanced properties in samples of silver oxide doped $YBa_2Cu_3O_x$. Appl Phys Lett. 1988;52:2066.

[11] Brandt EH. Levitation in physics. Science. 1989;243:349.

[12] Bomemann HJ, Tonoli A, Ritter T, Urban C, Zaitsev O, Weber K, Rietschel H. Engineering prototype of a superconducting flywheel for long term energy storage. IEEE Trans on Appl Supercond. 1995;5:618.

[13] Wang J, Wang S, Zeng Y, Huang H, Luo F, Xu Z, et al. The first man-loading high temperature superconducting Maglev test vehicle in the world. Physica C. 2002;809:378–381.

[14] Moon FC. Superconducting levitation. New York, Wiley; 1994.

[15] Hull JR. Superconducting bearings. Supercond Sci Technol. 2000;13(2):R1.

[16] Ma KB, Postrekhin YV, Chu WK. Superconductor and magnet levitation devices. Rev Sci Instruments. 2003;74:4989.

[17] Murakami M. Materials developments and applications of bulk Re–Ba–Cu–O superconductors. In: Narlikar A, editor. Frontiers in superconducting materials. Berlin, Springer Verlag; 2005. pp. 869–884.

[18] Werfel FN, Floegel-Delor U, Rothfeld R, Riedel T, Goebel B, Wippich D, Schirrmeister P. Superconductor bearings, flywheels and transportation. Supercond Sci Technol. 2012;25:014007.

[19] Wang JS. Proceedings of International Workshop on HTS Maglev. ISMAGLEV'2002; 2002 June 25–27; Chengdu, P. R. China; 2002.

[20] Wang JS, Wang SY. Synthesis of bulk superconductors and their properties on permanent magnet guideway. In: Narlikar A, editor. Frontiers in superconducting materials, Springer Verlag, Germany; 2005. pp. 885–912.

[21] Song HH, Wang JS, Wang SY, Ren ZY, Wang XR, de Haas O, Fuchs G, Schultz L. Studies of YBCO electromagnetic properties for high-temperature superconductor Maglev technology. In: Martins BP, editor. New topics in superconductivity research. Nova Science Publishers, Inc.; New York, 2006. pp. 107–156.

[22] Wang JS, Wang SY. High temperature superconducting Maglev measurement system. In: Kr Sharma M, editor. Advances in measurement systems. InTech; Vienna, Austria, 2010. pp. 51–80.

[23] Ma GT, Wang JS, Wang SY. 3-D finite-element modelling of a Maglev system using bulk high-$T_c$ superconductor and its application. In: Luiz AM, editor. Applications of high-$T_c$ superconductivity. InTech; Vienna, Austria, 2011. pp. 119–146.

[24] Werfel FN, Floegel-Delor U, Riedel T, Wippich D, Goebel B, Rothfeld R, Schirrmeister P. HTS magnetic bulk superconductors towards large-scale application. Nova Science Publishers; New York, 2012. pp. 183–201.

[25] Deng ZG, Wang JS, Zheng J, Lin QX, Wang SY. Quasi-static optimization of HTS Maglev systems with a focus on bulk superconductor part. In: Miryala M, editor. Superconductivity: recent developments and new production technologies. Nova Science Publishers; New York, 2012. pp. 215–239.

[26] Wang JS, Wang SY. High temperature superconductive Maglev vehicle. In: Fan TD, editor. Endless quest, Peking University Press; 2014.

[27] Kadyrov E, Gurevich A, Larbalestier DC. High critical current densities in Nb47%Ti multilayers with a planar copper flux pinning nanostructure. Appl Phys Lett. 1996;68:1567.

[28] Wilson MN. NbTi superconductors with low ac loss: A review. Cryogenics. 2008;48:381.

[29] Godeke A, Jewell MC, Fischer CM, Squitieri AA, Lee PJ, Larbalestier DC. The upper critical field of filamentary $Nb_3Sn$ conductors. J Appl Phys. 2005;093909:97–9.

[30] Godeke A, Cheng D, Dietderich DR, Ferracin P, Prestemon SO, Sabbi G, Scanlan RM. Limits of NbTi and $Nb_3Sn$, and development of W&R Bi-2212 high field accelerator magnets. IEEE Trans on Appl Supercond. 2007;17:1149.
[31] Reitz JR. Forces on moving magnets due to Eddy currents. J Appl Phys. 1970;41:2067.
[32] Borcherts RH, Davis LC. Force on a coil moving over a conducting surface including edge and channel effects. J Appl Phys. 1972;43:2418.
[33] Guderjahn CA, Wipf SL, Fink HJ, Boom RW, MacKenzie KE, Williams D, Downey T. Magnetic suspension and guidance for high speed rockets by superconducting magnets. J Appl Phys. 1969;40:2133.
[34] Coffey HT, Chilton F, Barbee TW, Jr. Suspension and guidance of vehicles by superconducting magnets. J Appl Phys. 1969;40:2161.
[35] Miericke J, Urankar L. Forces on null-flux magnetic levitation systems. Appl Phys. 1973;2:201.
[36] Urankar L, Miericke J. Theory of electrodynamic levitation with a continuous sheet track-part II. Appl Phys. 1974;3:67.
[37] Hieronymus H, Miericke J, Pawlitschek F, Rudel M. Experimental study of magnetic forces on normal and null flux coil arrangements in the inductive levitation system. Appl Phys. 1974;3:359.
[38] Iwasa Y. High speed magnetically levitated and propelled mass ground transportation. In: Foner S, Schwartz BB, editors. Superconducting machines and devices: large systems applications. New York, Plenum Press; 1974. pp. 347–399.
[39] Kolm HH, Thornton RD. Magneplane: guided electromagnetic flight. Proceedings of the 5th International Applied Superconductivity Conference, 1972 May 1–3; New York. pp. 76–85.
[40] Kolm H, Thornton R. Electromagnetic flight. Sci Amer. 1973;229:17.
[41] Reitz JR, Davis LC. Force on a rectangular coil moving above a conducting slab. J Appl Physics. 1972;43:1547.
[42] Coffey HT. Magnetic Suspension studies for high-speed vehicles. In: Advances in cryogenic Engineering. New York, Plenum Press; 1974. pp. 137–153.
[43] Lever JH. Technical assessment of Maglev system concepts: final report by the Government Maglev System Assessment Team. Chicago, 1998.
[44] Lever JH. Technical assessment of Maglev system concepts: final report by the Government Maglev System Assessment Team. October 1998. Also see final report on the National Maglev Initiative, September, 1993.
[45] Proise M, Deutsch L, Gran R, Herbermann R, Kalsi S, Shaw P. System concept definition of the Grumman Superconducting Electromagnetic Suspension (EMS) Maglev Design. 13th International Conference on Maglev, Argonne National Laboratory, 1993 May 19–21. pp. OS4–4.
[46] Kalsi SS. Superconducting Electromagnetic Suspension (EMS) system for Grumman Maglev concept. Second International Symposium on Magnetic Suspension Technology, Part 1, 1994. pp. 197–211.
[47] Kolm H, Thornton RD, Iwasa Y, Brown WS. Magneplane system. Cryogenics. 1975;15:377.
[48] Thome RJ, Radovinsky A, Montgomery B. EDS Levitation and guidance using sheet guideways. 16th International Conference on Maglev; Proceeding of the Maglev' 2000. pp. 236–241.
[49] Montgomery DB. Overview of the 2004 magplane design. Proc of the Maglev'2004. Shanghai, China; 2004, pp. 106–113.
[50] Powell JR, Danby GT. Transport by magnetic levitation. 21st Science & Technology Magazine. 2003:43–57.
[51] Rhodes RG, Mulhall BE, Hawell JP, Abel E. The Wolfson Maglev project. IEEE Trans on Magnetics. 1974;10:398–401.

[52] Eastham AR, Atherton DL. Superconducting maglev and LSM development in Canada. IEEE Trans on Magnetics. 1975. MAG-11: 627–632.
[53] Atherton DL, Eastham AR, Belanger PR, Eggleton PL, Rudback NE. High speed magnetically levitated transport development in Canada. J Adv Transp. 1980;14:73–105.
[54] Gutberlet HG. The German magnetic transportation program. IEEE Trans on Magnetics. 1974;MAG-10:417–420.
[55] Glatzel K, Khurdok G, Rogg D. The development of the magnetically suspended transportation system in the Federal Republic of Germany. IEEE Trans Veh Technol. 1980;VT-29:3–17.
[56] Rogg D. Development of magnetic levitation transport systems in the Federal Republic of Germany: survey, present state, prospects and reasons. International Conference on Maglev'85;1985 Sep 17–19; Tokyo, Japan. Washington, DC: The National Academies of Sciences, Engineering, and Medicine. p. 1–11.
[57] Yamamura S. Magnetic levitation technology of tracked vehicles present status and prospects. IEEE Trans on Magnetics. 1976;12:874–878.
[58] Sawadi Kazuo. Superconducting Maglev developed by RIRI and JR central. Japan railway & transport review. 2000;25:58–60.
[59] http://en.wikipedia.org/wiki/File:JNR_ML500_1.jpg.
[60] https://3dwarehouse.sketchup.com/model.html?redirect=1&mid=6dfe0a02e0809cbc666bcaed61061f38&ct=lc&hl=zh-CN.
[61] Shirakuni N, Terai M, Watanabe K, Takahashi K. The status of development and running tests of superconducting Maglev. Maglev 2006; Germany, Sep. 2006, p. 27–34.
[62] Ono M, Koga S, Ohtsuki H. Japan's superconducting Maglev train. IEEE Instrum Meas Mag. March 2002;5(1):9–15.
[63] Sawada K. Outlook of the superconducting Maglev. Proceedings of the IEEE. 2009;97: 1881–1885.
[64] Sakamoto S, Watanabe H, Takizawa T, Suzuki E, Terai M. Development of a Maglev superconducting magnet for the Yamanashi test line in Japan: vibration characteristics and analysis for design. IEEE Trans on Appl Supercond. 1997;7:3791–3796.
[65] Terai M, Igarashi M, Kusada S, Nemoto K, Kuriyama T, Hanai S, Yamashita T, Nakao H. The R&D project of HTS magnets for the superconducting Maglev. IEEE Trans on Appl Supercond. 2006;16:1124–1129.
[66] Zhang Y, Xu SG. In: Lin LZ, Shen GL, Yan LG, editor. Proceeding of Fifteenth International Conference on Magnet Technology, Science Press; 1998. p. 763–766.
[67] Schultz L, de Haas O, Verges P, Beyer C, Rohlig S, Olsen H, Kuhn L, Berger D, Noteboom U, Funk U. Superconductively levitated transport system – the SupraTrans project. IEEE Trans on Appl Supercond. 2005;15:2301–2305.
[68] Kovalev KL, Koneev SMA, Poltavec VN, et al. Magnetically levitated high-speed carriages on the basis of bulk HTS elements. Pro. 8th Intern Symp Magn Susp Technol. (ISMST'8), Dresden; 2005. p. 51.
[69] Stephan RM, Nicolsky R, Neves MA, Ferreira AC, de Andrade R, Jr., Cruz Moreira MA, Rosario MA, Machado OJ. A superconducting levitation vehicle prototype. 2003 IEEE Intern Sym Indust Electro. 2004;1:206–209.
[70] Sotelo GG, Dias DH, Machado OJ, David ED, de Andrade R, Jr., Stephan RM, Costa GC. Experiments in a real scale Maglev vehicle prototype. J Phys Conf Ser. 2010;234:032054.
[71] Deng Z, Wang J, Zheng J, Zhang Y, Wang S. An efficient and economical way to enhance the performance of present HTS Maglev systems by utilizing the anisotropy property of bulk superconductors. Private communication. 2013.
[72] Okano M, Iwamoto T, Furuse M, Fuchino S, Ishii I. Running performance of a pinning-type superconducting magnetic levitation guide. J Phys Conf Ser. 2006;43:999.

[73] D'Ovidio G, Crisia F, Lanzara GA. "V" shaped superconducting levitation module for lift and guidance of a magnetic transportation system. Physica C. 2008;468:1036–1040.
[74] Kusada S, Igarashi M, Nemoto K, Okutomi T, Hirano S, Kuwano K, Tominaga T, Terai M, Kuriyama T, Tasaki K, Tosaka T, Marukawa K, Hanai S, Yamashita T, Yanase Y, Nakao H, Yamaji M. The project overview of the HTS magnet for superconducting Maglev. IEEE Trans on Appl Supercond. 2007;17:2111–2116.
[75] Kuwano K, Igarashi M, Kusada S, Nemoto K, Okutomi T, Hirano S, Tominaga T, Terai M, Kuriyama T, Tasaki K, Tosaka T, Marukawa K, Hanai S, Yamashita T, Yanase Y, Nakao H, Yamaji M. The running tests of the superconducting Maglev using the HTS magnet. IEEE Trans on Appl Supercond. 2007;17:2125–2128.
[76] Zhou T, Xing HW. Controller design and test results for a four axis HTS coil based Maglev system. Physica C. 2007;454:20–26.
[77] Ogata M, Miyazaki Y, Hasegawa H, Sasakawa T, Nagashima K. Basic study of HTS magnet using 2G wires for Maglev train. Physica C. 2010;470:1782–1786.
[78] Mizuno K, Ogata M, Nagashima K. Fabrication of 5 T magnet using 2G wires directed at Maglev application. IEEE Trans on Appl Supercond. 2013;23:3600104–3600104.
[79] Lee J, Bae DK, Sim K, Chung YD, Lee YS. Characteristics on electodynamic suspension simulator with HTS levitation magnet. Physica C. 2009;469:1820–1824.
[80] Sass F, Henrique D, Dias N, Sotelo GG, de Andrade R, Jr. Superconducting levitation using coated conductors. IEEE Trans on Appl Supercond. 2014;24:3600905–3600905.
[81] Meissner W, Ochsenfeld R. Short initial announcements. Naturwiss. 1933;33:787–788.
[82] Simon I. Forces acting on superconductors in magnetic fields. J Appl Phys. 1953;24:19–24.
[83] Buchhold TA. Three-phase motor with superconductive bearings. Cryogenics. 1965;5: 216–219.
[84] Tachikawa K, Togano K. Potential methods for the fabrication of high-$T_c$ superconductors for wires and cables. Proc IEEE. 1989;77:1124–1131.
[85] Perini E, Giunchi G, Saglietti L, Albisetti A, Matrone A, Cavaliere V. Magnetic field trapping in $MgB_2$ bulks and inserts. IEEE Trans on Appl Supercond. 2011;21:2690–2693.
[86] Lüke B, Kiirschner HG, Lehndorff B, Lenkens M, Piel H. Levitation force of a stack of epitaxial $YBa_2Cu_3O_{7-\delta}$ films. Physica C. 1996;259:151–156.
[87] Johansen TH, Riise AB, Bratsberg H, Shen YQ. Magnetic levitation with high-$T_c$ superconducting thin films. J Supercond . 1998;11:519–524.
[88] Riise AB, Johansen TH, Bratsberg H, Koblischka MR, Shen YQ. Levitation force from high-$T_c$ superconducting thin-film disks. Phys Rev B. 1999;60:9855–9861.
[89] Wang JS, Wang SY. Synthesis of bulk superconductors and their properties on permanent magnet guideway. In: Narlikar A, editor. Frontiers in superconducting materials. Springer Verlag; 2005. 885–912.
[90] Wang JS, Wang SY, Zeng YW, Deng CY, Ren ZY, Wang XR, Song HH, Wang XZ, Zheng J, Zhao Y. The present status of the high temperature superconducting Maglev vehicle in China. Supercond Sci Technol. 2005;18:S215.
[91] Wang JS, Wang SY. A superconducting Maglev train system. Chinese patent No. ZL01128867.1, 2001.
[92] Wang JS, Wang SY, Ren ZY, Jiang H, Zhu M, Wang XR, Shen XM, Song HH. High temperature superconducting Maglev equipment on vehicle. Physica C. 2003;386:531–535.
[93] Tomita M, Murakami M. High-temperature superconductor bulk magnets that can trap magnetic fields of over 17 tesla at 29 K. Nature. 2003;421:517–520.
[94] Durrell JH, Dennis AR, Jaroszynski J, Ainslie MD, Palmer KGB, Shi YH, Campbell AM, Hull J, Strasik M, Hellstrom EE, Cardwell DA. A trapped field of 17.6T in melt-processed, bulk Gd-Ba-Cu-O reinforced with shrink-fit steel. Supercond Sci Technol. 2014;27:082001.

[95] David EG, Stephan RM, Costa GC, Andrade R, Jr., Nicolsky R. Feasibility study of an HTS-Maglev line at the Federal University of Rio de Janeiro. Maglev 2006, Dresden, Germany.

[96] Sotelo GG, Dias DH, Machado OJ, David ED, de Andrade R, Jr., Stephan RM. Experiments in a real scale Maglev vehicle prototype. J Phys Conf Ser. 2010;234:032054.

[97] Stephan RM, de Andrade R, Jr., Ferreira AC, Machado OJ, dos Santos MDA, Sotelo GG, Dias DH. The first full scale functional prototype of the superconducting magnetically levitated vehicle MagLev-Cobra. The 21st International Conference on Maglev (MAGLEV 2011), Daejeon, Korea,10–13 October 2011, pp. DPO-8.

[98] Werfel FN, Floegel-Delor U, Rothfeld R, Riedel T, Goebel B, Wippich D, Schirrmeister P. Recent up-scaling in HTS magnetic device technology. IEEE Trans on Appl Supercond. 2011;21:1473–1476.

[99] Werfel FN, Floegel-Delor U, Riedel T, Wippich D, Goebel B, Rothfeld R, Scirrmeister P. HTS Flywheel from R&D to Pilot Energy Storage System. The Superconductivity Centennial Conference EUCAS/ICEC/ICMC2011 (SCC2011), Hague, the Netherlands, Sep. 2011, 18–23, No. 4-LB-010.

[100] Stephan R, David E, Andrade R, Jr., Machado O, Dias D, Sotelo GG, de Haas O, Werfel F. A full-scale module of the Maglev-Cobra HTS-superconducting vehicle. Proc 20th International Conference on Maglev, San Diego, Dec. 2008, paper 27.

[101] Schultz L, Kuehn L, de Hass O, Berger D, Holzapfel B. The Superconductivity Centennial Conference, EUCAS/ICEC/ICMC2011 (SCC2011), Hague, the Netherlands, Sep. 2011, 18–23, Paper No. 4-LB-12.

[102] Stephan RM, de Andrade R, Jr., Ferreira AC. Superconducting light rail vehicle: a transportation solution for highly populated cities. IEEE Vehi Technol Mag. 2012;12:122–127.

[103] Oster D. Evacuated tube transport (ETT), Private correspondence and www.et3.com.

[104] Okano M, Iwamoto T, Fuchino S, Tamada N. Feasibility of a goods transportation system with a superconducting magnetic levitation guide – load characteristics of a magnetic levitation guide using a bulk high-$T_c$ superconductor. Physica C. 2003;386:500–505.

[105] Wang JS, Wang SY, Deng CY, Zeng YW, Song HH, Zheng J, Wang XZ, Huang HY, Li F. Design consideration of a high temperature superconductor Maglev vehicle system. IEEE Transa on Appl Supercond. 2005;15:2273–2276.

[106] Wang JS, Wang SY, Deng CY, Zeng YW, Song HH, Huang HY. Design consideration of a high temperature superconductor Maglev vehicle system Sign In or Purchase. Int J Mod Phys B (IJMPB). 2005;19:399.

[107] Wang JS. Past, Present and Future R&D of HTS Maglev in China. The 21st International Conference on Maglev, Daejeon, Korea, 2011 Oct 10–13; Invitation Keynote Speech No. 2.

[108] Deng ZG, Wang JS, Wang SY, Zheng J, Lin QX, Zhang Y. Research and development status of high temperature superconducting magnetic bearings. Transactions of China Electrotechnical Society. 2009;24:1–8. (In Chinese).

[109] Deng ZG, Wang JS, Wang SY, Zheng J, Ma GT, Lin QX, Zhang Y. Status of high $T_c$ superconducting flywheel energy storage system. Transactions of China Electrotechnical Society. 2008;23:1–10. (In Chinese).

[110] Wang J-S, et al. Design of 5 kWh HTS magnetic levitation flywheel energy storage. Unpublished.

[111] Yen F, Li J, Zheng SJ, Liu L, Ma GT, Wang JS, Wang SY, Liu W. A single-sided linear synchronous motor with a high temperature superconducting coil as the excitation system. Supercond Sci Technol. 2010;23:105015–105018.

[112] Wang JS. The written application of the National Natural Science Foundation in China, in March, 2008. (In Chinese).

[113] Lee C, Lee J, Jo J, Park C, Ryu W, Chung Y, Hwang Y, Ko T, Oh S, Lee J. Conceptual design of superconducting linear synchronous motor for 600-km/h wheel-type railway. IEEE Trans on Appl Supercond. 2014;24:3600304.
[114] Ohsaki H, Lv Z, Matsushita N, Sekino M, Koseki T, Tomita M. Superconducting power cable application in DC electric railway systems. IEEE Trans on Appl Supercond. 2013;23:3600705.
[115] Tomita M, Muralidhar M, Fukumoto Y, Ishihara A, Suzuki K, Kobayashi Y, Akasaka T. Design and development of superconducting DC cable for railway applications. IEEE Trans on Appl Supercond. 2013;23:3601504.

Su-Yu Wang, Jia-Su Wang, Yi-Yun Lu, and Wei Liu

# 5 HTS Maglev experimental methods and set-up

Experimental study data are the foundation of scientific research. Before the design and development of the first manned HTS Maglev vehicle, a foundation of reliable scientific evidence must first be obtained by excellent measurement methods and test devices. Therefore, Jia-Su Wang and Su-Yu Wang propose some new test and analytical methods, and the successful development of the measurement devices. This chapter will introduce the measurement methods and devices.

## 5.1 Introduction

The potential engineering applications mentioned above are based on high-quality HTS bulk samples, and it is especially important to investigate the levitation properties between the YBCO bulk and the PM. Preparation methods and material properties for HTS bulk are described in details by Krabbes et al. [1].

The measurements of the levitation forces can be realized by directly measuring the forces between a YBCO bulk and a cylindrical PM. Normally, the YBCO bulk is placed in the liquid nitrogen, while the levitation forces are measured directly with a cylindrical PM moving up and down. The liquid nitrogen vessel is normally placed below the PM. However, if the liquid nitrogen vessel is placed above the PM, the measurement results can be more accurate since the PM is not affected by the variation of temperature. This requires that the bottom wall thickness of the liquid nitrogen vessel must be sufficiently thin. For accurate, rapid, and convenient measurement of the levitation forces, a special thin bottom liquid nitrogen vessel has been designed. It will be introduced later.

The measurement of levitation force is not the only important issue, measurements of many other parameters of HTS Maglev are equally important. For example, the guidance force as well as the cross stiffness, dynamic rigidity, relaxation time, etc. are important. Synchronous measurements of multiple parameters are also required. In order to meet these requirements, several special HTS Maglev measurement devices have been designed and developed.

In order to investigate the levitation properties (levitation force, guidance force, trapping flux, etc.) of the HTS Maglev vehicle over the PMG, the SCML-01 HTS Maglev measurement system [2] was developed at Jia-Su Wang's group in 1999. This measurement system has the measured HTS bulk over the PMG, thus, a special cryogenic vessel is needed. Before the SCML-01, previous HTS Maglev testing systems had the PM over the HTS bulk. The YBCO was put inside a simple cryostat container for liquid nitrogen, and a cylindrical PM was moving above.

https://doi.org/10.1515/9783110538434-005

This measurement system included a liquid nitrogen vessel, a PMG, data acquisition and processing, a mechanical drive and Autocontrol software. The thickness of the bottom wall of the vessel was 3.5 mm. The PMG has a length of 920 mm and a magnetic induction up to 1.2 T. The measuring process was controlled by a computer.

The SCML-01 measurement system is capable of performing real time measurements of Maglev properties through a combination of one or several YBCO bulks and one or two PMGs. This set-up is usually employed with the on-board HTS Maglev equipment over one and two PMGs. The on-board Maglev equipment includes a rectangular-shaped liquid nitrogen vessel containing YBCO bulk superconductors.

Based on the original research results [3, 4] from SCML-01, the first manned HTS Maglev test vehicle in the world was successfully developed [5] in 2000 in China. After 2004, several HTS Maglev vehicle prototypes over a PMG were created in Germany [6], Russia [7], Brazil [8], Japan [9], and Italy [10].

Given the limited measurement functions and measurement precision of the SCML-01, 5 years later, a new HTS Maglev Measurement System (SCML-02) with more functions and higher precision was developed [11] to extend the investigation of the Maglev properties of YBCO bulks over a PM or PMG. The new features in this measurement system are unique and include higher measurement precision, instant measurements upon the movement of the measured HTS sample, automatic measurements of both levitation and guidance forces, dynamic rigidity, three-dimensional simultaneous movement of the HTS sample, relaxation measurements of both levitation and guidance forces, etc.

All these experimental parameters are very helpful to evaluate the load ability of the HTS Maglev vehicle. But the running performance over a PMG cannot be measured by these mentioned measurement systems.

To further develop the HTS Maglev vehicle for engineering applications, the dynamic properties should be clearly understood. In order to investigate the dynamic characteristics of the HTS Maglev, a HTS Maglev dynamic measurement system (SCML-03) was designed and successfully developed [12]. The PMG was fixed along the circumference direction of a big circular disk with a diameter of 1500 mm. The maximum linear velocity of the PMG was about 300 km/h when the circular disk rotated around the central axis at 1280 rpm. The liquid nitrogen vessel with the HTS bulk samples was placed above the PMG during the dynamic testing. The liquid nitrogen vessel was made not rigid along the three principal axes, but instead measurement sensor devices were attached. These sensors can detect weak changes of force along the three principal directions.

In order to investigate HTS Maglev bearing and flywheel energy storage technology, a HTS Maglev bearing measurement system [13] was designed and successfully developed in 2014.

The principles, methods, structure, functions, and specifications of the several HTS Maglev measurement systems [14] will be discussed in detail in this chapter.

## 5.2 Key components

The basic structure and main function of any measurement methods and devices generally include mechanical and electrical drive, control system, data collection and processing, etc. However, HTS Maglev measurement devices need some special requirements, for instance, the bottom wall thickness of the liquid nitrogen vessel and various PMG. They are very important for the engineering applications of the HTS Maglev vehicle.

### 5.2.1 Permanent magnet guideway (PMG)

Figure 5.1 shows the cross-section drawings, the distribution of magnetic field lines, magnetic field along the transverse directions, and magnetic field along the vertical directions of the several PMGs. The parameters of NdFeB PMs are magnetic energy product of 40 MOGe, coercivity $H_c$ of 979 kA/m, and relative permeability $\mu$ of 1.049. The PMGs are composed of normal PMs and iron plates. The arrows represent magnetic poles where the arrowheads represent north. The first PMG in the table was used in the first manned HTS Maglev test vehicle in the world.

The magnetic field in the center of the PMG is stronger than that at any other position, and it decreases rapidly with the increasing of the gap from the surface of the PMG. The surface magnetic field of the single pole normal PM is about 0.45 T, while the surface concentrated magnetic flux density of the first PMG is up to 1.2 T. The magnetic flux density is 0.4 T at 20 mm above the surface of the PMG and is about equivalent to the surface magnetic field of a single PM.

### 5.2.2 Liquid nitrogen vessel with thin bottom

We have developed a thin bottom liquid nitrogen vessel in order to realize real-time, accurate measurements of Maglev performance of HTS bulk over PMG. This is a very important technique in the HTS Maglev measurements over a PMG. Generally, the wall thickness of liquid nitrogen vessel is not limited, and its main specification is a low vapor rate. However, YBCO bulks are levitated over a PMG in the HTS Maglev measurement system, therefore a liquid nitrogen vessel with a thin bottom is very necessary in order to get a maximum net levitation gap, i.e. the gap between the outside wall of the vessel and the surface of the guideway.

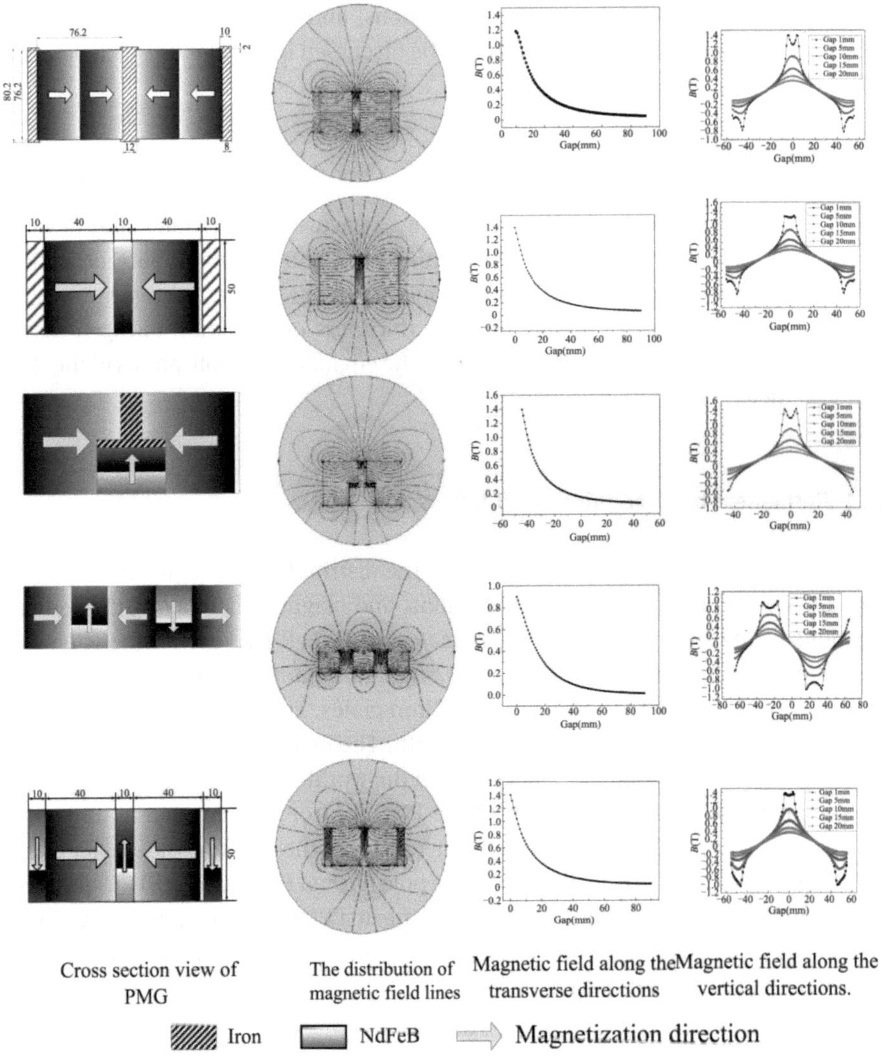


**Fig. 5.1:** The properties of PMG for the investigation of magnetic levitation properties of HTS YBCO bulk. (From Chang-Qing Ye and Dong-Hui Jiang).

A small columnar liquid nitrogen vessel with thin bottom has been developed [15, 16] in order to meet the needs of testing device and to verify the possibility of large-size liquid nitrogen vessel with thin wall. The bottom thickness is only 3.5 mm (The design thickness is 3 mm, but after the molding process the thickness is 3.5 mm). The schematic, photograph, and vapor rate of of the small columnar liquid nitrogen vessel with thin bottom are shown in Figs. 5.2–5.4, respectively. The outside and

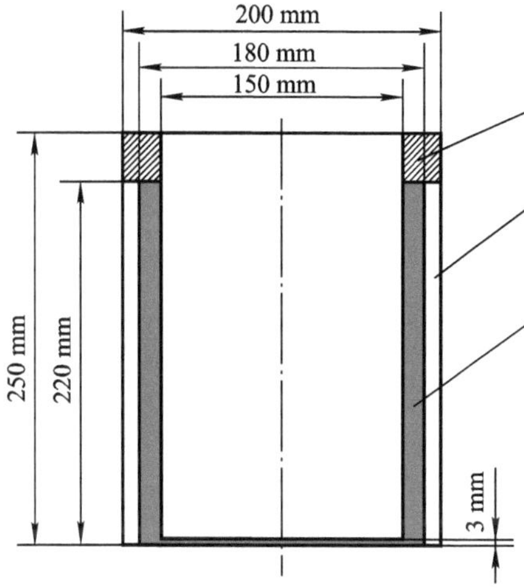


**Fig. 5.2:** Schematic of the columnar liquid nitrogen vessel with thin bottom. (1) Flange. (2) Vacuum. (3) Super-insulation [15].

**Fig. 5.3:** Photograph of the cylindrical liquid nitrogen vessel with thin bottom.

inside diameters of the vessel are 200 and 150 mm, and the height is 250 mm. The liquid nitrogen vessel can operate continuously for over 16 hours, and can hold 7 YBCO samples of 30 mm in diameter. After 14 years the liquid nitrogen vessel still has excellent constant temperature performance. The vessel has been used successfully to measure the levitation forces of YBCO bulk over the PMGs. During the experiment, the YBCO is fixed to the bottom of the columnar liquid nitrogen vessel. It also provides experience for the development of the on-board rectangular liquid nitrogen vessel with a thin bottom and large size.

In order to avoid a collision between the vessel's bottom surface and the upper surface of the PMG, there is a gap of 1.5 mm when the vessel moves down to the lowest point. Thus, the minimum distance between the sample and the PMG surface is 5 mm. The YBCO is cooled in a zero magnetic field with liquid nitrogen, and it can move up and down at different velocities.

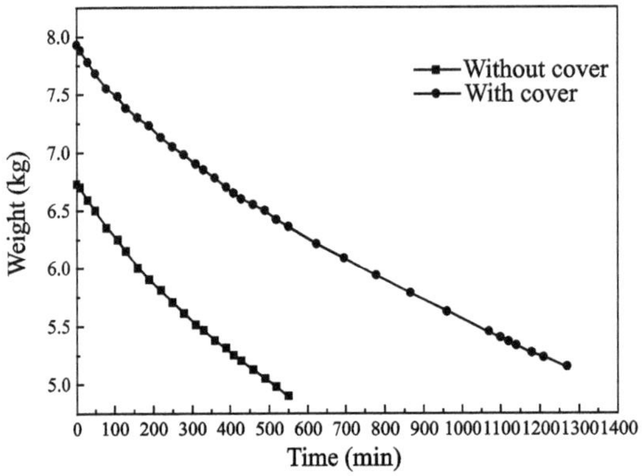


**Fig. 5.4:** Vapor rate of the columnar liquid nitrogen vessel with thin bottom [15].

## 5.3 HTS Maglev measurement system [2]

Figure 5.5 shows the schematic diagram of the HTS Maglev measurement system. During the experiment, the YBCO is placed inside the cylindrical liquid nitrogen vessel which is positioned above the PMG. The YBCO is zero field cooled, and the vessel is allowed to move up and down at different speeds. The horizontal drive platform is used to measure the guidance force (stable equilibrium force along lateral orientation of PMG). The drive device of three dimensions can make scanning measurements of the magnetic field of the PMG and trapped flux inside a HTS.

The specifications of the SCML-01 measurement system are: vertical maximal displacement of 200 mm, ±0.1 mm precision, ±2000 N vertical maximal support force, 0.2% precision, 100 mm PMG horizontal maximal displacement, ±0.1 mm precision, 1000 N of horizontal maximal support force, and 0.1% precision. The trapped flux of high-$T_c$ superconductors and the magnetic induction of the PMG can be scanned in the range of 100 mm × 100 mm.

In the measurement, the YBCO HTS bulk sample is fixed at the bottom of the thin wall liquid nitrogen vessel and cooled to the superconducting state in a zero magnetic field. Secondly, the vessel is fixed at a connecting fixture with a servo electromotor. In order to avoid collision between the bottom of the vessel and the surface of PMG, there is still a gap of 1.5 mm left between the bottom of the vessel and the surface of the PMG when the vessel is lowered to the lowest point, so that minimum gap is 5 mm between the bottom of the sample and the surface of the PMG. The vessel first moves downward, after reaching the lowest point, then moves upward at a speed of 2 mm/s, and the computer samples data every 0.5 s. The system can make real-time measurements of one or many superconductors. The measurement process is controlled by a computer. The main interfaces for the measurement results of the magnetic levitation forces, guidance forces, and scanning magnetic field of a HTS trapping flux are shown in Fig. 5.6.

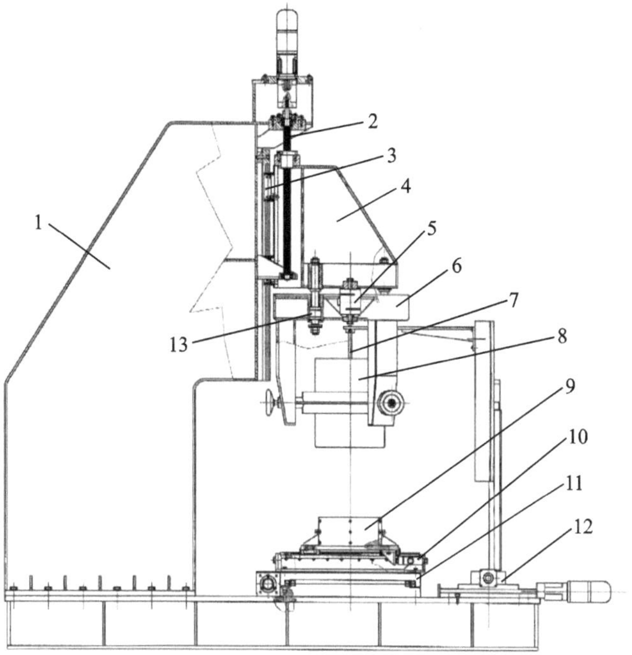


**Fig. 5.5:** The overall design of HTS Maglev measurement system (from You-Wen Zeng). (1) Square column arm. (2) Ball screw. (3) Rolling guideway. (4) Vertical motion unit. (5) Vertical force measuring unit. (6) Cryogenic vessel fixed frame. (7) Trapped flux scanning unit. (8) Liquid nitrogen vessel. (9) PMG. (10) Horizontal force measuring unit. (11) Horizontal motion unit. (12) Three-dimensional rolling platform. (13) Linear ball bearing.

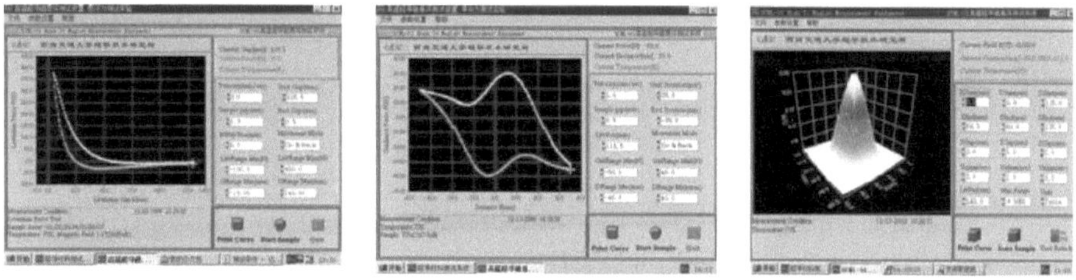

(a) Levitation forces (b) Guidance forces (c) Trapping flux profile

**Fig. 5.6:** Main interfaces for the measurement Results [2].

The SCML-01 measurement system is capable of making real time measurement of Maglev properties with one or many YBCO pieces and with a PM or PMGs. Figure 5.7a shows a columnar liquid nitrogen vessel over a PMG. The columnar vessel includes a YBCO bulk or an array of 2–7 YBCO bulks. This set-up was employed in on-board HTS Maglev equipment over the PMGs (Fig. 5.7b). The on-board Maglev equipment includes a rectangular-shaped liquid nitrogen vessel and an array of 43 YBCO bulk

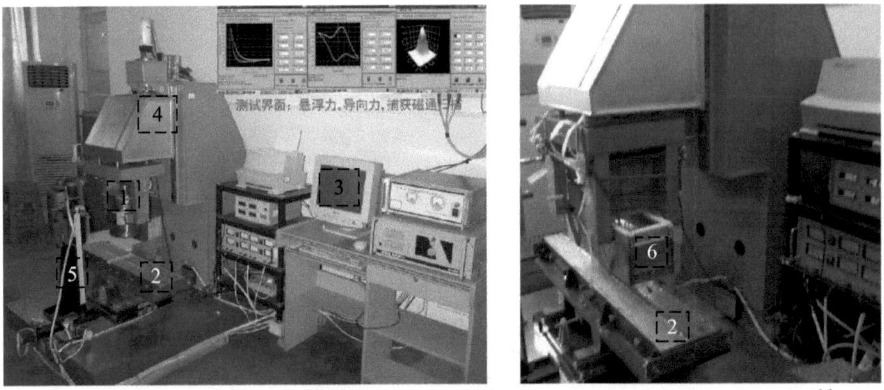


(a) Single columnar vessel[17] (b) Single rectangular vessel[16]

**Fig. 5.7:** Photos of SCML-01 HTS Maglev measurement system. (1) Columnar liquid nitrogen vessel. (2) Permanent magnet guideway (PMG). (3) Data collection and processing. (4) Mechanical drive and autocontrol. (5) Scan magnetic flux. (6) Rectangular vessel.

**Fig. 5.8:** Measurement for two rectangular vessels over two PMGs.

superconductors. Figure 5.8 shows the measurement system for two on-board HTS Maglev equipments over two parallel PMGs.

Based on the original results [3, 4] from the SCML-01 HTS Maglev measurement system, the first manned HTS Maglev test vehicle in the world was successfully developed in 2000 [5].

Many of these research results [17–27] (see chapters 6, 8, and 9) were also obtained with the SCML-01 HTS Maglev measurement system.

In order to further investigate the Maglev properties of HTS bulks with PM or PMG, an improved HTS Maglev test system (SCML-01) was designed and successfully developed [28]. The recent improvements included an integration of displacement, temperature, force, torque, and magnetic field testing. The upgraded SCML-01 system (Fig. 5.9) has augmented circular type, slant-type and z-type mechanical movement

**Fig. 5.9:** Photo of upgraded SCML-01 HTS Maglev measurement system [28].

functions, i.e. a two-dimension motion mechanism and a three-dimension motion mechanism. Thus, this system can accurately locate and simultaneously measure both levitation force and guidance force, as well as both levitation force and torque. The improved SCML-01 system has augmented new functions:

(1) Measured HTSCs specimens can be moved with the combination of any two dimension directions of three dimension directions to operate in circular, slant, and z-type movements.
(2) Measurements of displacement, temperature, levitation force, torque at the same time.
(3) Measurements of levitation force and guidance force simultaneously.
(4) Measurements of relaxation time of levitation force and guidance force simultaneously.
(5) Measurements of relaxation time of the torque.

The measurement precisions are calibrated experimentally and the augmented new functions are verified. These results show that the development of the system is completely successful.

## 5.4 Guidance forces of the HTS Maglev vehicle [29–31]

The guidance force defines the lateral stability of the Maglev vehicle when either at rest or moving. The lateral guidance force is dependent on the trapped flux in the bulk superconductors, so the larger the amount of trapped flux, the stronger the guidance force. This is a distinctive characteristic of the bulk HTS Maglev vehicle. This sort of Maglev vehicle with bulk HTS does not need any lateral stability control systems, which makes it superior to other conventional Maglev vehicle systems. The guidance

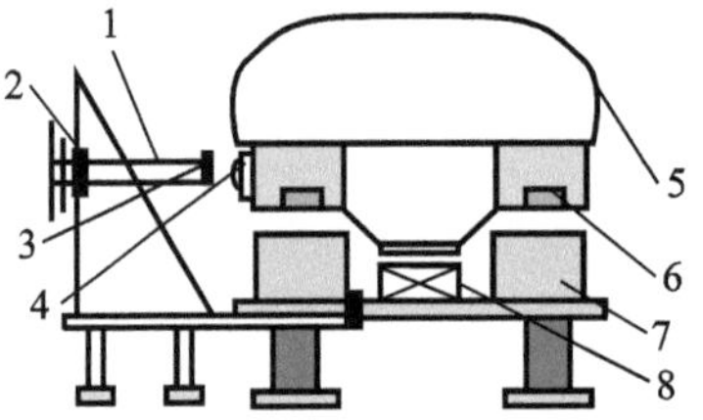


**Fig. 5.10:** Sketch of guidance force measuring equipment for the entire HTS Maglev vehicle [29]. (1) Horizontal propulsion system. (2) Vertical propulsion system. (3) Screw for adjusting zero. (4) Force sensor. (5) Vehicle body. (6) HTS bulks. (7) PMG. (8) Linear motor.

**Fig. 5.11:** Photograph of guidance force of the entire HTS Maglev vehicle [29].

forces are large and sufficient enough to guide the vehicle when large levitation forces are present.

The measuring equipment of the guidance forces for the entire HTS Maglev vehicle is depicted in Figs. 5.10 and 5.11 [29]. The set-up includes two probing levers and two sets of force sensors. The force sensors are fixed on the vehicle. Each set of the propulsion system can move in both the horizontal and vertical directions so that they can measure the guidance forces of the entire vehicle at different levitation gaps. Two sets of propulsion systems are connected by a chain with a synchronization precision of 0.5 mm. The moving range of the propulsion system along the horizontal direction is 0 to 20 cm with a moving precision of 1 mm and along the vertical direction a range from 0 to 10 cm with a vertical moving precision of 1 mm.

The experimental results of the guidance forces under a maximum lateral distance of 20 mm will be shown in Fig. 6.70. The data show that the lateral guidance forces have a large hysteresis effect. The guidance forces rise rapidly when the vehicle leaves its initial position, and the increase rate becomes smaller as the vehicle move further away from its original position (see section 6.18). The guidance forces drop rapidly when the vehicle moves back to its initial position from the maximum lateral displacement of 20 mm and vanishes at about 10 mm. This indicates that the range of effective lateral displacement is smaller than 10 mm. It can be seen from Fig. 6.70 that the lateral guidance forces of the entire HTS Maglev vehicle under different field cooling heights (FCH) are sufficiently large. The guidance forces with displacements

20 mm are 4407, 2908, and 1980 N for field cooling heights of 12, 26, and 42 mm, respectively.

The HTS Maglev vehicle can return to its initial position after a lateral displacement from 0 to 6 mm, but this did not happen when the lateral displacement was from 7 to 20 mm. For example, there was a zero guidance force at the displacement 9 mm when FCH was 26 mm. The measured guidance pull force, which makes the vehicle return to the initial rest position, was 1214 N when the FCH was 26 mm. Again, this force is sufficiently large enough to keep the vehicle laterally stable.

## 5.5 Measurement system with more functions and higher precision [11]

Although the first manned HTS Maglev vehicle in the world [5] was developed successfully with the research results from the SCML-01 measurement system [2], the measurement functions and precision of the SCML-01 platform needed to be improved. Therefore, to make more thorough and careful investigations on the magnetic levitation properties of the HTS Maglev vehicle over a PMG, the SCML-02 HTS Maglev measurement system [11] with several special functions and high precision was successfully developed.

### 5.5.1 System description

The SCML-02 HTS Maglev measurement system includes the experiment system, control hardware and control processing software.

The experiment system is shown in Fig. 5.12. Four vertical support posts are fixed on an optical bedplate The upside liquid nitrogen vessel with HTSCs is placed above a PM or PMG The underside liquid nitrogen vessel with HTS bulks is placed under a PM. In this way, the experiment system can be fit for different measurements, especially measurements of Maglev properties of superconductor samples on a PMG. The inertial force effect (tension) of movement parts is decreased by four force balance chain wheels and corresponding counterweights. The *x*-*y* electromotion seat is fixed onto the optical bedplate Horizontal directions along the *x* and *y* axis can be moved together or separately. Therefore, the measured HTS bulk samples can be moved in three dimensions.

The vertical levitation forces and the lateral guidance forces are measured by four force sensors. The vertical and horizontal sensors are isolated from each other, and it is possible to avoid interference between the two kinds of force sensors. Therefore, the SCML-02 Maglev measurement system can measure the levitation forces and the guidance forces at the same time. Figure 5.13 shows a photograph of the SCML-02 HTS Maglev measurement system. Figure 5.14 shows that researchers are testing and calibrating the accuracy of the SCML-02 measurement system.

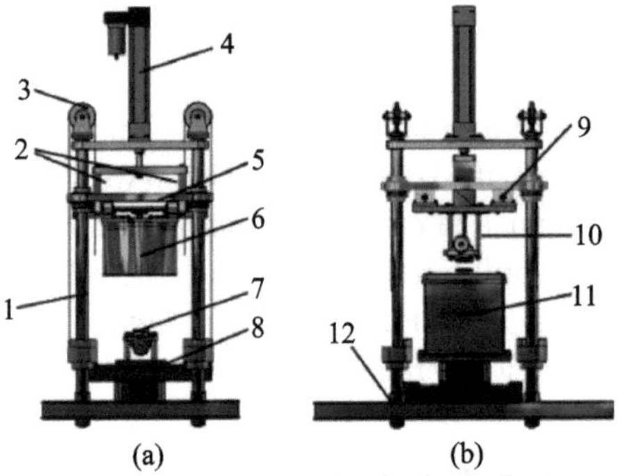


**Fig. 5.12:** The design scheme of the SCML-02 HTS Maglev measurement system: (a) upside liquid nitrogen vessel; (b) underside liquid nitrogen vessel [11]. (1) Vertical support post. (2) Vertical force sensor. (3) Force balance chain wheel. (4) Servo motor. (5) Slip set. (6) Upside liquid nitrogen vessel. (7) Underside clamp of cylinder PM or PMG. (8) *x*-*y* electromotion seat. (9) Horizontal force sensor. (10) Underside clamp of cylinder PM or PMG. (11) Underside liquid nitrogen vessel. (12) Optical bedplate.

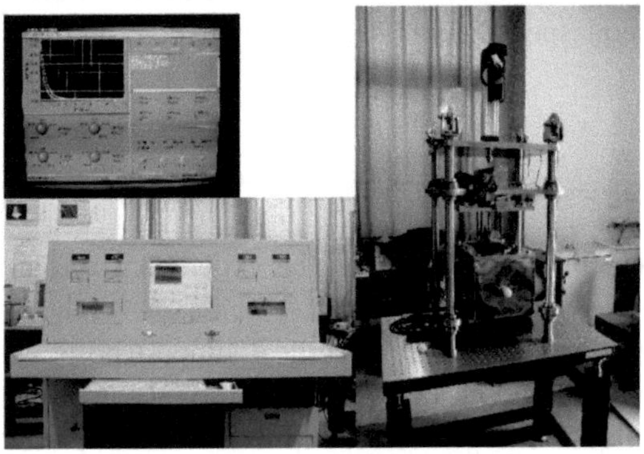

**Fig. 5.13:** Photo of the SCML-02 HTS Maglev measurement system [11]: levitation force measurement interface (top left corner); console control desk (under left corner); experiment system (right).

## 5.5.2 The function and technical specification

The system includes a liquid nitrogen vessel, cylindrical PM (or PMG), etc. The special functions of the SCML-02 Maglev measurement system can be summarized as follows:

1. High-precision measurements.
2. Measurements of both a single or a number of HTS bulk samples.
3. Measured HTS bulk samples and the PM can move along all three dimensions at one time.
4. The location of the HTS bulk and PM can be interchanged.

Fig. 5.14: Researchers are testing and calibrating accuracy of SCML-02 measurement system.

5. Maglev measurements between the HTS bulk specimens and a cylindrical PM or a PMG.
6. Simultaneous measurements of the levitation forces and guidance forces for which the sample can be displaced vertically and horizontally at the same time.
7. Synchronous measurements of not just the levitation and guidance forces, but also the cross stiffness of the levitation forces or the guidance forces in a real-time independent measurement along the vertical or horizontal.
8. Interaction of dynamic rigidity of levitation forces and guidance forces synchronous measurements.
9. Relaxation time of levitation forces and guidance forces measurements along the vertical or horizontal directions.

The SCML-02 Maglev measurement system's main technical specifications are as follows:
1. Vertical maximal displacement of 150 mm.
2. Horizontal maximal displacement of 100 mm.
3. Position precision of 0.05 mm.
4. Vertical maximal support force of 1000 N.
5. Vertical force precision of 2.
6. Horizontal maximal support force of 500 N.
7. Horizontal force precision of 1.

### 5.5.3 Measurement precision calibration

In order to calibrate the measurement precision of the SCML-02 HTS Maglev measurement system, two cylindrical PMs were used. The two PMs have the same strength

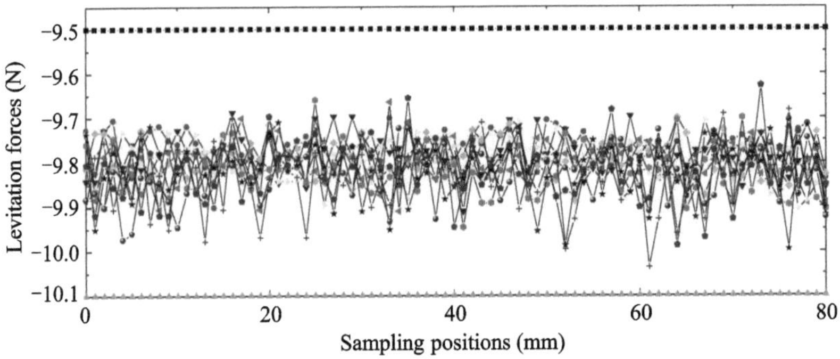


**Fig. 5.15:** Test results of levitation forces at the standard weight of 9.8 N at different sampling velocities (the vertical move velocities are 0.1 to 1.0 mm/s (velocity partition 0.1 mm/s)) [11].

and dimensions. During the levitation forces test, standard weights of 9.8, 18.62, and 24.95 N, and vertical moving velocities of 0.1 to 1.0 mm/s (velocity partition 0.1 mm/s) were used. Figure 5.15 shows the test results of the levitation forces at the standard weight of 9.8 N and at different sampling velocities. At the standard weight of 9.8 N, the measurement precision of the levitation force is 99.23%–99.55%, and the error 0.453%–0.773% with the range of 0.00256–0.02924.

Thus, it can be seen from Fig. 5.16 that the levitation force change is independent of the standard weights of 9.8, 18.62, and 24.95 N. As the total change is between 0.00163–0.0292. The levitation force measurement error is found to be plus or minus 0.5%.

Figure 5.17 shows the test results of the levitation forces at different sampling velocities which justifies the good repeatability of the levitation force measurements of the SCML-02 HTS Maglev measurement system.

Figure 5.18 shows the synchronous test results of the levitation forces and guidance force at the sampling velocity of 0.5 mm/s. The curves with other sampling

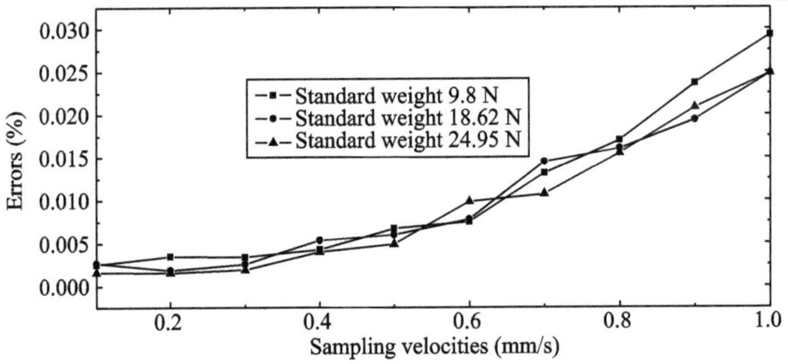


**Fig. 5.16:** Test errors of levitation forces at the different standard weights and sampling velocities [11].

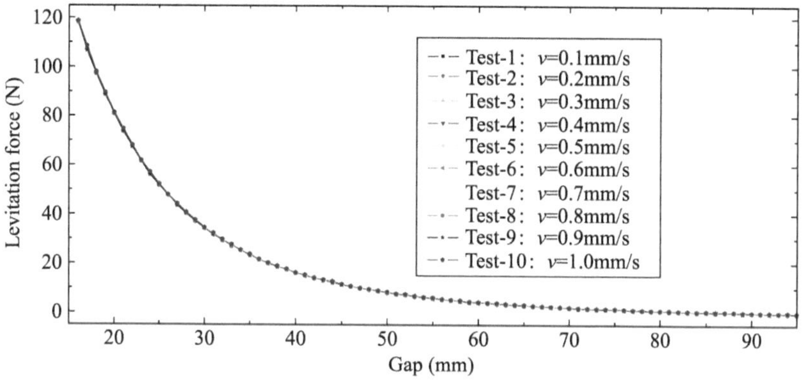


**Fig. 5.17:** Test results of levitation forces at different sampling velocities [11].

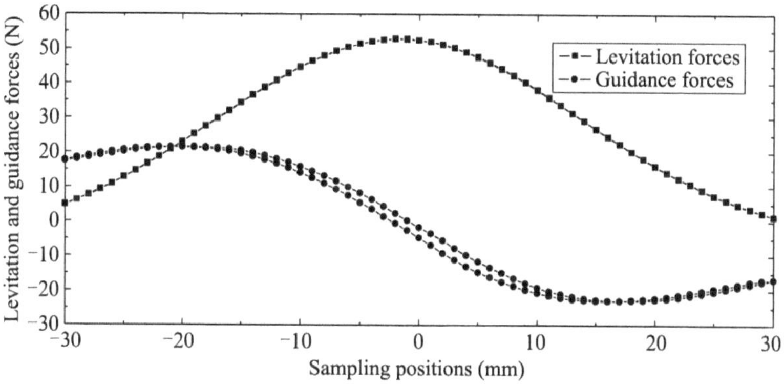


**Fig. 5.18:** Synchronous test results of the levitation forces and the guidance forces at sampling velocity of 0.5 mm/s [11].

velocities of 0.1 to 1.0 mm/s (velocity partition 0.1 mm/s) are also shown and found to fit well with the profile.

Figure 5.19 shows the stiffness test results of the levitation forces at the sampling velocity of 1.0 mm/s. Sampling point gap was 1.0 mm, and 61 sampling points were taken in the measurement range of 95–15 mm.

Figure 5.20 shows the stiffness test results of the guidance forces with the sampling velocity of 1.0 mm/s. The sampling point gap was 1.0 mm, and 61 sampling points were taken in the measurement range of −30 to 30 mm.

Figure 5.21 shows time-dependent levitation forces at the standard weights of 18.62 N. The sampling point interval was 0.1 s, and 1001 sampling points were acquired at the measurement range of 1–100 s.

The above magnetic levitation properties between the PMs were measured by the SCML-02 HTS Maglev measurement system. These measurement results validated the high precision and multi-functionality of the SCML-02.

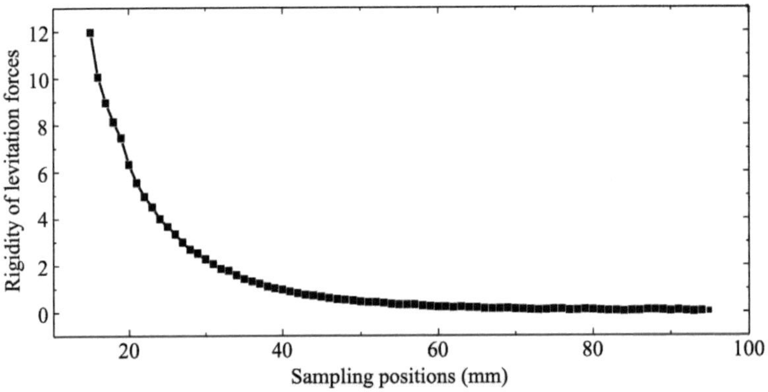


**Fig. 5.19:** Stiffness test results of the levitation forces at the sampling velocity of 1.0 mm/s [11].

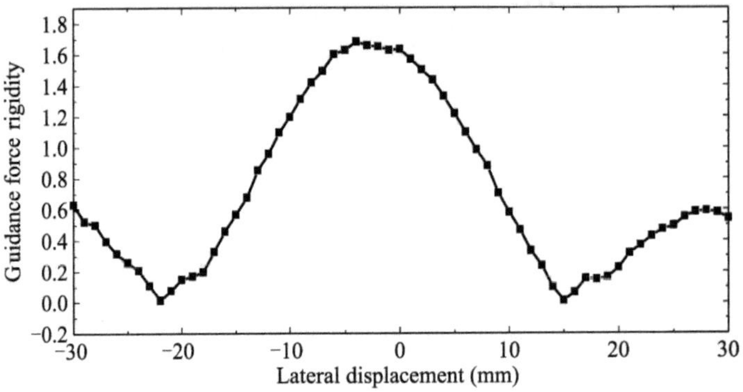


**Fig. 5.20:** Test results of levitation forces at the standard weight 9.8 N and different sampling velocities (the vertical move velocities are 0.1 to 1.0 mm/s (velocity partition 0.1 mm/s)) [11].

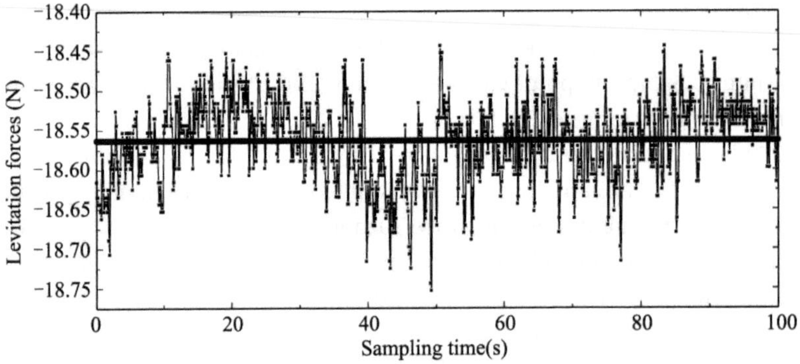


**Fig. 5.21:** Test errors of levitation forces at the different standard weight and sampling velocity [11].

### 5.5.4 Measurement results of HTS Maglev properties

The Maglev properties of YBCO bulk are measured on the basis of the above calibration results of SCML-02 HTS Maglev measurement system. The levitation forces of single YBCO bulk and an array of 7 YBCO bulks above a PMG are measured by the SCML-02 HTS Maglev measurement system. The array of 7 YBCO bulks are concentrically arranged, where three bulks share one axis along the axis line of the PMG.

The diameter of the single YBCO bulk is 48 mm, and the diameter of each of the array of 7 YBCO bulks is 30 mm.

Figure 5.22 shows the measurement results of the levitation forces of two YBCO bulks with diameter of 48 mm with a field cooling height (FCH) of 35 mm and vertical measurement ranges from 60 to 100 mm and backward.

Figure 5.23 shows the measurement results of the guidance forces along the lateral direction of the PMG. The YBCO samples have a diameter of 48 mm, field cooling height (FCH) of 35 mm, measurement height (MH) of 15 mm, and lateral displacement measurement range of –20 to 20 mm.

Figure 5.24 shows the measurement results of the levitation forces of three different arrangements of 7 YBCO samples above the PMG. These 7 YBCO bulk samples have a diameter of 30 mm and FCH of 35 mm.

Figure 5.25 shows the measurement results of the guidance forces of three arrangements of 7 YBCO bulks above a PMG. The MH for this experiment was of 10 mm, and the lateral measurement ranged from –10 to 10 mm.

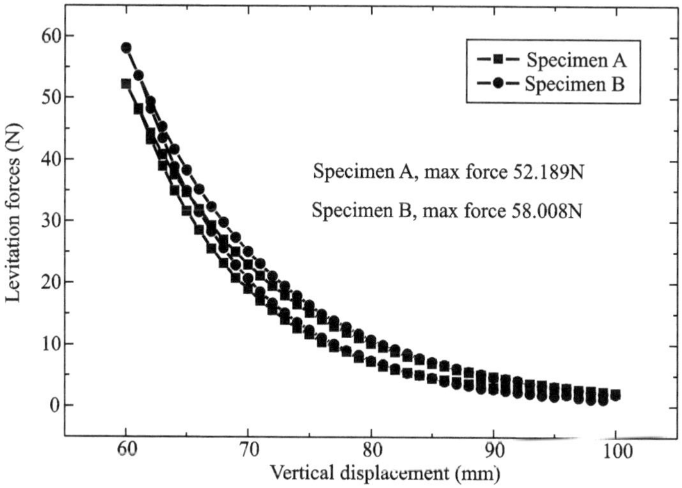


**Fig. 5.22:** Levitation forces of the two YBCO bulk samples with diameter of 48 mm (FCH of 35 mm, measurement range of 60–100 mm) [11].

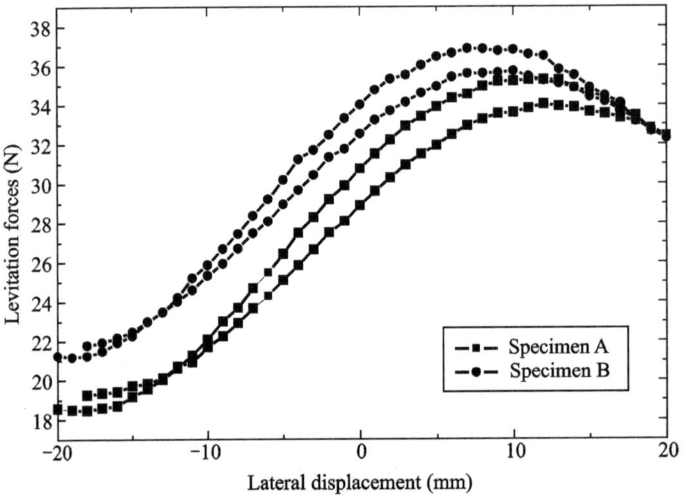


**Fig. 5.23:** Guidance forces of the two YBCO bulk samples with diameter of 48 mm (FCH of 35 mm, measurement height of 15 mm) [11].

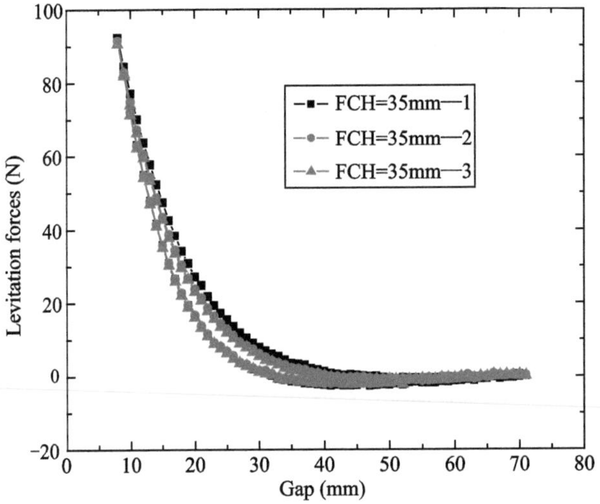


**Fig. 5.24:** Levitation forces of an array of 7 YBCO bulk samples with diameter of 30 mm (FCH 35 mm) [11].

Figure 5.26 shows the measurement results of the array of 7 YBCO samples above a PMG at different MHs. The FCH was set to 20 mm, the measurement heights ranged from 10 to 30 mm, and the lateral measurement ranged from –10 to 10 mm.

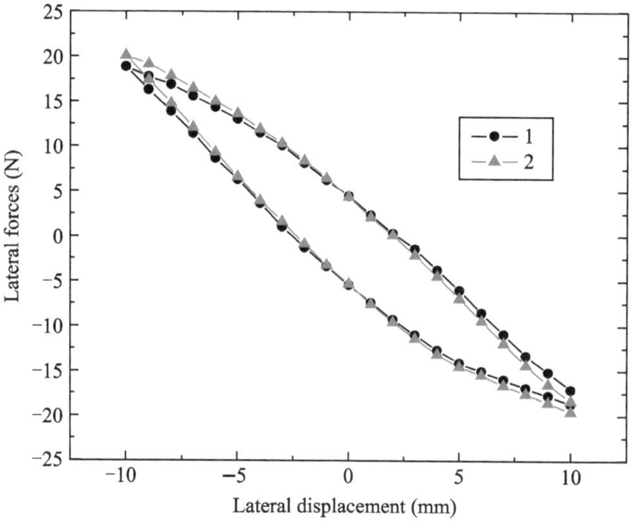


**Fig. 5.25:** Guidance forces of an array of 7 YBCO bulk samples of diameter 30 mm (FCH 35 mm, measurement height 10 mm) [11].

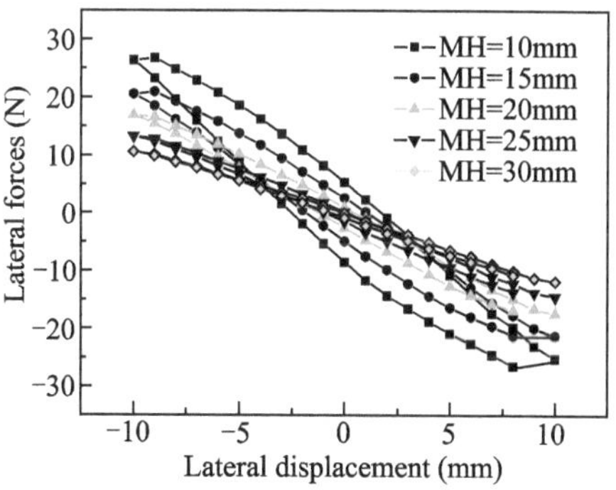


**Fig. 5.26:** The guidance forces of 7 YBCO bulks with diameter of 30 mm are measured by SCML-02 (FCH of 20 mm, MH range of 10–30 mm, lateral measurement range of –10 to 10 mm) [11].

The levitation forces and guidance forces of YBCO bulks were measured with the SCML-02 HTS Maglev measurement system. The measurement results of HTS Maglev properties validated the accuracy and reliability of the SCML-02.

The SCML-02 measurement system can make real-time measurements, and the data acquisition process is controlled by a computer. The main functions and specifications of the system are verified. A lot of precious experimental results have been obtained from this system, like the Maglev vehicle's levitation force, guidance force, their stiffness, and cross stiffness, etc. (see Chapters 8, 9, and 10). All these experimental parameters are very helpful to evaluate the load capability of the HTS Maglev vehicle.

## 5.6 HTS Maglev dynamic measurement system [12]

Although the HTS Maglev measurement system SCML-02 had more functions and higher precision than SCML-01, it cannot measure the running performance of bulk YBCO samples above a PMG. For the further engineering application of the HTS Maglev vehicle, the dynamic properties of the Maglev must be clearly understood. From this viewpoint, a HTS Maglev dynamic test system (SCML-03) [12] was then designed and successfully developed in ASCLab. The test process and results of the HTS Maglev dynamic test system will be reported in this section.

### 5.6.1 System description

When the HTS Maglev vehicle runs along the PMG, it is difficult to measure its dynamic properties. In SCML-03, the rotational motion of a circular PMG instead of the physical motion of the YBCO bulk is taken to be the equivalent measure of the dynamic interaction between the superconductor and the PMG. That is, the circular PMG can rotate to different speeds while the on-board HTS Maglev equipment is fixed above the PMG. This will simulate the fact that the superconductor is traveling above a PMG. SCML-03 is composed of a vertical load, horizontal load, three-dimensional measurement systems, liquid nitrogen vessel, circular PMG, drive device, data acquisition and processing, and autocontrol.

The main measurement functions include the dynamic stability of the HTS Maglev equipment (liquid nitrogen vessel including the HTS samples), the levitation force and guidance force of both single and multi HTS bulk samples, the levitation force and guidance force rigidity of both single and multi HTS bulk samples, the levitation force and guidance force change at the levitation gap, etc.

The main design scheme of SCML-03 is shown in Fig. 5.27. The airframe of the HTS Maglev dynamic test system SCML-03 (not including the measurement control desk) is shown in Fig. 5.28. The total dimensions of the principal part of the SCML-03 are 3.3 m long, 2.4 m wide and 3.15 m high. The total weight is 13.95 t which includes the circular PMG disk that weights 0.6 t. Figure 5.29 shows the measurement scene of the HTS Maglev dynamic test system. Figure 5.30 shows the panoramic photos of the HTS Maglev dynamic test system SCML-03.

A DC motor is used to rotate the circular PMG and control its rotational speed. The rotational direction of the DC motor is translated into the horizontal rotation direction of the circular PMG by a gear redirection case. The circular PMG is fixed along the circumferential direction of a big circular disk with a diameter of 1500 mm. The rotating imbalance of the big circular disk is less than 20 gm. The maximum linear velocity of the PMG is about 300 km/h when the circular disk rotates round the central axis at 1280 rpm. The rotation speed error of the circular disk is less than 3%.

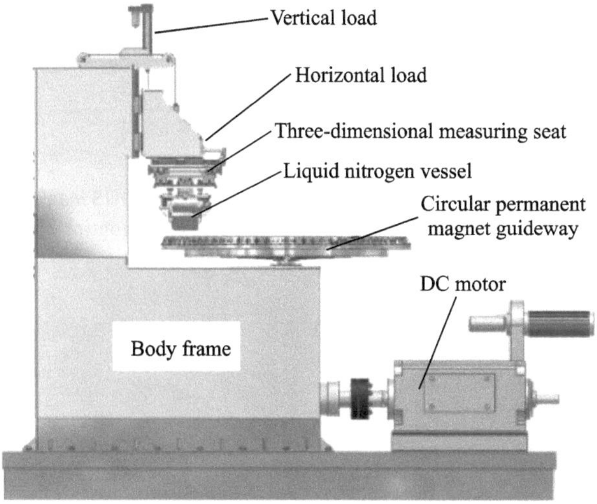


**Fig. 5.27:** The design scheme and principal parts of the HTS Maglev dynamic test system SCML-03 [12] (not including power supply and measurement control desk).

**Fig. 5.28:** Photo of the airframe of the HTS Maglev dynamic test system SCML-03 [12]: 3.3 m long, 2.4 m wide, and 3.15 m high, with a total weight of 13.95 t.

The three-dimensional measuring seat is fixed on the horizontal load. The seat can be moved along the horizontal direction which is perpendicular to the tangential direction of the circular PMG. Therefore, the HTS Maglev dynamic guidance force can be measured at the same time. Six force sensors are used to measure the vertical, transverse, and longitudinal directional forces of the liquid nitrogen vessel. To measure

**Fig. 5.29:** The first measurement scene of the HTS Maglev dynamic test system. On the right side is the control desk [12].

**Fig. 5.30:** Panoramic photos of the HTS Maglev dynamic test system SCML-03. In this picture, the parts below the laboratory ground shown by the square frame line in Fig. 5.28 are not visible.

the three-dimensional dynamic response, the liquid nitrogen vessel is connected to the three-dimensional measuring seat by elastomers.

The autocontrol of all components is completed by the total control desk. Both measurement and control software was developed on basis of Labview software by National Instruments. The servo-motor movement and the circular PMG speed are controlled by both the software program and a relative control card.

In order to calibrate the measurement precision of the dynamic measurement system, an 11-channel Noise&Vibration Measurement System made by B&K company is used. The dynamic vertical levitation force is measured when the circular PMG is rotated at different speeds.

### 5.6.2 The function and technical specification

The SCML-03 can measure the dynamic properties with opposite motion between the vehicle and the PMG. The main measurement functions include:

1. Dynamic stability measurements of the HTS Maglev equipment (liquid nitrogen vessel included HTS bulks) at different velocities.
2. Measurements of levitation force and guidance force of both single and multi HTS bulk samples.

3. Measurements of levitation force stiffness and guidance force stiffness of both single and multi HTS bulk samples.
4. Measurements of the levitation force and guidance force change at different levitation gaps.

Main technical specifications:
1. Diameter of the circular PMG 1500 mm.
2. Maximal linear speed of the circular PMG 300 km/h.
3. Rotation speed precision of the circular PMG ±3%.
4. Vertical maximal displacement 200 mm.
5. Horizontal maximal displacement ±50 mm.
6. Position precision ±0.05 mm.
7. Vertical maximal support force 3350 N.
8. Horizontal maximal support force 500 N.
9. Force sensor precision ±1%.

### 5.6.3 Stability of dynamic test system

The most important property of the dynamic test system is its self-stability. The rotating imbalance of the PMG circular disk was measured to be less than 20 gm. The rotating imbalance value satisfied the system measurement's needs. The other important parameter of the dynamic test system is the self-stability of the liquid nitrogen vessel and the body frame. In order to confirm the self-stability, an 11-channel Noise & Vibration measurement System was used. A 4507-004B accelerometer was fixed onto the body frame, and four 4507-004B accelerometers were fixed on the clamp device of the liquid nitrogen vessel. The vibrations of the perpendicular and horizontal directions on the clamp device were measured by two sensors. A picture of the first measurement scene of the SCML-03 is shown in Fig. 5.29. Figure 5.30 shows panoramic photo of whole measurement system. In this picture, it could not show the parts below laboratory ground which is shown by the square frame line in Fig. 5.28

The vibration spectra of the body frame and the clamp device of the liquid nitrogen vessel were measured at the rotation speeds of 50 rpm, 100 rpm, 200 rpm, 300 rpm, and 400 rpm. The vibration spectrum of the clamp device of the liquid nitrogen vessel includes both the perpendicular direction and the horizontal direction. The vibration spectrum of the body frame at 400 rpm is shown in Fig. 5.31. The vibration spectra of the perpendicular direction and the horizontal direction on the liquid nitrogen vessel at 400 rpm are shown in Figs. 5.32 and 5.33.

Figure 5.34 shows the vibration measurement results of the body frame and the clamp device of the liquid nitrogen vessel (the perpendicular and the horizontal direction) at a rotation speed of 50 rpm, 100 rpm, 200 rpm, 300 rpm, and 400 rpm. The left figure shows the measuring vibration results without the liquid nitrogen

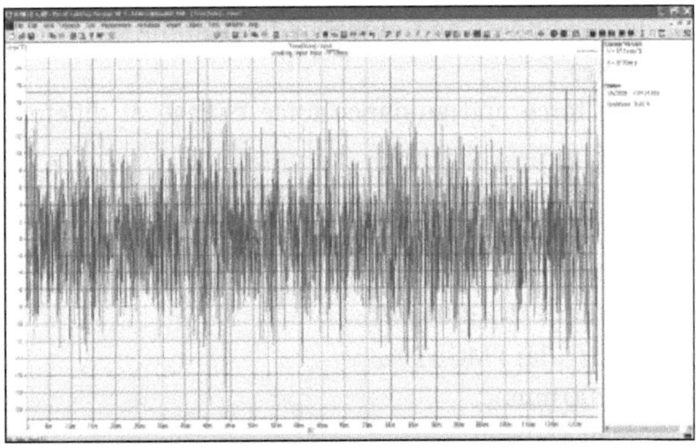

**Fig. 5.31:** Vibration spectrum of the body frame at 400 Rpm [12].

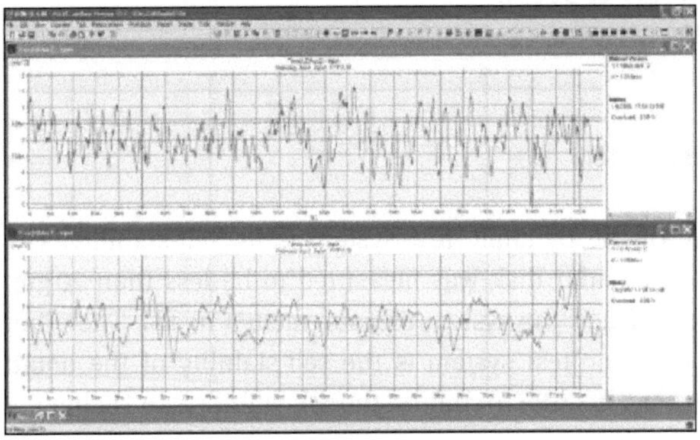

**Fig. 5.32:** Vibration spectrum along the perpendicular direction on the clamp device of the liquid nitrogen vessel at 400 rpm [12].

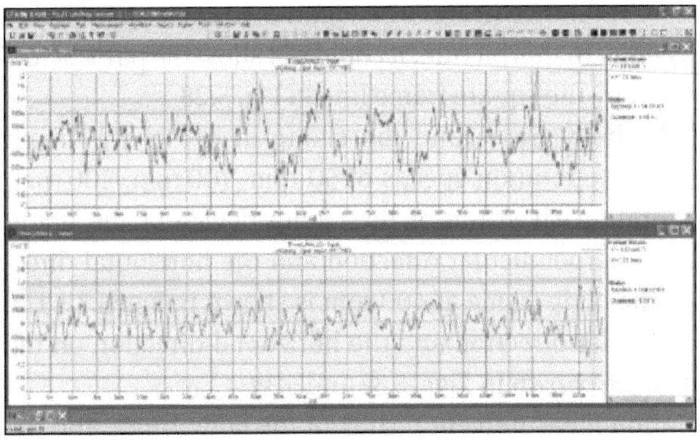

**Fig. 5.33:** Vibration spectrum along the horizontal direction on the clamp device of the liquid nitrogen vessel at 400 rpm [12].

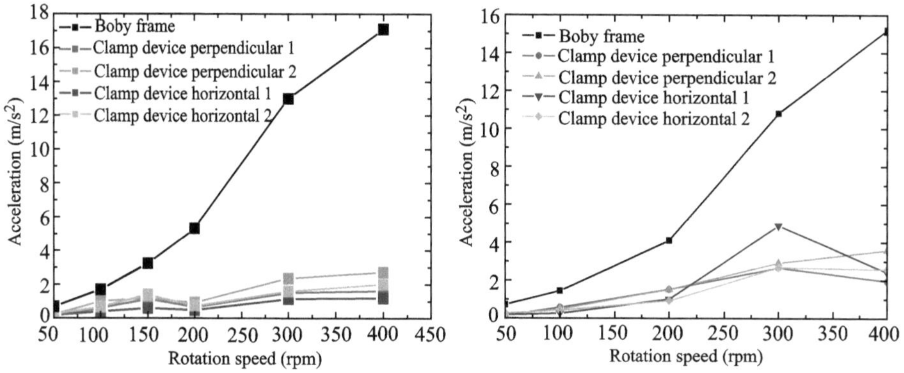


**Fig. 5.34:** Vibration measuring results of the body frame and the clamp device of the liquid nitrogen vessel (perpendicular and horizontal direction) at rotation speeds of 50 rpm, 100 rpm, 150 rpm, 200 rpm, 300 rpm, and 400 rpm. Left (right) figure shows the measuring results without (with) the liquid nitrogen vessel on the clamp device [12].

vessel on the clamp device. The right figure shows results with the liquid nitrogen vessel attached. After the YBCO superconductors in the vessel were cooled in the zero field, vibration properties were measured at the levitation gap of 20 mm and lateral displacement of 0 mm. Experimental results showed that the vibration of the body frame increases quickly with increased of rotation speed, and the vibration of the clamp device with or without the liquid nitrogen vessel also increased slowly with increased rotation speed. This indicated that the measuring seat had the function of vibration isolation. The vibration of the clamp device with the liquid nitrogen vessel satisfied measurement requirements.

### 5.6.4 Measurements of HTS Maglev properties

The main distinction between the SCML-03 [12] and previous HTS Maglev measurement systems [2, 11] is measurement of the Maglev performance under operating conditions, with the SCML-03, both levitation forces and guidance forces are measured experimentally at different velocities.

The experimental results of the levitation forces and the guidance forces of YBCO bulk measured with the SCML-03 are shown in Figs. 5.35 and 5.36, respectively, in the static state. The levitation forces of YBCO bulk were measured in zero field cooling (ZFC), and the guidance forces of YBCO bulk were measured at a 15 mm field cooling height above the circular PMG. Those measurement results verify that the SCML-03 can accurately measure the levitation and the guidance forces in the moving state.

In order to confirm the measurement results for the levitation forces and the guidance forces of YBCO bulks, the changes of the measurement results with different

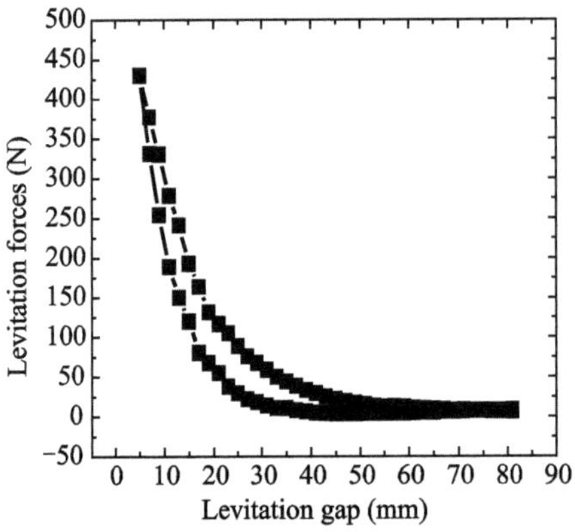


Fig. 5.35: Measurement results of the levitation forces of YBCO bulk in zero field cooling (ZFC) [12].

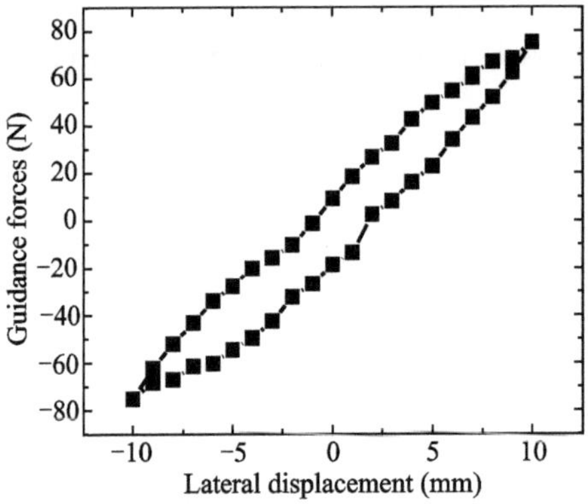


Fig. 5.36: Measurement results of the guidance forces of YBCO bulk at a 15 mm levitation gap for the field cooling (FC) [12].

measuring times and different rotational speeds were investigated experimentally. Rotational speeds of 50 rpm, 100 rpm, 200 rpm, 300 rpm, and 400 rpm, with time intervals of 12 s were tested and shown in Fig. 5.37. The experimental results were measured with a levitation gap of 10 mm.

Figure 5.37 shows that the changes of the measured levitation forces are small, especially at the 300 and 400 rpm speeds. The higher the PMG rotation speed, the smaller the levitation force changes. At a levitation gap of 10 mm the observed changes of the measured levitation forces in a 48 s period were 0.54 N at 300 rpm and 0.10 N at 400 rpm.

The levitation force decreased with increased rotational speed. The decrease in value of the levitation forces was 22.98 N when the rotation speed increased from 50 to 400 rpm at time intervals of 12 s, which a decrease of the levitation force of 7.4% from the original value. The reason for the decrease of the levitation force can be attributed to some subtle changes in the uniformity of the circular PMG's magnet field along the circumferential direction which at high speed is equivalent to subjecting the HTS to an AC magnetic field.

The experimental results of the guidance forces on field cooling YBCO superconductor bulks at time intervals of 12 s, levitation gap of 15 mm, and special rotation

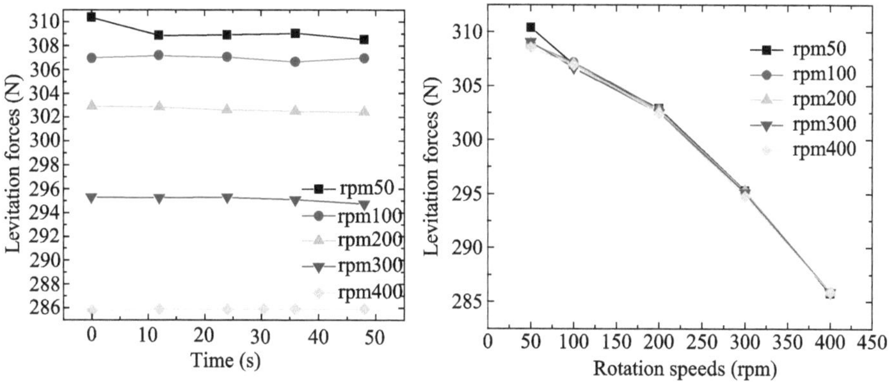


**Fig. 5.37:** Experimental results of the levitation forces for zero field cooling YBCO superconductor bulks at time intervals of 12 s and special rotation speeds (the levitation gap of 10 mm) [12].

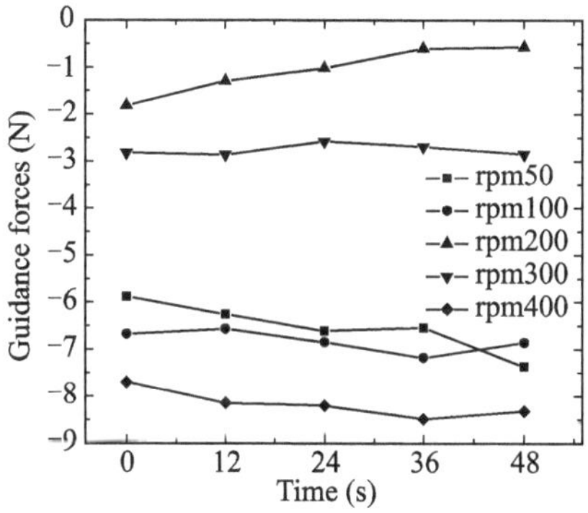


**Fig. 5.38:** Experimental results of the guidance forces for field cooling YBCO bulks at time intervals of 12 s, levitation gap was of 15 mm, and special rotation speeds [12].

speeds are shown in Fig. 5.38. It indicates that the changes of measured guidance forces at the time intervals of 12 s were small, especially at 300 and 400 rpm. The higher the PMG rotation speed, the smaller the guidance force changes. The experimental results for the guidance forces were not better than that of the levitation forces. The largest change of measured guidance forces in the 48 s time frame was 0.28 N at 300 rpm and 0.79 N at 400 rpm.

A lot of precious experimental results of dynamic tests were obtained from this system (see Chapters 8 and 9).

## 5.7 HTS Maglev bearing measurement system [13]

The unique operation characteristic of the HTS Maglev bearings is the heavy load capacity at a high rotational speed, which is very important to the practical application but has not been well studied in the relevant research fields.

The general or standard test equipment can only be used for static or low speed test, and are incapable of performing the test for a HTS Maglev bearing under heavy load and high speed. In order to study the effects of the dynamic loss of the HTS bulk in high-speed bearing, and also to investigate HTS Maglev bearing and flywheel energy storage technology, a HTS Maglev bearing measurement system was successfully designed and developed in Beijing Century Superconductivity Company limited, from 2011 to 2014. Figure 5.39 shows the three-dimensional design of main body for the HTS Maglev bearing measurement system. Figure 5.40 is the HTS Maglev bearing measurement system and its supporting units.

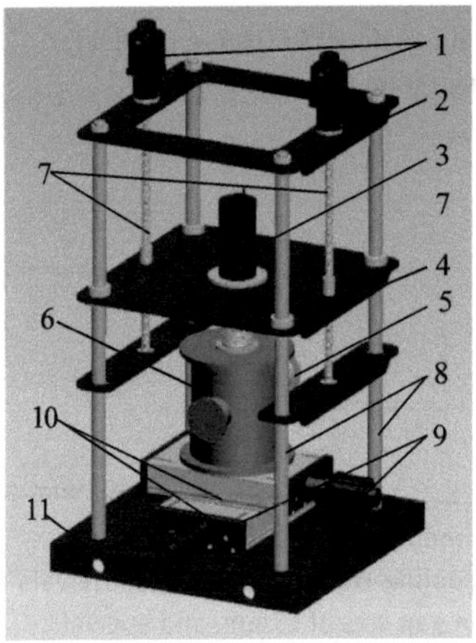


**Fig. 5.39:** The three-dimensional design of main body for HTS Maglev bearing measurement system [13]. (1) Servo motor. (2) Upper frame. (3) Electric spindle. (4) Slide. (5) Refrigerator cooling head. (6) Vacuum cover. (7) 2-Ball screw. (8) 4-Columns. (9) *X-Y* Step motor. (10) *X-Y* Sliding platform. (11) Base.

**Fig. 5.40:** The HTS Maglev bearing measurement system and supporting facilities. (1) Servo motor. (2) Upper frame. (3) Electric spindle. (4) Slide. (5) Refrigerator cooling head. (6) Vacuum cover. (7) Refrigeration compressor. (8) Molecular pump. (9) Circulating water cooling device. (10) System control platform. (11) Power supply cupboard. (This picture composition composed of four photos) [13].

This measurement system can accomplish the automatic test of various mechanical, electromagnetic, and temperature performances of HTS Maglev bearing at static and dynamic conditions, which will provide strong technical support for the development and engineering applications of HTS Maglev bearing and flywheel energy storage system (FESS).

The characteristics and special technical requirements for the HTS Maglev bearing measurement system are:

1. Basic platform with high positioning accuracy.
2. Improved machining of the PM rotor and the assembly accuracy, in order to ensure structural stiffness and safety.
3. Very high concentricity between one PM rotor and the three stators (high-speed motor, HTS bearing, and PM bearing stator).
4. The system dynamic balance compensation for the best balance.
5. Shock absorption and vibration isolation for the test platform.
6. The precision machining and assembly of the arc plane for HTS bulks, in order to ensure the accuracy of the HTS Maglev bearings.
7. The cryogenic system of wide temperature region for automatic control.
8. The vacuum dynamic sealing device at the high speed.

The HTS Maglev bearing measurement system is mainly used for levitation characteristic tests of the HTS Maglev bearing in static and dynamic operations, especially with heavy load and high-speed conditions. The main functions include:

1. Real-time monitoring and recording of rotation rate.
2. Setting and monitoring of the running state for the high-speed electric spindle.
3. Measurements of axial and radial forces of the HTS Maglev bearing under static test.
4. Measurements of axial and radial relaxation time of the HTS Maglev bearing under static test.
5. Vibration characterization of HTS Maglev bearing under dynamic conditions.
6. Force characterization tests of bearing system at low temperature.
7. Measurements of temperature variation of HTS bulk at dynamic test.
8. Dynamics balance characterization of HTS bulk bearing at dynamic test.
9. Studies on influence of external disturbance on bearing performance under dynamic test.

Specifications of the HTS Maglev bearing measurement system:

1. The base size, $1.1 \times 1.1$ m.
2. Rotating speed, 020,000 rpm.
3. PM rotor diameter, 220 mm.
4. Test force range, (axial) 1500 kgf, (radial) 800 kgf.
5. Test temperature, 40–300 K.
6. The position precision, ±0.05 mm (axial) and ±0.01 mm (radial).

7. Force measuring accuracy, ±0.1%.
8. Vertical test speed, 10 mm/s (max.), 0.1 mm/s (min.).
9. Horizontal test speed, 5 mm/s (max,), 0.01-0.03 mm/s (min.).
10. Vertical maximum displacement, ±150 mm.
11. Vertical minimum test distance, 0.1 mm.
12. Level $(x-y)$ the maximum displacement, ±50 mm.
13. Horizontal displacement minimum test distance, 0.01 mm.
14. Location: after each test back to the origin (0 points).
15. Power supply, AC 220 V, 15 A (max.).
16. Working temperature, 0°C–50°C
17. Humidity, 10%–90%.

The maximal testing axial force is 1500 kgf, which is designed with the consideration of the HTS Maglev bearing. This is a common knowledge in the last decade development in FESS, from the cost-effective view. If we want to increase the capacity, we should use assembly of several FESS. The ESC of 2–5 kWh can satisfy 90% of the requirements for FESS, which is based on a mass of market research. The flywheel rotor composed of metal or composite materials, with ESC of 5 kWh and rotation speed of 15,000 rpm, is about 450 kg by the simulation calculation. Because of the limitation of maximal line speed of metal materials, the weight of the flywheel rotor is about 750 kg if composed completely of metal at rotate speed of 15,000 rpm. This HTS Maglev bearing measurement system has a reserve testing capacity for a metal flywheel rotor of 1500 kg, in other words, a FESS with ESC of 10 kWh. If the 1500-kg rotor is composed of metal and composite materials, the maximal ESC for testing FESS will be about 20 kWh. From experimental experience, the maximal horizontal force of HTS Maglev system is always smaller than half of the maximal vertical force, so the maximal testing radial force is set to 800 kgf.

Because of risk in flywheel tests at high speed, all the processes are controlled remotely on the test platform. Some special technologies are applied to ensure the precision of testing signal, such as the transition between RS 232/optical fiber and the application of shielded wires. Furthermore, the key accessory equipment of the test system, such as the vacuum pump, the cryogenic refrigerator and the water cooling device, are also controlled on the test platform. Once a dangerous signal regarding the vacuum degree, the temperature or the cooling system appears, the control system will stop the ongoing experiment immediately to protect the safety of operators and facilities.

## References

[1] Krabbes G, Fuchs G, Canders WR, May H, Palka R. High temperature superconductor bulk materials: Fundamentals - Processing - Properties control - Application aspects. Weinheim: Wiley-VCH Verlag GmbH & Co. KGaA; 2006. p. 1–30.

[2] Wang JS, Wang SY, Lin GB, Huang HY, Zhang CF, Zeng YW, et al. High $T_c$ superconducting magnetic levitation measurement system. High technol lett. 2000;10:56–58+55. (In Chinese).

[3] Wang JS, Wang SY. Synthesis of bulk superconductors and their properties on permanent magnet guideway. In: Narlikar A, editor. Frontiers in superconducting materials. Berlin Heidelberg: Springer Verlag; 2005. p. 885–912.

[4] Song HH, Wang JS, Wang SY, Deng CY, Zeng YW. Studies of YBCO electromagnetic properties for high-temperature superconductor Maglev technology. In: Martins BP, editor. New topics in superconductivity research. New York: Nova Science Publishers; 2006. p. 107–156.

[5] Wang JS, Wang SY, Zeng YW, Huang H, Luo F. The first man-loading high temperature superconducting Maglev test vehicle in the world. Physica C Supercond. 2002;s378–381(1): 809–814.

[6] Schultz L, De Haas O, Verges P, Beyer C, Rohlig S, Olsen H, Kuhn L, Berger D, Noteboom U. Superconductively levitated transport system - the SupraTrans project. IEEE Trans on Appl Supercond. 2005;15(2):2301–2305.

[7] Kovalev KL, Koneev SM-A, Poltavec VN. Magnetically levitated high-speed carriages on the basis of bulk HTS elements. Proceedings of 8th International Symposium on Maglev Suspension Technology (ISMST'8), Dresden, Germany, September 26–28, 2005, p. 51.

[8] Sotelo GG, Dias DHN, Machado OJ, David ED, De Andrade R, Jr., Stephan RM, Costa GC. Experiments in a real scale maglev vehicle prototype. J Phys Conf Ser. 2010;234(3):032054.

[9] Okano M, Iwamoto T, Furuse M, Fuchino S, Ishii I. Running performance of a pinning-type superconducting magnetic levitation guide. J Phys Conf Ser. 2006;43:999.

[10] D'Ovidio G, Crisia F, Lanzara G. A "V" shaped superconducting levitation module for lift and guidance of a magnetic transportation system. Physica C Supercond & Its Appl. 2008;468(14):1036–1040.

[11] Wang S, Wang J, Deng C, Zheng J, Song H. Laboratory-scale high temperature superconducting Maglev launch system. IEEE Trans Appl Supercond. 2007;17(2):2091–2094.

[12] Wang J, Wang S, Deng C, Zeng Y, Zhang L. A high-temperature superconducting Maglev dynamic measurement system. IEEE Trans Appl Supercond. 2008;18(2):791–794.

[13] Liu W, Kang D, Zhang CN, Peng GH, Yang X, Wang S. Design of a high-$T_c$ superconductive Maglev flywheel system of 100 kW level. IEEE Trans Appl Supercond. 2016;26(4):5700805.

[14] Wang JS, Wang SY. High temperature superconducting Maglev measurement system. In: Kr Shanma M, editor. Advances in measurement systems. Vukovar: InTech; 2010. p. 51–80.

[15] Wang SY, Wang JS. A liquid nitrogen vessel with thin wall. Cryogenics and superconductivity. Cryogenics & Super conductivity. 1999;27(3):1–3. (In Chinese).

[16] Wang SY, Wang JS, Ren ZY, Zhu M, Jiang H, Wang XR, et al. High temperature superconducting Maglev equipment on vehicle. Physica C. 2003;386:531–535.

[17] Wang JS, Wang SY, Ren ZY, Zhu M, Jiang H, Tang QX. Levitation force of a YBaCuO bulk high temperature superconductor over a NdFeB guideway. IEEE Trans on Appl Supercond. 2001;11:1801–1804.

[18] Wang SY, Wang JS, Ren ZY, Jiang H, Zhu M, Wang XR, Tang QX. Levitation force of multi-block YBaCuO bulk high temperature superconductors. IEEE Trans on Appl Supercond. 2001;11:1808–1811.

[19] Ren ZY, Wang JS, Wang SY, Jiang H, Zhu M, Wang XR, Song HH. Influence of shape and thickness on the levitation force of YBaCuO bulk HTS over a NdFeB guideway. Physica C. 2003;384:159–162.

[20] Jiang H, Wang JS, Wang SY, Ren ZY, Zhu M, Wang XR, Shen XM. The magnetic levitation performance of YBaCuO bulk at different temperature. Physica C. 2002;s378–381:869–872.

[21] Zhu M, Ren ZY, Wang SY, Wang JS, He J, Tang QX. Proceedings of International Conference on Engineering and Technological Sciences 2000; 2000 Oct 11–13; Beijing, China; Advanced materials. 2000;1:578–581.
[22] Wang XR, Song HH, Ren ZY, Zhu M, Wang JS, Wang SY, Wang XZ. Levitation force and guidance force of YBaCuO bulk in applied field. Physica C. 2003;386:536–539.
[23] Song HH, De Haas O, Ren ZY, Wang XR, Zheng J, Wang XZ, Wang SY, Wang JS, Zhao Y. Magnetic interaction between multiple seeded YBCO bulks and the permanent magnet guideway. Physica C. 2004;407:82–87.
[24] Wang XZ, Wang SY, Wang JS. FEM analysis of a new hybrid superconducting magnetic levitation system. Int J Mod Phys B. 2005;19:403–405.
[25] Wang JS, Wang SY, Zeng YW, Deng CY, Ren ZY, Wang XR, et al. The present status of the high temperature superconducting Maglev vehicle in China. Supercond Sci Technol. 2005;18:S215.
[26] Wang JS. Proceedings of Int Workshop on HTS Maglev; 2002; Chengdu, P. R. China; 2002. p. 25–27.
[27] Wang JS, Wang SY. In: Narlikar A, editor. Synthesis of bulk superconductors and their properties on permannent magnet guideway. Frontiers in superconducting materials. vol. 36. Berlin Heidelberg: Springer Verlag; 2005. p. 885–912.
[28] Zeng R, Wang SY, Wang JS, Zheng J, Lin QX, Liao XL, et al. An improved High-$T_c$ superconducting Maglev measurement system with multi-parameter test and movement functions. IEEE Trans on Appl Supercond. 2013;23:9000904.
[29] Wang JS, Wang SY, Ren ZY, Wang XR, Zhu M, Tiang H, Song HH, Wang XZ, Zheng T. Guidance forces on high temperature superconducting Maglev test vehicle. IEEE Trans on Appl Supercond. 2003;13:2154–2156.
[30] Ren ZY, Wang JS, Wang SY, Wang XR, Song HH, Wang XZ. An equipment measuring guidance force of a HTS Maglev vehicle. Adv Technol Electral Eng Energ. 2003;22:77–80. (In Chinese).
[31] Wang XR, Ren ZY, Song HH, Wang XZ, Zheng J, Wang SY, Wang JS, Zhao Y. Guidance force in an infinitely long superconductor and permanent magnetic guideway system. Supercond Sci Technol. 2005;18:S99–S104.

Jia-Su Wang and Su-Yu Wang

# 6 First manned HTS Maglev vehicle in the world

The above chapters have introduced the basic issues about the superconducting Maglev systems, including the permanent magnet (PM) levitations and the normal conductive Maglev, the LTS Maglev, and the HTS Maglev. The HTS Maglev includes using the HTS wire and bulk Maglev.

The HTS bulk Maglev could realize stable levitation without any active control. This is especially attractive for practical applications of Maglev transportation and energy storage. The experts in the world are trying to complete the HTS Maglev vehicle test. The first manned HTS Maglev test vehicle in the world – "Century" was built on December 31, 2000 [1]. This chapter is devoted to introduce the application of HTS Maglev transportation. The focus is on the development process of the "Century". The other applications of HTS Maglev, for example, Maglev bearings, flywheel energy storage, Maglev electromagnetic launch technology, etc., will be discussed in later chapters.

## 6.1 Introduction

Since the discovery of the HTS bulk Maglev phenomenon, significant progresses have been made in the theory and application of HTS bulk Maglev, especially in the fields of transportation and energy storage.

Various Maglev trains and traditional railway transport have their own characteristics for ground transportation. The normal conductive Maglev trains with moderate/low speed or high speed and the LTS Maglev trains with high speed or ultra-high speed have started to enter the market and show their new features. In constant pursuit of the "perfect" transportation, the HTS bulk Maglev train might be a possible candidate.

The HTS bulk Maglev train is a transport whose levitation and guidance do not require an on-board power supply, and it is a self-stable system that does not require active control. The HTS bulk Maglev system can integrate superconducting levitation, guidance, and linear motor propulsion. The vehicle in the evacuated tube transportation (ETT) can achieve ultra-high speed (thousands of kilometers per hour), and is economical in energy consumption, safety and comfort.

International research zest on the manned HTS Maglev vehicle was aroused after the manned HTS Maglev vehicle was invented [2]. Two more manned HTS Maglev vehicles were developed in Germany [3] and Russia [4] in 2004. A full-scale HTS Maglev train as a substitute for urban light track train is under the development in Brazil [5, 6]. Moreover, research groups in Japan [7] and Italy [8], among others, also developed HTS Maglev prototypes.

https://doi.org/10.1515/9783110538434-006

At first, this chapter will introduce the key properties of HTS Maglev vehicle, the HTS Maglev vehicle project as part of the National High-tech R&D Program (National 863 Program), and supports and attentions from experts and organizations. Then, the research and development process on the "Century" vehicle before and after its birth will be reviewed.

There are some comprehensive review papers on HTS bulk Maglev which can be found elsewhere [9–20].

## 6.2 HTS Maglev in National 863 Program

Both rare-earth materials yttrium (Y) and neodymium (Nd) are used in HTS bulk Maglev trains, and these resources are abundant in China. In 1996, Jia-Su Wang and Su-Yu Wang discussed the problems about the manned HTS Maglev vehicle with Prof. Hong-Tao Ren of the GRINM (General Research Institute for Nonferrous Metals) in the Dalian Conference of Information Net of Cryogenics and Superconductivity in China. Prof. Ren finally acceded to the challenging project.

During the period of the ninth five-year plan, the Applied Superconductivity Laboratory (ASCLab) directed one of the National High-tech R&D Program (National 863 Program) in China. The research project of manned HTS Maglev vehicle (grant number 863-CD080000) was approved by National 863 Program in 1997. The Maglev performance of the HTS YBCO bulk over permanent magnet guideway (PMG) has been experimentally investigated using the special measurement equipment developed in ASCLab. They were among the first to establish a number of theoretical and basic technological results (Sections 6.3–6.16) [11, 12]. The first manned HTS Maglev test vehicle in the world, "Century", was unveiled in ASCLab on December 31, 2000 [1]. All of the on-board melt-textured YBCO bulk superconductors were purchased from Prof. Hong-Tao Ren's group in GRINM in Beijing, China. They were fixed in rectangle-shaped vessels with very thin soleplates.

The Maglev vehicle [1, 21] was 3.5 m long, 1.1 m wide, 0.9 m high and the PMG had a length of 15.5 m. The net levitation height was more than 20 mm when five people sat on the vehicle (about 630 kg including vehicle's weight). Thereafter, the R&D of the HTS Maglev vehicle had been continually receiving the investment from the National 863 Program.

On January 3, 2001, a press conference about "Century" was held in Southwest Jiaotong University.

The National Committee of Experts in Superconductivity of the National 863 Program reviewed and approved the completed Maglev vehicle project on February 11, 2001. According to the National Committee of Experts in Superconductivity of the National 863 Program, "This is a creative achievement via a long period of time of hardworking. It has expanded the possibility of crossing over the development about HTS Maglev technology ... By hardworking of the group, China made an important breakthrough of basic research in HTS Maglev application and other related technologies. Now we can play a leading role in this field in the world ... This original work also lays the foundation in applying HTS Maglev technology in transportation and other fields and provides guidance for long-term development in the subject". "Century" was awarded "Top 10 Advancements of Science and Technology of Chinese Colleges" in 2001 and was described as: "It possesses tremendous developing opportunity and commercial potential" (Acceptance Report of National Expert Committee on Superconductivity of 863 Program-internal unpublished document). ASCLab is also the recipient of the National Certificate of Merit for the May Day by All-China Federation of Trade Unions in 2001.

The HTS Maglev vehicle "Century" was under the spotlight in Review of 15 Years (1986–2000) for National High-tech R&D Program (National 863 Program) of China in Beijing, 2001. Five years later, the vehicle "Century" was the main attraction in the First Science and Technology Festival of Chengdu, 2005. The performance of the "Century" is still similar to its performance in 2001. "Century" makes passengers feel like they are not running above the railway.

The International Workshop on HTS Maglev (ISMAGLEV 2002) was held on June 25–27, 2002, in ASCLab of Southwest Jiaotong University, Chengdu, China. The topic was the HTS Maglev technology and its application [22].

On November 5, 2003, 14 Chinese Academy of Sciences members co-signed a letter to the State Council to petition for higher priority on further research and development of the HTS Maglev train technology and related it to domestic high-technology industry.

During the period of the ninth 5-year plan, the SCML-01 HTS Maglev measurement system was developed. Thereafter, the SCML-02 HTS Maglev measurement system with high precision and multi-functions and the SCML-03 HTS Maglev dynamic test system with the linear speed of 300 km/h were developed in 2004 and 2005. A laboratory prototype of the HTS Maglev launch system was completed in 2006. The HTS Maglev bearing measurement system was developed successfully in 2014 in Beijing Century Superconductivity Company.

The project proposal review meeting of the ultra-high speed ETT HTS Maglev train was held in the Southwest Jiaotong University on January 12, 2004. According to research results of the long-term stability (less than 5% of the levitation force decrease), Jia-Su Wang made the report on an ETT HTS Maglev train project which proposed a high speed of more than 600 km/h. The project discussion meeting with more than 50 experts lasted all day long. The meeting approved this project and suggested to launch R&D on the HTS Maglev vehicle. Unfortunately, the meeting did not approve the R&D program of the high-speed experiment of 600 km/h for the ETT HTS Maglev train.

In December 2004, the First Academicians Forum Meeting on the ETT-HTS high-speed transportation was held in the Sichuan Academy Advisory Centre. The CAS members and experts unanimously recommended strong support for the development and industrialization of China's first HTS Maglev vehicle after the author reported on the high-speed HTS Maglev vehicle technology.

"National 863 Program Strategy Development Report (Superconducting Materials and Technology)" pointed out: China had developed the manned HTS Maglev system technology, and it had expanded the possibility to cross over the development of HTS Maglev technology. HTS Maglev system technology had bright Chinese characteristic. This Maglev technology was a new transport vehicle which will have high energy efficiency, environment-friendly, safe, and comfortable.

## 6.3 Levitation forces of HTS bulk above PMG

### 6.3.1 Levitation forces of single HTS bulk above PMG

A HTS Maglev measurement system was developed (see Section 5.3) in order to realize the real-time, high-precise measurement of HTS Maglev performance and to develop the manned HTS Maglev vehicle. This measurement system (Fig. 5.7) is different from the previous one, i.e. measuring the levitation forces of HTS bulk over PMG (see Section 5.2.1). Thus, a special thin-bottom liquid nitrogen vessel [23] (see Section 5.2.2) was needed. The outside and inside diameters of the thin-bottom liquid nitrogen vessel were 200 and 150 mm, and the height was 250 mm. The liquid nitrogen vessel can operate continuously for over 16 hours and can hold a 7-YBLO-bulks unit, where each bulk shares the 30-mm diameter. The PMG was composed of NdFeB PMs and an iron plate whose function is concentrating field in the center of the surface was up to 1.2 and 0.4 T at a height of 20 mm above it. The length of the PMG was 920 mm.

A series of the magnetic levitation properties, for example, levitation force, guidance force, levitation stiffness, trapping flux, and influence of HTS bulk shape, thickness, etc., of the HTS YBCO bulk above PMG have been systematically investigated by this HTS Maglev measurement system SCML-01. These research will be presented in this chapter.

The levitation forces of a HTS YBCO bulk over the PMG was measured by the SCML-01 HTS Maglev measurement system (see Section 5.3). In our measurement, the measured YBCO bulks were fixed at the bottom of the thin-bottom liquid nitrogen vessel, then they were cooled into the superconducting state in zero magnetic field. Second, the vessel was fixed at a connecting fixture with a servo electromotor. In order to avoid collision between the bottom of the vessel and the surface of the PMG, there was still a gap of 1.5 mm between them when the vessel moved to the closest position, so the minimum gap was 5 mm between the bottom of the HTS bulk sample and the surface of the PMG. The vessel first moved downward, after reaching the lowest point of 1.5 mm, then moved upward at a speed of 2 mm/s, and the computer collected data every 0.5 second.

First, we measured the levitation forces of the HTS YBCO bulk over the single-NdFeB PM and the PMG. Figure 6.1 shows the measured results of the levitation forces over the NdFeB PMG (its surface magnetic flux density was 1.2 T) and over the

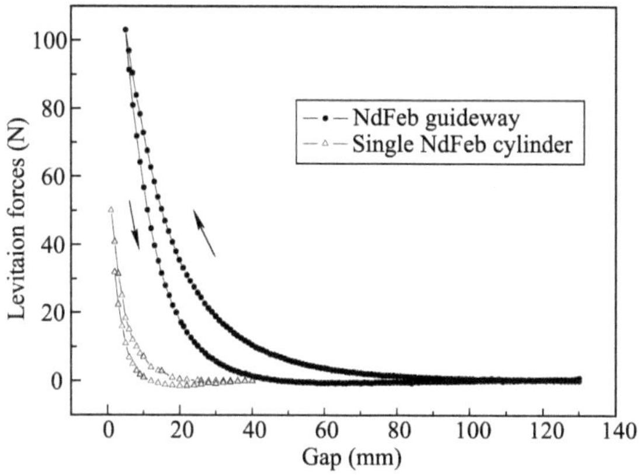


**Fig. 6.1:** Levitation forces of single-NdFeB PM cylinder and a NdFeB PMG in ZFC at 77 K [24].

single-NdFeB PM cylinder (its surface magnetic flux density was 0.38 T), respectively [24]. Clearly, the levitation forces of the NdFeB PMG were much higher than that of the single-NdFeB PM cylinder.

Table 6.1 lists the levitation forces at different position above the surface of the magnets. $F_{LG}$ is the levitation force over the PMG magnet and $F_{LS}$ is the levitation force over the single-NdFeB PM cylinder. The levitation force $F_{LG}$ between YBCO ($\phi$ 30 × 17 mm thickness) and the PMG is 25 N, but the levitation force $F_{LS}$ between the same YBCO and the single-NdFeB PM cylinder is 0.1 N when the levitation gap is 25 mm. It is clear that the levitation force over the PMG is much larger than that over a single-NdFeB PM cylinder. It is interesting that there is still a large levitation force at higher levitation gaps.

Another melt-textured HTS YBCO bulk sample made in the Beijing General Research Institute for Nonferrous Metals was subsequently measured. The YBCO bulk sample had a diameter of 30 mm and a thickness of 14 mm. In this case, there was a 103.4-N (14.6 N/cm$^2$) levitation force at the 5-mm gap between the YBCO sample and the NdFeB PMG. At that time, the levitation force was relatively high at the same.

**Tab. 6.1:** Levitation force vs. levitation gap over the single-NdFeB PM [$F_{LS}$(N)] and the PMG [$F_{LG}$(N)] [24].

| Levitation gap (mm) | $F_{LG}$(N) | $F_{LS}$(N) |
|---|---|---|
| 5.0 | 94.0 | 18.4 |
| 10.0 | 68.9 | 7.0 |
| 15.0 | 47.9 | 2.8 |
| 20.0 | 34.4 | 0.5 |
| 25.0 | 25.0 | 0.1 |
| 30.0 | 18.1 | 0.0 |

### 6.3.2 Levitation forces of a four-HTS-bulk array [25]

The different arrays of four YBCO bulks are shown in Fig. 6.2. The levitation forces over PMG for arrays 1, 2, 3, and 4 at 5-mm levitation gap were 267.0, 221.1, 203.8, and 279.5 N, respectively (Fig. 6.3a). The levitation forces over the PMG for arrays 1, 2, 3, and 4 at 15 mm levitation gap were 139.7, 119.6, 126.3, and 152.2 N, respectively (Fig. 6.3a). Figure 6.3 shows that the levitation force of array 4 was the largest and that of array 3 was the smallest. Table 6.2 lists the levitation forces of different arrays.

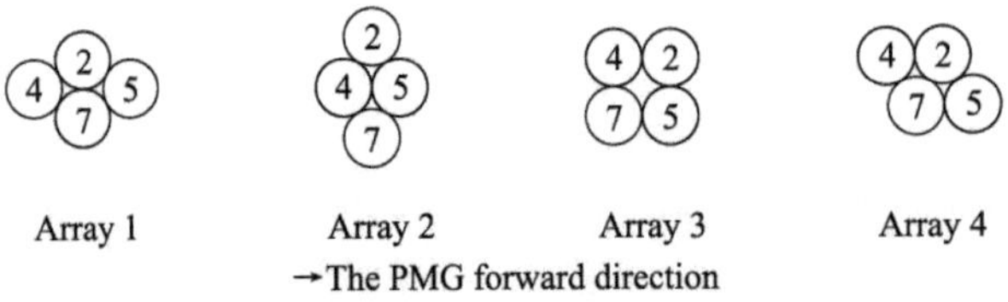


**Fig. 6.2:** Different array of four-block YBCO HTS bulks [25].

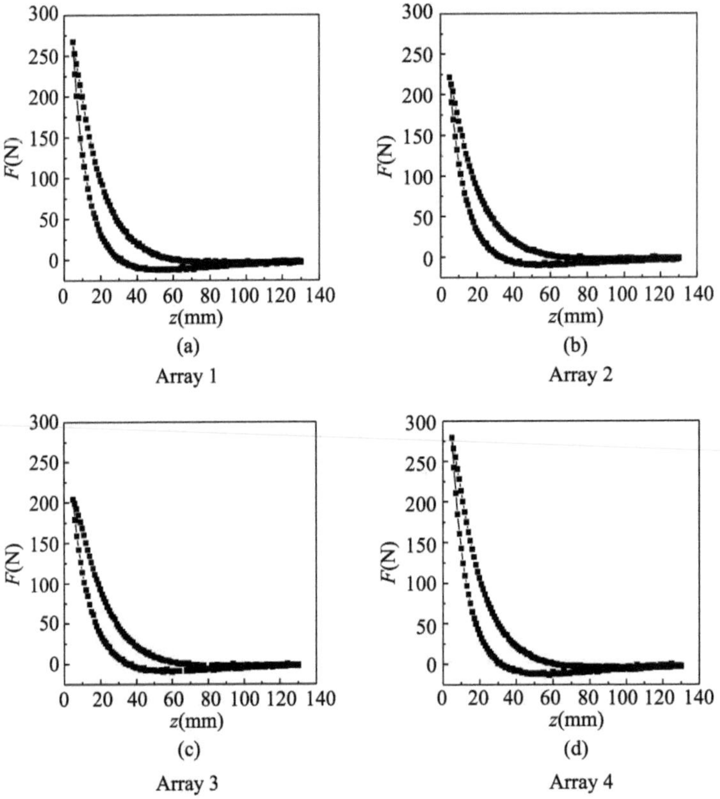


**Fig. 6.3:** The levitation forces dependence on the gap of four different arrays [25].

Tab. 6.2: Levitation forces of different arrays of four YBCO [25].

| Gap/mm | Levitation forces (N) | | | |
|---|---|---|---|---|
| | Array 1 | Array 2 | Array 3 | Array 4 |
| 5 | 267.0 | 221.1 | 203.8 | 279.5 |
| 10 | 200.0 | 167.5 | 168.4 | 213.4 |
| 15 | 139.7 | 119.6 | 126.3 | 152.2 |
| 20 | 96.7 | 84.2 | 92.8 | 106.2 |
| 25 | 67.0 | 59.3 | 66.0 | 73.7 |
| 30 | 45.0 | 41.2 | 46.9 | 49.8 |
| 35 | 28.7 | 28.7 | 32.5 | 35.4 |
| 40 | 19.1 | 21.1 | 22.1 | 22.0 |

With the increase of the levitation gap, more and more YBCO bulks were in a lower magnetic field region, and the effective areas of the different arrays were approximately the same. At 5-mm gap, the effective area of array 4 was maximum, arrays 1 and 2 had smaller effect area, and array 3 had the smallest. Those effective areas were in accordance with the levitation forces at a 5-mm gap, that is, the levitation force of array 4 was maximal. At 40-mm gap, the levitation forces were very close and consistent with the almost same effective areas for different arrays.

### 6.3.3 Levitation forces of seven-HTS bulks array [25]

The levitation forces of seven-block HTS YBCO over a PMG were investigated. The melt-textured YBCO bulks were made in the Beijing General Research Institute for Nonferrous Metals. The samples shared a diameter of 30 mm and a thickness of 14 mm. The YBCO bulks are arrayed in Fig. 6.4. In this measurement, the HTS YBCO bulks were cooled into superconducting state in a zero magnetic field. The measurement results for levitation forces of each YBCO sample and the total levitation forces of a seven-block YBCO sample are shown in Figs. 6.5 and Fig. 6.6, respectively. The levitation forces of each sample were similar to each other. The best measurement results of total levitation forces of seven HTS YBCO bulks are shown in Fig. 6.7.

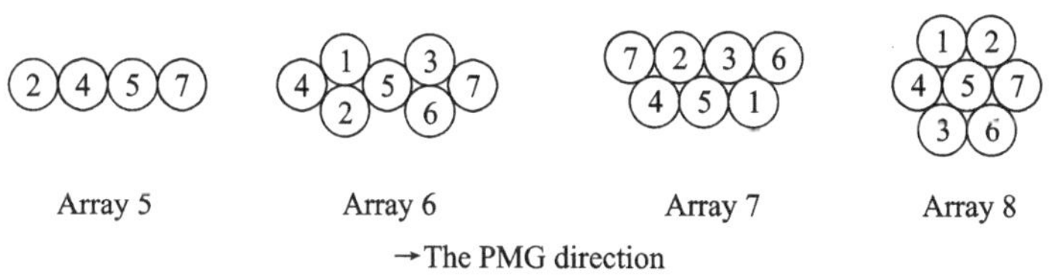


Fig. 6.4: Different array of multi-block YBCO HTS bulks [25].

Table 6.3 lists the levitation forces (N) of different seven-block YBCO samples. The levitation force of 7 YBCO in array 8 over PMG was 264.1 N at 10-mm gap and the total levitation force of 7-block-alone YBCO was 500.6 N at 10-mm gap [24]. The total levitation force of array 8 was smaller than the sum of the levitation forces of the seven single pieces of YBCO, since four pieces in array 8 were not placed at the center position along the PMG lengthways direction. In order to get the optimum levitation force of the system, selecting array 8 was more appropriate.

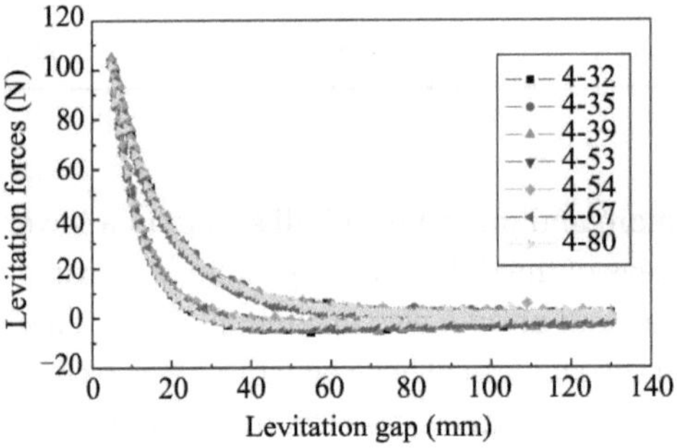


**Fig. 6.5:** Levitation forces of different seven-block YBCO [25].

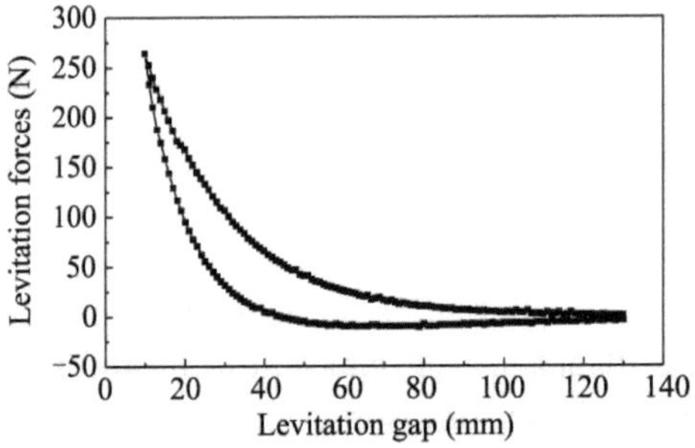


**Fig. 6.6:** Total levitation forces of seven-block YBCO in array 8 [25].

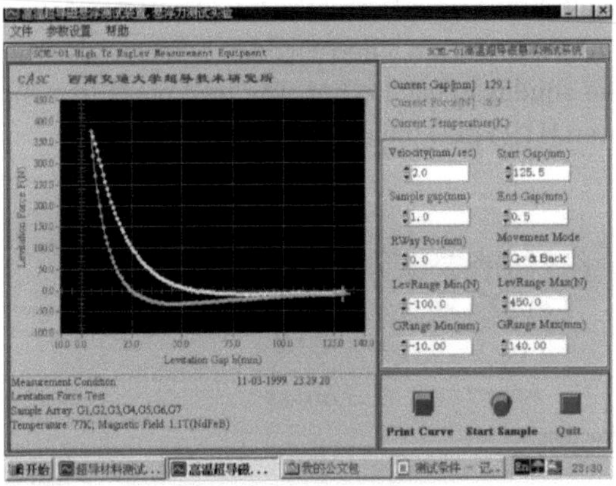


**Fig. 6.7:** The best total levitation forces of the seven-block YBCO in array 8 [24].

Tab. 6.3: Levitation force (N) of different seven-block YBCO [24].

| Sample | Gap (mm) | | | | | |
|---|---|---|---|---|---|---|
| | 5 | 10 | 15 | 20 | 25 | 30 |
| 4-32 | 103.3 | 73.8 | 52.9 | 35.2 | 24.9 | 18.2 |
| 4-35 | 101.4 | 71.0 | 48.6 | 33.3 | 23.9 | 17.2 |
| 4-39 | 101.6 | 69.6 | 47.7 | 33.8 | 23.9 | 17.2 |
| 4-53 | 103.3 | 72.9 | 48.6 | 34.3 | 23.9 | 17.2 |
| 4-54 | 103.3 | 72.9 | 51.0 | 34.3 | 24.9 | 18.2 |
| 4-67 | 101.6 | 69.8 | 47.7 | 32.6 | 23.0 | 19.2 |
| 4-80 | 101.6 | 70.6 | 49.9 | 35.3 | 24.9 | 18.2 |
| Array* | – | 264.1 | 206.7 | 167.5 | 133.0 | 106.2 |

*Array is total magnetic levitation of a seven-block YBCO.

### 6.3.4 Comparison of levitation forces per unit length [26]

The levitation forces of different arrays of multi-block HTS YBCO were investigated over the specially designed NdFeB PMG. Different arrays of multi-block YBCO HTS are shown in Figs. 6.2 and 6.4. The levitation forces per unit length of different arrays are listed in Tab. 6.4. The data in Tab. 6.4 was measured at 20-mm gap between HTS YBCO and the NdFeB PMG.

Table 6.4 shows that the levitation forces of per unit length of array 8 was the largest. It merits our attention that the data were measured under given conditions, for example, the thin-wall liquid nitrogen vessel, the NdFeB PMG whose concentrating field in the center of the surface was up to 1.2 T and still 0.4 T at a position of 20 mm above it, and so forth.

The test results of different arrays of YBCO bulks show the different levitation forces between each array and the NdFeB PMG and can be used as a reference for the design of a Maglev vehicle. The measured results were different if the PMG was not the same. Because the external magnetic field for various bulks array was not completely identical, it may not be precise for the system design, but it provides a reference.

### 6.3.5 Levitation forces of multiple-seeded melt-growth YBCO bulk [27]

Song et al. [27] reported the experimental results of magnetic levitation forces on top-seeded melt-growth (TSMG), multi-seeded melt-growth (MSMG) YBCO bulks above the PM guideway in parallel and perpendicular mode between the length directions of a

Tab. 6.4: Levitation forces of per unit length (N/mm) of different arrays of YBCO bulk [26].

| Array No. | 1 | 2 | 3 | 4 | 5 | 6 | 7 | 8 |
|---|---|---|---|---|---|---|---|---|
| Levitation force (N) | 105 | 123 | 136 | 118 | 145 | 193 | 177 | 163 |
| N/mm | 1.75 | 1.50 | 2.27 | 1.97 | 1.20 | 1.44 | 1.68 | 2.32 |
| Figure of merit | 4 | 6 | 2 | 3 | 8 | 7 | 5 | 1 |

PM guideway. All the bulk samples were fabricated by IFW, Dresden, Germany [28]. Each trisected part corresponds to one growth domain with one seeding crystal. They were stacked up as one bulk array (90 mm length, 35 mm width, and 15 mm thickness). Moreover, there was a piece of MSMG bulk with the dimension (90 × 36 × 15 mm). The levitation forces were measured with the HTS Maglev measurement system (see Section 5.3).

The levitation force of the MSMG bulk and the stacked array in parallel (Fig. 6.8) and perpendicular (Fig. 6.9) mode were compared. In the parallel mode, the hysteresis loss of the MSMG bulk was clearer than that of the stacked array in Fig. 6.8, and the weak-link grain boundaries gave rise to hysteresis behavior. However, their maximum levitation forces were slightly different. The difference between them was 7.0 N with the percentage of 4.68%. However, the difference rose up to 36.1 N with the percentage of 21.40% in the perpendicular mode, but their maximum forces dropped to 204.8 and 168.7 N, respectively.

## 6.4 Levitation stiffness [29]

Levitation force and stiffness of HTS bulk over a PM guideway are key parameters which are equally important. The levitation stiffness directly represents the change of

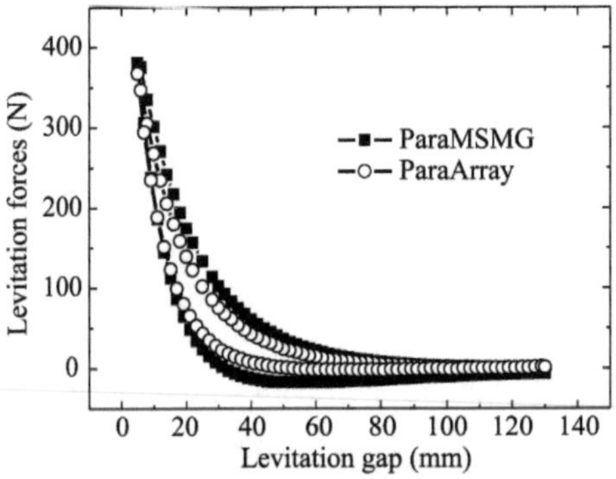


**Fig. 6.8:** Levitation forces of the MSMG bulk and the stacked array consisting of TSMS bulk S01, S02, and S03 in the parallel mode [27].

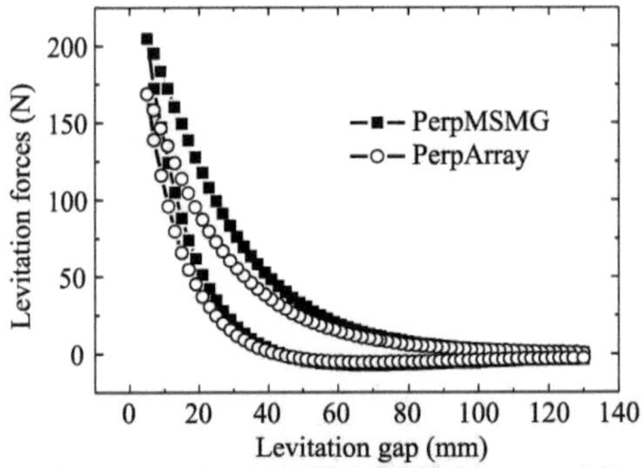


**Fig. 6.9:** Levitation forces of the MSMG bulk and the stacked array consisting of TSMS bulk S01, S02, and S03 in perpendicular mode [27].

**Tab. 6.5:** Levitation stiffness at a gap of 20 mm with the FC height of 40 mm [29].

| Displacement (mm) | Stiffness (N/mm) | Displacement (mm) | Stiffness (N/mm) |
|---|---|---|---|
| 0 | 2.29 | 50 | 0.32 |
| 5 | 2.21 | 55 | 0.37 |
| 10 | 1.76 | 60 | 0.45 |
| 15 | 1.29 | 65 | 0.55 |
| 20 | 0.98 | 70 | 0.69 |
| 25 | 0.64 | 75 | 0.74 |
| 30 | 0.48 | 80 | 0.76 |
| 35 | 0.38 | 85 | 0.71 |
| 40 | 0.35 | 90 | 0.58 |
| 45 | 0.34 | 95 | 0.38 |

the levitation height when the load varies on the suspended carrier. When the load changes in the Maglev system, bigger levitation stiffness corresponds to a smaller change of the levitation height.

Generally, levitation force and stiffness at an operation gap of 20 mm is essential for the practical operation of the vehicle "Century". Song et al. [29] measured the minor loops upon the major hysteresis loop from the gap of 22–18 mm in field cooling (FC). The rectangle-shaped single-domain melt-textured bulk ($30 \times 35.6 \times 15.5$ mm) was fabricated by IFW. According to the minor loops upon the major hysteresis loop, the levitation stiffness in both symmetrical and unsymmetrical applied field was calculated. The associated levitation stiffness in the descending and ascending processes (from 22 mm to 18 mm and back to 22 mm) of minor loops were 2.63 and 2.75 N/mm, respectively. These in the descending and ascending processes of the major loop (from 22 to 18 mm and back to 22 mm again) were 2.28 and 2.18 N/mm. For the same minor loop, the stiffness in the descending process was slightly different from the stiffness for the ascending process. The stiffness upon the descending process of the major hysteresis loop was larger than that in the ascending process of the major hysteresis loop. The levitation stiffness was measured with the HTS Maglev measurement system (see Section 5.3).

Table 6.5 lists the levitation stiffness of the rectangular bulk at a gap of 20 mm with the FC height of 40 mm. Figure 6.10 shows the levitation force versus gap for the rectangle bulk at a displacement of 5 mm.

## 6.5 Levitation forces vs. temperature [30–32]

In order to study the influence of temperature on levitation force, Jiang et al. [30] designed a test device (Fig. 6.11) using the GM refrigerator in the HTS Maglev measurement system SCML-01 (see Section 5.3), and they compiled a temperature control program.

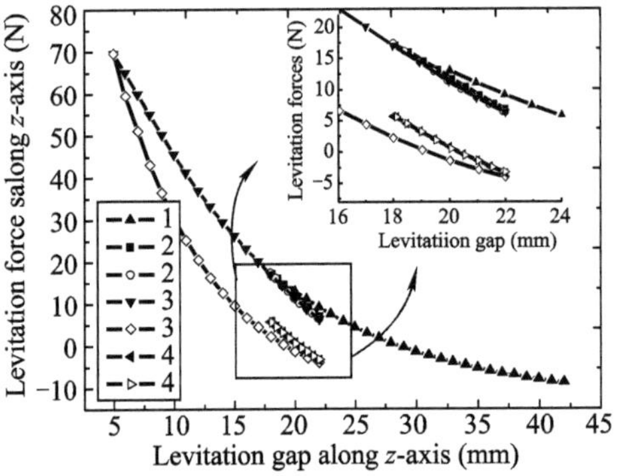


**Fig. 6.10:** Levitation forces vs gap at a displacement of 5 mm. The top right is the minor hysteresis loops conducted upon major loop from gaps of 22–18 mm [29].

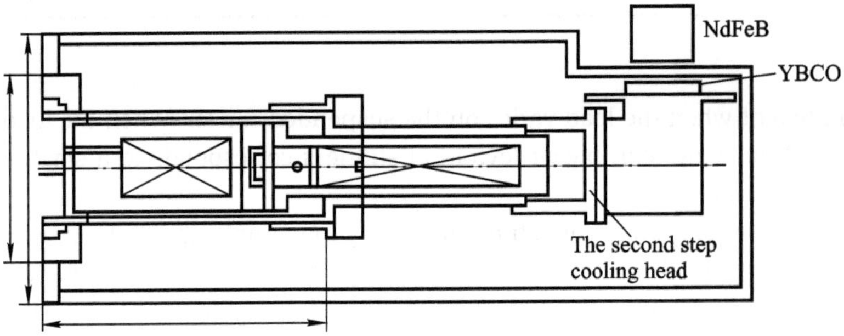


**Fig. 6.11:** Schematic drawing of the test device using the GM refrigerator for measurements of levitation forces at 38–77 K [33].

In order to measure the levitation force between the NdFeB disk and the YBCO bulk, the permanent magnet NdFeB disk of $\phi$ 30 × 15 mm was fixed on the vertical moving part of the SCML-01, and the YBCO bulk of $\phi$ 30 × 14 mm was fixed on the second step cooling head of the refrigerator. The levitation force between YBCO and NdFeB could be measured at different gaps and at different temperatures. The maximum magnetic fields were 0.42, 0.219, and 0.0535 T at the NdFeB surface, 5 mm, and 20 mm above, respectively. The minimum gap between YBCO and NdFeB was 5 mm.

Figure 6.12 shows the photograph of the test device using the G-M refrigerator at 38–77 K with the SCML-01 platform [32].

After the YBCO bulk was cooled down to 40 K by the refrigerator, the NdFeB was fixed on the vertical moving part of the SCML-01, which moved down and up. The gap between YBCO and NdFeB changed from 150 to 5 mm, the levitation force data were sampled by the computer at the same time, which was used to control the measurement system and to sample the measurement data. Figure 6.13 shows one of the measurement results at 38.4 K, which indicates the relationship between the levitation force and the gap. Table 6.6 shows the measured maximum levitation force at different temperature. The relationship between the maximum levitation force at the same (minimal) gap and temperature is shown as Fig. 6.14 [30].

**Fig. 6.12:** Photograph of the test device using the G-M refrigerator at 38–77 K in the SCML-01 platform [32].

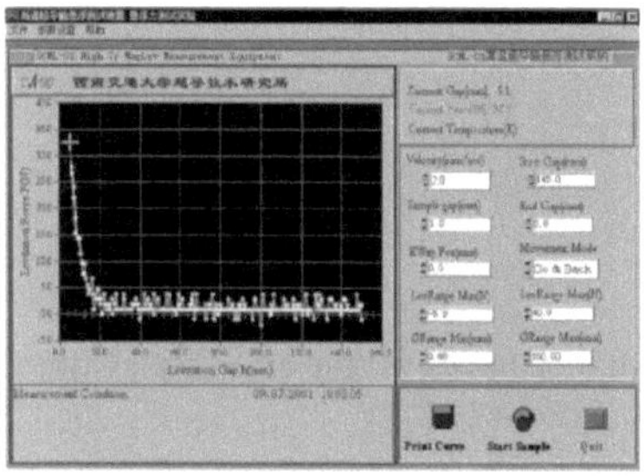

**Fig. 6.13:** Photograph of the measurement curve for levitation forces at 38.4 K [30].

**Tab. 6.6:** The maximum levitation forces at different temperatures [30].

| Levitation forces (N) | Temperatures (K) |
|---|---|
| 32.5 | 38.4 |
| 33.5 | 48.5 |
| 32.5 | 50.0 |
| 32.5 | 54.3 |
| 30.6 | 59.0 |
| 29.7 | 64.5 |
| 29.6 | 67.6 |
| 28.7 | 70.5 |
| 25.8 | 75.5 |
| 21.1 | 82.0 |
| 9.6 | 88.0 |
| 0 | 93.0 |
| 0 | 100.0 |

On the basis of the above studies, Wang et al. [31] further studied the relationship between the levitation force and the temperature. A single YBCO bulk ($\phi$ 30 × 17 mm) was fixed on the second stage cooling head of the G-M refrigerator, rather than being immerged in liquid nitrogen (77 K), whose temperature can be controlled by the

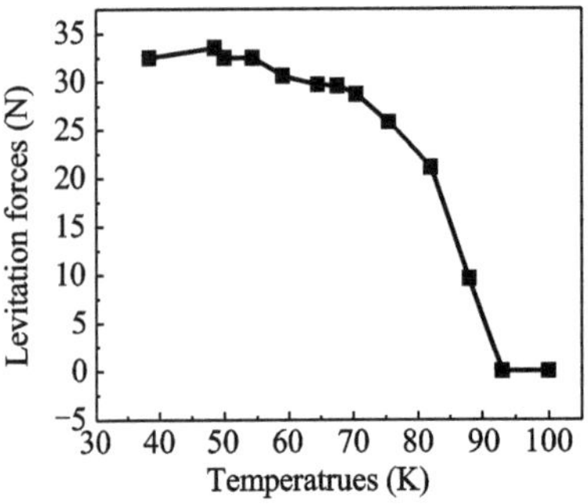


**Fig. 6.14:** Relationship between the levitation force and the temperature [30].

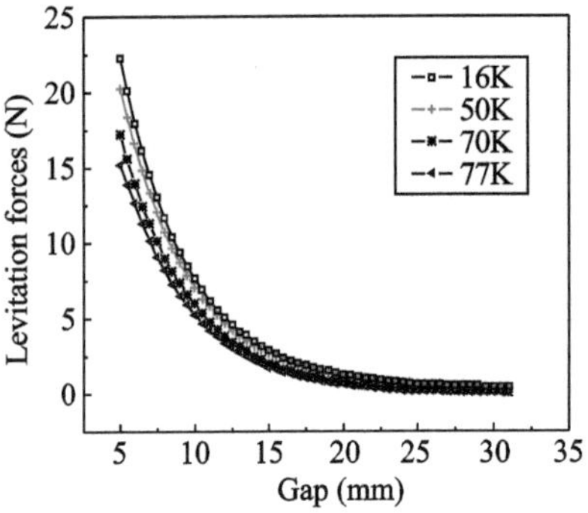


**Fig. 6.15:** Levitation forces vs. gap at the temperature of 16, 50, 70, and 77 K [31].

deep cryogenic measurement system, ranging from 77 to 16 K. A single-NdFeB PM disk ($\phi$ 30 × 15 mm) could be lifted vertically above the HTS YBCO bulk with the gap between 100 and 5 mm. Levitation forces of YBCO bulk at different gaps were obtained at different temperatures under the control of the HTS Maglev measurement system (see Section 5.3).

Figure 6.15 shows the levitation forces of a single YBCO bulk under a single permanent magnet disk at 16, 50, 70, and 77 K, respectively. It is clear that temperature has a notable effect on the levitation force at small levitation gaps, for example, 22.3 N at 16 K and 15.2 N at 77 K for the same gap of 5 mm. However, the effect of temperature on the levitation force is not obvious at larger levitation gaps. Therefore in practical applications, if the levitation gap is larger, it is not economic to have a cryogenic temperature lower than 77 K.

Song et al. [32] reported the temperature dependence of the HTS YBCO bulk at different levitation gaps. Results showed that the levitation force at lower temperature was larger than that at higher temperature near the critical temperature of the YBCO. Figure 6.16 shows that the levitation force at a gap of 5 mm increased with decreasing temperature. The levitation performance of the YBCO below 35 K was stable with maximum levitation force of 22.3 N (16 K). When the gap was 10 mm, improvement of the levitation performance versus the temperature was not notable.

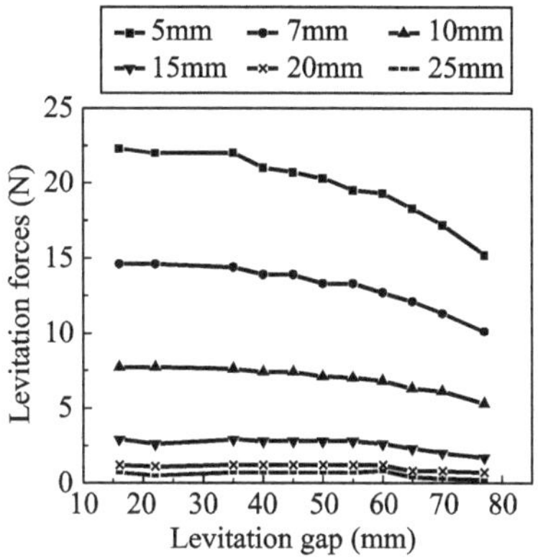


**Fig. 6.16:** Temperature dependence of levitation forces at the gaps of 5, 7, 10, 15, 20, and 25 mm [32].

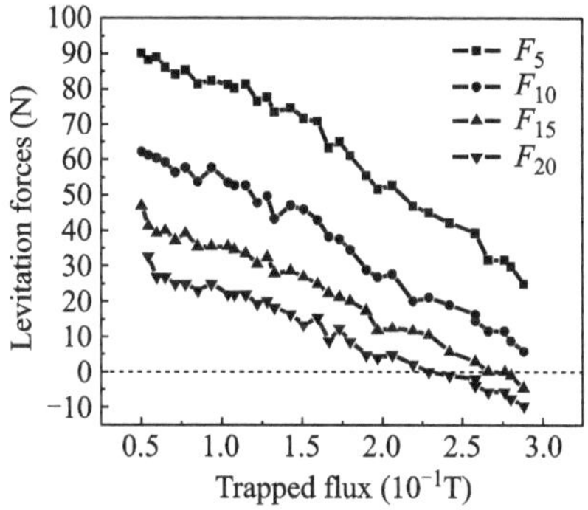


**Fig. 6.17:** Levitation forces vs. trapped flux at different gaps [31].

## 6.6 Levitation forces vs. trapped flux [31]

In the experiment of the dependence of levitation forces on the trapped flux, the same YBCO bulk was used. First, YBCO bulk was cooled by liquid nitrogen in applied field, which was provided by a permanent magnet. Trapped flux of YBCO bulk was controlled by changing the gap between the sample and the permanent magnet. Secondly, the levitation forces of the YBCO bulk above the PM guideway were measured by the HTS Maglev measurement system (see Fig. 5.7). The central surface concentrating magnetic field of the PM guideway was 1.2 T and the field distribution in space was symmetrical about its central vertical plane.

Figure 6.17 gives the experimental results of levitation forces of a single YBCO bulk with different trapped flux above the PM guideway. Figure 6.17 shows that the levitation force decreases almost linearly with increasing trapped flux. It is well known that the trapped flux ensures stable levitation of YBCO bulk in applied field, but it weakens the levitation force simultaneously. This Phenomenon can be seen clearly from Fig. 6.17. Thus, an optimum trapped flux is very important in practical levitation systems. This Phenomenon allows YBCO bulk to provide both a larger levitation force and a larger guidance force.

## 6.7 Guidance forces of HTS bulks over PMG

### 6.7.1 Guidance forces of HTS bulks over single PMG [34]

The guidance forces refer to the forces that compels the carrier to return to its original location of magnetic balance, after a lateral displacement. These forces are from the pinning forces caused by trapped flux. Maglev guidance forces are measured with the relative translational motion between a Maglev carrier and the external magnetic field (PMG or PM). The moving element of the relative motion can be the Maglev carrier or also the external magnetic field. The test process of guidance forces is similar to the levitation force measurements.

The difference is that the PMG moves in the horizontal plane while the guidance force (lateral force) is measured.

In order to test guidance forces over a PM guideway, Wang et al. [34] designed and produced two types of special test apparatus. One was the special test apparatus of the HTS Maglev measurement system SCML-01 (see Fig. 5.7). The PMG was composed of NdFeB PMs and iron plate whose concentrating field in the center the surface was up to 1.2 T and remains 0.4 T at the position of 20 mm above it. The length of the PMG was 920 mm (see Section 6.14).

A sketch of the apparatus measuring the guidance forces is shown in Fig. 6.18. The guidance forces of the YBCO bulk above the PMG were measured with the SCML-01. A circular liquid nitrogen vessel (see Section 5.2.2) with the HTS bulks fixed inside was hung over the PMG. The vessel could only swing freely around the point O in the $x$-$z$ plane (see Fig. 6.18a). The bottom center of the liquid nitrogen container was taken as the origin of the Cartesian coordinate.

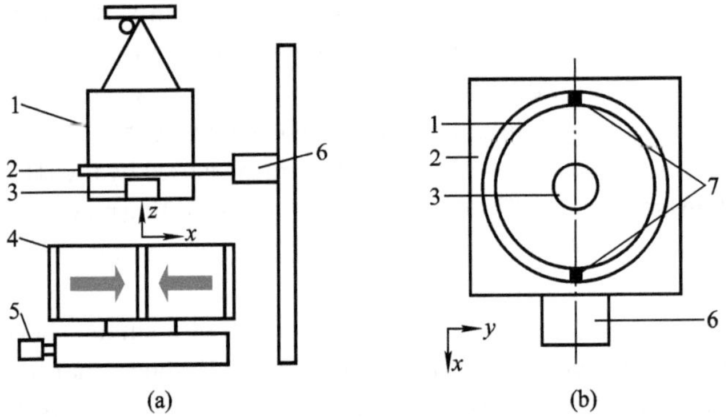


**Fig. 6.18:** (a) Sketch of the guidance force test apparatus of the HTS bulk over the single PM guideway. (b) Overview of the epoxy ring and the cryogenic vessel [34]. (1) Liquid nitrogen vessel. (2) Epoxy ring. (3) Single HTS bulk. (4) PM guideway. (5) Step motor. (6) Force transducer. (7) Epoxy wedges.

During the experiment, when the PMG was moving along the $x$-axis, the guidance forces acting on the HTS bulks might make the vessel swing in the $x$-$z$ plane. We rang the vessel by a rigid epoxy ring with a little larger diameter than that of the vessel. Two epoxy wedges inserted between the ring and the vessel, as shown in Fig. 6.18b. The wedges prevented the motion of the vessel and the HTS bulks remained motionless in the experiment. The guidance forces acting on the HTSCs were transferred to the force transducer connected to the epoxy ring. The precision of the transducer was ±0.1 N.

The PMG was made of NdFeB and iron yokes, as shown in Fig. 6.18a. It was controlled by a PC-based measurement system (see Sections 5.3 and 5.4); the velocity was 1 mm/s. The displacement precision of the PMG was ±0.1 mm.

The YBCO bulks were about 30 mm in diameter and 18 mm in thickness. The HTS bulks were arranged into five arrays (Fig. 6.19) [34]. The $c$-axis of each HTS bulk was perpendicular to the top surface of the guideway. The HTS bulks were all field cooled. The distance between the bottom surface of the HTS bulk and the top surface of the PMG was 10 mm in both the cooling procedure and the measurements.

The guidance forces were sampled automatically as a function of lateral displacement of the PMG during the experiment. First, the PMG moved in the positive $x$-direction until the maximum lateral displacement (MLD) was achieved. Then, a complete cycle was performed by moving the PMG in the negative $x$-direction for a displacement of −2 MLD and reversing. Finally, the PMG returned to its original position shown in Fig. 6.18. For each HTS bulk C or HTS bulks array, we measured the guidance forces over lateral displacements with different MLDs, i.e. 0.5, 1, 2, 3, 4, 5, 7, 10, 12, 15, 17, and 20 mm. After all these MLDs were finished, the HTS bulk or the

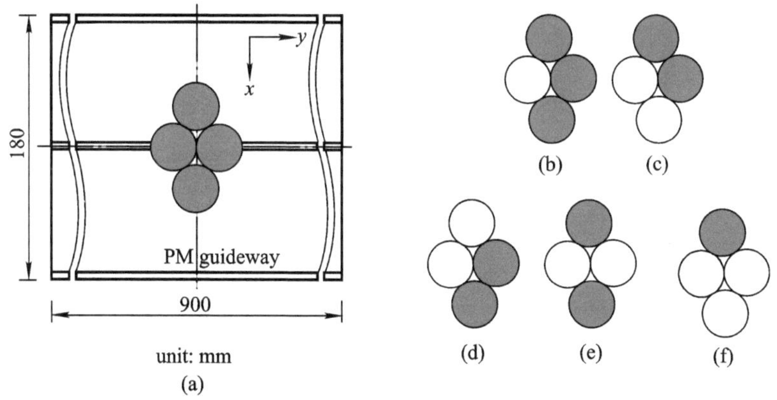


**Fig. 6.19:** Arrangements of the HTS bulks. Of the four circles in each figure, those black and numbered are HTSCs, with the rest epoxy bulks [34]. (a) Arrays 2, (b) 2, (c) 4, (d) 4, and (e) 5 and (f) an example arrangement for a single HTS bulk measurement.

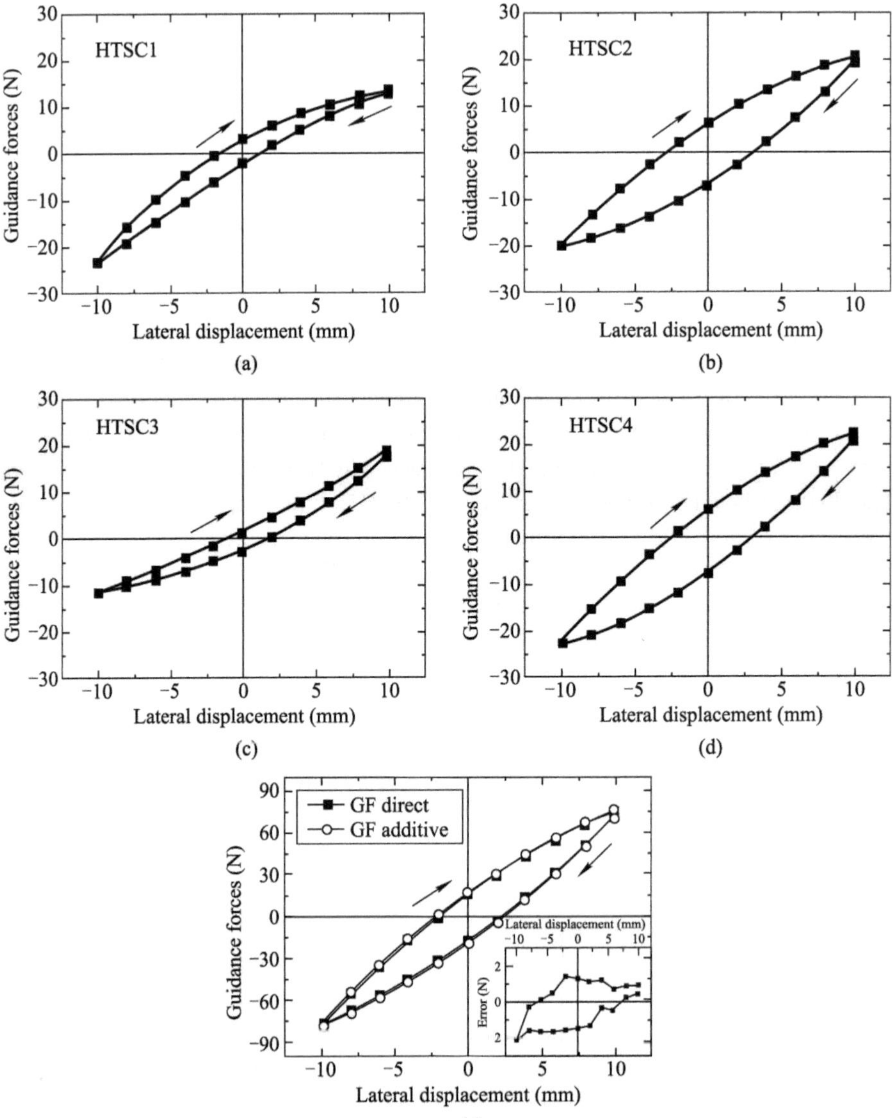


**Fig. 6.20:** Guidance force of array 1 when the maximum lateral displacement is 10 mm. (a) HTS bulks 1, (b) 2, (c) 3, and (d) 4 and (e) the results of the additive (epoxy bulks) method and the direct method. The inset shows the difference between the two methods as a function of lateral displacement of the single PMG [34].

HTS bulks array were quenched, then the measurements with another HTS bulk or HTS bulks array were started.

We used rigid epoxy bulks to replace those HTS bulks that should be absent while measuring the guidance forces of the HTS bulks arrays (see Fig. 6.20b–e). Because the epoxy bulk had the same geometry as the HTS bulk, the position of each HTS bulk was kept the same in every measurement so that the measured results of the additive (epoxy bulks) method could be reasonably compared with that of the direct method.

The guidance forces of each HTS bulk in array 1 are drawn as a function of lateral displacement in Fig. 6.20a–d [34]. Figure 6.20e shows the results of the additive (epoxy bulks) method and the direct method. With respect to each lateral displacement of the PMG where the guidance forces are sampled, we subtract $GF_{additive}$ from $GF_{direct}$ and the difference, denoted by d, is shown as a function of lateral displacement in the inset of Fig. 6.20e.

### 6.7.2 Guidance forces of YBCO bulks over two parallel PMG [31]

In order to test the guidance forces over the two parallel PM guideway, Wang et al. [31] designed and produced another special test apparatus. It was the special apparatus composed of two cryogenic vessels over two parallel PMGs. This test apparatus was equivalent to the frame of a quarter of the "Century" vehicle on the PMG. It could hold two onboard HTS Maglev equipments (see Section 6.16) in this framework.

A sketch of the apparatus measuring guidance forces is shown in Fig. 6.21 [31]. The guidance forces of the YBCO bulk over the PMG were measured using the measuring equipment of the guidance forces for the entire HTS Maglev vehicle (see Section 5.4). Here, the measuring equipment of the guidance forces was split into two parts, and they were placed on both sides of the PMG. The bottom center of the liquid nitrogen vessel was aligned to the center line along the PMG length. During the experiment, the onboard HTS Maglev equipments were moving component relative to the PMG, and the guiding force was measured directly.

The guidance forces (lateral forces returning to the initial position) depend on the field-cooling height (FCH). Guidance forces for a single YBCO bulk over two parallel PMG were measured by the guidance force test apparatus (Fig. 6.21) in two FCH cases.

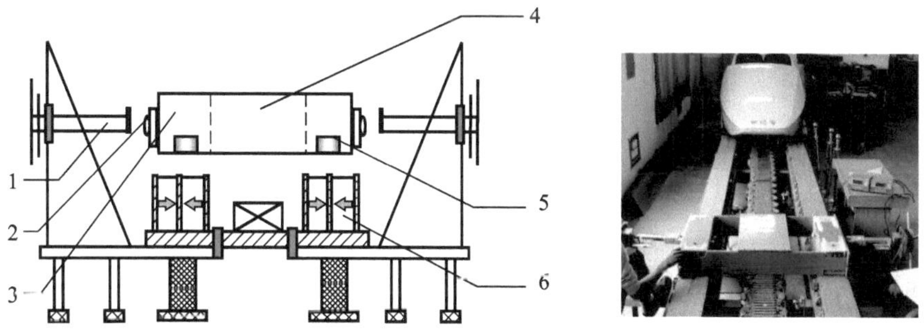


**Fig. 6.21:** Schematic diagram and photograph of guidance force test apparatus of YBCO arrays on two parallel PM guideways [31]. (1) Horizontal propulsion arm. (2) Force transducer. (3) Liquid nitrogen vessel. (4) Rigid link. (5) HTS arrays. (6) PMG.

In the first FCH case, YBCO bulk was cooled by liquid nitrogen at 50 mm above the center of the PMG, where the applied field was 0.1575 T. Guidance forces of the single YBCO bulk were measured at two different heights, i.e. 10 and 20 mm. In the second FCH case, YBCO bulk was cooled at 8 mm above the center of the PMG, where the applied field was 0.6874 T, and then the guidance forces of the YBCO bulk were measured at two different heights mentioned above.

Figure 6.22 shows the guidance forces of a single YBCO bulk above the PMG in the first FC case, whose trapped flux was 0.092 T. The arrows indicate the moving direction of the YBCO bulk during the experiments. As shown in Fig. 6.22, the guidance forces exhibit strong hysteresis; moreover, the hysteresis increases with the decrease of the measurement height (MH). It is obvious that guidance force at MH 10 mm is much larger than that at MH 20 mm, especially near the center of the guideway. This is because guidance forces do not depend only on trapped flux in the YBCO bulk, but also depend on the applied field. Although the trapped flux was the same, the applied field at MH 10 mm was stronger than that at MH 20 mm. Another possible reason is that the strong magnetic field at MH 10 mm increases the amount of trapped flux. Especially in the center of YBCO bulk, the pinning effect was not strong. In addition, it implied that YBCO bulk had better lateral stability at a lower position above the PM guideway.

Figure 6.23 shows the guidance forces for the same sample above the same normal permanent magnet (NPM) guideway in the second FC case, whose trapped flux was 0.184 T. Although its tendency is similar to that in the first FC case, there are also differences between them. For example, the change of the guidance forces is monotonous in Fig. 6.23; however, it is not monotonous in Fig. 6.22, especially at MH 10 mm. There was no difference except for the trapped flux in YBCO bulk for the two groups of experimental results. Thus, trapped flux is very important for investigation of the guidance forces.

The guidance forces of two YBCO bulk arrays above two parallel PMG were measured. Each array included 9 YBCO bulks arranged in a line. Each sample was 30 mm in diameter and 17 mm in thickness, and was fixed at the bottom of the rectangular vessel (Fig. 6.21).

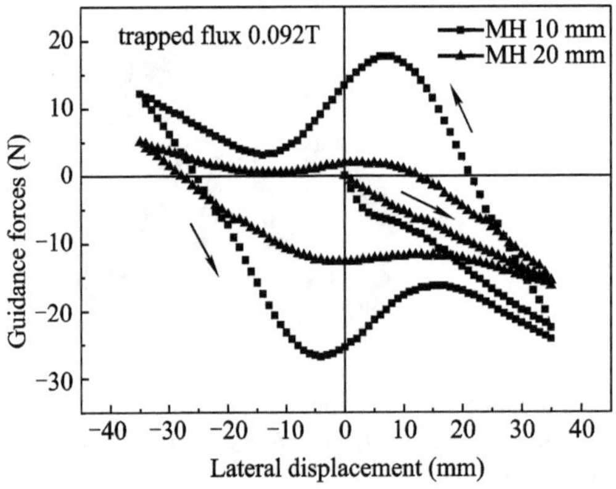


**Fig. 6.22:** Guidance forces vs. lateral displacement in 0.092 T trapped flux [31].

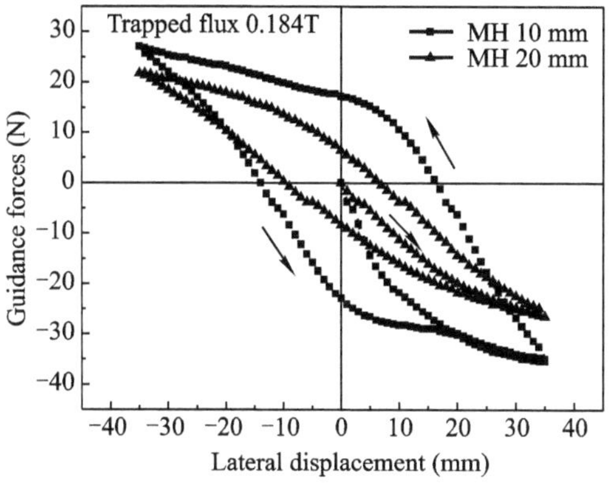


**Fig. 6.23:** Guidance forces vs. lateral displacement in 0.184 T trapped flux [31].

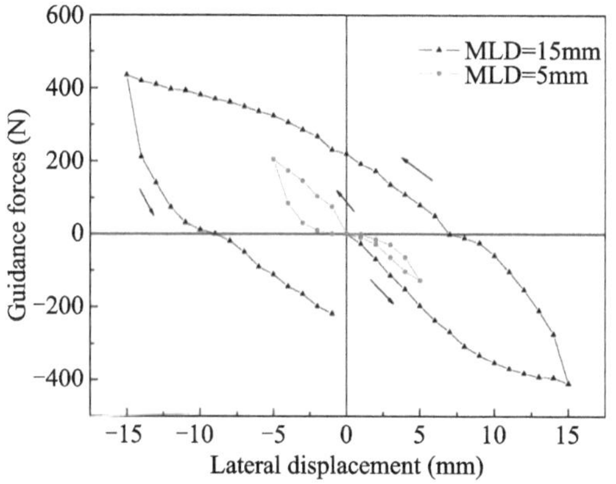


**Fig. 6.24:** Guidance forces of YBCO bulk arrays above two parallel PMG with different maximum lateral displacement [31].

Figure 6.24 shows guidance forces on the YBCO bulk arrays above two parallel PMG when the maximum lateral displacement (MLD) was 5 and 15 mm, respectively. When MLD was 5 mm, even though guidance forces were small, the place corresponding to zero guidance force was always the original position, which implied that the YBCO bulk arrays can return to its original place freely after they have been displaced, i.e. the center of the PMG. Compared with the guidance force at MLD = 5 mm, when MLD was 15 mm, the guidance forces were larger and increase faster with the increase of the lateral displacement, but the guidance forces decreased faster to zero before it returned to its original position. This is not beneficial to lateral stability of the levitation system. Thus, the effective lateral displacement, which is defined as the maximum lateral displacement corresponding to the levitation system being able to return to its original position after being displaced, is very important for the design of practical levitation systems.

On the basis of the experimental research, Wang et al. [35, 36] also completed a numerical simulation and programmed a calculation of the guidance forces.

### 6.7.3 Guidance forces of multiple seeded melt growth YBCO [27]

The guidance forces of the multiple seeded melt growth (MSMG) YBCO bulk over the PMG were measured by the HTS Maglev measurement system SCML-01 (see Fig. 5.7). First, the YBCO bulk was cooled rightly above the center of the PMG and the field cooling height (FCH) was 20 mm. Then, it moved along the *x*-axis from a maximum lateral displacement (+20 mm) and finally returned to the start point of this lateral movement (similar to Fig. 6.18). The speed was 2 mm/s in the cases of both vertical and horizontal movement, so the interaction between PMG and MSMG might be regarded as quasi-static.

The guidance forces of the MSMG bulk and the array of three SMG bulk were compared in both parallel and perpendicular modes (Figs. 6.25 and 6.26). Their maximum guidance forces were 70.1 and 69.3 N in the parallel mode, while 56.3 and 56.2 N in the perpendicular mode. The difference between them was 13.8 and 13.1 N in the two different modes. Although both the placement and movement direction of the MSMG bulks were extremely different in the parallel and perpendicular modes, their maximum guidance forces almost did not change. This phenomenon seems to be controversial with the conclusion that the guidance forces are dependent on the trapped field of the HTS bulk, because the trapped fields in the different modes are significantly different. In fact, the guidance forces are the total pinning forces determined by both the trapped field and the applied field during the magnetic interaction process.

### 6.7.4 Influence of two tilted PMG on the guidance forces [37]

In order to measure the lateral guidance forces of HTS Maglev equipment over tilted PMGs, the Maglev measurement device of tilted PMGs was manufactured (Figs. 6.27

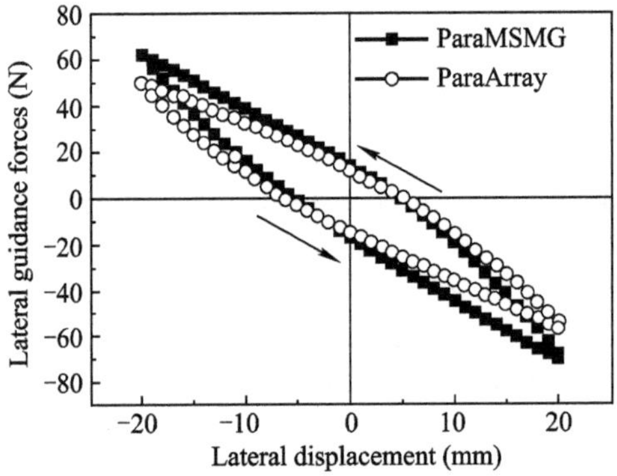


**Fig. 6.25:** Guidance forces of the MSMG bulk and the array of three SMG bulk in the parallel mode [27].

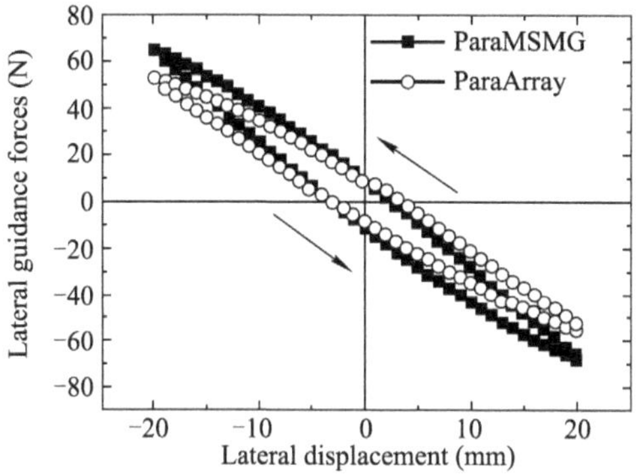


**Fig. 6.26:** Guidance forces of the MSMG bulk and the array of three SMG bulk in the perpendicular mode [27].

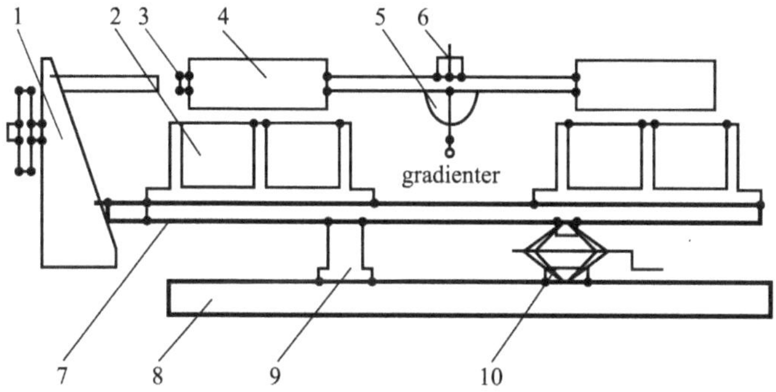


**Fig. 6.27:** Schematic diagram of the HTS Maglev measuring system of tilted PMGs [37]. (1) Propulsion. (2) NdFeB PMGs. (3) Sensor. (4) Rectangular-shaped liquid nitrogen vessels. (5) Gradienter. (6) Load. (7) Beam. (8) Support. (9) Fulcrum. (10) Jack.

and 6.28). The tilted PMGs were fixed in the SCML-01. The entire measuring equipment included a pair of onboard HTS Maglev equipments (see Section 6.24) and PMGs which can be tilted. Experimental results showed that tilted guideways helped to improve the guidance forces.

## 6.8 Guidance forces stiffness

Wang et al. [38] also investigated the cooling height dependence of the lateral magnetic stiffness. Defined as the slope of the line connecting the two acute corners of the hysteresis loop, the concept of effective stiffness was adopted to study the lateral magnetic stiffness of the HTS bulks in the lateral motion. Results are shown in Fig. 6.29. It is clear from the figure that larger lateral displacement leads to less effective stiffness. We could also increase the effective stiffness by field cooling the HTS bulks at lower heights.

Fig. 6.28: Photograph of the measuring system of tilted PMGs [37].

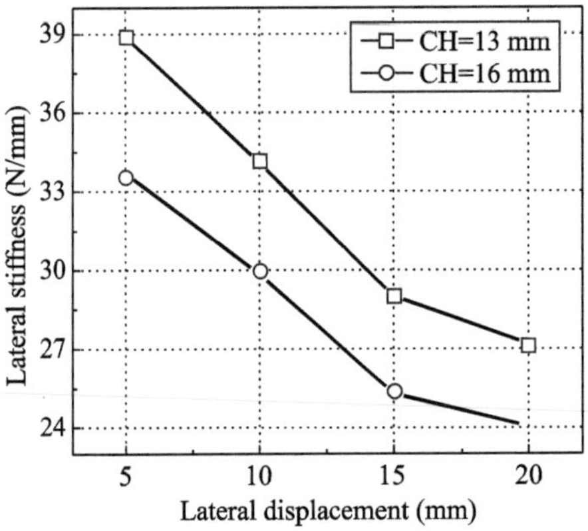


Fig. 6.29: Cooling height and lateral displacement dependence of effective stiffness [38].

## 6.9 Influence of HTS bulk geometry on the properties

### 6.9.1 Influence of HTS bulk shape on the properties

Influence of HTS bulk shape on the levitation properties were investigated experimentally by Zhu et al. [39] and Ren et al. [40] The trapped flux and levitation forces of all HTS bulk shapes were measured at 77 K. The levitation forces of all shapes were measured using the HTS Maglev measurement system SCMI-01 (see Fig. 5.7).

**Tab. 6.7:** Influence of HTS bulk shape on the levitation forces.

| **Sample** | **Cylinder 1** | **Hexagon 1** | **Square** | **Cylinder 2** |
|---|---|---|---|---|
| Shape | 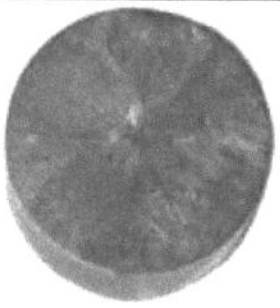 |  |  |  |
| Weight (g) | 45.35 | 38.89 | 29.37 | 23.3 |
| Diameter (mm) | 29.5 | 29.5*<br>27*** | 22.0** | 22.0 |
| Thickness (mm) | 10.8 | 11 | – | – |
| Area ($cm^2$) | 6.83 | 5.65 | 4.84 | 3.80 |
| Trapped flux (T) | 0.168 | 0.163 | 1.39 | 1.04 |
| $f_0$ ($N/cm^2$) | 10.10 | 8.67 | 6.53 | 3.00 |
| $F_5$ (N) | 69.0 | 52.7 | 31.6 | 11.4 |
| $F_{10}$ (N) | 52.7 | 40.1 | 23.9 | 9.5 |
| $F_{15}$ (N) | 22.0 | 21.1 | 18.2 | 7.6 |
| $F_{20}$ (N) | 15.3 | 16.3 | 14.3 | 6.7 |

*Diagonal length, **side length, ***side to side, $F_{15}$ (*N*) is the levitation force of gap of 15 mm.

The research results are listed in Tab. 6.7. It is well known that the magnetic levitation forces are related to the critical current density $J_c$, the diameter of screening current $d$ and magnetic field gradient $dB/dz$. These measurement results also revealed the physical essence of the HTS Maglev. Both diameter of cylinder 1 and the diagonal length of hexagon 1 were equal, and the levitation forces at a gap of 15 mm [$F_{15}$ (N)] on cylinder 1 and hexagon 1 were 22.0 and 21.1 N, respectively. Both the diameter of cylinder 2 and the side length of the square were equal, and the levitation forces at a gap of 15 mm [$F_{15}$ (N)] on cylinder 1 and hexagon 1 were 18.2 and 7.6 N, respectively. The two groups of data above indicated that the path size of the ring shielding current depended on the levitation forces on the different shaped bulks, for the same conditions of the critical current density $J_c$ and magnetic field gradient $dB/dz$. In order to fully utilize the magnetic field of the PMG and not allow the magnetic field energy leakage, the HTS bulks should generally be arranged to be gapless. Thus, the hexagon and square were thought to be the best shapes, but experimental results showed that the levitation forces on the hexagon and square HTS bulks over the PMG did not increase, or it even was smaller sometimes. The experimental results implied that a cylinder is better than a hexagon or square, because the cylinder has a longer shielding current path than the hexagon and square. The longer the shielding current path, the greater is the levitation force.

Song et al. [41] proved experimentally the correctness of this choice. He measured the levitation forces of cylindrical and rectangular shaped HTS YBCO bulks over the PMG (see Section 6.14) using the HTS Maglev measurement system SCMI-01 (see

Section 5.3). The average force densities (AFD) were calculated and compared. The melt-textured single-domain YBCO cylinder bulk Beijing 207 ($17 \times \phi 30$ mm) was fabricated by the Beijing General Research Institute for Nonferrous Metals. Another rectangular bulk TG 208-7 ($30 \times 35.6 \times 15.5$ mm) was made by IFW. In all the experiments, the *c*-axis of the HTS bulk was oriented perpendicular to the surface of the PMG.

The actual vehicle "Century" was designed based on the above measurement results. For the full-scale HTS Maglev vehicle "Century", 43 pieces of cylindrical YBCO bulks were used in each rectangular-shaped on-board the HTS Maglev equipment (see Section 6.15) in ASCLab. The inside size of the liquid nitrogen vessel equipment was 102 mm wide, 470 mm long, and 168 mm high. The levitation forces of two HTS Maglev vessels composed of rectangular and cylindrical bulks were calculated. The results indicated that an asymmetric field applied about the axes of bulks had more of an effect on the average force density of the rectangular bulk than on the cylindrical bulk. Particularly for the full-scale HTS Maglev vehicle "Century", the total levitation force was increased from 1092 N to 1282 N at a gap of 20 mm, if a rectangular bulk was employed rather than cylindrical bulk. Arranging rectangular bulks rather than cylindrical bulks in the same vessel equipment increased the number of YBCO bulks, thus strengthened the levitation force to 1320 N at a gap of 20 mm [41]. These measured data had great significance for practical application. The conclusion of this experiment study has also been used in subsequent development of the HTS Maglev vehicle.

### 6.9.2 Influence of HTS bulk size on the properties

Ren et al. [42] researched the influence of the size of the HTS bulks on the levitation forces. Three top-seeded MTG YBCO bulks were used in the levitation force measurement. These samples were made in the Beijing General Research Institute for Nonferrous Metals. All the levitation forces were measured at 77 K by the HTS Maglev measurement system SCML-01 (see Fig. 5.7). Table 6.8 lists the parameters and levitation forces over the PMG in ZFC at 77 K at different gaps.

Table 6.8 lists the levitation forces of the three samples at different gaps. The diameters of samples 2 and 3 are 16.6% and 33.3% larger than the diameter of sample 1, respectively, but their levitation forces are increased 30.5% and 69.9% at a gap of 5 mm and 39.4% and 90.1% at a gap of 20 mm, respectively, than that of sample 1. This implies that the larger the sample, the more rapidly levitation forces increase for larger gaps. Figure 6.30 shows the levitation forces over the PMG of samples 1, 2, and 3 in ZFC at 77 K. As shown in Fig. 6.30, levitation forces of sample 3 are obviously larger than that of sample 2, but the difference between the thicknesses of the two samples is very small. The influence of thickness on levitation forces in this case is negligible. However, the thickness must reach a certain value (see Section 6.9.3).

Fourteen years ago, "Century" was built with cylindrical shaped HTS bulks, because that was what we had. Of course, the larger the size of YBCO HTS bulks, the

**Tab. 6.8:** Parameters and levitation forces over the PMG in ZFC at 77 K and different gaps [42].

| Sample No. | Sample 1 | Sample 2 | Sample 3 |
|---|---|---|---|
| Diameter (mm) | 3.0 | 3.5 | 4.0 |
| Weight (g) | 77.6 | 101.4 | 133.2 |
| Height (mm) | 17.5 | 18.6 | 19.1 |
| $F_5$ (N) | 103.0 | 134.5 | 175.0 |
| $F_{10}$ (N) | 72.8 | 97.0 | 129.5 |
| $F_{15}$ (N) | 50.4 | 69.2 | 92.9 |
| $F_{20}$ (N) | 35.5 | 49.5 | 67.5 |

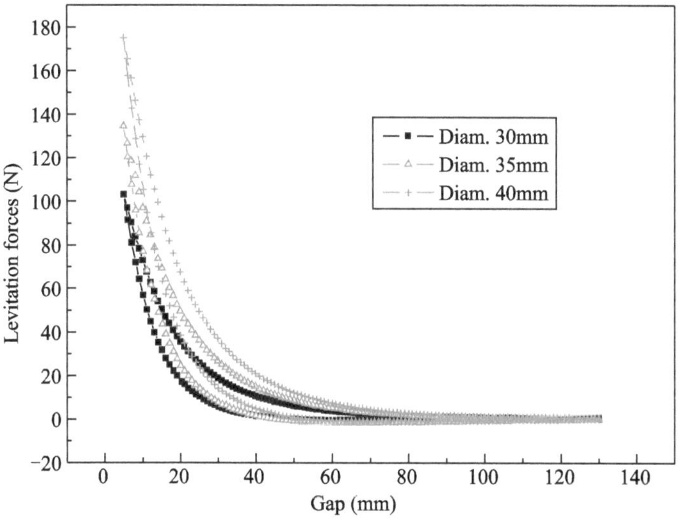


**Fig. 6.30:** Levitation forces over the NdFeB PMG of sampled 1, 2, and 3 vs. gap in ZFC at 77 K [42].

better, since the larger shielding current area gives the greater levitation force. The test results for the levitation forces of 30 and 26 mm diameter showed that the levitation force on the 30-mm diameter was much larger than that of the 26-mm diameter one. The batch preparation of HTS bulks of diameter larger than that of 30 mm was difficult when "Century" was developed. Therefore, we selected the sample of diameter 30 mm.

Ren [43] researched the influence of HTS bulk size on the guidance force. In order to study the influence of HTS bulk size on the guidance forces, three different cylindrical samples with different diameters (Tab. 6.9) were used in the experiment. The field cooling height and the measured height was 15 mm. The maximum lateral displacement was 20 mm, and the sampling interval was 1 mm. The guidance forces were measured at 77 K with the HTS Maglev measurement system SCML-01.

**Tab. 6.9:** Parameters measured for HTS bulks over the PMG [43].

| Samples | diameter (mm) | Thickness (mm) | Mass (g) | Samples from |
|---|---|---|---|---|
| TG186/20 | 26.6 | 9.2 | 32.2 | IFW |
| BJ218 | 30.0 | 16 | 69.7 | GRINM |
| TG177/6 | 49.0 | 14.8 | 169.3 | IFW |

Figure 6.31 shows the measured results of the guidance forces of each sample for a 15-mm FCH [43]. The arrow direction represents the testing process from the original location to the maximum lateral displacement, 20 mm, from which it would return to the starting position. Guidance forces increase rapidly with increasing sample diameter, while guidance forces hysteresis become more and more obvious, but the return distance decreases with the increase of diameter.

Taking into account the three samples with different thicknesses, the influence of sample size on the experimental results of guidance forces is not accurately reflected by Fig. 6.31. In addition, another factor reducing the accuracy of conclusions based on these results was inconsistent variations in the physical properties of the samples.

### 6.9.3 Influence of HTS bulk thickness on the properties

Zhu et al. [39] investigated experimentally the influence of HTS YBCO bulk thickness on the levitation properties. In this experiment, a HTS bulk sample with a diameter

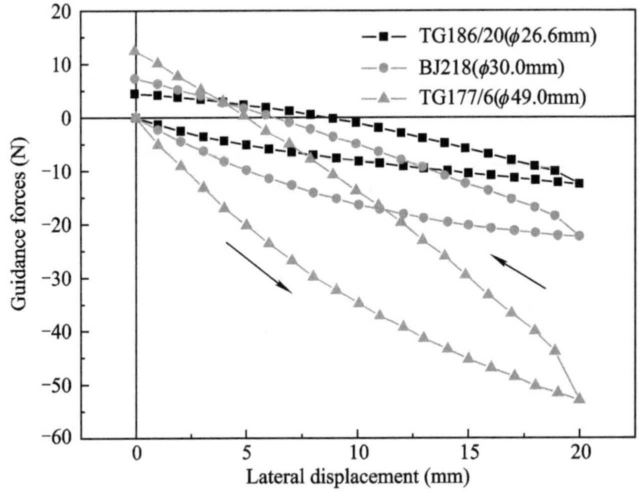


**Fig. 6.31:** Guidance forces vs. lateral displacement of three samples over the NdFeB PMG at 77 K [43].

of 30 mm and thickness of 13.7 mm was ground, and the levitation forces of HTS bulk were measured with each 0.5-mm-thickness removal of material. During the grinding, many cracks, holes, and impurities were present. In one case, a mass of copper which was not oxidized appeared (d). The appearance of the YBCO upper surface (a) and inside surface that was ground (b–d) is shown in Fig. 6.32. From (a) and (b), we can see clearly that when the bulks was ground thin enough, the appearance of the upper surface and the grinding surface was very similar except for a small quantity of holes that were scattered on the ground surface. All of these defects must affect the properties of the YBCO material. If the internal structure can be improved, the property of the HTS bulks can be greatly improved. Figure 6.32 shows the influence of thickness of HTS YBCO bulk using the grinding method on the levitation forces.

Table 6.10 lists the relationship between levitation forces and thickness of HTS bulk [40], and the levitation force of cylinder 2 over the PMG was much larger than that of cylinder 1. We expected that the levitation forces would increase with increasing thickness of the sample in an applied magnetic field, especially in a high applied magnetic field. When the thickness exceeded a certain value, the thickness dependence of the levitation forces was not clear (see Fig. 6.33).

The levitation forces on part of a ground YBCO were not measured in the grinding method of reducing thickness [40]. Cutting the sheet to reduce the thickness can measure the levitation forces on section cut [43]. The HTS bulk sample in this cutting experiment was an MTG YBCO bulk with a diameter of 30 mm and thickness of 18 mm. The sample was cut sequentially removing 3.2, 3.4, and 3.4 mm from face not seeded. The residual thickness of the sample was 14.3, 10.1, and 6 mm, respectively. Figures 6.34 and 6.35 show the measured results of the levitation forces on the slice

(a) Upper surface (b) 2 mm (c) 7.5 mm (d) 9 mm

**Fig. 6.32:** Appearance Of The YBCO upper and grinding inside surface [39]. The appearance of YBCO bulk (a) upper surface, ground (b) 2 mm, (c) 7.5 mm and (d) 9 mm.

**Tab. 6.10:** Relationship between thickness and levitation forces of YBCO bulk samples [40].

| Parameters | Cylinder 1 | Cylinder 2 |
|---|---|---|
| Thickness (mm) | 12 | 14 |
| Diameter (mm) | 30 | 30 |
| $F_{15}$ (N) | 28.8 | 41.2 |
| $F_{20}$ (N) | 22.9 | 29.8 |

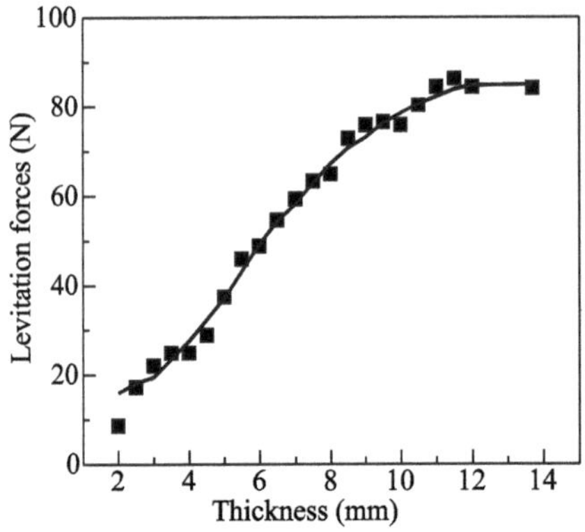


**Fig. 6.33:** Levitation forces vs. thickness [39].

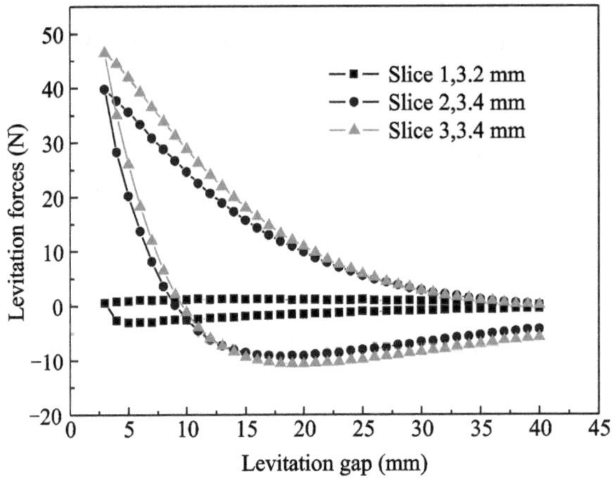


**Fig. 6.34:** Levitation forces of slice thickness of YBCO sample [43].

thickness and the residual thickness of this HTS YBCO bulk, respectively. It can be seen from Fig. 6.34 that the influence of YBCO sample thickness on levitation force is larger when the levitation gap is smaller, but the levitation forces of different thickness YBCO samples are almost the same when the levitation gap is bigger than 20 mm.

The influence of sheet cutting method to reduce thickness on the guidance forces was investigated by Ren [43]. The field cooling height and measured height were 20 mm. The maximum lateral displacement was 10 mm, and the sampling interval was 1 mm. The guidance forces were measured at 77 K with the HTS Maglev measurement system SCML-01. The results show that when the lateral displacement is small, the effect of the HTS sample thickness on the guidance forces is not clear. With the increase of lateral displacement, the guidance forces increase rapidly with increase of sample thickness and tend to saturation. However, due to the restrictions on the sample thickness, the guidance forces cannot be studied on thicker samples.

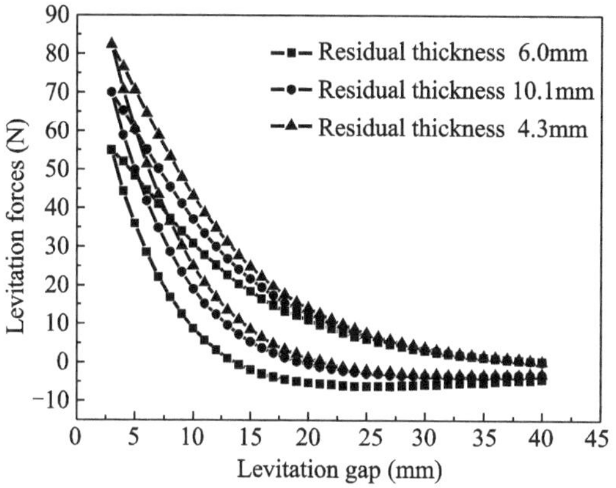


**Fig. 6.35:** Levitation forces of residual thickness of YBCO sample [43].

## 6.10 Levitation forces and guidance forces of ring HTS bulks [43]

The levitation forces of ring-shaped YBCO bulks above an NPM guideway were studied. The investigation focused on the influence of the ring HTS YBCO bulk on the levitation forces. In this experiment, three cylindrical YBCO bulks, one with a hole of 8 mm in diameter was made in sample BJ214, one with a hole of 12 mm in diameter for sample BJ215 and one with a hole of 18 mm in diameter for sample BJ216, were used (Fig. 6.36). The outer diameter was 30 mm, and the thickness was 16 mm. The levitation forces of each bulk were measured above the PMG at FCH 40 mm.

Figure 6.37 shows the levitation forces comparison for each of these bulk as shown in Fig. 6.37a and b, when the hole diameter was 8 or 12 mm, there was hardly any difference in the levitation forces between with or without a hole. As shown in Fig. 6.37a and b, when the hole diameter was 8 and 12 mm, it hardly had any influence on the levitation forces in the whole measuring range, i.e. gap from 40 to 3 mm. When the hole was enlarged to 18 mm in diameter, the levitation forces began to decrease at a lower levitation gap, but still hardly influenced the levitation forces at a larger levitation gap. For example, gaps more than 20 mm are shown in Fig. 6.37c. This proved that the outer part of YBCO bulk contributes more to the levitation forces

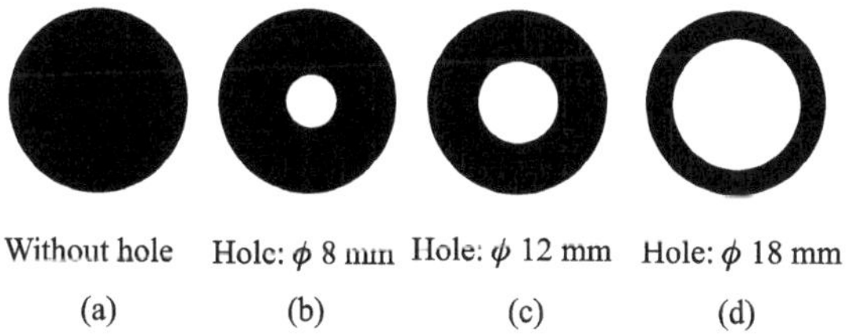


**Fig. 6.36:** Ring-shaped YBCO bulk with different holes [43].

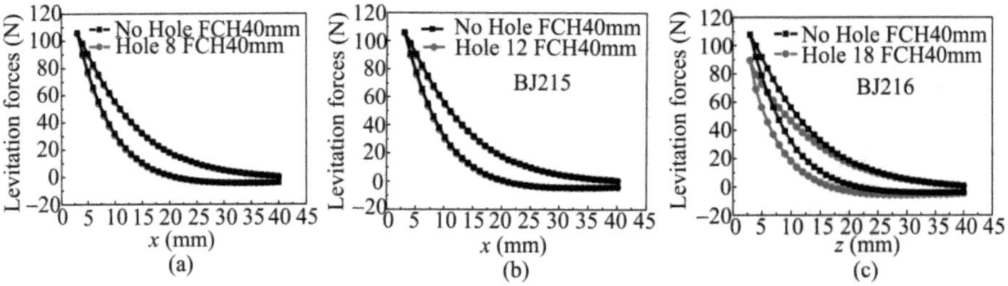


**Fig. 6.37:** Levitation forces comparison of YBCO bulk with different hollow holes [43]. (a) A hole with 8 mm in diameter. (b) A hole with 12 mm in diameter. (c) A hole with 18 mm in diameter.

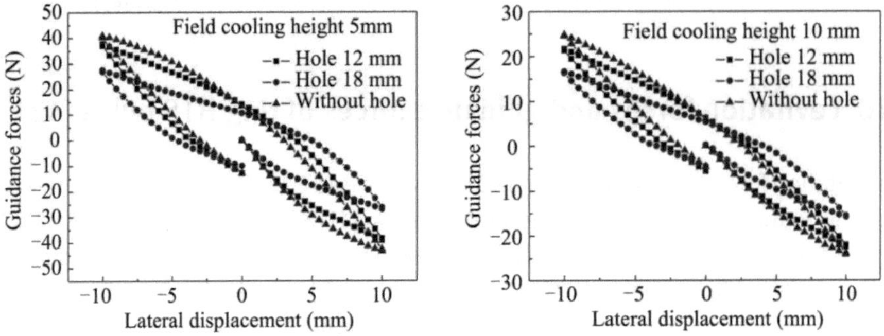


**Fig. 6.38:** Guidance forces vs. lateral displacement with different holes in FCH 5 mm (left) and 10 mm (right) [43].

than the inner part. In some cases, we can substitute a cylindrical YBCO bulk with a ring-shaped YBCO bulk in order to achieve a lightweight Maglev system.

While the hole did not cause much difference in the levitation forces, it had much more influence on the guidance forces. When the hole was small enough, i.e. a hole with diameter of 4 mm, the influence on guidance forces was not clear. As the diameter of the hole increased, the guidance forces decreased rapidly (Fig. 6.38).

The reason was that the guidance forces originates from the interaction between trapped flux and applied field. The trapped flux depended on the pinning centers. As the hole enlarged, the number of pinning center decreased, which resulted in smaller trapped flux and led to a smaller guidance force.

## 6.11 Early scheme consideration

Both rare-earth materials yttrium (Y) and neodymium (Nd) are used in a HTS bulk Maglev system, and these resources are abundant in China. After the HTS magnetic levitation phenomenon was discovered in 1988, the idea of the Maglev train using a HTS bulk Maglev was considered. In 1987, the breakthrough in HTS material was followed by the rapid development in China, which has greatly strengthened and promoted the original ideas about HTS Maglev. In 1992, Jia-Su Wang and Su-Yu Wang started to read the relevant literature, electromagnetic suspension, electrodynamic suspension, normal conductive Maglev, superconducting Maglev, high $T_c$ superconductivity etc., and then gradually formed some tentative plans.

In 1994, the president of Southwest Jiaotong University, Xiang Sun, approved the report on the superconducting Maglev submitted by Jia-Su Wang and Su-Yu Wang. In 1996, Jia-Su Wang and Su-Yu Wang discussed the problems about the manned HTS Maglev vehicle with Prof. Hong-Tao Ren of GRINM (General Research Institute for Nonferrous Metals) in the Dalian Conference of Information Net of Cryogenics and Superconductivity. At first, Ren believed that the project was too risky, but finally accepted this high-risk project. Continued technical progress steps for the HTS Maglev had been achieved through the efforts of the experts, which included improvement of the HTS bulk performance, suspension of a sumo wrestler in Japan, successful demonstration of a HTS Maglev model on a small orbit during the 100th anniversary of Southwest Jiaotong University), and the Maglev demonstration model using a linear motor achieved through Chinese and German cooperation. These vigorously promoted the manned HTS Maglev vehicle development process.

In the early days, the superconducting Maglev vehicle only referred to LTS magnetic levitation [44]. The concept of a HTS Maglev vehicle [45] was gradually formed in the 1990s. Jia-Su Wang and Su-Yu Wang [45–48] have presented several Maglev vehicle schemes using HTS YBCO bulks in the 1990's. First, the YBCO bulk superconductors were arranged above the guideway using NdFeB PMs. The second scheme was the electromagnetic suspension (EMS) Maglev vehicle using HTS PM or HTS wire. The third was the EMS Maglev using HTS PM.

## 6.12 Maglev of HTS bulk above magnets

At first, we considered several HTS Maglev versions of HTSCs over an applied magnetic field using PM or electromagnets [45, 48]. In Fig. 6.39a, [45] PM poles are arranged alternately, and in Fig. 6.39b, the iron between both opposite PM poles produces a concentrated magnetic field. The latter has higher magnetic flux density, and it can attain a greater magnetic levitation force. Figure 6.39c shows the electromagnet guideway.

There is a large levitation force between two normal PMs, but this levitation force is unstable. The HTS bulks in the condition of field cooling (FC) have a high stability. It perhaps is a valid scheme to combine the large levitation forces between both normal PMs and the high stability of HTS bulk superconductors. Combining

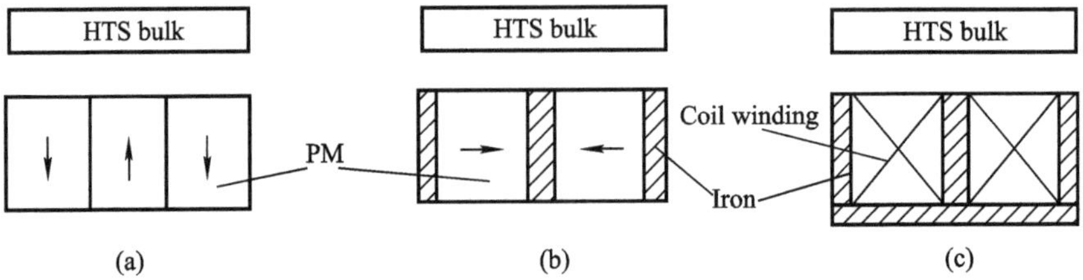


**Fig. 6.39:** Several HTS Maglev versions of HTSCs over an applied magnetic field from PM or electromagnet [45]. Three versions of HTS Maglev that HTSCs levitated above, respectively, (a) three vertical magnetized PMs, (b) two antiparallel magnetized PMs assembled with iron, and (c) electromagnet with iron, where the arrows in PMs show the magnetization directions.

levitation schemes using both normal PMs and HTS bulk superconductors provides many possibilities:

1. All levitation forces can be generated by the interaction between normal PMs. The guidance forces are provided by HTS bulks in field cooling (FC).
2. The levitation forces can be generated by interaction between two normal PMs and by the interaction between a normal PM and bulk superconductors in zero field cooling (ZFC). The guidance forces can be provided by HTS bulks in field cooling.
3. Both levitation and guidance forces can be achieved by HTS bulk over a PMG.

An understanding of the first scheme is relatively easy. The levitation using normal PM is very attractive all the time, since its levitation forces are high. The levitation forces and the guidance forces are independent. An advantage of this scheme is that the levitation forces are higher than the other schemes when the quality of the HTS bulk superconductors is not very good.

The second scheme is complex and perhaps is theoretically superior. However, it is difficult to design the best combination of both normal PMs and bulk superconductors.

Ren et al. [49] presented a hybrid Maglev vehicle using PMs and HTS bulks that was similar to the second scheme. Some PMs were installed in the vehicle together with HTS bulks. The HTS bulks provided guidance forces and a partial levitation force. On-board PMs provided additional levitation forces.

Figure 6.40 shows the schematic diagram of the hybrid Maglev vehicle using PMs and HTS bulks. HTS bulks were over the center of two parallel PMs (see Section 6.14) so that a larger levitation force and guidance force could be created. The on-board PMs were positioned over the two sides of two parallel PMs. These provided additional levitation forces. The total levitation forces were generated by the on-board HTS bulks and PMs. The hybrid Maglev vehicle did not need an on-board power supply and achieved stable levitation, but the total magnetic levitation forces could be greater than that of one of the HTS bulks over the PMG.

The levitation forces between a single PM and a single YBCO bulk and one between two arrays of PMs and HTS bulks over a PMG were studied. Research results showed

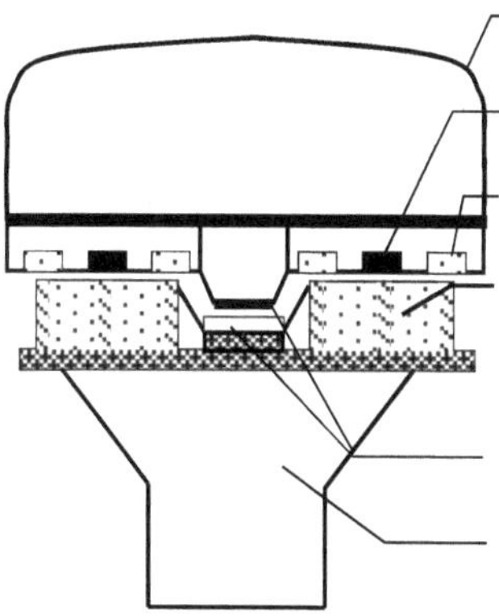


**Fig. 6.40:** Schematic diagram of a hybrid Maglev vehicle using PMs and HTS bulks [49].

that both the levitation forces and guidance forces of the hybrid Maglev vehicle provided larger levitation forces and stiffness.

The third scheme (Fig. 6.39b) was the implementation plan of the HTS Maglev vehicle "Century". It will be discussed in detail in the following sections. This section describes some of the early schemes.

The guidance forces of HTS bulk superconductors were used in each of the above-mentioned schemes, because of the important characteristics of HTS bulk superconductors. All three schemes had an important defect, i.e. the levitation height of these schemes was limited. Both height and forces of the levitation are crucial for a Maglev vehicle.

Jia-Su Wang et al. [48] reported the magnetic levitation of HTS bulk above the PM or the electromagnets.

The electrical magnet can be controlled in sections along a road. Only the section where the vehicle was located was charged with electricity. This system of supplying power was analogous to that of a linear synchronous motor (LSM), but the electrical magnets were powered by direct current. The magnetic flux density of an electrical magnet with a soft iron core achieves a field of least 1.0 T, which can generate a large levitation force. Its shortcoming was the need for a complex guideway and high electrical energy. Figure 6.41 shows a scheme for a Maglev vehicle system using the levitation between the HTS bulks and the electrical magnets [48]. The HTS bulks bathed in liquid nitrogen generate a high levitation force. The propulsion system which used a linear induction motor (LIM) is shown in Fig. 6.41. This system was only suitable for experimental research. For actual use, it should have a LSM.

## 6.13 Maglev vehicle using HTS PMs

The rare-earth PMs with a high energy product and a high intrinsic coercive force have been widely used and have brought great technical and economic benefit. The HTS

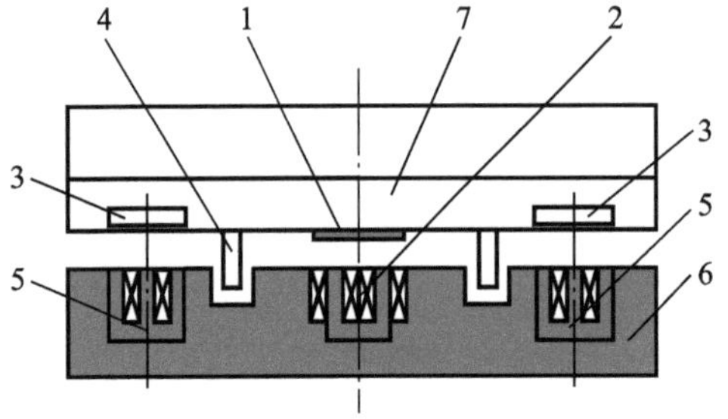


**Fig. 6.41:** A scheme for Maglev vehicle system using the HTS bulk and the electromagnets [48]. (1) Induction plate. (2) Induction motor stator. (3) HTS Bulks. (4) Slide for landing. (5) Electrical magnets. (6) Road pier. (7) Vessel of liquid nitrogen.

bulks can trap enough flux to become a new PM. The HTS PMs have a much higher energy product than the rare-earth PMs, and their potential industrial applications are very promising. The HTS PMs can realize an energy product of 1.8 MJ/m$^3$ [50]. This value is 3.5 times higher as NdFeB. The maximum trapped field $B_0$ is 17.24 T in a bulk YBCO sample of 2.65 cm diameter at 29 K [51]. A production run of 60 melt-textured YBCO trapped field magnets, 2 cm in diameter, was recently reported by Sawh et al. [52] The resulting average trapped field at the center of the seed-side surface was 2.04 T at 77 K. It can also be applied to EMS systems because of the simple structure and large levitation forces that can be obtained. Figure 6.42 shows a full-superconducting scheme with HTS PMs [48]. In Fig. 6.42, the control coils are attached to the bulk superconductors bathed in liquid nitrogen. When the trapped field of the HTS PMs is changed by inherent or external fluctuations, the field change can be compensated by the control coils. The coils of the LSM are used for the propulsion. The special geometry of the liquid nitrogen vessel containing the bulk superconductors is designed for exciting the field. It is very important for the HTS PMs to excite the field. The applied magnetic field is not only high, but also both its size and orientation are restricted.

Another scheme for a Maglev vehicle system using the HTS PMs is shown in Fig. 6.43 [48]. This scheme requires that the HTS PMs have very high performance. The HTS PMs play the role of the suspension, the propulsion, and the guidance. The HTS

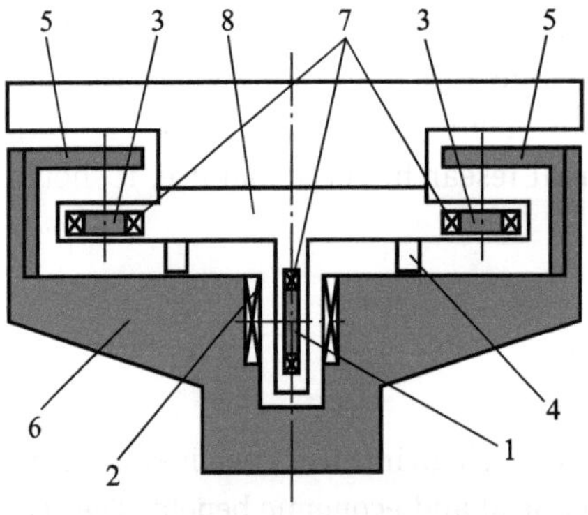


**Fig. 6.42:** Scheme A for Maglev vehicle systems using HTS PMs [48]. (1) HTS PMs for LSM. (2) Coil of LSM. (3) HTS PMs for the suspension. (4) Slide for landing. (5) Iron guideway. (6) Road pier. (7) Control coil. (8) Vessel of liquid nitrogen.

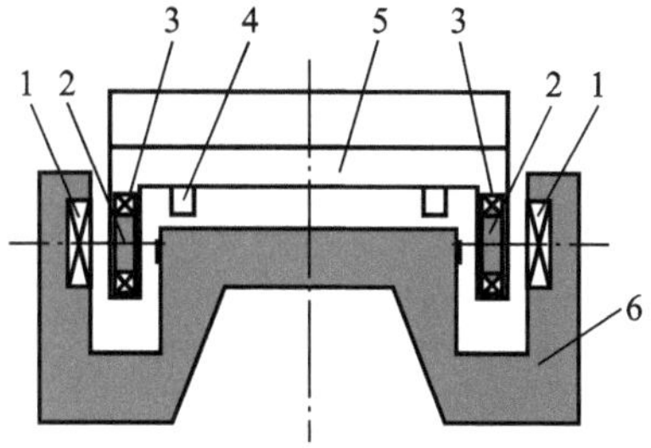


**Fig. 6.43:** Scheme B for Maglev vehicle systems using the HTS PMs [48]. (1) Coil of LSM. (2) HTS PMs. (3) Control coils. (4) Slide for landing. (5) Vessel of liquid nitrogen. (6) Road pier.

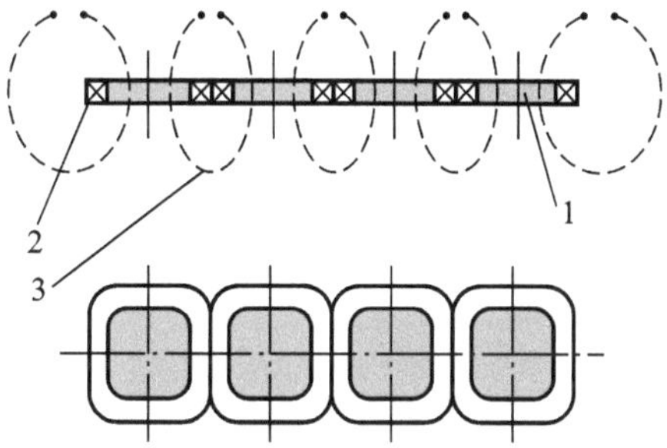


**Fig. 6.44:** Arrangement scheme of HTS PMs [48]. (1) HTS PMs. (2) Control coils. (3) Flux line.

PMs are arranged in a row (see Fig. 6.44). This scheme cannot be realized until a stable, high field is reached in the HTS PM. The current EMS system using normal conductors is called the Transrapid-07 (TR-07) in Germany. One of the principal drawbacks of the TR-07 EMS system is the small gap (10 mm) between the vehicle and the guideway rail. The Japanese have developed the electrodynamic suspension (EDS) system using LTS magnets at 4.2 K. The EDS system uses a large gap (100 mm), but the operating temperature is very low, and it is not a static suspension. The all-superconducting scheme with HTS PMs possesses a great number of the merits [48]:

1. It is a new EMS system. In other words, not only it is a static suspension, but also it has a larger gap between the vehicle and the guideway rail.
2. The speed can reach over 500 km/h.
3. The restriction for the tolerance in the guideway rail is lower.
4. Both the production cost and the operating cost are lower because the low-temperature system is cooled with liquid nitrogen.
5. The energy consumption is lower.
6. The environmental pollution is very small.

The above advantages predict success using the HTS PMs for the Maglev vehicle systems. Unfortunately, the trapped field is limited by the mechanical tensile strength of the HTS YBCO superconductors, and the stability of the trapped field is lower. In addition, both the magnetic circuit and the exciting mode are not suitable for practical application. Therefore, a model of a hybrid Maglev transportation system using HTS PMs is presented.

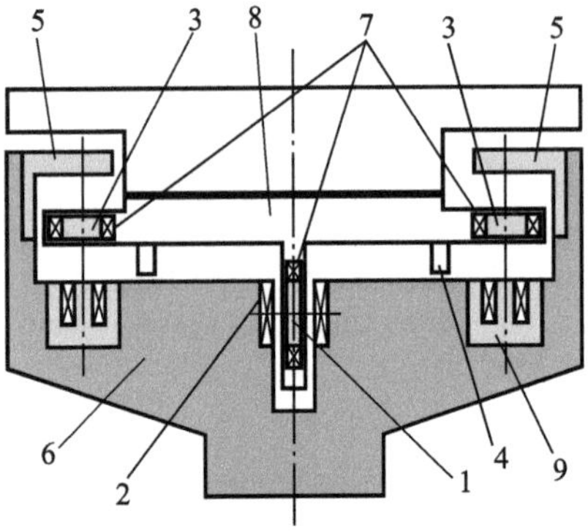


**Fig. 6.45:** A scheme for a hybrid Maglev vehicle system using HTS PMs [48]. (1) HTS PMs for LSM. (2) Coil for propulsion. (3) HTS PMs for the suspension. (4) Slide for landing. (5) Iron guideway. (6) Road pier. (7) Control coil. (8) Vessel of liquid nitrogen. (9) Electrical magnet for suspension.

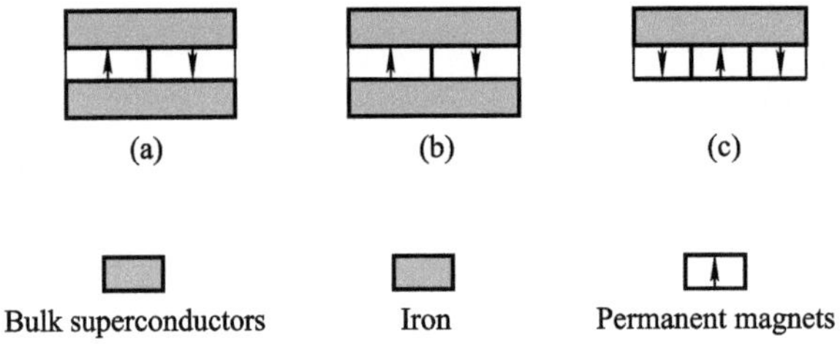


**Fig. 6.46:** The PMs are joined to bulk superconductors [48].

The model of a hybrid Maglev vehicle system using HTS PMs is shown in Fig. 6.45 [48]. The principal parts of this scheme are the same as those for the full-superconducting scheme with HTSCs in Fig. 6.42. One difference is that additional electrical magnets are used for suspension. They provide the suspension when HTS PMs are not suitable. The other difference is the use of the induction plates for propulsion part of the LIM. Those plates are fixed to the outside of the liquid nitrogen vessel of and the opposite the propulsion coil. In addition, HTS PMs can be replaced by PMs or be joined HTS bulks (Fig. 6.46) [48] Stability of HTS PM increases with the decrease of temperature. Therefore, to achieve high and stable trapped flux, a hybrid Maglev vehicle system using HTS PMs is suitable for research of the Maglev vehicle, since the present capabilities of HTS PMs do not achieve a practical level.

The schematic diagram of key components of the EMS Maglev using HTS PM is shown as in Figs. 6.47 and 6.48 [46]. The scheme requires the HTS PM to have high performance. It is reasonable to choose a trapped magnetic field of 2 T for the HTS PM, because the peak field in the iron is greater than 2 T.

There is no gap between the iron and the HTS PM in the whole cooling (Fig. 6.47), and the magnetic resistance is null. The geometry gap between the rail and the magnetic pole area of iron core is reduced because of the thickness of the cryogenic vessel.

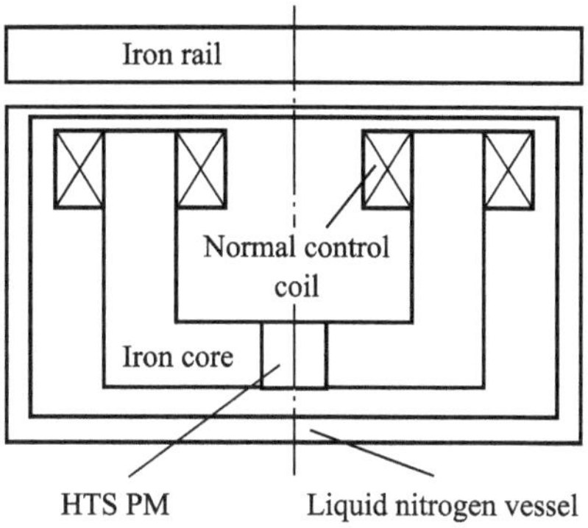


**Fig. 6.47:** Scheme A for the EMS Maglev vehicle using the HTS PM [46].

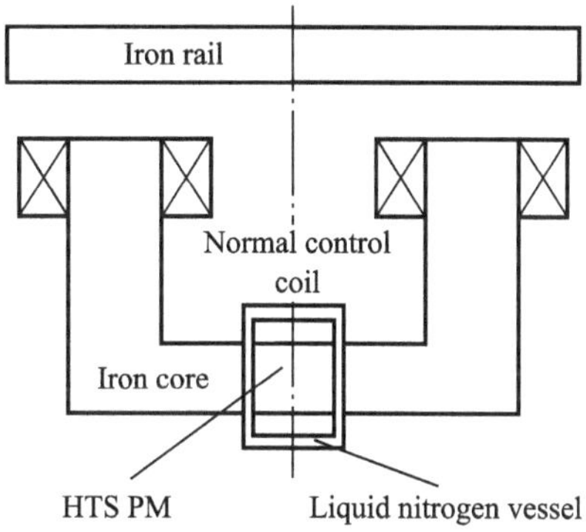


**Fig. 6.48:** Scheme B for the EMS Maglev vehicle using the HTS PM [46].

In Fig. 6.48, the net gap (the suspension height) between the rail and the magnetic pole area of iron core is higher, because there is no cryogenic vessel near the magnetic pole area. In contrast, there is a gap between the iron and the HTS PM, and the magnetic resistance is not null. It is difficult to make a thin-wall cryogenic vessel. In any case, this scheme can be realized, and its many problems should be thoroughly and carefully studied.

This scheme has the possibility to be achievable because the trapped field of the HTS bulk materials has achieved the value of 2 T [52]. In practical application, it is also very important to excite the field of the HTS PMs. Both the method and process of the exciting the field should be further researched. In the cases of both Figs. 6.47 and 6.48, the method of exciting field is completely not identical, and the exciting field of whole cooling in Fig. 6.48 is the most complex. The applied magnetic field is not only high, but also both size and orientation are restricted.

The scheme of the EDS Maglev using HTS PMs is discussed in Section 4.5.

The scheme of the EDS Maglev using HTS wires (1G or 2G HTS wires) is discussed in Section 4.6.

These early schemes of the HTS Maglev had been discussed, before we decided on the "Century" plan. After several years of research, we finally decided to choose the scheme of Maglev vehicle of HTS bulk above PMs (see Section 6.26).

The following will introduce the technical details in the manned HTS Maglev vehicle "Century".

## 6.14 Permanent magnet guideway

The disadvantages of an electrical magnet are the complexity of the guideway and high electric energy consumption. If the applied magnetic field is provided by an NPM, a larger levitation force between HTS bulk and the applied magnetic field can be achieved. The magnetic field of the PMG can be increased by the application of a flux concentration scheme in the magnet arrangement. In order to increase the field on the top of the PMG, ferromagnetic materials are introduced as magnetic conductors between two NPMs [53].

Figure 6.1 and Tab. 6.1 show the measured results of the levitation forces with the single-NdFeB PM cylinder (surface magnetic flux density is 0.38 T) over the NdFeB PMG (surface magnetic flux density is 1.2 T) [24]. Clearly, the levitation forces of the NdFeB PMG are much higher than those of single-NdFeB PM cylinder.

The NdFeB PMs has magnetic energy product of 40 MOGe, coercivity $H_c$ of 979 kA/m, and relative permeability $\mu$ of 1.049. The transverse distribution of magnetic field of a PMG is shown in Fig. 6.49. The PMG is composed of normal PMs and iron plate. In Fig. 6.49, the magnetic field at the center of the PMG is stronger than that at any other position, and it decreases rapidly with the increase of the gap from the surface of the PMG.

Figure 6.50 shows the measured and calculated results of the magnetic field distribution of the PMG along the transverse and the vertical direction a PMG along the vertical direction [35, 43]. Figure 6.50b shows the transverse field distribution at 15 mm above PMG surface. The calculated 1 and calculated 2 correspond to the finite element method and the equivalent current method, respectively. Table 6.11 lists the measured magnetic flux densities of PMG center along the vertical direction.

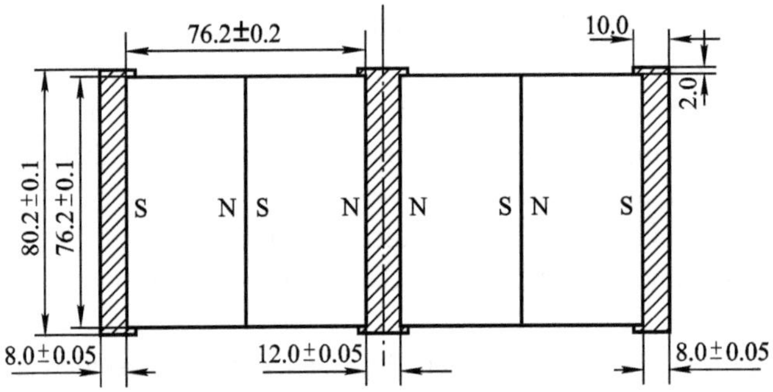


**Fig. 6.49:** Cross-sectional view of the PMG.

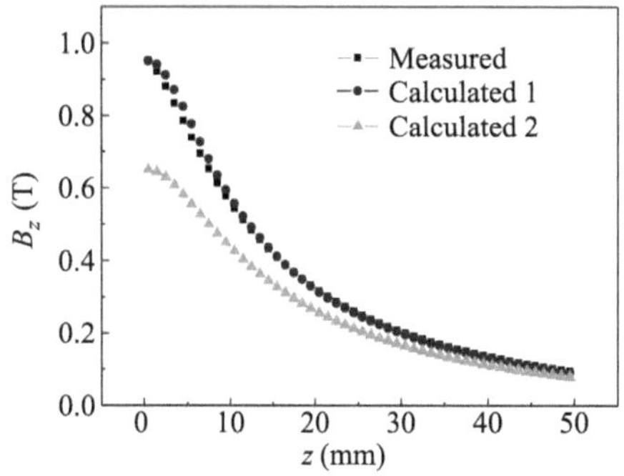

(a) The vertical field distribution of PMG

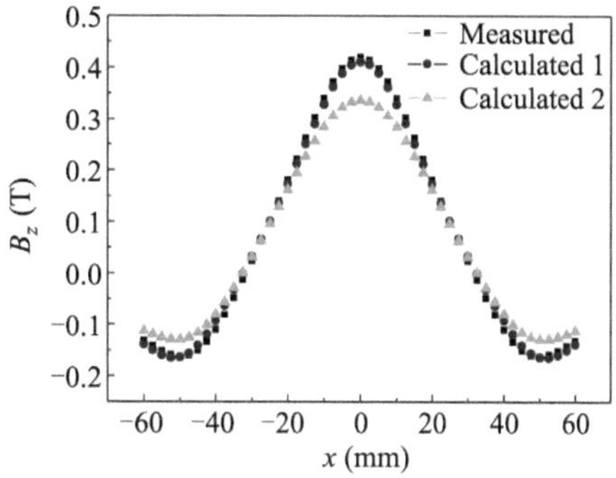

(b) The transverse field distribution at 15 mm above PMG surface

**Fig. 6.50:** Measured and calculated results of the magnetic field distribution of the PMG along the transverse and the vertical direction. Calculated 1 and calculated 2 correspond to the finite element method and the equivalent current method, respectively. [35, 43]

**Tab. 6.11:** Measured magnetic flux densities at PMG center along the vertical direction.

| Height (mm) | Measured $B_z$ (T) | Height (mm) | Measured $B_z$ (T) |
|---|---|---|---|
| 0 | 1.2125 | 20 | 0.432 |
| 5 | 0.8388 | 25 | 0.363 |
| 10 | 0.6481 | 30 | 0.3089 |
| 15 | 0.5209 | | |

**Fig. 6.51:** Photograph of the PMG. The surface concentrating magnetic flux density of the PMG is up to 1.2 T. The magnetic flux density is 0.4 T at 20-mm gap from the surface of the PMG.

The PMG (Fig. 6.51) has a length of 920 mm and width of 178 mm. The cross-section area of the PMG is $76.2 \times 76.2 \times 2 = 116\ \text{cm}^2$. The surface magnetic field of a single permanent magnet is about 0.45 T, and the surface concentrating magnetic flux density of the PMG is up to 1.2 T. The magnetic flux density is 0.4 T at 20 mm gap from the surface of the PMG which is close to the surface magnetic field of the NdFeB PM cylinder. The PMG is used not only in the measurement of HTS Maglev properties but also in the HTS Maglev vehicle system. [1]

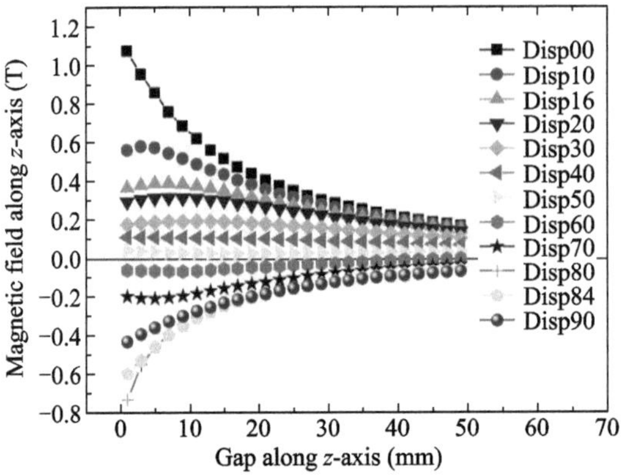


**Fig. 6.52:** Displacement dependence of the magnetic field for a gap of 5 mm to 50 mm over the PMG. Disp00 represents the magnetic field along the *z*-axis at the displacement of 0 mm; Disp10, 10 mm; Disp16, 16 mm; and so forth [41].

Figure 6.52 shows the displacement dependence of the magnetic field from a gap of 5 to 50 mm over the PMG. Disp00 represents the magnetic field along the *z*-axis at the displacement of 0 mm, similarly for Disp10, Disp16, etc [41].

## 6.15 On-board HTS Maglev equipment [54]

Based on the experience with the small columnar liquid nitrogen vessel (see Section 5.2.2), the on-board rectangular-shaped liquid nitrogen vessel was developed. The bottom thickness of the rectangular-shaped vessel is thinner than that of the small columnar vessel. It is only 3 mm. The scheme of the rectangular-shaped thin bottom liquid nitrogen vessel is shown in Fig. 6.53. Its outside and inside dimensions are $150 \times 516 \times 168$ mm and $102 \times 470 \times 168$ mm, respectively. This liquid nitrogen vessel can operate continuously over 6 hours. Figure 6.54 shows the evaporation rate of a new rectangular liquid nitrogen vessel with a thin bottom [21]. The vessel was successfully installed on the HTS Maglev test vehicle "Century".

The on-board HTS Maglev equipment is the key component of the HTS Maglev vehicle system. The photographs of the on-board HTS Maglev equipment are shown in Fig. 6.55.

The on-board Maglev equipment is composed of 43 pieces of YBCO bulks and rectangular-shaped liquid nitrogen vessel. There are 43 pieces of YBCO bulks in each vessel. Figure 6.55a shows the installed bulks without the seal box. The YBCO bulks are 30 mm in diameter and 17–18 mm in thickness. The YBCO bulks are arranged at the bottom of the liquid nitrogen vessel. In order to avoid degradation of YBCO bulks, all 43 pieces of YBCO bulks are sealed by a special technology. Figure 6.55b shows the upper and lower seal box cover (above) and the sealed welded box with the 43 pieces of YBCO (below). In order to ensure that YBCO bulks are not damaged by the high-temperature

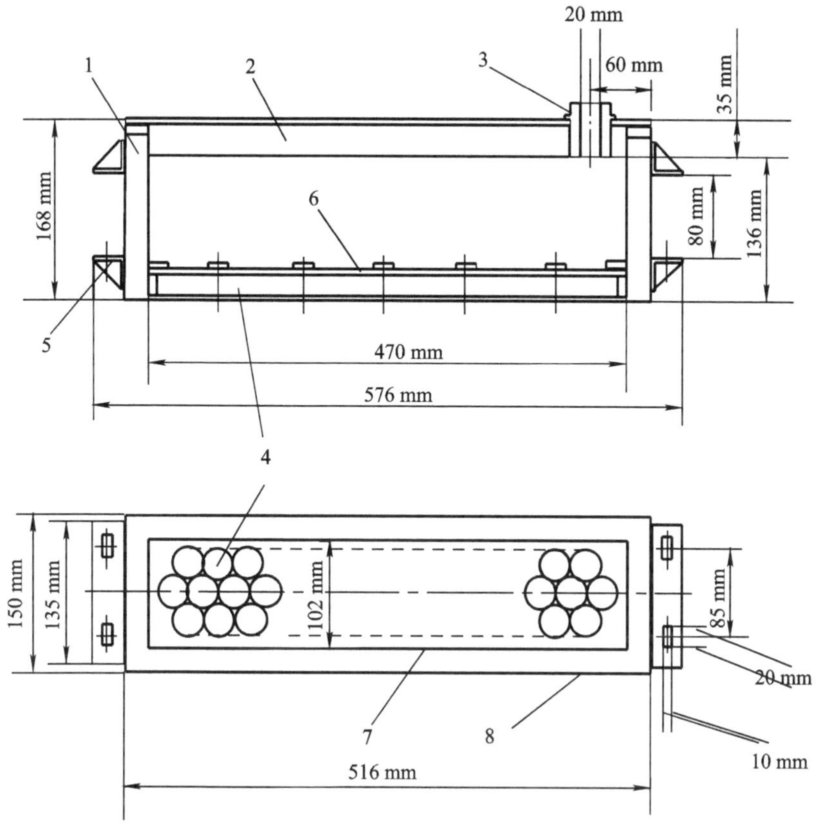


**Fig. 6.53:** Structure diagrams of the on-board HTS Maglev equipment. (1) Thermal insulation. (2) Vessel cover. (3) Liquid nitrogen entrance. (4) YBCO. (5) Pedestal connecting with vehicle. (6) Press board of YBCO. (7) Inner vessel container. (8) Outer vessel container.

welding, the YBCO temperature must be limited during welding. After leak detection of the sealed can, dry nitrogen gas of slightly more than one atmospheric pressure is pumped into the can and it is permanently sealed. After 10 years of operation, the can has been opened and YBCO bulks have been removed. Test results show that the performance of the YBCO bulks remain unchanged (see Section 6.20).

The sealed cans are fixed firmly in the liquid nitrogen vessel. Figure 6.55c shows the on-board HTS Maglev equipment completely assembled. The on-board Maglev equipment is connected rigidly to the two sides of the vehicle body.

## 6.16 Experimental results of the on-board Maglev equipment

The levitation forces on a single on-board HTS Maglev equipment above the PMG were measured with the SCML-01 HTS Maglev measurement system. The measurement method is described in Section 6.3. During the experiment, the on-board HTS Maglev

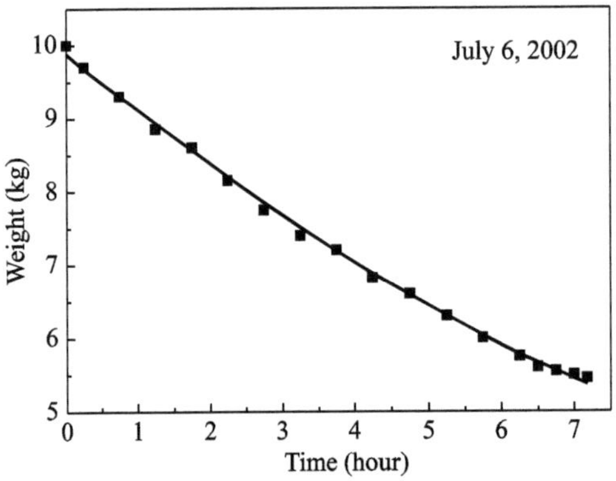


**Fig. 6.54:** Evaporation rate of the new thin-bottom liquid nitrogen vessel [22].

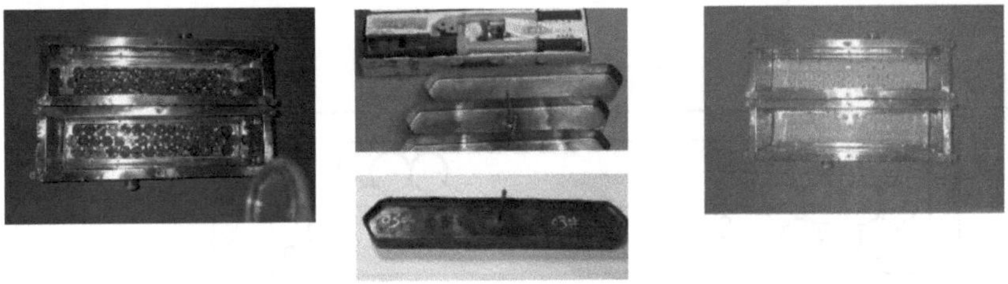

(a) 43 pieces of HTS YBCO bulks in the rectangular thin bottom vessel, not sealed.

(b) The upper and lower sealing box covcr(abovc) and the sealed welded box with 43 pieces of YBCO(below).

(c) The on-board HTS Maglev equipment is completed assembly with the sealed welded box and the thin bottom vessel.

**Fig. 6.55:** Photographs of the on-board HTS Maglev equipment.

equipment was fixed to the HTS Maglev measurement system SCML-01. The on-board Maglev equipment was above the PMG, and it could move up and down at different velocities. Figure 5.7b shows the measuring system for the levitation forces on a single on-board HTS Maglev equipment over a single-NdFeB PMG [54].

The first levitation forces of single on-board HTS Maglev equipment over the PMG were measured successfully on December 2, 2000. Figure 6.56 shows the measurement interface for the levitation forces on the single on-board HTS Maglev equipment over a NdFeB PMG (surface magnetic flux density was 1.2 T). The main data of the levitation forces of the single on-board HTS Maglev equipment are listed in Tab. 6.12. As shown in Fig. 6.56 and Tab. 6.12, the levitation force of single on-board HTS Maglev equipment over the PMG was 1202 N at a levitation gap (between PMG surface and superconductors) of 15 mm. The levitation force was 1724 N at a levitation gap of 8 mm. Moreover, the levitation force was up to 1823 N at a levitation gap of 7 mm.

The levitation forces of single on-board HTS Maglev equipment above the PMG were measured again on December 24, 2001 [26]. Figure 6.57 shows the levitation forces of single on-board HTS Maglev equipments above the PMG. Figure 6.58 shows

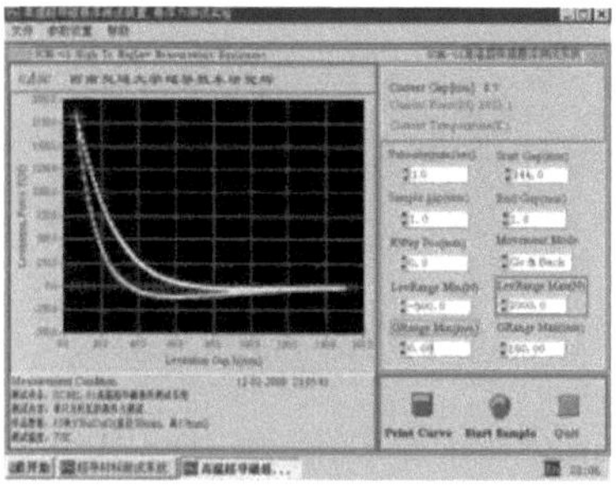

**Fig. 6.56:** Measurement results for the single on-board HTS Maglev equipment over a single-NdFeB PMG on December 2, 2000 [54].

**Tab. 6.12:** Levitation forces of the best single on-board HTS Maglev equipment on December 2, 2000 [54].

| Gap (mm) | Forces (N) | Gap (mm) | Forces (N) |
|---:|---:|---:|---:|
| 60 | 44.9 | 8 | 1591.5 |
| 50 | 122.5 | 9 | 1422.1 |
| 40 | 265.1 | 10 | 1277.6 |
| 30 | 506.3 | 11 | 1152.2 |
| 25 | 685.2 | 12 | 1031.6 |
| 20 | 913.0 | 13 | 928.3 |
| 15 | 1202.0 | 14 | 830.7 |
| 14 | 1267.1 | 15 | 746.4 |
| 13 | 1341.7 | 20 | 421.1 |
| 12 | 1409.7 | 25 | 216.2 |
| 11 | 1484.3 | 30 | 87.0 |
| 10 | 1561.8 | 35 | 3.8 |
| 9 | 1638.4 | 40 | −46.8 |
| 8 | 1723.6 | 50 | −96.6 |
| 7 | 1823.1 | 60 | −105.2 |

the levitation force measuring results for each of the eight on-board HTS Maglev equipments over the PMG, and Tab. 6.13 gives the levitation forces of each on-board Maglev equipment at levitation gaps of 8, 10, and 15 mm, respectively [26]. It shows that the levitation force of the No. 2 Maglev equipment was the largest. Its levitation forces were 1418, 1354, and 1127 N at 8, 10, and 15 mm, respectively. The smallest levitation force was different for different levitation gaps. The levitation forces of the No. 4 were the smallest at 10- and 15-mm gaps, 1166 and 964 N, respectively. However, as far as the levitation forces for the 8-mm gap were concerned, that of No. 3 was the smallest of 1249 N. A reason for this variation is perhaps that the YBCO bulks were produced in different batches.

Figure 6.59 shows that the levitation forces for each of the eight on-board Maglev equipments above the PMG differed slightly [22]. For example, the levitation forces of Maglev equipment No. 7 were 1493 and 1227 N for levitation gaps of 10 and 15 mm, respectively. The levitation forces of Maglev equipment No. 3 were 1091 and 902 N for

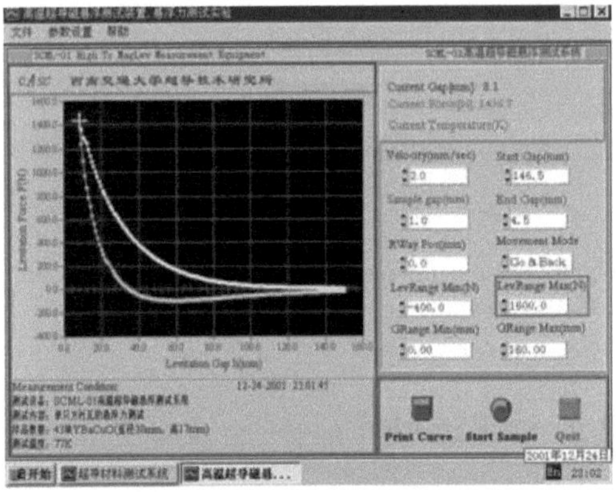


**Fig. 6.57:** Measurement interface of the levitation forces of a single on-board HTS Maglev equipment on December 24, 2001 [26].

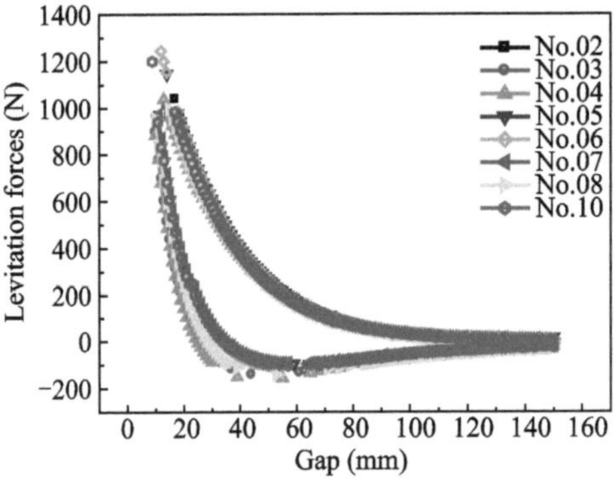


**Fig. 6.58:** Levitation force results of each of the eight on-board HTS Maglev equipments above the PMG [26].

**Tab. 6.13:** Levitation forces (N) of each on-board HTS Maglev equipment [26].

| Gap | No. 2 | No. 3 | No. 4 | No. 5 | No. 6 | No. 7 | No. 8 | No. 10 |
|---|---|---|---|---|---|---|---|---|
| 8 mm | 1418 | 1249 | 1252 | 1403 | 1399 | 1437 | 1269 | 1385 |
| 10 mm | 1354 | 1170 | 1166 | 1279 | 1291 | 1303 | 1195 | 1330 |
| 15 mm | 1127 | 974 | 964 | 1104 | 1113 | 1099 | 998 | 1078 |

levitation gaps of 10 and 15 mm, respectively. Figure 6.60 shows the total levitation forces of the eight on-board Maglev equipment assemblies. The total levitation forces of these equipment assemblies yielded levitation forces of 10,431 and 8486 N at levitation gaps of 10 and 15 mm, respectively.

All measurement results of each on-board HTS magnetic levitation equipment had some differences. The reason may be ascribed to the different properties of each YBCO bulk or 43 pieces of YBCO integration, the different thin-bottom liquid nitrogen vessel, different degrees of accuracy in installing them on the measurement system

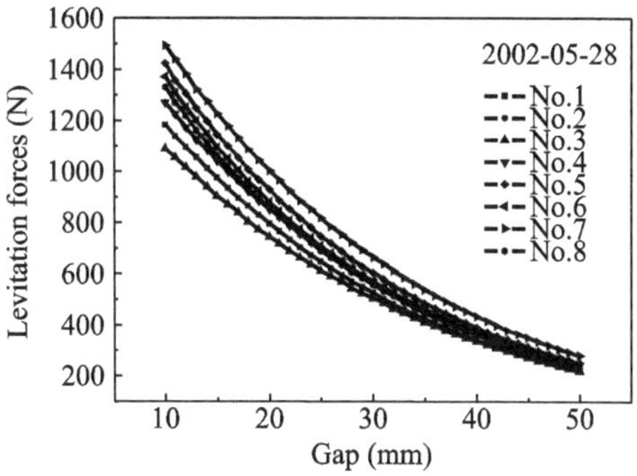


**Fig. 6.59:** Measured results of each of the eight on-board HTS magnetic levitation equipments (rectangular-shaped vessels included 43 pieces of YBCO bulks) [22].

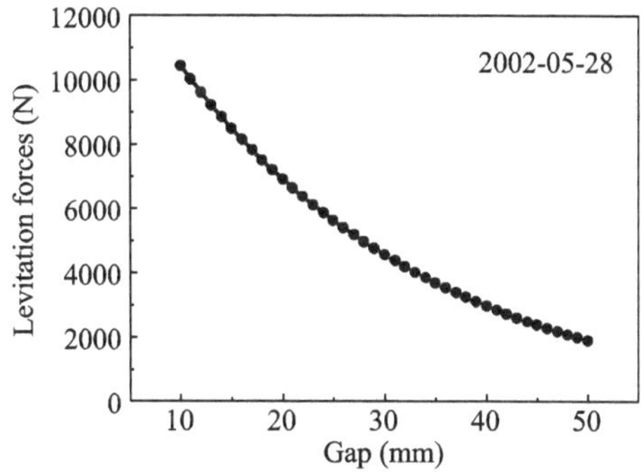


**Fig. 6.60:** Total levitation forces of the eight on-board HTS Maglev equipments above the PMG [22].

**Fig. 6.61:** Summary of the measurement results of the eight HTS Maglev equipments.

SCML-01, and especially variations in the distances between the eight on-board HTS magnetic levitation equipments and the center of the PMG. The accuracy of the measurement system SCML-01 was high. Although there were some differences between the measurement process and the results, the general measurement results were reliable. Figure 6.61 shows the summary of the measurement results of the eight HTS Maglev equipments.

In addition, in order to simulate the actual situation for the Maglev vehicle, the levitation forces of two parallel on-board HTS Maglev equipments were measured with the SCML-01 because the levitation forces of the Maglev vehicle system on a

PMG cannot be precisely measured. Figure 5.8 shows the measuring system for the levitation forces of two on-board HTS Maglev equipments over two NdFeB PMGs. In order to accurately test the levitation forces of two parallel on-board HTS Maglev equipments, a three-dimensional force sensor was used. The interference of the two force components in the horizontal plane must be eliminated at the same time of the measurement. The measurement results showed that the total levitation forces of two parallel on-board HTS Maglev equipments were in agreement with the sum of two single on-board HTS Maglev equipments. We confirmed that the vehicle performance could be achieved completely through this test before the completion of the HTS Maglev vehicle – the "Century" system.

## 6.17 First manned HTS Maglev vehicle in the world – "Century" [1, 2]

One of the prospective applications of HTS YBCO bulk superconductors is the superconducting Maglev vehicle [45–59]. The Maglev concept using HTS bulk had been demonstrated by the Beijing Small Model [55, 58] through the cooperation between China and Germany. There was no report about manned HTS Maglev vehicle at that moment, since the HTS Maglev vehicle is a large system rather than a small model. At that time, developing a manned HTS Maglev was very challenging.

With the improvement of the preparation technology of YBCO bulk, melt-textured YBCO bulk was made that could achieve higher critical current density and trapped magnetic flux density. The HTS Maglev measurement system SCML-01 had been developed successfully (see Chapter 5), and a series of valuable research results had been obtained with this test device. The PMG, the rectangular liquid nitrogen vessel with large size and thin bottom, and the on-board HTS Maglev equipment had been developed successfully. Substantial progress in theoretical and experimental research had finally made the development of the manned HTS Maglev vehicle possible.

In accordance with the requirements of the National 863 Program, after the research results of HTS Maglev had met the prerequisite conditions, the manned HTS Maglev vehicle system project was approved. In January 14–15, 2000, Zhong-Xian Zhao presided over the conference for the HTS Maglev vehicle demonstration program in Southwest Jiaotong University. Experts confirmed that the early research results had reached the expected target and unanimously approved the development scheme for a manned vehicle system proposed by Jia-Su Wang et al. The R&D of the Maglev vehicle system formally started.

Before the breakthroughs in HTS theory and basic technology on HTS Maglev over a PMG, the design of the HTS Maglev vehicle system was considered. Once the previous research results were recognized by our fellow scientists, immediately we began to implement the system development plan. The design, processing, assembly, and debugging for the manned HTS Maglev vehicle system were completed in 2000.

(a) Installing the permanent magnetic guideway (PMG) and linear motor

(b) Measuring the distribution of magnetic field around the PMG

(c) The Maglev system including the vehicle body frame on the PMG.

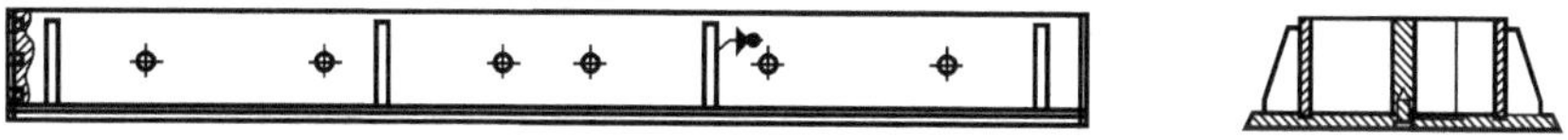

**Fig. 6.62:** The PMG of the total length of 15.5 m had a magnetic flux density of about 0.4 T at height of 20 mm (see Section 6.14).

First, the supporting beam for the PMG was installed. Then each segment of the PMG was mounted on the beam. The PMG consisted of two parallel PM tracks, whose concentrated magnetic field at the height of 20 mm is about 0.4 T. The total length of the PMG was 15.5 m. The linear motor, power supply, and control unit, liquid pressure elevation equipment, photoelectric positioning device, etc. were assembled. Figure 6.62 shows the PMG (see Section 6.14). Figure 6.62a shows the installed PMG and linear motor, Fig. 6.62b shows the measuring distribution of the magnetic field around the PMG, and Fig. 6.62c shows the Maglev system with the vehicle body frame on the PMG.

Each on-board HTS Maglev equipment was 550 mm long, 150 mm wide, and 160 mm high. The bottom wall thickness was only 3 mm. There were 43 pieces of HTS YBCO bulks in each rectangular-shaped liquid nitrogen vessel. The vehicle was driven by a LIM and controlled by a ground control device. There was no control device on the vehicle. The control device only controlled linear motor drive system. The levitation and guidance systems were not required control. The stator of the linear motor was positioned between two PMGs. The rotor, an induction plate, was attached on the bottom of the vehicle. The magnetic field distribution between two the PMGs and the linear motor was determined by theoretical calculation and experimental test. There was no interaction when the linear motors operated. The whole stator was 15.5 m in length, separated into six segments with each segment provided electric power separately. No electric power was required in the vehicle, except for the stator of the linear motor on the support beam. When the stator was powered, it produced a traveling magnetic wave. The interaction of the traveling magnetic wave and the induction plate generated an electromagnetic force, which drove the vehicle forward. The function of suspension was separated from the function of propulsion. The HTS Maglev vehicle stably moved forward and backward using the linear motor.

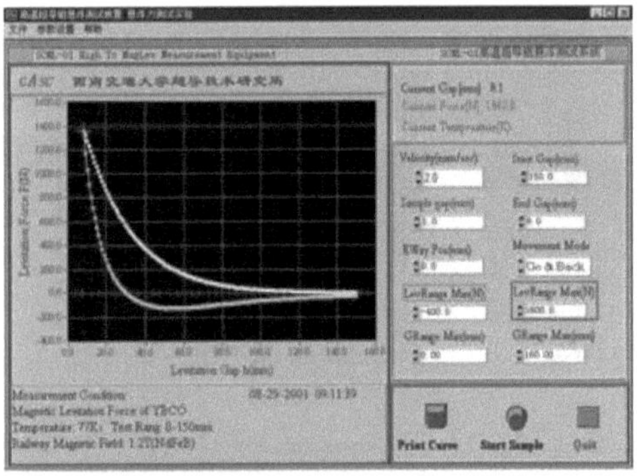


**Fig. 6.63:** Levitation forces of a single on-board HTS Maglev equipment over a single-NdFeB PMG on August 29, 2000 [1].

**Tab. 6.14:** Total levitation forces of eight on-board HTS Maglev equipments over the PMG.

| | 10 | 15 | 20 | 25 | 30 |
|---|---|---|---|---|---|
| 2000/12/24 | 10,884.5 | 9221.7 | 7562.6 | 6185.3 | 5061.2 |
| 2001/07/21 | 10,943 | 8940.4 | 7270.6 | 5889.7 | 4790.7 |

To obtain both larger levitation force and guidance force, HTS YBCO bulks had to be cooled in an appropriate applied magnetic field. First, the vehicle body was lifted with a liquid pressure elevation equipment until the bottom of the on-board HTS Maglev equipment was 42 mm away from the surface of the PMG, then YBCO bulks were cooled with liquid nitrogen. After the YBCO bulks were in the superconducting state, the vehicle body was released from the elevation equipment. The HTS Maglev provide inherent stable forces both in levitation and for guidance. The net gap between the bottom of the on-board HTS Maglev equipment and the PMG face was 35 mm. After the cryogenic vessel was filled with liquid nitrogen, the vehicle could continuously run for about 6 hours.

Figure 6.63 shows the measurement result of the single on-board HTS Maglev equipment above a single-NdFeB PMG from August 29, 2000 [1]. The levitation forces on the single rectangular-shaped liquid nitrogen vessel were 1055 and 1362 N at levitation gaps (between the PMG face and HTS bulks) of 15 and 8 mm, respectively. In addition, the levitation force increased to 1823 N at a levitation gap of 7 mm in another experiment over the NdFeB PMG.

Figure 6.64 and Tab. 6.14 show the total levitation forces of the eight on-board HTS Maglev equipments over the PMG. The total levitation forces of the eight on-board HTS Maglev equipments over the NdFeB PMG were 9000 and 11,000 N at levitation gaps of 15 and 10 mm, respectively. The total levitation forces of the eight on-board HTS Maglev equipments over the NdFeB PMG were 8000 and 10,500 N at net levitation gaps (deduct the 3 mm bottom thickness of liquid nitrogen vessel) of 15 and 8 mm, respectively.

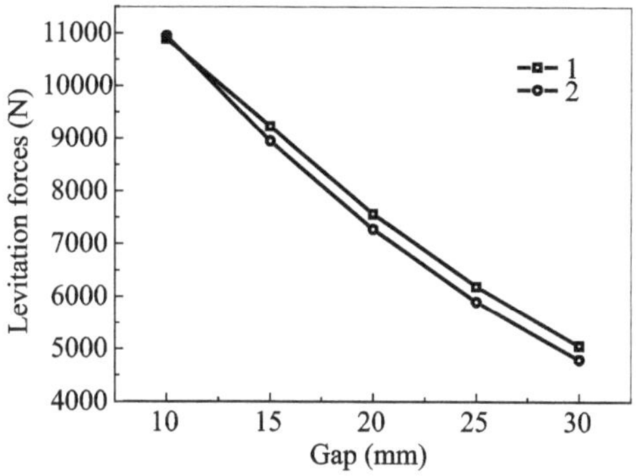


Fig. 6.64: Total levitation forces of the eight on-board HTS Maglev equipments over the PMG [1].

(a) (b) (c)

Fig. 6.65: The first manned HTS Maglev vehicle in the world [1] was tested successfully at 2:26 pm, December 31, 2000, at ASCLab, Southwest Jiaotong University, P. R. China. (a) The first manned HTS Maglev vehicle in the world—vehicle body framework without an outer shell. (b) First pass perfusion of liquid nitrogen to the on-board HTS Maglev equipments on December 31, 2000 [21]. (c) The net levitation gap of the HTS Maglev vehicle was more than 20 mm when five people stood on the vehicle [1].

The first manned HTS Maglev vehicle in the world [1] (Fig. 6.65) was tested successfully at 2:26 pm, December 31, 2000, at ASCLab, Southwest Jiaotong University, China. The on-board HTS Maglev equipment was fixed to both sides of the vehicle body frame. Figure 6.65a shows the vehicle body frame without the outer shell of first manned HTS Maglev vehicle in the world: 2268 mm long, 1038 mm wide, and 120 mm high. The external size of the vehicle "Century", with an outer shell of 3.5 m long, 1.2 m wide, and 0.8 m high and there were four seats in the HTS Maglev vehicle (Fig. 4.10b).

Figure 6.65b is the first-pass perfusion of liquid nitrogen to the on-board HTS Maglev equipments on December 31, 2000 [21]. Figure 6.65c shows that the net levitation gap of the HTS Maglev vehicle was more than 20 mm when five people stood on the vehicle. [1]

During this test run, five people were standing on the vehicle body and the total weight was about 530 kg. The net levitation gap was more than 20 mm. Accurate measurements later showed that the total levitation forces of the entire Maglev vehicle were 6351 and 7850 N at net levitation gaps of 20 and 15 mm, respectively, in July 2001. The net levitation gap was the distance between the PMG's upper surface and the liquid nitrogen vessel's bottom.

The essential parameters for the first manned HTS Maglev test vehicle in the world – “Century” are listed in Tab. 4.1.

In order to improve the performance of the manned HTS Maglev vehicle system, in-depth and detailed theoretical and experimental research have been carried out unceasingly. Through continuous study for more than 10 years, it was found that the cross-sectional area of the PMG could be further reduced to the about half to third of the original size. On the one hand, the corresponding PMG was designed according to different HTS bulk properties. On the other hand, the shape and size of the HTS bulks were prepared for a small area PMG. Consequently, it is possible to improve the performance of the levitation system and reduce the manufacturing and operating cost greatly.

In January 3, 2001, the press conference about “Century” was held in Southwest Jiaotong University, Chengdu, China (Fig. 6.66).

The Committee of Experts in Superconductivity of National 863 Program reviewed and approved the completed Maglev vehicle project on February 11, 2001 (Fig. 6.67) in Southwest Jiaotong University, Chengdu, China. “Century” was highlighted in the Review of 15 Years (1986–2000) for National 863 Program of China in Beijing, 2001 (Fig. 4.10c). Five years later, “Century” was the main attraction in the First Science and Technology Festival of Chengdu, 2005 (Fig. 6.68). The performance of “Century” was still similar to its performance in 2001.

**Fig. 6.66:** On January 3, 2001, the press conference was held at Southwest Jiaotong University. (From the *People’s Daily Overseas Edition*.)

**Fig. 6.67:** The Committee of Experts in Superconductivity of National 863 program reviewed and approved completed Maglev vehicle project on February 11, 2001.

**Fig. 6.68:** The vehicle "Century" on the First Science and Technology Festival of Chengdu in 2005.

## 6.18 Guidance forces of the entire HTS Maglev vehicle [60, 61]

Both guidance and levitation forces are very important parameters for the HTS Maglev vehicle. The levitation forces between HTS bulks and PMGs are directly proportional to the gradient of the magnetic field over PMG. The guidance forces provide the lateral stability of the Maglev vehicle. Both levitation and guidance forces depend on the flux pinning capability of HTS bulks.

The lateral guidance forces are dependent on the trapped flux in the HTS bulks and the applied magnetic fields. This Maglev vehicle with bulk HTS does not need any lateral stability control system, which makes it competitive compared with all other conventional Maglev vehicle systems. The guidance forces are large and sufficient in guiding the vehicle while large levitation forces are also achieved.

The guidance forces of the entire HTS Maglev vehicle are not only different from those in Section 6.7.1, but also not the same as those in Section 6.7.2. The guidance forces of HTS bulks over a single PMG are discussed in Section 6.7.1. Although Section 6.7.2 describes the guidance forces of HTS bulks over parallel two PMG, it is not the guidance forces of the entire vehicle. The guidance forces are also measured at forces sensor on both sides of the suspension carrier. The guidance forces of the entire vehicle at two parallel force sensors are measured at a force sensor on one side of the vehicle.

The measuring equipment of the guidance forces of the entire HTS Maglev vehicle is shown in Fig. 11. The set-up included two horizontal propulsion systems and two sets of force sensors which were fixed on the vehicle. Each set of the propulsion system could move in both the horizontal and vertical directions so that they could measure the guidance forces of the entire vehicle at different levitation gaps. A chain with a synchronization precision of 0.5 mm connected two sets of propulsion systems. The moving range of the propulsion system along the horizontal direction was 0 to 20 cm and the moving precision is 1 mm, and the range along the vertical direction was 0 to 10 cm with vertical moving precision of 1 mm. A photograph of the measuring equipment of guidance forces is shown in Fig. 5.12. In order to ensure the accuracy of the guidance force measurements at different levitations of the vehicle, we ordered eight pull and push force sensors with measuring ranges of 200, 500, 700, and 1000 kg of ±15% accuracy. In order to test the feasibility of the guidance force measurement

**Fig. 6.69:** The measuring scene of the guidance forces of the HTS aglev vehicle.

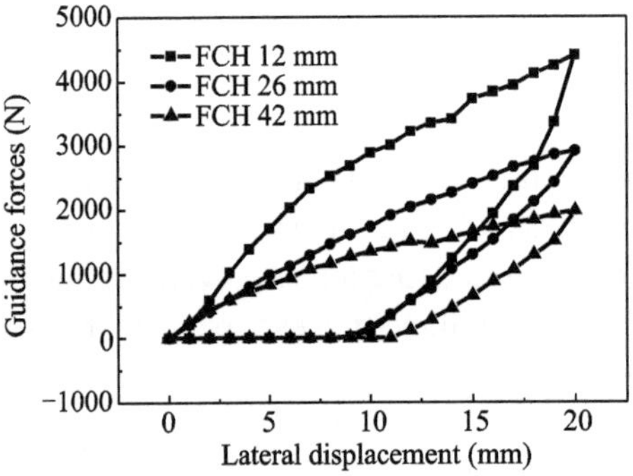


**Fig. 6.70:** Results of the guidance forces measured for the entire HTS Maglev vehicle [60].

**Fig. 6.71:** Firsthand experience of the guidance forces of the Maglev vehicle framework.

equipment, the guidance forces of the HTS Maglev vehicle were measured at the no-load condition and in the manned condition. Measurements need several people simultaneously performing the operation (Fig. 6.69).

Under no-load condition, the on-board Maglev equipments were fully cooled at the FCH of 42 mm. Henceforth, vehicle support was removed, and the net levitation height (work height) was 26 mm for the free levitation vehicle. The net levitation height was the gap between the PMG surface and the bottom surface in on-board Maglev equipments.

The experimental results of the guidance forces are listed in Tab. 6.15 and Fig. 6.70. It can be seen that the lateral guidance forces of the entire HTS Maglev vehicle were large enough (Fig. 6.71). The guidance forces at a 20-mm displacement were 4407, 2908, and 1980 N for FCH of 12, 26, and 42 mm, respectively. The guidance forces at 11 mm displacement were 3003, 1913, and 1420 N for FCHs of 12, 26, and 42 mm, respectively. The experimental results for the guidance forces under the maximum lateral displacement of 20 mm are shown in Fig. 6.70. The lower the FCH, the higher the guiding force. This is because the trapped flux was more at the lower position.

**Tab. 6.15:** Guidance force under different FCHs [60]. (Disp., displacement; N, forces unit (newton); FCH, field cooling height).

| Disp. (mm) | FCH (mm) | | | Disp. (mm) | FCH (mm) | | |
|---|---|---|---|---|---|---|---|
| | 12 | 26 | 42 | | 12 | 26 | 42 |
| 1 | 215 | 207 | 250 | 19 | 3364 | 2422 | 1515 |
| 2 | 592 | 414 | 450 | 18 | 2682 | 2120 | 1295 |
| 3 | 1026 | 603 | 600 | 17 | 2361 | 1834 | 1070 |
| 4 | 1391 | 802 | 725 | 16 | 1932 | 1533 | 885 |
| 5 | 1713 | 987 | 835 | 15 | 1580 | 1301 | 665 |
| 6 | 2033 | 1122 | 945 | 14 | 1244 | 1073 | 470 |
| 7 | 2330 | 1289 | 1085 | 13 | 886 | 767 | 295 |
| 8 | 2521 | 1465 | 1165 | 12 | 590 | 606 | 120 |
| 9 | 2685 | 1610 | 1275 | 11 | 354 | 375 | 0 |
| 10 | 2883 | 1734 | 1355 | 10 | 110 | 190 | 0 |
| 11 | 3003 | 1913 | 1420 | 9 | 20 | 24 | 0 |
| 12 | 3220 | 2038 | 1510 | 8 | 0 | 0 | 0 |
| 13 | 3350 | 2151 | 1475 | 7 | 0 | 0 | 0 |
| 14 | 3418 | 2263 | 1565 | 6 | 0 | 0 | 0 |
| 15 | 3726 | 2399 | 1660 | 5 | 0 | 0 | 0 |
| 16 | 3834 | 2517 | 1740 | 4 | 0 | 0 | 0 |
| 17 | 3941 | 2659 | 1785 | 3 | 0 | 0 | 0 |
| 18 | 4117 | 2744 | 1845 | 2 | 0 | 0 | 0 |
| 19 | 4246 | 2851 | 1930 | 1 | 0 | 0 | 0 |
| 20 | 4407 | 2908 | 1980 | 0 | 0 | 0 | 0 |

The experimental data showed that the lateral guidance forces had a large hysteretic effect. The guidance forces rose rapidly when the vehicle left its initial position, and the rate of increase became smaller as the vehicle moved further away from its original position as shown in Fig. 6.70. The guidance forces dropped rapidly when the vehicle moved back to its initial position from the maximum lateral displacement of 20 mm, and vanished at about 11 mm for a FCH of 42 mm. This showed that the trend of the vehicle to return to the original position had disappeared at 11 mm. Thus, it was just a kind of pseudo-guiding force. The true meaning of guidance force was to direct the suspension carrier back to the original point, i.e. the guidance forces should be restored to the original center position without an additional external force. This result implied that the maximum lateral displacement is equal to or less than 11 mm, for the vehicle to be able to return to the initial point. When the vehicle is pushed away from the center position 11 mm (lateral displacement), it should return to the origin position. The maximum guidance force is equal to or less than 1420 N for a FCH of 42 mm with a free levitation height of 26 mm.

These measurement results were obtained under the condition of a slow measurement rate. Because of the large hysteretic effect in guidance force measurements,

**Tab. 6.16:** Horizontal magnetic field distribution at height of 25 mm above the PMG [29].

| | | | | | | | |
|---|---|---|---|---|---|---|---|
| Horizontal position (mm) | −2 | 0 | 2 | 4 | 6 | 8 | 10 |
| Field *B* (T) | 0.354 | 0.354 | 0.354 | 0.349 | 0.342 | 0.332 | 0.322 |
| Horizontal position (mm) | 12 | 14 | 16 | 18 | 20 | 22 | 24 |
| Field *B* (T) | 0.310 | 0.295 | 0.281 | 0.266 | 0.251 | 0.220 | 0.205 |

the above measurement may not be accurate. The maximum guidance force should be an instantaneous explosive force which maintained the original position. When a sudden impact force acted on the Maglev vehicle system, the guidance forces were not reflected with the above slow measurement rate. Perhaps the relaxation had ceased to exist in this emergency, because the impact had already ended before relaxation can occur.

Trapped flux was 0.220 T for a FCH of 42 mm. If both the trapped flux and surface magnetic field of the PMG did not change during the measuring process of the guidance forces, the measured guidance forces were the magnetic attractive forces between the trapped flux of 0.220 T and magnetic field of PMG. In a certain range, the guidance forces gradually increased with the increase of horizontal displacement, but beyond this range it was not attractive, i.e. the measurement guidance force was zero (in fact, the guidance force of the original position was an objective existence value). Table 6.16 lists the horizontal magnetic field distribution at a levitation height of 25 mm above the PMG. The levitation height of 25 mm can be used as a reference in the analysis of the experimental data at the levitation height of 26 mm. However, the surface magnetic field of the PMG was smaller and smaller with the increase of the horizontal displacement. The effect of the horizontal displacement on the guidance forces was much larger than that of the horizontal field change on the PMG surface.

In order to research the true effective guidance forces at certain lateral displacement, the guidance forces were measured while gradually increasing the maximum lateral displacement [61]. Figure 6.72 shows the experimental results of the guidance forces of the entire vehicle under different maximum lateral displacement.

Figure 6.72 shows that the vehicle could return to the initial position after the horizontal lateral forces were removed at the maximum lateral displacement of 2 mm. The vehicle could not return to the initial position after the horizontal lateral forces were removed at the maximum lateral displacement of 4 mm. There was separation between the force sensor and the propulsion system at a lateral displacement of 1 mm. In the operating state, the effective lateral displacement of the vehicle was less than 4 mm. The guidance force at a vehicle lateral displacement of 4 mm was 377 N. The guidance force at a vehicle lateral displacement of 6 mm was 524 N. Thus, it can be seen that the guidance forces increased with the increase of lateral displacement; however, these are the above-mentioned pseudo-guidance forces.

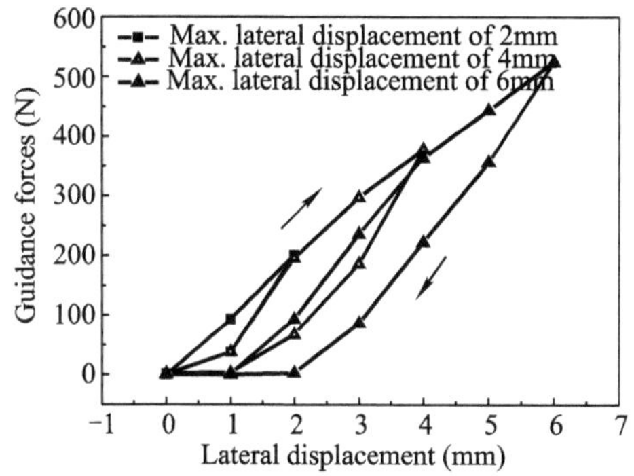


**Fig. 6.72:** Experimental results of guidance forces of the whole vehicle under different maximum lateral displacement [61].

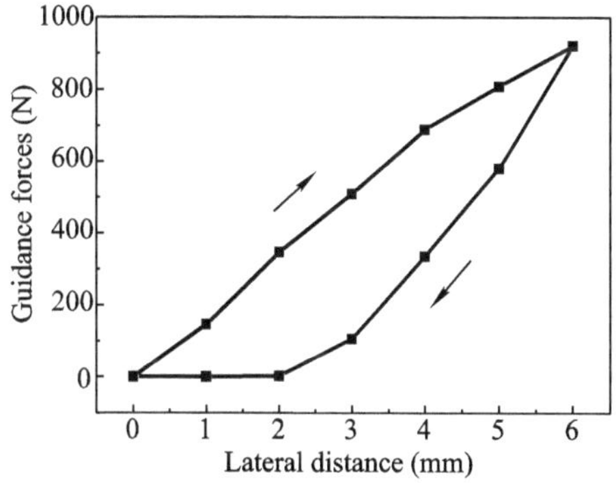


**Fig. 6.73:** Experimental results of guidance forces with three people loaded [60].

Figure 6.73 shows the experimental results for the guidance forces with three people on the vehicle. The total loading weight was about 500 kg with a levitation height (work height) of 20 mm. It can be seen that the guidance forces rapidly increased with the decrease of the levitation height due to the increase in weight. The guidance force at a vehicle lateral displacement of 6 mm was 921 N, and it was about twice that in the no-load condition. However, the vehicle was not centered by 2 mm after the removal of the horizontal lateral forces, and it was like the no-load condition. According to these measured results, the previous conclusion about the guidance force of 1980 N may be incorrect.

The previous guidance forces were defined as the forces that would compel the carrier to return to its original location of the magnetic suspension balance after lateral displacement of the carrier over the PMG. All of the above experiments were conducted according to this definition.

In fact, the guidance force should be defined as the maximum lateral force for which it does not leave the original position after applying lateral force. It is not precise to identify the guidance forces from which its position cannot return to the origin. For example, it is not appropriate that the maximum guidance force was affirmed to be 1420 N according to the lateral forces vanishing at about 11 mm for a FCH of 42 mm.

Guidance force research is quite complex. Not only larger relaxation phenomenon can occur, but also the trapped flux may change in the testing process. Further research is very necessary. In order to provide simultaneously both the levitation forces and sufficient guidance forces for the Maglev system, it is necessary to find an effective way to measure these forces more precisely.

Ma et al. [62] reported the investigation of the lateral guidance forces characteristics of an analogous HTS Maglev vehicle system of YBCO bulks above a PMG by measuring its resonant frequency ($f_{RF}$) after being subject to a lateral displacement. In conditions of FCH at 45 mm and working height (WH) of 25 mm, the measured maximum restorable lateral displacement was 2.5 mm. The experimental conditions were almost the same as for the above guidance force measurement of the entire Maglev vehicle (FCH of 42 mm and WH of 26 mm). The measured maximum restorable lateral displacement of 2.5 mm was identical with that of 2 mm in Fig. 6.72.

## 6.19 Long-term stability of the HTS Maglev vehicle in 2001–2003

Long-term stability of the HTS Maglev vehicle is very important. 46 days After the "Century" was born, it was loaded onto two trucks and transported to Beijing for exhibition at the Review of 15 Years for National High-tech R&D Program (National 863 Program) of China. In 2001, Tanaka, the chairman of the 14th International Symposium on Superconductivity in Japan invited Jia-Su Wang to report about the title of the first manned HTS Maglev vehicle in the world. At the meeting, Jia-Su Wang reported the research results of "Century" and also published the first results [1] (see Fig. 6.64) on the stability studies of the HTS Maglev vehicle.

The results of the long-term stability of the HTS Maglev vehicle were measured with SCML-01 HTS Maglev measurement system in July 2001, December 2001, May 2002, and March 2003 [1, 12, 21, 26]. Table 6.17 lists the total levitation forces on the eight on-board HTS Maglev equipments over the PMG at different levitation gaps.

**Tab. 6.17:** Total levitation forces of eight on-board HTS Maglev equipments over the PMG (N) in 2000–2003.

| Gap (mm) | 2000/12/24 | 2001/07/21 | 2001/12/24 | 2002/05/28 | 2003/03/05 |
|---|---|---|---|---|---|
| 10 | 10,885 | 10,943 | 10,091 | 10,431 | – |
| 15 | 9222 | 8940 | 8457 | 8486 | 8633 |
| 20 | 7563 | 7271 | 6927 | 6908 | 7094 |
| 25 | 6185 | 5890 | 5656 | 5618 | 5782 |
| 30 | 5061 | 4791 | 4685 | 4561 | 4693 |

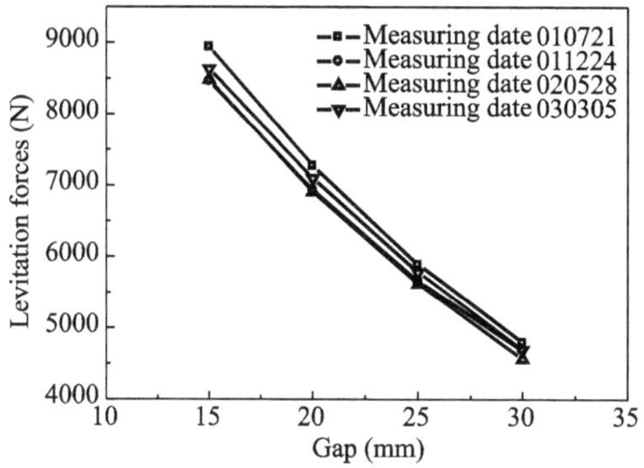


**Fig. 6.74:** Variation of the total levitation forces of entire HTS Maglev vehicle over the PMG [12].

In December 2000, the total levitation forces of the eight on-board HTS Maglev equipments over the PMG were 9222 and 7563 N at levitation gaps of 15 and 20 mm, respectively. In July 2001, the total levitation forces were 8940 and 7271 N at levitation gaps of 15 and 20 mm, respectively. The total levitation forces were 8633 and 7049 N at levitation gaps (subtract the 3-mm bottom thickness of the liquid nitrogen vessel) of 15 and 20 mm, respectively, in March 2003.

Figure 6.74 shows the variation for the total levitation forces of the eight on-board HTS Maglev equipments (entire HTS Maglev vehicle) over the PMG from 2001 to 2003 [12].

The total levitation forces of the entire HTS Maglev vehicle were 8486 and 6908 N at levitation gaps of 15 and 20 mm, respectively, in May 2002. At a gap of 30 mm, there was a 46% decrease in levitation force compared to the gap of 15 mm.

The data set of total levitation forces for March 2003 was slightly higher than that of May 2002. This was because the center lines of the on-board HTS Maglev equipments and the PMG were laid to overlap more accurately in March 2003. The overlap accuracy of both center lines directly impacts the magnitude of the levitation forces.

The comparison of total levitation forces of the entire HTS Maglev vehicle during the 10-month period from July 2001 to May 2002 showed that the levitation force decreased by 5.1%, 5.0%, 4.6%, and 4.8% at levitation gaps of 15, 20, 25, and 30 mm, respectively. All data were nearly the same, i.e. each levitation force decrease was about 5.0% at the different levitation gaps.

The levitation forces became lower after 2 years. Up to September 2009, over 40,000 passengers had ridden on the Maglev vehicle, and the total recorded shuttle distance was about 400 km. Experimental results indicate that the long-term stability of the HTS Maglev vehicle is better in the static mode than when the vehicle is operating at low speeds. HTS Maglev can be used in high-speed traffic, but the pinning behavior must be at a higher level, which is better adapted to high-speed needs. Therefore, the investigation of the dynamic properties of the HTS Maglev vehicle at the high-speed operation is extremely important.

## 6.20 Long-term stability of YBCO bulks in 2001–2009 [63]

In order to evaluate the long-term stability of the single HTS bulks for practical application, the levitation performance of single bulk YBCO samples were investigated with the SCML-02.

The 43 YBCO bulk pieces were packaged in the No. 6 on-board HTS Maglev equipment in "Century" in 2000. The YBCO bulk samples were not opened, but they were measured independently in separate tests while running for nine years. After opening the seal cover and removing the YBCO bulks, the levitation forces of each YBCO bulk were measured in 2009. These YBCO samples were signed $A_{15}$, $B_{14}$, and $C_{14}$, and included 15 pieces over the center of the PMG and 14 pieces over both sides of the center of the PMG. Table 6.18 lists the average values of measurements for $A_{15}$, $B_{14}$, and $C_{14}$ of 45 YBCO bulks.

From the test results, the average levitation forces of the three groups, $A_{15}$, $B_{14}$, and $C_{14}$ YBCO bulk, were close. The $A_{15}$ group was slightly higher than that of the other two groups. The better YBCO bulks in the same batch were located in the strong magnetic field at the PMG center line in 2000. This was consistent with the measured results. This showed that, after the long-term operation in about 9 years, the performance of $A_{15}$ group bulk were not degraded at the work environment of the high magnetic field.

The single piece of A5 YBCO bulk over the center of the PMG had the maximum levitation force. The YBCO bulk 07-1 was produced in 2007, and old-1 was an old YBCO bulk produced in 2001 and not loaded for operation. Table 6.19 lists the comparison of long-term stability of single YBCO bulk. It shows that the maximum levitation forces of the A5 bulk and the old-1 bulk sample at a gap of 10 mm were very close. However, the levitation force of the A5 bulk and the old-1 bulk were higher than the 07-1 bulk. This was due to the smaller thickness of the 07-1 bulk.

**Tab. 6.18:** Average values of $A_{15}$, $B_{14}$, and $C_{14}$ [63].

| | $A_{15}$ | $B_{14}$ | $C_{14}$ |
|---|---|---|---|
| Average levitation forces (N) | 30.95 | 29.10 | 28.54 |
| Maximum levitation force $F_{max}$ (N) | 33.59 | 31.9 | 30.2 |
| Piece number of $F_{max}$ | A5 | B1 | C4 |

**Tab. 6.19:** Comparison of long-term stability of single YBCO bulk [63].

| | A5 | 07-1 | Old-1 |
|---|---|---|---|
| Diameter (mm) | 31 | 31 | 30 |
| Thickness (mm) | 17.5 | 13 | 18.5 |
| $F_{max}$ (N) | 33.59 | 28.5 | 33.25 |
| Levitation forces of unit volume (N/mm$^3$) | $2.544 \times 10^{-3}$ | $2.906 \times 10^{-3}$ | $2.544 \times 10^{-3}$ |

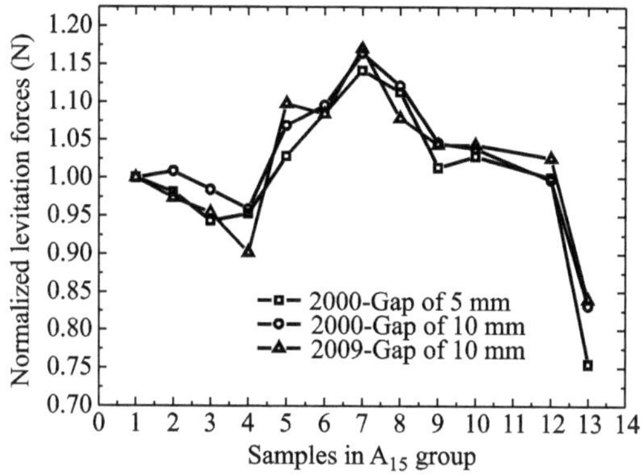


**Fig. 6.75:** Measurement results of YBCO bulk properties of some samples for $A_{15}$ group in 2000 and 2009 [63].

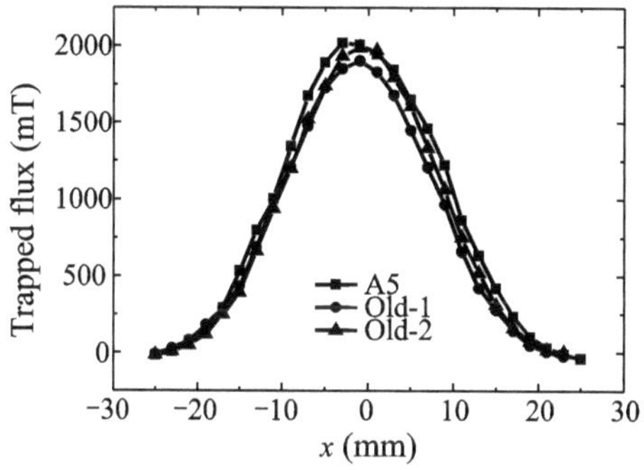


**Fig. 6.76:** Measurement results of trapped flux for several YBCO bulks in 2000 and 2009 [63].

In addition, the properties of the YBCO bulks from different periods for some samples in the $A_{15}$ group were compared. Figure 6.75 shows the measurement results of YBCO bulk properties for some samples from the $A_{15}$ group in 2000 and 2009 which were in close agreement. The figure demonstrates that the difference between levitation forces for different YBCO bulks was small after long-term use.

In order to confirm the influence of the vehicle operation on pinning in the YBCO bulk, the trapped flux of both bulk A5 and bulk old-1 was measured using constant magnetic field magnetization method. The old-2 bulk was another sample from the same A5 batch purchased, but not used in the loading operation.

The constant magnetic field was provided by an EM4-VC electromagnet from Lakeshore, duration was 15 minutes, and the magnetic field was 1.16 T at a current of 60 A. Magnetic field scan at gap of 6 mm above the YBCO bulk seeded surface was measured with 460 Gaussmeter and Hall probe. Figure 6.76 shows the measurement results for the trapped flux in several YBCO bulks.

As you can see from Fig. 6.76, the trapped flux in the A5 bulk and the old-2 bulk was close, and both were slightly higher than that with the old-1 bulk. In addition, the maximum trapped flux of A5, old-1, and old-2 were 2270, 2084, and 2132 mT, respectively. The measurement results showed that the trapped magnetic

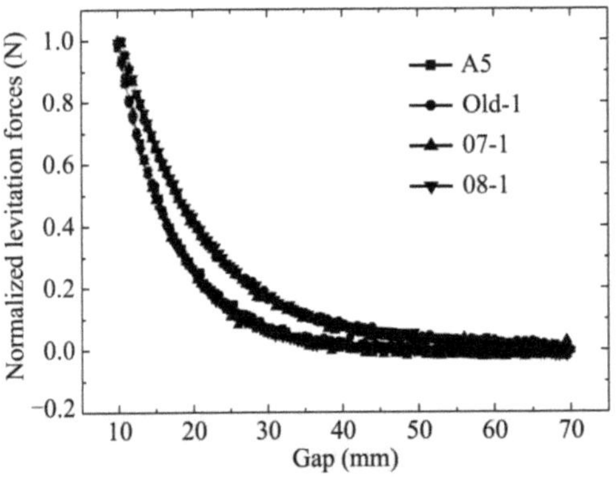


**Fig. 6.77:** Measurement results of levitation forces.

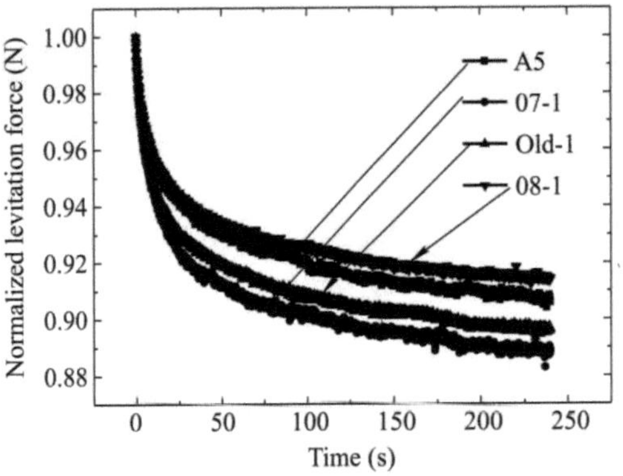


**Fig. 6.78:** Measurement results of levitation force relaxation.

ability, namely the pinning properties, did not decrease because of long-term loading operations.

The levitation force (Fig. 6.77), hysteresis, and relaxation (Fig. 6.78) of YBCO bulks, which were cooled at the liquid nitrogen temperature 77 K, were measured with the SCML-02 HTS Maglev measurement system in June 2009 [63]. (A5) is And on-board bulk YBCO from 2000, (old-1) is bulk YBCO from 2000 but not loaded, (07-1) was new in 2007; (08-1) was new in 2008 [63]. This was to analyze the variation of levitation performance of different bulk YBCO samples during different periods after fabrication of the YBCO.

Figure 6.77 is the comparison of the levitation forces of different bulk YBCOs fabricated during 2000–2008, and Fig. 6.78 is the comparison of the levitation force relaxation of these YBCO bulks. The square YBCO bulk 08-1 was produced in 2008.

These experimental results validated that the performances of the bulk YBCO samples above the PMG did not change after nearly 9 years. Our research results can be applied not only to the manned HTS Maglev vehicle, but also to other superconducting applications such as HTS Maglev bearings and superconducting flywheel energy storage devices.

## References

[1] Wang J, Wang S, Zeng Y, Huang H, Luo F, Xu Z, et al. The first man–loading high temperature superconducting Maglev test vehicle in the world. Physica C. 2002;378–381:809–814.

[2] Wang JS, Wang SY. A superconducting Maglev system. Chinese patent, No. ZL 01128867.1, 2001.

[3] Schultz L, de Haas O, Verges P, Beyer C, Rohlig S, Olsen H, et al. Superconductively levitated transport system – the SupraTrans project. IEEE Trans on Appl Supercond. 2005;15(2):2301–2305.

[4] Kovalev KL, Koneev SM-A, Poltavec VN, et al. Pro 8th Int Symp Magn Susp Technol. (ISMST'8), Dresden, Germany; 2005. p. 51.

[5] Stephan RM, Ferreira AC, de Andrade R, Neves MA, Jr., Cruzmoreira MA, Rosario MAP, et al. A superconducting levitated small scale vehicle with linear synchronous motor. IEEE Int Sym Indust Electro. 2004;1:206–209.

[6] Sotelo GG, Dias DH, Machado OJ, David ED, de Andrade R, Jr., Stephan RM, et al. Experiments in a real scale Maglev vehicle prototype. J Phys Conf Ser. 2010;234:032054.

[7] Okano M, Iwamoto T, Furuse M, Fuchino S, Ishii I. Running performance of a pinning–type superconducting magnetic levitation guide. J Phys Conf Ser. 2006;43:999–1002.

[8] D'Ovidio G, Crisia F, Lanzara G. A. "V" shaped superconducting levitation module for lift and guidance of a magnetic transportation system. Physica C. 2008;468:1036–1040.

[9] Moon FC. Superconducting levitation: applications to bearings and magnetic transportation. New York: Wiley; 1994.

[10] Murakami M. Materials developments and applications of bulk Re–Ba–Cu–O superconductors. In: Narlikar A, editor. Frontiers in superconducting materials, Springer Verlag; 2005. pp. 869–884.

[11] Wang JS, Wang SY. Synthesis of bulk superconductors and their properties on permanent magnet guideway. In: Narlikar A, editor. Frontiers in superconducting materials, Springer Verlag; 2005. pp. 885–912.

[12] Wang JS, Wang SY, Zeng YW, Deng CY, Ren ZY, Wang XR, et al. The present status of the high temperature superconducting Maglev vehicle in China. Supercond Sci Technol. 2005;18(2):S215.

[13] Song HH, Wang JS, Wang SY, Ren ZY, Wang XR, de Haas O, Fuchs G, Schultz L. Studies of YBCO electromagnetic properties for high-temperature superconductor Maglev technology. In: Martins BP, editor. New topics in superconductivity research. Nova Science Publishers Inc.; 2006. pp. 107–156.

[14] Wang JS, Wang SY. High temperature superconducting Maglev measurement system. In: Kr sharma M, editor. Advances in measurement systems. InTech; 2010. pp. 51–80.

[15] Ma GT, Wang JS, Wang SY. 3-D Finite-element modelling of a Maglev system using bulk high-$T_c$ superconductor and its application. In: Luiz AM, editor. Applications of high-$T_c$ superconductivity. InTech; 2011. pp. 119–146.

[16] Deng ZG, Wang JS, Zheng J, Lin QX, Wang SY. Quasi-static optimization of HTS Maglev systems with a focus on bulk superconductor part. In: Miryala M, editor. Superconductivity: recent developments and new production technologies. Nova Science Publishers; 2012. pp. 215–239.

[17] Wang JS, Wang SY, Zheng J. Recent development of high temperature superconducting Maglev system in China. IEEE Trans on Appl Supercond. 2009;19(3):2142–2147.

[18] Wang JS, Wang SY, Zheng J, Deng ZG, Lu YY, Ma GT, et al. Recent developments of high temperature superconducting Maglev system. Chinese J Low Temp Phys. 2009;31(5):142–150. (In Chinese).

[19] Wang JS. Past, present and future R&D of HTS Maglev in China. The 21st International Conference on Maglev, Daejeon, Korea, October 10–13, 2011, Invitation keynote speech No. 2.

[20] Wang JS, Wang SY. High temperature superconductive Maglev vehicle. In: Fan JD, editor. Endless quests. Peking University Press; 2014.

[21] Wang SY, Wang JS, Wang XR, Ren ZY, Zeng YW, Deng CY, et al. The man–loading high–temperature superconducting Maglev test vehicle. IEEE Trans Appl Supercond. 2003;13(2):2134–2137.

[22] Wang JS. Proceedings of the International Workshop on HTS Maglev (ISMAGLEV 2002), Chengdu, P. R. China, June 25–27, 2002.

[23] Wang SY, Wang JS. A liquid nitrogen vessel with thin wall. 1999;27(3):1–3. (In Chinese).

[24] Wang JS, Wang SY, Ren ZY, Zhu M, Jiang H, Tang QX. Levitation force of multi–block YBaCuO bulk high temperature superconductors. IEEE Trans on Appl Supercond. 2001;11(1):1801–1811.

[25] Wang SY, Wang JS, Ren ZY, Zhu M, Jiang H, Tang QX. Levitation force of a YBaCuO bulk high temperature superconductor over a NdFeB guideway. IEEE Trans on Appl Supercond. 2001;11(1):1801–1804.

[26] Wang JS, Wang SY, Ren ZY, Jiang H, Zhu M, Wang XR, et al. High temperature superconducting Maglev equipment on vehicle. Physica C. 2003;386:531–535.

[27] Song HH, De Haas O, Ren ZY, Wang XR, Zheng J, Wang XZ, et al. Magnetic interaction between multiple seeded YBCO bulks and the permanent magnet guideway. Physica C. 2004;407(1–2):82–87.

[28] Schäzle P, Krabbes G, Stover G, Fuchs G, Schlafer D. Multi–seeded melt crystallization of YBCO bulk material for cryogenic applications. Supercond Sci Technol. 1999;12(2):69–76.

[29] Song HH, Wang JS, Wang SY, De Haas O, Ren ZY, Wang XR, et al. The relationship between levitation force and stiffness in symmetrical and unsymmetrical applied fields. Supercond Sci Technol. 2005;18(2):S95.

[30] Jiang H, Wang JS, Wang SY, Ren ZG, Zhu M, Wang XR, et al. The magnetic levitation performance of YBaCuO bulk at different temperature. Physica C. 2002;378–381:869–872.

[31] Wang XR, Song HH, Ren ZY, Zhu M, Wang JS, Wang SY, et al. Levitation force and guidance force of YBaCuO bulk in applied field. Physica C. 2003;386(8):536–539.

[32] Song HG, Jiang H, Wang JS, Wang SY, Ren ZG, Wang XR, Wang XZ, Zhu M, Shen XM. Proceedings of the International Workshop on HTS Maglev (ISMAGLEV 2002), Chengdu, P. R. China, June 25–27, 2002. pp. 94–99.

[33] Jiang H, Tang QX, Wang SY, Ren ZG, Zhu M, Wang JS, Wang XR. Proceedings of the Beijing International Conference on Cryogenics (Beijing-ICC), Beijing, Oct 5–8, 2000. pp. 223–226.

[34] Wang XR, Wang JS, Wang SY, Ren ZG, Song HG, Wang XZ, et al. The relationship of guidance force between single and multiple cylindrical Y–Ba–Cu–O superconductors. Physica C. 2003;390:113–119.

[35] Wang XR. Master's degree thesis, Southwest Jiaotong University, China, 2003.

[36] Wang X, Ren Z, Song H, Wang X, Zheng J, Wang S, et al. Guidance force in an infinitely long superconductor and permanent magnetic guideway system. Supercond Sci Technol. 2005;18(2):S99.

[37] Wang XG, Wang JS, Wang SY, Ren ZG, Wang XR, Song HG. Proceedings of the International Workshop on HTS Maglev (ISMAGLEV 2002), Chengdu, China, June 25–27, 2002. pp. 100–104.

[38] Wang XR, Wang SY, Wang JS, Ren ZY, Song HH, Wang XZ, et al. Guidance force and lateral magnetic siffness of YBaCuO superconductors in array. Unpublished.

[39] Zhu M, Ren ZG, Wang SY, Wang JS, Jiang H. Influence of thickness and shape of a YBaCuO bulk high temperature superconductor on levitation force. Chinese J Low Temp Phys. 2002;24(3):213–217. (In Chinese).

[40] Ren ZG, Wang JS, Wang SY, Jiang H, Zhu M, Wang XR, et al. Influence of shape and thickness on the levitation force of YBaCuO bulk HTS over a NdFeB guideway. Physica C. 2003;384(1–2):159–162.

[41] Song HH, Wang JS, Wang SY, De Hass O, Ren ZY, Wang XR, et al. Levitation force and average force density between HTS YBCO bulks and permanent magnet guideway. Unpublished.

[42] Ren ZY, Wang JS, Wang SY, Wang XR, Song HG, Wang XG, et al. Enhancing of Levitation force of YBCO bulk in applied magnetic field. Chinese J Low Temp Phys. 2003;25(3):202–208.

[43] Ren ZY. Doctorate thesis, Southwest Jiaotong University, China, 2004.

[44] Wang SY, Wang JS, Zhang QG, Lian JS. Research and development of superconducting Maglev train. Superconductivity and Cryogenics. 1995;23(4):1–7. (In Chinese).

[45] Wang SY, Wang JS, Lian JS. High temperature superconducting Maglev. Superconductivity and Cryogenics. 1997;25(1):17–22. (In Chinese).

[46] Wang JS, Wang SY, Ren ZG, Lian JS, Lin GB, Zhang CF, et al. A scheme of Maglev vehicle using high $T_c$ bulk superconducters. IEEE Trans on Appl Supercond. 1999;9(2):904–907.

[47] Wang SY, Wang JS. High temperature superconducting electromagnetic levitation. 1999;27(4):8–12. (In Chinese).

[48] Wang SY, Wang JS, Lian JS. Some considerations using high $T_c$ bulk superconductors for Maglev transportation systems. In: Lin LZ, Shen GL, Yan LG, editors. Proceeding of the Fifteenth International Conference on Magnet Technology (Part one). Science Press; 1998. pp. 767–769.

[49] Ren ZY, Wang JS, Wang SY, Jiang H, Zhu M, Wang XR, et al. A hybrid Maglev vehicle using permanent magnets and high temperature superconductor bulks. Physica C. 2002;378–381:873–876.

[50] Wipf SL, Laquer HL. Superconducting permanent magnets. IEEE Trans on Magnetics. 1989;25(2):1877–1880.

[51] Tomita M, Murakami M. High–temperature superconductor bulk magnets that can trap magnetic fields of over 17 tesla at 29 K. Nature. 2003;421(6922):517–20.

[52] Sawh RP, Weinstein R, Carpenter K, Parks D, Davey K. Production run of 2 cm diameter YBCO trapped field magnets with surface field of 2 T at 77 K. Supercond Sci Technol. 2013;26(10):105014.

[53] Wang JS, Wang SY, Lin GB, Huang HY, Zhang CF, Zeng YW, et al. High $T_c$ superconducting magnetic levitation measurement system. High technol lett. 2000;10:55–58. (In Chinese).

[54] Wang SY, Wang JS, Ren ZY, Zhu M, Jiang H, Wang XR, et al. High temperature superconducting Maglev equipment on vehicle. Physica C. 2003;386:531–535.

[55] Zhang Y, Xu SG. In: Lin LZ, Shen GL, Yan LG, editors. Proceeding of the Fifteenth International Conference on Magnet Technology, Science Press; 1998. pp. 763–766.

[56] Weh H, Pahl H, Hupe H, Steingrover A, May H. Proceedings Maglev'95 14th International conference on Maglev; 1995. pp. 217–222.

[57] Weh H. In: Lin LZ, Shen GL, Yan LG, editos. Proceeding of the Fifteenth International Conference on Magnet Technology, Science Press; 1998. pp. 833–838.

[58] Wang JR, Zhang PX, Zhou L, Wu MZ, Gawalek W, Gornert P, Weh H. 15th International Conference on Magnet Technology: Beijing, October 20–24, 1997.

[59] Fujimoto H, Kamijo H, Higuchi T, Nakamura Y, Nagashima K, Murakami M, et al. Preliminary study of a superconducting bulk magnet for the Maglev train. IEEE Trans on Appl Supercond. 1999;9(2):301–304.

[60] Wang JS, Wang SY, Ren ZG, Wang XR, Zhu M, Jiang H, et al. Guidance forces on high temperature superconducting Maglev test vehicle. IEEE Trans on Appl Supercond. 2003;13(2):2154–2156.

[61] Ren ZG, Wang JS, Wang SY, Wang XR, Song HG, Wang XG. An equipment measuring guidance force of a HTS Maglev vehicle. Advanced technology of electric engineering and energy. 2003;22(1):77–80. (In Chinese).
[62] Ma GT, Wang JS, Lin QX, Liu MX, Deng ZG, Li XC, et al. Lateral restorable characteristics of the high-$T_c$ superconducting Maglev vehicle above the permanent magnet guideway. Physica C. 2009;469(21):1954–1957.
[63] Liu W, Wang SY, Lin GX, Yang P, Liu HF, Ma GT, et al. Research on the long—term operation stability of the bulk YBCO on vehicle. Chinese J Low Temp Phys. 2009;31(5):85–89. (In Chinese).

Guang-Tong Ma and Yi-Yun Lu

# 7 Numerical simulations of HTS Maglev

## 7.1 Introduction

Numerical simulation is a counterpart of the direct measurement in scientific research. It has the advantage of low cost and being flexible to tune the geometrical and material characteristics of the studied object, as well as being able to understand and verify the macroscopic observations by direct measurements. Before running the simulation, one should establish a theoretical model to mathematically characterize the physical properties of the studied object and then convert the model into computer language. Therefore, modeling is the theoretical basis for performing the simulations.

The theoretical basis for conducting simulations toward HTS Maglev is to properly establish the partial differential equations (PDEs) for governing the electromagnetic behavior of HTSC, which has properties being different from the conventional conductors, in the presence of non-uniform magnetic fields. In the early phase after the discovery of the HTSC with $T_c$ over 77 K in 1987, the theoretical models were basically explored in the framework of the mirror-image principle by considering the HTSC as a perfect diamagnet [1–3]. This is easy to implement, but the applicable scope is limited due to the dipole approximation, although its extended form to reflect the hysteresis of levitation forces has been suggested [4]. This sort of models cannot be used to study the Maglev performance of the HTSCs above the PMG, which has a large scale so that the dipole approximation does not hold.

Understanding the HTSC working in the mixture state, a number of electromagnetic models, directly derived from the Maxwell's equations, have been established and verified in both two- and three-dimensional levels [5–14] to investigate the Maglev performance of the HTSCs in a more realistic sense. Using these models, the geometrical and material effects on the levitation performance [15], and the distributions of the magnetic field and the induced supercurrent [7, 11, 14, 16]. have been studied. The effect of the mutual interaction among the superconducting constituents of a multi-blocked HTS unit on the Maglev performance has also been identified [14], apart from the influence of the lateral movement on the levitation forces [17, 18]. Also, the propitious configuration of PMG for improving the levitation capability has been estimated [8, 19, 20], and miscellaneous strategies for the practical design of such levitation systems have been suggested [8, 21–25].

In this chapter, our emphasis is to present the efforts and achievements of numerical modelings and simulations towards the HTS Maglev system with a configuration of translational symmetry in J. S. Wang group. The mathematical foundations for governing the electromagnetic behavior of HTSC subjected to magnetic excitation,

https://doi.org/10.1515/9783110538434-007

along with the analytical model to calculate the magnetic field generated by PMG, will be introduced at both two- and three-dimensional levels, after a brief description of the macroscopic electromagnetic properties of HTSC and the basic Maxwell's equations for the electromagnetic field. Making recourse to these and using the finite element method (FEM), we have developed numerical codes, on the basis of which some practically interesting cases of HTS Maglev have been numerically simulated to provide the scientific criteria for designing and optimizing a HTS Maglev system with translational symmetry, i.e. for the Maglev transportation application.

## 7.2 Maxwell's equations

Maxwell's equations [26–28] are a set of four independent integral or differential expressions that mathematically characterize the generation, interaction, and propagation of the electric and magnetic fields in the nature. These equations, originally proposed by Maxwell 150 years ago, unify the mathematical expression of the time-steady and time-dependent electromagnetic fields via the introduction of a displacement current in Ampère's law and constitute the theoretical foundation of the subject of electromagnetism.

### 7.2.1 Ampère's law with Maxwell's addition

This equation describes the relationship between the magnetic field and its sources, viz., the current and/or the changing electric field. It can be expressed in either the differential form

$$\nabla \times \boldsymbol{H} = \boldsymbol{J} + \frac{\partial \boldsymbol{D}}{\partial t}, \tag{7.1}$$

or the integral form

$$\oint_C \boldsymbol{H} \cdot \mathrm{d}\boldsymbol{l} = \int_S \left( \boldsymbol{J} + \frac{\partial \boldsymbol{D}}{\partial t} \right) \cdot \mathrm{d}\boldsymbol{s}, \tag{7.2}$$

where $\boldsymbol{H}$ is the magnetic field intensity, $\boldsymbol{J}$ is the current density, and $\partial \boldsymbol{D}/\partial t$ is the displacement current added by Maxwell to state that a changing electric field can induce a magnetic field, even without a current – the converse of Faraday's law. For situations with a low frequency so that the wavelengths are much longer that the configurations of interest, the term of displacement current can be reasonably dropped, which leads to a quasi-static approximation that will be used in the following simulations.

In the absence of a free surface current density, the tangential component of the magnetic field intensity will be continuous in passing through the surface from Region 1 ($\boldsymbol{H}_1$) to Region 2 ($\boldsymbol{H}_2$), as expressed by

$$\boldsymbol{n}_{12} \times (\boldsymbol{H}_2 - \boldsymbol{H}_1) = 0. \tag{7.3}$$

### 7.2.2 Faraday's law

This equation describes how a time-varying magnetic field generates an electric field in a stationary object. It can be expressed in either the differential form

$$\nabla \times \boldsymbol{E} = -\frac{\partial \boldsymbol{B}}{\partial t}, \tag{7.4}$$

or the integral form

$$\oint_C \boldsymbol{E} \cdot \mathrm{d}\boldsymbol{l} = -\int_S \frac{\partial \boldsymbol{B}}{\partial t} \cdot \mathrm{d}s. \tag{7.5}$$

The tangential component of the electric field $\boldsymbol{E}$ is always continuous in passing through the surface from Region 1 ($\boldsymbol{E}_1$) to Region 2 ($\boldsymbol{E}_2$), i.e.

$$\boldsymbol{n}_{12} \times (\boldsymbol{E}_2 - \boldsymbol{E}_1) = 0. \tag{7.6}$$

### 7.2.3 Gauss's law

This equation presents the relationship between the electric field and the electric charge that generates the electric field. It can be expressed in either the differential form

$$\nabla \cdot \boldsymbol{D} = \rho, \tag{7.7}$$

or the integral form

$$\int_S \boldsymbol{D} \cdot \mathrm{d}\boldsymbol{s} = \int_V \rho \mathrm{d}v. \tag{7.8}$$

In the absence of a free surface charge density, the normal component of the electric displacement will be continuous in passing through the surface from Region 1 ($\boldsymbol{D}_1$) to Region 2 ($\boldsymbol{D}_2$), as expressed by

$$\boldsymbol{n}_{12} \cdot (\boldsymbol{D}_2 - \boldsymbol{D}_1) = 0. \tag{7.9}$$

### 7.2.4 Conservation of magnetic flux density

This equation states that the magnetic flux lines have no beginning or ending, and any magnetic flux line that enters an object must exit that object. It can be expressed in either the differential form

$$\nabla \cdot \boldsymbol{B} = 0, \tag{7.10}$$

or the integral form

$$\int_S \boldsymbol{B} \cdot \mathrm{d}\boldsymbol{s} = 0. \tag{7.11}$$

The normal component of the magnetic flux density $\boldsymbol{B}$ is always continuous in passing through the surface from Region 1 ($B_1$) to Region 2 ($B_2$), i.e.

$$\boldsymbol{n}_{12} \cdot (\boldsymbol{B}_2 - \boldsymbol{B}_1) = 0. \tag{7.12}$$

## 7.3 Macroscopic electromagnetic properties of HTSC

### 7.3.1 Nonlinear constitutive equation

HTSC is a nonlinear conducting medium, with the resistivity (conductivity) being dependent on the parameters of the local electromagnetic field, which requires the constitutive law between the electric field $\boldsymbol{E}$ and current density $\boldsymbol{J}$ in HTSCs to be represented by a nonlinear formulation, rather than a linear one of the traditional Ohm's law.

The typical model of the nonlinear $\boldsymbol{E}$-$\boldsymbol{J}$ constitutive law in HTSCs is Bean's model of the critical state [29], which postulates that the current density in HTSCs is limited by a critical value $J_c$, and until this threshold is reached, the electric field is zero. This assumption results in a discontinuous expression and is very difficult to represent in numerical simulation. The general form of Bean's model used in simulations is that proposed in [30]

$$\boldsymbol{J} = \begin{cases} J_c \dfrac{\boldsymbol{E}}{|\boldsymbol{E}|} & (|\boldsymbol{E}| \neq 0), \\ \dfrac{\partial \boldsymbol{J}}{\partial t} & (|\boldsymbol{E}| = 0). \end{cases} \tag{7.13}$$

In modeling the HTS Maglev, this constitutive law is not appropriate for investigating problems such as force relaxation [31] and drift under vibration [32] due to the assumption that the current flowing in the HTSCs is constant with time and the lack of material-related parameters.

The second model of the nonlinear **E-J** constitutive law in HTSCs is the so-called power law [33, 34], which is expressed as

$$\boldsymbol{E} = E_c \left( \frac{|\boldsymbol{J}|}{J_c} \right)^n \frac{\boldsymbol{J}}{|\boldsymbol{J}|}, \tag{7.14}$$

with $1 \leqslant n \leqslant \infty$. This power law corresponds to a logarithmic current dependence of the activation energy $U(J) = U_c \ln(J_c/J)$, which inserts into an Arrhenius law yielding $E(J) = E, \exp(-U/\kappa\Theta) = E_c(J/I_c)^n$. Index $n$ is defined as $n = U_0/\kappa\Theta$, where $U_0$ is the pinning potential of the HTSC at an absolute temperature $\Theta$ and $\kappa$ is the Boltzmann constant. At the extremes, the power law reduces to the Ohm's law with $n = 1$ and to the Bean's model with $n \to \infty$.

The third model of the nonlinear **E-J** constitutive law in HTSCs is the so-called flux flow and creep model, which was derived from the Anderson's theory [35] and has the form [31],

$$\boldsymbol{E} = \begin{cases} 2\rho_c J_c \sinh\left(\dfrac{U_0 |\boldsymbol{J}|}{\kappa\Theta J_c}\right) \exp\left(-\dfrac{U_0}{\kappa\Theta}\right) \dfrac{\boldsymbol{J}}{|\boldsymbol{J}|}, & 0 \leqslant |\boldsymbol{J}| \leqslant J_c, \\ E_c + \rho_f J_c \left(\dfrac{|\boldsymbol{J}|}{J_c}\right) \dfrac{\boldsymbol{J}}{|\boldsymbol{J}|}, & |\boldsymbol{J}| > J_c, \end{cases} \tag{7.15}$$

where $\rho_c$ and $\rho_f$ are the creep and flow resistivity, respectively, and $E_c = \rho_c J_c[1 - \exp(-2U_0/\kappa\Theta)] \approx \rho_c J_c$.

Recently, a smoothed Bean's model of the critical state in the hyperbolic tangent approximation [12, 36],

$$\boldsymbol{J} = J_c \tanh\left(\frac{|\boldsymbol{E}|}{E_0}\right) \frac{\boldsymbol{E}}{|\boldsymbol{E}|} \tag{7.16}$$

was proposed to describe the nonlinear **J-E** constitutive law in HTSCs, where $E_0$ is a characteristic electric field that determines the electrical conductivity $\sigma_0$ at zero electrical field by $\sigma_0 = J_c/E_0$. The continuity of the current density around zero electric field makes the associated numerical simulation stable and it is not necessary to introduce a residual resistivity for numerical stability.

Additionally, the field-dependent property of the critical current density is generally characterized by the Kim's model [37],

$$J = J_{c0} \left( \frac{B_0}{|\boldsymbol{B}| + B_0} \right), \tag{7.17}$$

where $J_{c0}$ is the zero-field critical current density depending on the prescribed electric field criterion $E_c$, and $B_0$ represents a critical magnetic flux density for which the critical current density is halved.

### 7.3.2 Anisotropy

The special microstructure of cuprate-based oxide HTSCs, which consists of the alternating stack of superconductive $CuO_2$ layers and almost insulating block layers, results in a remarkably anisotropic behavior in the present HTSC family [38]. Due to this anisotropic behavior arising from intrinsic pinning and other defects in the HTSCs, flux-line curvature will always occur in HTSCs in the presence of a magnetic field. Consequently, $J_c$ in HTSCs is anisotropic and strongly dependent on the orientation of the applied magnetic field [39]. Namely, the value of the critical current density flowing in the *a*-*b* plane, $J_c^{ab}$, is larger than that along the *c*-axis, $J_c^c$ [40].

## 7.4 Calculation of the magnetic field of PMG

### 7.4.1 Two-dimensional case [14]

In the two-dimensional case, it is assumed that the PMG extends infinitely along the forward direction of the HTS Maglev vehicle, the non-uniform magnetic field generated by the PMG can be calculated through the combined contribution of each PM therein, via an analytic model established by resorting to the surface current model [41]. Using a vertically magnetized PM as a representative, the cross-sectional view of the surface current model for the designated PM of width $2w$ and thickness $d$ is schematically drawn in Fig. 7.1, with two sheet currents counterflowing near the marginal parts. If the PM is supposed to be magnetized uniformly with magnetization $\boldsymbol{M} = M_0\boldsymbol{z}$, the volume current density is null due to the zero gradient of $\boldsymbol{M}$ across the *y*-*z* plane and only the surface current density, estimated by $\boldsymbol{j}_s = \boldsymbol{M} \times \boldsymbol{n}$ remains. The surface current density is equal to $M_0\boldsymbol{x}$ and $-M_0\boldsymbol{x}$, respectively, for the left and right parts, as marked in Fig. 7.1. In this way, the problem is reduced to solving for the magnetic field of two infinitely long current sheets of height $d$ with opposite direction and separated by a dimension of $2w$. In this case, the magnetic vector potential $\boldsymbol{A}$, defined as $\boldsymbol{B} = \nabla \times \boldsymbol{A}$, at point $(y, z)$ due to the combined contribution of the sheet current element $dI_1 = j_s dz'$ of the left one and $dI_2 = -j_s dz'$ of the right one is given by

$$\mathrm{d}\boldsymbol{A} = \left( \frac{\mu_0 M_0}{4\pi} \ln \frac{(y-w)^2 + (z-z')^2}{(y+w)^2 + (z-z')^2} \mathrm{d}z' \right) \boldsymbol{x}. \tag{7.18}$$

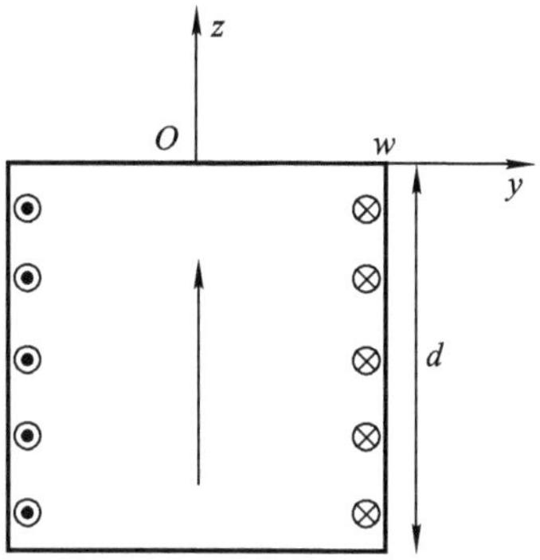


**Fig. 7.1:** Cross-sectional view of the surface current model for a vertically magnetized PM of width $2w$ and thickness $d$ and $b$ infinite along the $x$-direction (normal to the paper) of a Cartesian coordinate system $x$, $y$, $z$.

Through the integral operation upon Eq. (7.18) from $z' = 0$ and $z' = -d$, one can formally arrive at the analytic equation to calculate the magnetic vector potential $A_x$ at point $(y, z)$ generated by PM in Fig. 7.1,

$$A_x = \frac{\mu_0 M_0}{4\pi} \left[ (z - z') \ln \frac{(y + w)^2 + (z - z')^2}{(y - w)^2 + (z - z')^2} + 2(y + w) \arctan \frac{z - z'}{y + w} - 2(y - w) \arctan \frac{z - z'}{y - w} \right]_{-d}^{0}. \tag{7.19}$$

Once the expression of the magnetic vector potential is obtained, the two components of the magnetic flux density, $B_y$ and $B_z$, can be deduced to be

$$B_y = \frac{\mu_0 M_0}{4\pi} \ln \frac{[(y + w)^2 + z^2][(y - w)^2 + (z + d)^2]}{[(y + w)^2 + (z + d)^2][(y - w)^2 + z^2]},$$
$$B_z = \frac{\mu_0 M_0}{2\pi} \left( \arctan \frac{z}{y - w} + \arctan \frac{z + d}{y + w} - \arctan \frac{z}{y + w} - \arctan \frac{z + d}{y - w} \right), \tag{7.20}$$

according to the relations that $B_y = \partial A_x / \partial z$ and $B_z = -\partial A_x / \partial y$.

The contribution of the PM with other magnetization directions and locations in a certain PMG can be estimated by the geometrical operations of translation and/or rotation.

### 7.4.2 Three-dimensional case [42]

The 3D model is realistic and can take into account the non-uniformity of the magnetic field along the forward direction, which may be caused by the existent gap between the adjacent PM elements in the PMG, into account. According to the Biot-Savart law, the elemental flux density d$\boldsymbol{B}$ can be described as,

$$\mathrm{d}\boldsymbol{B} = \frac{\mu_0}{4\pi}\frac{I\mathrm{d}\boldsymbol{l}\times\boldsymbol{l}\times\boldsymbol{r}}{r^2}, \tag{7.21}$$

where $\mu_0$ is the magnetic permeability of vacuum, $\boldsymbol{r}$ is a coordinate vector from the element of length d$\boldsymbol{l}$ to an observation point $P$, and $I$ is the magnitude of the current.

Therefore, the three components of the d$\boldsymbol{B}$ in the $x$, $y$, and $z$ directions produced by the element current $I$d$y$ of the element area ABCD in Fig. 7.2 can be expressed as

$$\mathrm{d}B_{x1} = \frac{\mu_0 M_0 \mathrm{d}y'}{4\pi}\int_{-t_{\mathrm{PM}}}^{0}\frac{(y'-y)\mathrm{d}z'}{((x-(l+g))^2+(y-y')^2+(z-z')^2)^{3/2}}, \tag{7.22}$$

$$\mathrm{d}B_{y1} = \frac{\mu_0 M_0 \mathrm{d}y'}{4\pi}\int_{-t_{\mathrm{PM}}}^{0}\frac{(x-(l+g))\mathrm{d}z'}{((x-(l+g))^2+(y-y')^2+(z-z')^2)^{3/2}}, \tag{7.23}$$

$$\mathrm{d}B_{z1} = 0, \tag{7.24}$$

where $M_0$ is the magnetization of the PM, $g$ is the half of the gap between the adjacent PMs. d$B_{xi}$, d$B_{yi}$, and d$B_{zi}$ ($i = 1, 2, 3$) can be obtained from the element areas CDEF, EFGH, and GHAB with the same method. We also introduce the following expressions to simplify the description as done in [43]

$$K = \frac{\mu_0 M_0}{4\pi}, \quad \psi_i(\phi_1,\phi_2,\phi_3) = \frac{\phi_i}{(\phi_1^2+\phi_2^2+\phi_3^2)^{3/2}} (i = 1, 2, 3), \tag{7.25}$$

$$\Gamma(\gamma_1,\gamma_2,\gamma_3) = \ln\frac{\sqrt{\gamma_1^2+(\gamma_2-y')^2+\gamma_3^2}-\gamma_3}{\sqrt{\gamma_1^2+(\gamma_2-y')^2+\gamma_3^2}+\gamma_3}, \tag{7.26}$$

$$\varphi(\phi_1,\phi_2,\phi_3) = \begin{cases} \arctan\left[\dfrac{\phi_1}{\phi_3}\dfrac{\phi_2-y'}{\phi_1^2+(\phi_2-y')^2+\phi_3^2}\right] & \phi_3 \neq 0, \\ 0, & \phi_3 = 0. \end{cases} \tag{7.27}$$

Therefore, we have

$$\begin{aligned}\mathrm{d}B_x = \sum_{i=1}^{4}\mathrm{d}B_{xi} = K\mathrm{d}y'\int_{-t_{\mathrm{PM}}}^{0}&[\psi_2(x-(L+g), y'-y, z-z')\\ &+\psi_2(x-g, y'-y, z-z')]\mathrm{d}z',\end{aligned} \tag{7.28}$$

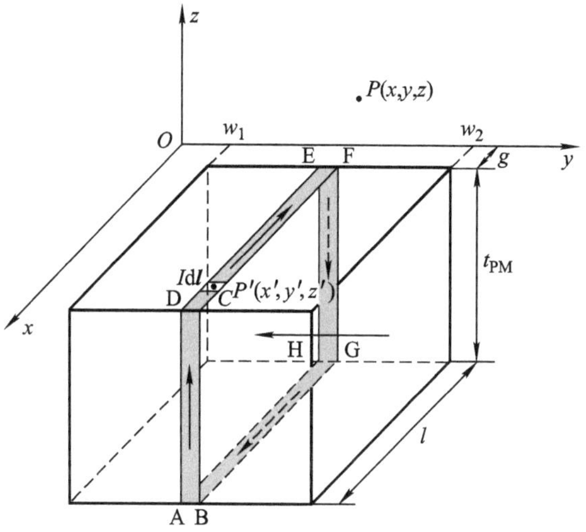


**Fig. 7.2:** Schematic drawing of the equivalent surface current of a rectangular PM ($g$: half of the gap between the adjacent PMs; $l$: length of the PM; $t_{PM}$: height of the PM).

$$\begin{aligned}
dB_y = \sum_{i=1}^{4} dB_{yi} = K dy' \Big( \int_{-t_{PM}}^{0} &(\psi_1(x-(L+g), y-y', z-z') \\
&+ \psi_1(g-x, y-y', z-z'))dz' + \int_{g}^{L+g} \\
&(\psi_3(x-x', y-y', z) + \psi_3(x-x', y-y', -z-t_{PM}))dx' \Big),
\end{aligned} \tag{7.29}$$

$$\begin{aligned}
dB_z = \sum_{i=1}^{4} dB_{zi} = K dy' \int_{g}^{L+g} &[\psi_2(x-x', y'-y, z) \\
&+ \psi_2(x-x', y-y', z+t_{PM})]dx'.
\end{aligned} \tag{7.30}$$

Furthermore, the 3D analytical expressions in Fig. 7.2 can be given by the following form, taking into account Eqs. (7.26) and (7.27),

$$\begin{aligned}
B_x = \frac{K}{2}(&\Gamma(x-(L+g), y-y', z+t_{PM}) - \Gamma(x-(L+g), y-y', z) \\
&- \Gamma(x-g, y-y', z+t_{PM}) + \Gamma(x-g, y-y', z))\Big|_{w_1}^{w_2},
\end{aligned} \tag{7.31}$$

$$\begin{aligned}
B_y = K(&\psi(z, y-y', x-(L+g)) - \psi(z+t_{PM}, y \quad y', x-(L+g)) \\
&- \psi(z, y-y'.x-g) + \psi(z+t_{PM}, y-y', x-g) \\
&+ \psi(x-(L+g), y-y', z) - \psi(x-g, y-y', z) - \psi(x-(L+g), y-y', z+t_{PM}) \\
&+ \psi(x-g, y-y', z+t_{PM}))\Big|_{w_1}^{w_2},
\end{aligned} \tag{7.32}$$

$$B_z = \frac{K}{2}(\Gamma(z, y - y', x - g) - \Gamma(z, y - y', x - (L + g)) - \Gamma(z + t_{PM}, y - y', x - g) + \Gamma(z + t_{PM}, y - y', x - (L + g)))\Big|_{w_1}^{w_2} \quad (7.33)$$

The analytical expressions for the designated PMG can be obtained by translating the PM along the $y$-axis with the direction of the magnetization considered, and then the final expressions can be established by translating the segment of the PMG along the $x$-axis with the total number of the segments taken into account.

## 7.5 Two-dimensional modelings and simulations

### 7.5.1 Prigozhin's model [6, 44, 45]

#### 7.5.1.1 Mathematical fundamentals

This model states that the electromagnetic properties of the HTSC in the framework of Bean's critical state model can be ruled by a complicated system of equations and inequalities, which leads to a free boundary problem. It has the shape of variation formulation as follows,

$$\mu_0 \int_0^T \int_{R^3} \frac{\partial \boldsymbol{H}}{\partial t} \cdot \boldsymbol{\varphi} + \int_0^T \int_{\Omega_{HTSC}} \rho(\nabla \times \boldsymbol{H}) \cdot (\nabla \times \boldsymbol{\varphi}) = 0, \quad (7.34)$$

where $R^3$ denotes the whole computational domain including the HTSC and the coolant domain, whereas $\Omega_{HTSC}$ represents the superconducting domain only. The function $\boldsymbol{\varphi}$ is a "test" vector function whose curl vanishes in the domain outside the HTSC and whose tangential components are continuous across the boundary $\Gamma$ between the HTSC and the coolant.

Now, we introduce a new variable, $\boldsymbol{h} = \boldsymbol{H} - \boldsymbol{H}_e$, satisfying

$$\begin{aligned} &\nabla \times \boldsymbol{h} = 0, && \text{in } \Omega_{Air}, \\ &|\nabla \times \boldsymbol{h}| \leqslant J_c(|\boldsymbol{h} + \boldsymbol{H}_e|), && \text{in } \Omega_{HTSC}, \\ &[\boldsymbol{h}_\Gamma] = 0, && \text{on } \Gamma. \end{aligned} \quad (7.35)$$

Let us define the set of functions

$$K(\boldsymbol{h}) = \{|\nabla \times \boldsymbol{\varphi}| \leqslant J_c(|\boldsymbol{h} \times \boldsymbol{H}_e|), \text{in } \Omega_{HTSC}\} \quad (7.36)$$

which is dependent on $\boldsymbol{h}$. Moreover, since

$$\nabla \times \boldsymbol{h} = \nabla \times \boldsymbol{H}, \quad |\nabla \times \boldsymbol{H}| \leqslant J_c(\boldsymbol{H}), \quad \text{in } \Omega_{HTSC}, \quad (7.37)$$

the formulation may be yielded as

$$\int_0^T \int_{\Omega_{HTSC}} \rho |\nabla \times \boldsymbol{h}|^2 = \int_0^T \int_{\Omega_{HTSC}} \rho J_c^2(|\boldsymbol{h} + \boldsymbol{H}_e|). \tag{7.38}$$

Correlating the above four functions (7.34), (7.36), (7.37), and (7.38), we obtain

$$\begin{aligned} \mu_0 \int_0^T \int_{R^3} \frac{\partial(\boldsymbol{h} + \boldsymbol{H}_e)}{\partial t} \cdot (\boldsymbol{\varphi} - \boldsymbol{h}) &= -\int_0^T \int_{\Omega_{HTSC}} \rho \nabla \times \boldsymbol{h} \cdot \nabla \times (\boldsymbol{\varphi} - \boldsymbol{h}) \\ &\geqslant \int_0^T \int_{\Omega_{HTSC}} \rho (J_c^2(|\boldsymbol{h} + \boldsymbol{H}_e|) - |\nabla \times \boldsymbol{\varphi}||\nabla \times \boldsymbol{h}|) \geqslant 0. \end{aligned} \tag{7.39}$$

This proves that $\boldsymbol{h}$ is a solution of the problem

$$\begin{aligned} &\text{Find function } \boldsymbol{h} \in K(\boldsymbol{h}) \text{ such that } \left( \frac{\partial(\boldsymbol{h} + \boldsymbol{H}_e)}{\partial t}, (\boldsymbol{\varphi} - \boldsymbol{h}) \right) \geqslant 0 \\ &\text{for any } \boldsymbol{\varphi} \in K(\boldsymbol{h}), \boldsymbol{h}|_{t=0} = \boldsymbol{h}_0, \end{aligned} \tag{7.40}$$

where $(\boldsymbol{u}, \boldsymbol{w}) = \int_0^T \int_{R^3} \boldsymbol{u} \cdot \boldsymbol{w}$ is the scalar product of two vector functions, with $\boldsymbol{h}_0 = \boldsymbol{B}_0/\mu_0 - \boldsymbol{H}_e|_{t=0}$ denoting the initial condition of the problem.

In the two-dimensional case, the magnetic field has only two components and it can be expressed as

$$\boldsymbol{H} = H_x \hat{\boldsymbol{x}} + H_y \hat{\boldsymbol{y}}. \tag{7.41}$$

The current flowing in the HTSC has only the $z$ component, which is derived as

$$\boldsymbol{J} = \begin{bmatrix} \hat{\boldsymbol{x}} & \hat{\boldsymbol{y}} & \hat{\boldsymbol{z}} \\ \frac{\partial}{\partial x} & \frac{\partial}{\partial y} & \frac{\partial}{\partial z} \\ H_x & H_y & H_z \end{bmatrix} = \left( \frac{\partial H_y}{\partial x} - \frac{\partial H_x}{\partial y} \right) \hat{\boldsymbol{z}}. \tag{7.42}$$

Since there is no imposed current in the HTSC, the net current induced by the variation of the external magnetic field must satisfy

$$\int_{\Omega_{HTSC}} J = I_{total}(t) = 0. \tag{7.43}$$

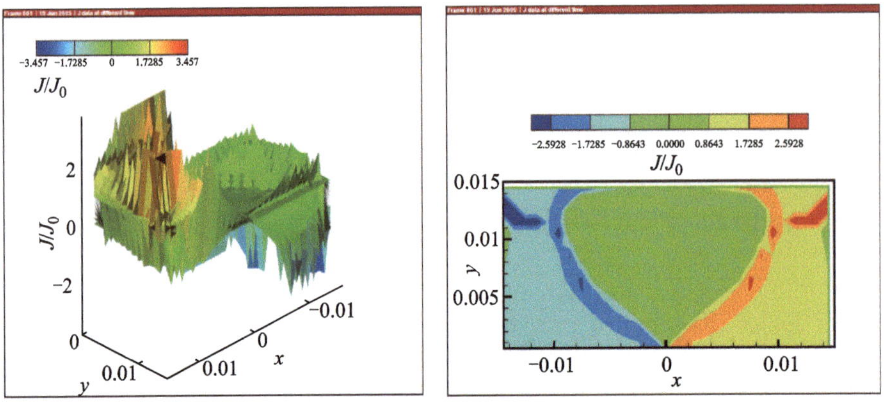


(a) 3D (b) 2D

**Fig. 7.3:** Cross-sectional distribution of the current in the HTSC in symmetrical applied field with levitation gap of 5 mm drawn in 3D (a) and 2D (b) views.

The magnetic forces generated by the interaction between the induced current and the external magnetic field of PMG are calculated by the Lorentz equation,

$$\boldsymbol{F} = \int \boldsymbol{J} \times \boldsymbol{B} \mathrm{d}V. \tag{7.44}$$

#### 7.5.1.2 Example of applications

Assuming a HTSC with 30 mm in width and 15 mm in height, numerical simulations using this method have been performed by quadratic programming [5].

When the HTSC is centrally above the PMG, the applied magnetic field is symmetrical about the axis of the HTSC. Once the HTSC is off the central position, the applied magnetic field will become asymmetrical. Supposing the HTSC is subjected to a ZFC condition at a position of 130 mm above the PMG, Figs 7.3 and 7.4 present the current distribution on the vertical cross section in $x$-$y$ plane (the infinite direction is on the $z$-axis in thesedrawings) when the HTSC moves to the height of 5 mm, i.e. the smallest levitation gap. As shown in Fig. 7.3, the current distribution is antisymmetrical about the center plane of the HTSC in the symmetrical applied field. However, the current distribution becomes asymmetrical when a lateral displacement of 30 mm is applied to the HTSC, resulting from the titled field in asymmetrical applied field as depicted in Fig. 7.4.

Figures 7.5 and 7.6 compare the calculated and experimental levitation forces in both symmetrical and asymmetrical applied fields. The details of the measurement rigs used in the experiment have been described in Section 5.3. The sample of HTSC had a dimension of 36 mm in length, 30 mm in width and 15 mm in thickness. After the sample was cooled in liquid nitrogen with the absence of the magnetic field, the levitation forces were collected real-time while the sample was moved from 130 mm

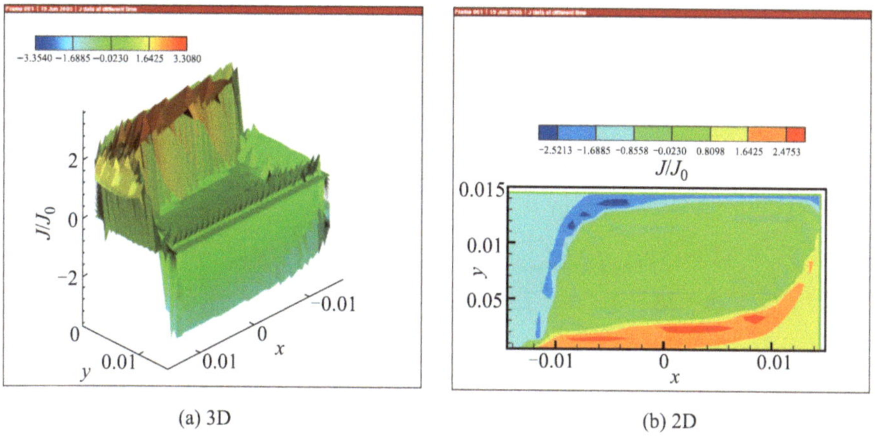


**Fig. 7.4:** Cross-sectional distribution of current in the HTSC in asymmetrical applied field with levitation gap of 5 mm and lateral displacement of 30 mm drawn in 3D (a) and 2D (b) views.

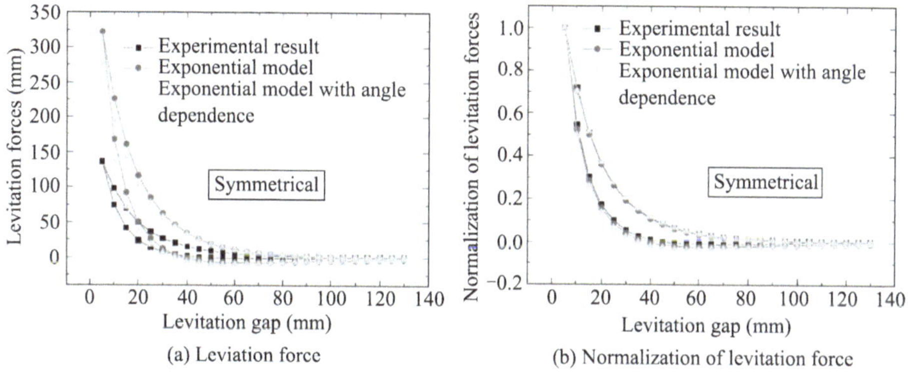


**Fig. 7.5:** Comparison between the calculated and experimental levitation forces as functions of levitation gap in the symmetrical field. Levitation force curve using actual (a) and normalized (b) data.

to 5 mm above the PMG and went back to the original position. When the sample vertically moved centrally above the PMG, the applied field was symmetrical and the relationship between the levitation force and gap was approximately linear. Once the HTSC sample was off, the center plane, the applied field became asymmetrical. The difference between two levitation force curves in the symmetrical and asymmetrical fields was obvious. In an asymmetrical applied field, the descending part of the hysteresis curve was convex [6].

We have two cases for the use of the critical exponential description to take the field-dependent properties of critical current density into account: without and with the angle dependence (details can be found in Ref. 6). There are differences between

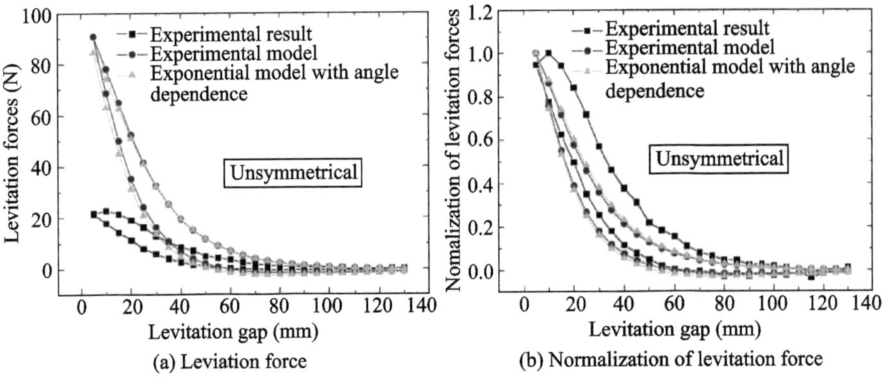


**Fig. 7.6:** Comparison between the calculated and experimental levitation forces as functions of levitation gap in the asymmetrical field. Levitation forces curve using actual (a) and normalized (b) data.

the calculated results in these two cases, which indicate that the angle of the applied field does have the influence on the critical current distribution. Moreover, both the calculated results are about twice more than the experimental results. Qualitatively, the calculated results agree well with the experimental results in the symmetrical applied field as the force normalization curves show in Fig. 7.5b. However, there is both qualitative and quantitative difference between them in the asymmetrical applied field, as shown in Fig. 7.6.

The guidance forces were carried out in the FC condition 20 mm above the PMG. Keeping MH at a gap of 20 mm, the sample was horizontally moved from the center to the displacement of –20 mm and across the center back to the displacement of 20 mm, and back to the starting position. The same happens to the guidance forces in Figs 7.7a and b, i.e. the difference between results without and with angle dependence still exists and the calculated hysteresis curve does not agree with the experimental curve, because the HTSC sample during the lateral movement experienced a changing field, from symmetrical to asymmetrical or vice versa.

## 7.5.2 Generalized magnetic vector potential

### 7.5.2.1 Mathematical fundamentals

With the introduction of a generalized magnetic vector potential and using the Ampère's law within the quasi-static approximation as the state equation, we have established the PDE that governs the electromagnetic field of the levitation system with single or multiple HTSCs. The prominent advantage of this course is that, for a two-dimensional problem, only the vector potential along the direction of translational invariance ($x$-axis in a Cartesian coordinate shown in Fig. 7.8) needs to be defined

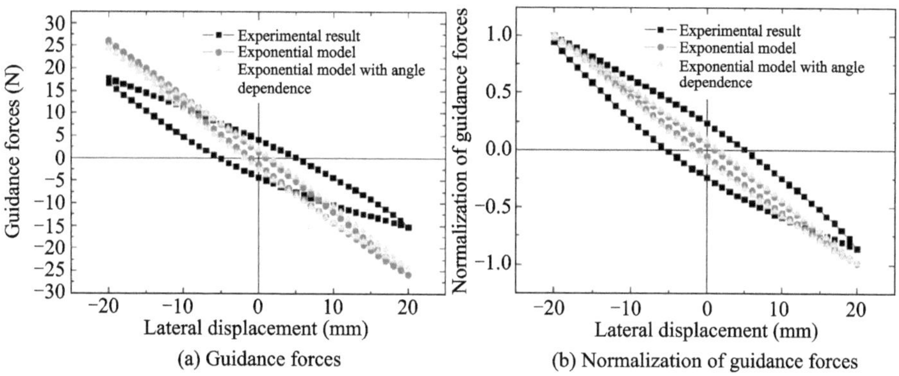


**Fig. 7.7:** Comparison between the calculated and experimental guidance forces during the lateral movement. Guidance forces curve using actual (a) and normalized (b) data.

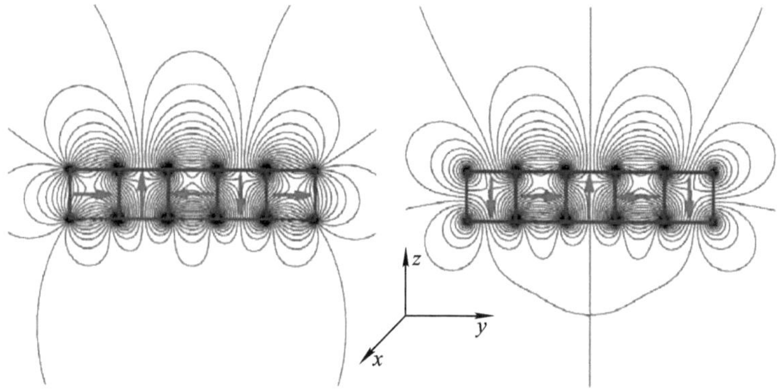


**Fig. 7.8:** Contour lines of magnetic flux generated by the two different PMGs (marked as PMG_A and PMG_B from left to right in this section), all with a geometrical configuration invariant along the $x$-direction and symmetric in terms of the $z$-direction of a Cartesian coordinate system. Shown for both plots is the horizontal component of magnetic flux density $B_y$. Each PM element, represented by a rectangle with blue color in both PMGs, has an identical width and thickness of 32 mm. A magnetization $M_0$ of $8.753 \times 10^5$ A/m was assigned to all PM elements, intending to approximately reproduce the performance of a N35 magnet.

and solved, which is rather profitable in terms of reducing the number of degrees of freedom when adopting the FEM technique to discretize the spatial domain including the HTSCs and the surrounding coolant as well.

With Faraday's law [Eq. (7.4)] and the definition of $\boldsymbol{B} = \nabla \times \boldsymbol{A}$, the nontrivial component of the induced electric field $E_x$ for a single HTSC can be expressed as

$$E_x = -\frac{\partial(A_{\mathrm{sc},x} + A_{\mathrm{ex},x})}{\partial t} - (\nabla V)_x, \tag{7.45}$$

with the magnetic vector potential along the $x$-direction $A_x$ being decomposed into $A_{sc,x}$ and $A_{ex,x}$, where $A_{sc,x}$ represents the vector potential induced by the supercurrent, and $A_{ex,x}$ serves as the magnetic excitation. The gradient of the electric scalar potential $\nabla V$ is invariant across the $y$-$z$ plane and only depends on time $t$, as both $\boldsymbol{E}$ and $\boldsymbol{A}$ are nontrivial merely along the $x$-axis and independent of the $x$-axis [46]. If we denote the value of $(\nabla V)_x$ at an arbitrary time $t_n$ as $C(t_n)$, Eq. (7.45) can be rewritten as

$$E_x = -\frac{\partial\left(A_{sc,x} + \int_0^{t_n} C(t)\mathrm{d}t + A_{ex,x}\right)}{\partial t}. \tag{7.46}$$

Making use of Ampère's law [Eq. (7.1)] within the quasi-static approximation and exploiting Eq. (7.46), the electromagnetic properties in the HTSC are governed by

$$-\frac{1}{\mu_0}\left(\frac{\partial^2 A_{sc,x}}{\partial y^2} + \frac{\partial^2 A_{sc,x}}{\partial z^2}\right) + \sigma\frac{\partial(A_{sc,x} + \int_0^{t_n} C(t)\mathrm{d}t)}{\partial t} + \sigma\frac{\partial A_{ex,x}}{\partial t} = 0, \tag{7.47}$$

where we have assumed that magnetic field intensity and magnetic flux density in the HTSC are related linearly with the vacuum permeability $\mu_0$.

Since $\int_0^{t_n} C(t)\mathrm{d}t$ is independent of $y$ and $z$, it is expedient to adopt a generalized vector potential $A'_{sc}$, which is defined as

$$A'_{sc,x} = A_{sc,x} + \int_0^{t_n} C(t)\mathrm{d}t, \tag{7.48}$$

to make the integral term in Eq. (7.46) implicit. Eq. (7.47) is thus reduced to

$$-\frac{1}{\mu_0}\left(\frac{\partial^2 A'_{sc,x}}{\partial y^2} + \frac{\partial^2 A'_{sc,x}}{\partial z^2}\right) + \sigma\frac{\partial A'_{sc,x}}{\partial t} + \sigma\frac{\partial A_{ex,x}}{\partial t} = 0, \tag{7.49}$$

where $A'_{sc,x}$ is the unknown to solve. It is worth noting that the contribution of the electric scalar potential will dominate in the computational region far from the HTSC, where the vector potential $A_{sc,x}$ due to the supercurrent is negligible, i.e. $A'_{sc,x} \approx \int_0^{t_n} C(t)\mathrm{d}t$ holds (if the HTSC is subjected to a symmetric magnetic field, $C(t_n) \equiv 0$ stands, and it is not needed to introduce the generalized vector potential).

For a superconducting unit containing $m$ constituents, the magnetic excitation to an arbitrary constituent $i$ is generated by the external source as well as by all the others, which requires modification of the electromagnetic master equation for the constituent $i$ to be,

$$-\frac{1}{\mu_0}\left(\frac{\partial^2 A'_{\mathrm{sc},x,i}}{\partial y^2}+\frac{\partial^2 A'_{\mathrm{sc},x,i}}{\partial z^2}\right)+\sigma\left(\frac{\partial}{\partial t}\sum_{j=1}^{m} A'_{\mathrm{sc},x,j}-\frac{\partial}{\partial t}\sum_{j=1,j\neq i}^{m} A'_{\mathrm{sc},x,j,\Gamma}\right)$$
$$+\sigma\frac{\partial A_{\mathrm{sc},x}}{\partial t}=0, \tag{7.50}$$

Here $A'_{\mathrm{sc},x,j,\Gamma}$ refers to the boundary unknown of the constituent $j(j\neq i)$, where $A'_{\mathrm{sc},x,j,\Gamma}\approx(\nabla V)_{x,j}$ stands. In this case, $m$ different unknowns, $A'_{\mathrm{sc},x,1}, A'_{\mathrm{sc},x,2},\cdots, A'_{\mathrm{sc},x,m}$ should be defined at each finite element node, and they are coupled through the second term of the left-hand side of Eq. (7.50).

The coolant to provide a cryogenic environment for the HTSC is supposed to be dielectric, and the second term of the left-hand side of Eqs. (7.49) and (7.50) thus vanishes in the coolant. The electromagnetic master equation in the coolant is actually represented by the Laplace's equation. The continuity of $A'_{\mathrm{sc},x}$ as well as its normal derivative applies to the interior bounds between the superconductor and the coolant, and the value of $A'_{\mathrm{sc},x}$ on the outer bounds is not constrained.

#### 7.5.2.2 Example of applications: mutual effect among the superconducting constituents in a levitation system with translational symmetry [14]

The PDEs governing the electromagnetic properties of HTSCs as well as the surrounding coolant, illustrated above, were numerically discretized by resorting to the FEM technique and FDM scheme, in the spatial and temporal domain. Considering the two different configurations of PMG shown in Fig. 7.8, both of which were derived from the Halbach array, we had applied this model to study the mutual effect among the superconducting constituents in a HTS unit composed of three HTSCs.

In these simulations, the dimension of each HTS constituent was 30 mm in width and 15 mm in thickness, and the gap between the constituents was 2 mm. We hade used $J_{c0}=1.5\times10^8$ A/m$^2$, $E_0=\times10^{-6}$ V/m, and $B_0=0.25$ T, to roughly represent the material properties of bulk Y–Ba–Cu–O at 77 K. According to the geometrical and material characteristics of the PMGs in Fig. 7.8, the starting position of vertical movement was 120 mm to approach the ZFC condition over both PMGs, where the strength of applied magnetic field was negligible. The minimum vertical distance between the HTSC and the PMG was 5 mm, and the speed of the HTSC was 1 mm/s to create a quasi-static state for all cases estimated in this section.

Two cases will be estimated and compared. One is the actual case with all constituents present, and the other is an envisaged case of considering one constituent

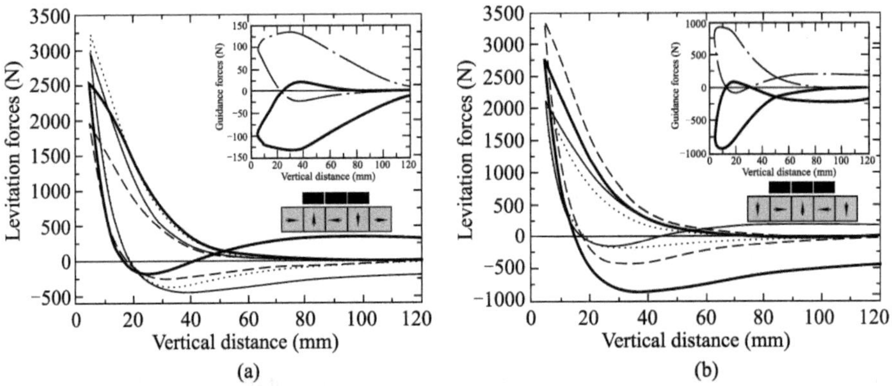


**Fig. 7.9:** Hysteretic loops of levitation forces on the left (thin solid line) and middle (thick solid line) constituent of a superconducting unit with three constituents inside while moving in the non-uniform magnetic field generated by PMG_A (a) and PMG_B (b). Also shown for comparison by the corresponding dashed line is that obtained in the envisaged case. The concurrent hysteretic loops of guidance forces on each constituent (left: solid line, middle: dash line, right: dash-dot line) of the actual case is plotted as an inset.

with all the rest absent. The envisaged case is used to identify how the mutual effect among the constituents acts.

#### 7.5.2.2.1 Mutual effect on the levitation forces

Figure 7.9 portrays the hysteretic loops of the levitation forces on each individual constituent for the actual case as well as for the envisaged case. The levitation forces of the right constituent in both cases are invisible due to the symmetry to the left one. The development of the guidance forces as the vertical movement on each individual constituent of the actual case is also plotted as an inset in each figure. The main findings from this figure include (i) the levitation forces on the left constituent over PMG_A and on the middle constituent over PMG_B, whose domains are dominated by the vertical component of magnetic field, are reduced due to the mutual effect, whereas the levitation forces on the middle constituent over PMG_A and on the left constituent over PMG_B, whose domains are dominated by the horizontal component of magnetic field, are enhanced; (ii) as a consequence of the mutual effect, the levitation forces on both constituents over both PMGs tend to be a finite value but not zero during the movement to its initial position, which is obviously different from that of the envisaged case. These residual levitation forces are generated by the magnetic interaction of the induced supercurrent in one constituent with the induced magnetic field of the rest; (iii) regardless of the configuration of the PMG, the guidance forces of the middle constituent are null but the left and right constituents undergo a finite guidance force with opposite direction. The superconducting unit as a whole can achieve a null guidance force.

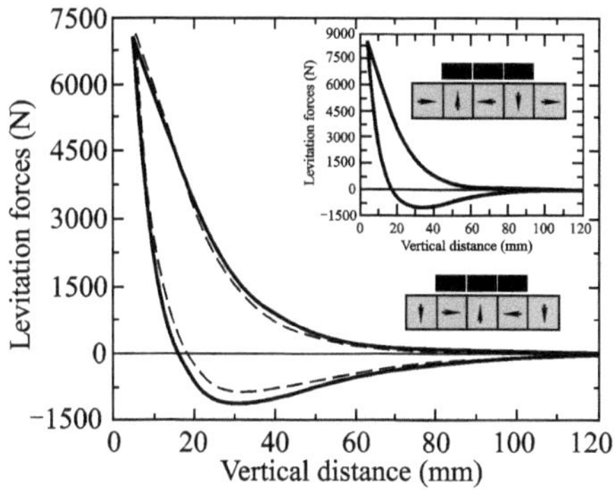


**Fig. 7.10:** Hysteretic loops of the total levitation forces on a superconducting unit with three constituents inside while moving in the non-uniform magnetic field generated by respectively PMG_A (inset) and PMG_B. Also shown for comparison is the sum of levitation forces on the individual superconductor obtained in the envisaged case (dash line).

Although the mutual effect brings out a pronounced alteration of the levitation forces on each individual constituent, the levitation forces as a whole, whose hysteretic loops of over both PMGs are plotted in Fig. 7.10, are influenced negligibly due to the mutual effect only a slight increase of the hysteretic loop over PMG_B found. The levitation forces for the envisaged case in Fig. 7.10 were obtained by summing the contributions from each HTS constituent calculated individually.

#### 7.5.2.2.2 Mutual effect on the distribution of magnetic flux density

The distributions of magnetic flux density $B$ inside the HTSCs, at the time instant when the smallest vertical distance is achieved during the calculations of the hysteretic loops of levitation forces in Fig. 7.9, is visualized and presented in Fig. 7.11 for both PMGs in the ZFC condition. The intensity of the red color is proportional to the strength of the local flux density.

Generally, the distributions reveal a physical symmetry of magnetic flux density in terms of the perpendicular bisector of the middle constituent in each case, and there exists a region, for all constituents of both actual and envisaged cases, where no magnet flux penetrates, i.e. a flux-free region.

Specifically speaking, (i) for the case with superconductors over PMG_A shown in Fig. 7.11a, the mutual effect on the middle constituent results in a notable decrease of the crescent-shaped flux-free region and extends the strong flux-penetration region at the bottom slightly upwards. Due to the mutual effect, the strong flux-penetration region localized at the lower-inner corners of the side constituents is slightly suppressed, and the flux-free region of the side constituents formed in the envisaged case is modulated to be a quasi-onion-shaped one by mostly altering the flux-penetrated region at the inner parts. The flux lines, mostly in the adjacent region between the constituents, are distorted by the mutual effect. (ii) For the case with superconductors over PMG_B shown in Fig. 7.11b, as a consequence of the mutual effect, the onion-shaped flux-free region formed in the middle constituent of the envisaged case is markedly reduced and the strength of the flux-penetration region

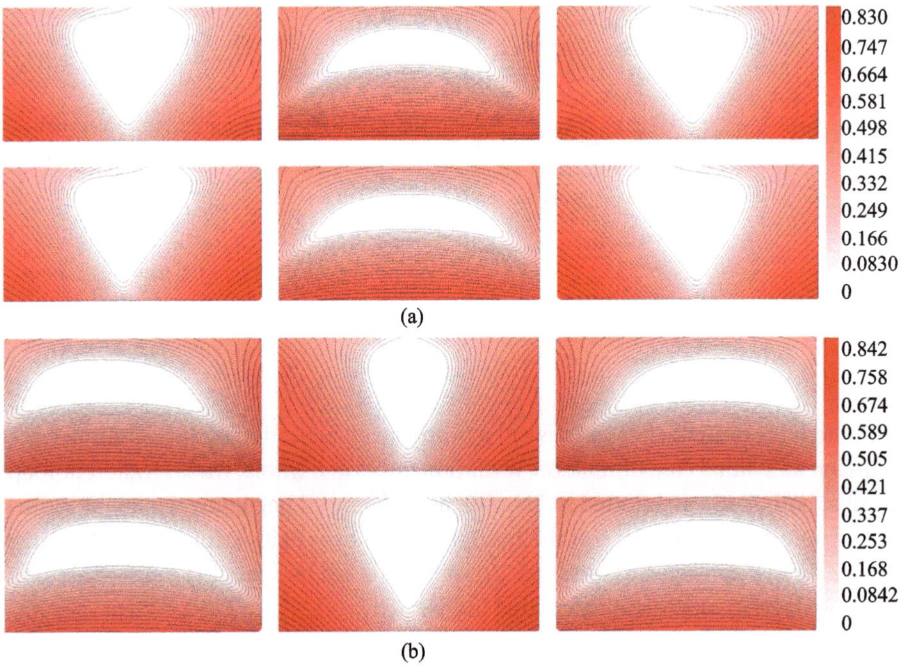


**Fig. 7.11:** Snapshots of the distributions of magnetic flux density $B$ (T) inside the HTS constituents at the time instant when the smallest vertical distance (5 mm) was achieved during the calculations of the hysteretic loops of levitation forces in Fig. 7.9 over PMG_A (a) and PMG_B (b), referring to the proposed PMGs shown in Fig. 7.8. The lower part of each picture, shown for comparison, represents the distributions inside each individual HTSC in the envisaged case.

localized at the lower-corners is weakened. The mutual effect on the side constituents gives rise to a broader region of flux penetration, or in other words, a smaller flux-free region, and slightly extends the strong flux-penetration region at the bottom upwards. The flux lines, mostly in the adjacent region between the constituents, are also distorted by the mutual effect, similar to that found in the case of over PMG_A.

#### 7.5.2.2.3 Mutual effect on the distribution of supercurrent density

The distributions of supercurrent density $J_x$ inside the HTSCs, at the time instant when the smallest vertical distance was achieved during the calculations of the hysteretic loops of levitation forces in Fig. 7.9, are visualized and presented in Fig. 7.12 for both PMGs in the ZFC condition. The intensity of the red and blue colors is proportional to the strength of the positive and negative supercurrents, respectively.

Generally, the distributions reveal a physical symmetry of supercurrent density in terms of the perpendicular bisector of the middle constituent over PMG_A and

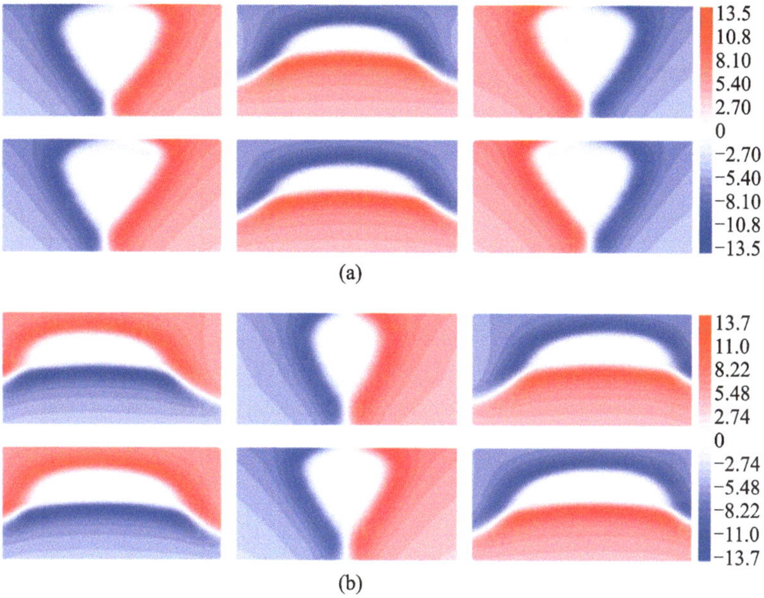


**Fig. 7.12:** Snapshots of the distributions of supercurrent density $J_x$ ($10^7$ A/m$^2$) inside the HTS constituents at the time instant when the smallest vertical distance (5 mm) was achieved during the calculations of the hysteretic loops of the levitation forces in Fig. 7.9 over PMG_A (a) and PMG_B (b), referring to the proposed PMGs shown in Fig. 7.8. The lower part of each picture, shown for comparison, represents the distributions inside each individual superconductor in the envisaged case.

a physical antisymmetry for the case over PMG_B. In each picture, there exists a region where no supercurrent flows in concert with the flux-free region in Fig. 7.11, and the whole domain of each constituent in the actual and envisaged cases is naturally divided into two regions with the flowing supercurrents opposite to each other. The two bands with the highest density in each constituent, one supercurrent flowing the opposite to the other, are always formed near the flux-free region where the variation of the magnet flux is intense according to Fig. 7.11.

Specifically speaking, (i) for the case with HTSCs over PMG_A shown in Fig. 7.12a, the mutual effect shortens the band with the highest density in the middle constituent and results in a reduced current-free region. Due to the mutual effect, the current-free regions formed in the side constituents of the envisaged case tends to be onion-shaped by mostly redistributing the positive supercurrent in the inner part of each side constituent. (ii) For the case with HTSCs over PMG_B shown in Fig. 7.12b, the mutual effect on the middle constituent gives rise to a reduced current-free region and extends the bands with the highest density inward. As a result of the mutual effect, the current-free region of the side constituents is slightly suppressed, as found in the middle constituent of over PMG_A.

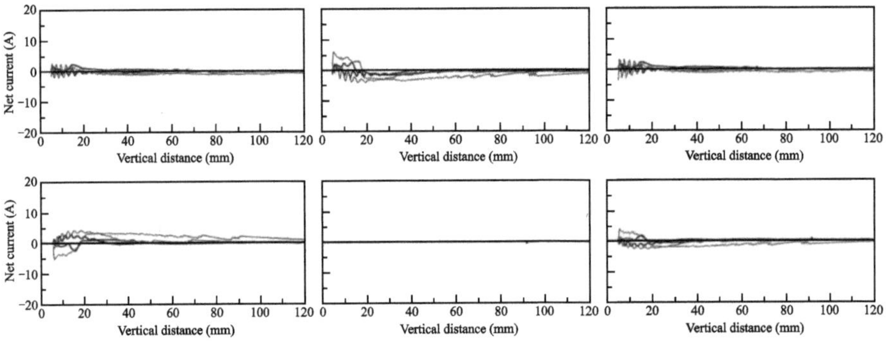


**Fig. 7.13:** Evolution of the net current inside the three constituents as a function of the vertical distance for both actual (thick solid line) and envisaged cases (thin solid line) over PMG_A (a) and PMG_B (b).

The evolution of the net current for the three constituents in both actual and envisaged cases as a function of the vertical distance over both PMGs is sketched and presented in Fig. 7.13. This figure indicates that the induced supercurrent of each constituent in actual case, estimated by integrating the supercurrent density across the respective cross section, is naturally zero, or very close to zero due to numerical errors, over both PMGs, like that found in the envisaged case. This demonstrates that the electrically isolated connection among the constituents is virtually left unchanged with the mutual effect present. This finding also proves that with the mathematical foundations derived above, zero net current can automatically be fulfilled without the constraint of supercurrent used in Prigozhin's model [46].

#### 7.5.2.2.4 Mutual effect on the distribution of levitation force density

The distributions of levitation force density inside the HTSCs at the time instant when the smallest vertical distance was achieved during the calculations of the hysteretic loops of levitation forces in Fig. 7.9 is visualized and presented in Fig. 7.14 for both PMGs in the ZFC condition. The intensity of the red and blue color is proportional to the strength of the positive and negative force density, respectively. In the following description, each HTSC in Fig. 7.14 is divided into two domains: one is the repulsive domain with a positive density, positively contributing to the levitation forces, and the other is the attractive domain with a negative density, negatively contributing to the levitation forces.

Generally, the distributions reveal a physical symmetry of the levitation forces in terms of the perpendicular bisector of the middle constituent in each case. Regardless of the PMG, there exists a force-free region, where the contribution to the levitation forces is null, for both the actual and envisaged cases, and the attractive domain is

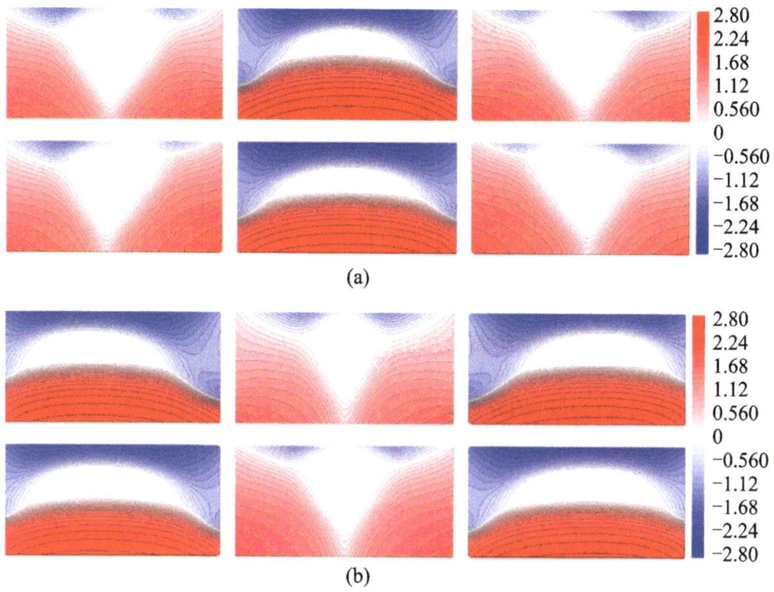


**Fig. 7.14:** Snapshots of the distributions of levitation force density ($10^7$ N/m$^2$) inside the HTS constituents at the time instant when the smallest vertical distance (5 mm) was achieved during the calculations of the hysteretic loops of levitation forces in Fig. 7.9 over PMG_A (a) and PMG_B (b), referring to the proposed PMGs shown in Fig. 7.8. The lower part of each picture, shown for comparison, represents the distributions inside each individual HTSC in the envisaged case.

usually formed at the upper part of the HTSC. The highest density of both the repulsive and attractive domains is always achieved in the HTSCs subjected to a magnetic field with the horizontal component dominating.

Specifically speaking, (i) for the case with superconductors over PMG_A shown in Fig. 7.14a, the mutual effect on the middle constituent gives rise to a reduced force-free region, and owing to the mutual effect, the strength of the attractive and repulsive domain is degraded and upgraded, respectively. Due to the mutual effect, the force-free region in the side constituents is reshaped and the contour lines of forces, mostly in the adjacent region between the constituents, are distorted by the mutual effect. (ii) For the case with HTSCs over PMG_B shown in Fig. 7.14b, the force-free region in the middle constituent slightly shrunk as a result of the mutual effect and the attractive domain is gently broadened. Similar to those found over PMG_A, the force-free region in the side constituents is reduced due to the upward expansion of the repulsive domain and the inward extension of the attractive domain. The contour lines of forces, mostly in the adjacent region between the constituents, are also distorted by the mutual effect.

## 7.6 Three-dimensional modeling and simulations

### 7.6.1 H-formulation [9]

We have developed a 3D-modeling numerical method using the dynamic boundary of the H-formulation. The resolving code is compiled with FORTRAN language using the FEM technique. It aims at simulating the electromagnetic properties of a HTS magnetic levitation system. As shown in Fig. 7.15, the levitation system is composed of one HTSC and one PMG, which generates the applied magnetic field for the HTSC. When the HTSC is moved vertically from position $A$ to position $B$ and then moved back to $A$, the levitation forces of the HTSC could be calculated with the 3D modeling numerical method. The geometrical and material parameters of HTSC in the calculations of this section is collected in Tab. 7.1.

#### 7.6.1.1 Mathematical fundamentals

In this section, we will give a 3D model and one PDE that is used to describe the electromagnetic behavior in HTSC. The geometric configuration of this 3D model is

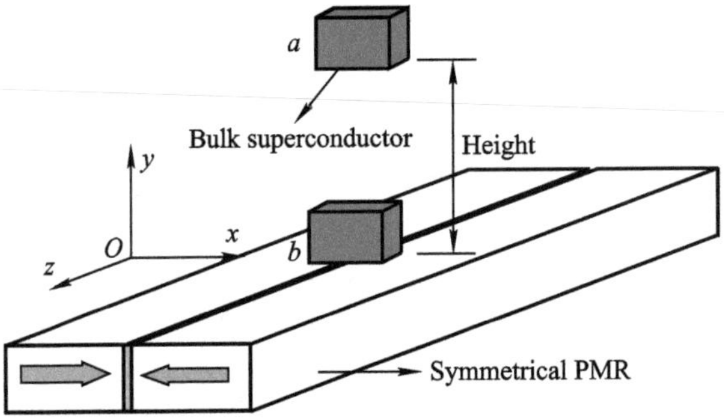


**Fig. 7.15:** Schematic sketch of levitation system composed of a bulk HTSC and PMG.

**Tab. 7.1:** Parameters of the levitation system used in the simulation.

| Parameter | Value |
|---|---|
| Dimension of HTSC (length × width × height) | 30 mm × 30 mm × 15 mm |
| Magnetization of PMG ($M_0$) | 96,000 A/m$^2$ |
| Vertical moving height | 50 mm |
| Velocity | 1 mm/s |
| Mini gap between the HTSC and the PMG | 3 mm |
| Critical current density $J_c$ | $8.0 \times 10^7$ A/m$^2$ |
| Material parameter $E_c$ of the HTSC | $1.0 \times 10^{-4}$ V/m |
| Dimension of PMG n(length × height) | 110 × 50 mm |

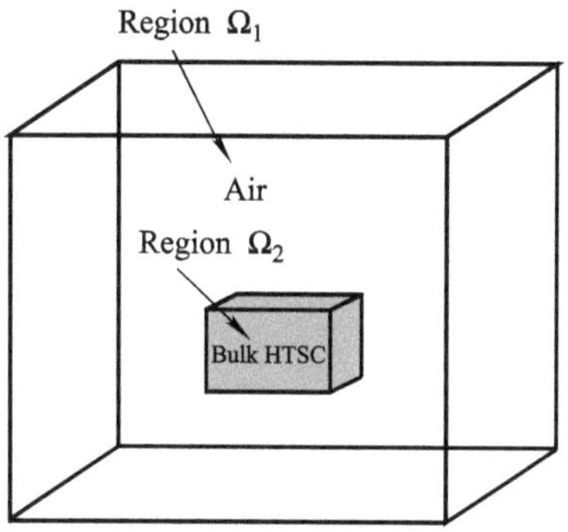


**Fig. 7.16:** Geometrical modeling of the 3D simulation of the HTS Maglev system.

described in Fig. 7.16. The virtual conductivity of the HTSC is calculated by

$$\frac{1}{\sigma_{sc}} = \frac{|\boldsymbol{E}_{sc}|}{|\boldsymbol{J}_{sc}|}, \tag{7.51}$$

where $\boldsymbol{E}_{sc}$ and $\boldsymbol{J}_{sc}$ are the electric field and current density of the HTSC, respectively. According to the $\boldsymbol{E}$-$\boldsymbol{J}$ exponent power law (see Eq. 7.14), the expression to calculate the conductivity is

$$\frac{1}{\sigma_{sc}} = \frac{|\boldsymbol{E}_{sc}|}{|\boldsymbol{J}_{sc}|} = \frac{E_{c0}}{J_c}\left(\frac{|\boldsymbol{J}_{sc}|}{J_c}\right)^{n-1}. \tag{7.52}$$

Based on Ampère's law [see Eq. (7.1)], we have

$$J_{sc,x}\hat{\boldsymbol{x}} + J_{sc,y}\hat{\boldsymbol{y}} + J_{sc,z}\hat{\boldsymbol{z}} = \left(\frac{\partial H_z}{\partial y} - \frac{\partial H_y}{\partial H_z} - \frac{\partial H_y}{\partial H_z}\right)\hat{\boldsymbol{x}} + \left(\frac{\partial H_x}{\partial z} - \frac{\partial H_z}{\partial x}\right)\hat{\boldsymbol{y}}$$
$$+\left(\frac{\partial H_y}{\partial x} - \frac{\partial H_x}{\partial y}\right)\hat{\boldsymbol{z}}. \tag{7.53}$$

Therefore, the three components of the supercurrent can be expressed in terms of the magnetic field as follows:

$$\begin{cases} J_{sc,x} = \dfrac{\partial H_z}{\partial y} - \dfrac{\partial H_y}{\partial z}, \\ J_{sc,y} = \dfrac{\partial H_x}{\partial z} - \dfrac{\partial H_z}{\partial x}, \\ J_{sc,z} = \dfrac{\partial H_y}{\partial x} - \dfrac{\partial H_x}{\partial y}. \end{cases} \tag{7.54}$$

The anisotropic feature of the critical current density in the HTSC is considered in the 3D modeling by assuming that the critical current density along the $c$-axis $J_c^c$ is one third of the critical current density within the $a$-$b$ plane $J_c^{ab}$ [47]. In order to describe this characteristic in simulations, we assume that the resistance of the currents flowing along the $c$-axis are three times larger than that within the $a$-$b$ plane, viz.,

$$\begin{cases} J_{\mathrm{sc},x} \leqslant J_\mathrm{c}, \\ J_{\mathrm{sc},y} \leqslant J_\mathrm{c}/3, \\ J_{\mathrm{sc},z} \leqslant J_\mathrm{c}. \end{cases} \tag{7.55}$$

Eq. (7.51) can be rewritten as

$$\boldsymbol{E}_\mathrm{sc} = \begin{bmatrix} E_{\mathrm{sc},x} \\ E_{\mathrm{sc},y} \\ E_{\mathrm{sc},z} \end{bmatrix} = \frac{1}{\sigma_\mathrm{sc}} \left\{ \begin{bmatrix} J_{\mathrm{sc},x} \\ J_{\mathrm{sc},y} \\ J_{\mathrm{sc},z} \end{bmatrix} + \begin{bmatrix} 0 \\ 2J_{\mathrm{sc},y} \\ 0 \end{bmatrix} \right\} = \frac{1}{\sigma_\mathrm{sc}} (\boldsymbol{J}_\mathrm{sc} + \boldsymbol{Q}), \tag{7.56}$$

where $\boldsymbol{Q} = (0, 2J_{\mathrm{sc},y}, 0)^\mathrm{T}$. By this method, the HTSC is mathematically composed of two parts, one is a homogeneous HTSC as bulk **A** and the other has only the $c$-axis conductivity which is not equal to zero, as bulk ***B***. By substituting Eq. (7.56) and $\boldsymbol{B} = \mu \boldsymbol{H}$ into Faraday's law [Eq. (7.4)], we have

$$\nabla \times \boldsymbol{E}_\mathrm{sc} = \nabla \times \frac{1}{\sigma_\mathrm{sc}} \boldsymbol{J}_\mathrm{sc} + \nabla \times \frac{1}{\sigma_\mathrm{sc}} \boldsymbol{Q} = -\frac{\partial}{\partial t} (\mu \boldsymbol{H}). \tag{7.57}$$

If we further substitute Ampère's law [Eq. (7.1)] into Eq. (7.57), we find the PDE that governs the electromagnetic properties in the HTSC,

$$\frac{\partial}{\partial t} (\mu \boldsymbol{H}) + \nabla \times \frac{1}{\sigma_\mathrm{sc}} \nabla \times \boldsymbol{H} + \nabla \times \frac{1}{\sigma_\mathrm{sc}} \boldsymbol{Q} = 0. \tag{7.58}$$

Taking all computational regions shown in Fig. 7.16 into account and considering that $J_y = \partial H_x / \partial z - \partial H_z / \partial x$, Eq. (7.58) can be reshaped as follows by defining an auxiliary parameter $\lambda$ to identify the medium involved,

$$\frac{\partial}{\partial t} (\mu \boldsymbol{H}) + \nabla \times \frac{1}{\sigma} \nabla \times \boldsymbol{H} + \lambda \nabla \times \frac{1}{\sigma} \boldsymbol{Q} = 0, \tag{7.59}$$

where $\sigma = \sigma_\mathrm{air}$, $\lambda = 0$ for region $\Omega_1$ and $\sigma = \sigma_\mathrm{sc}$, $\lambda = 1$ for region $\Omega_2$ (see Fig. 7.16), and $\boldsymbol{Q}$ is defined as

$$\boldsymbol{Q} = \left[0, 2\left(\frac{\partial H_x}{\partial z} - \frac{\partial H_z}{\partial x}\right), 0\right]^\mathrm{T}. \tag{7.60}$$

A stable resolution of PDE needs some suitable boundary conditions. For Eq. (7.59), Dirichlet and Neumann boundary conditions are used to describe the electromagnetic situations. Considering the situation of a cubic-shaped HTSC in a non-uniform applied magnetic field generated by the PMG. The boundary condition between the HTSC and the dielectric subdomain $\Omega_1$ region is continuity:

$$\mu_1 H_{1n} = \mu_2 H_{2n}. \tag{7.61}$$

The outer boundary of subdomain $\Omega_2$ is a dynamic boundary described by a time-dependent function:

$$\boldsymbol{H}_{\Omega_2}(r, t) = f_{\Omega_2}(r, t), \tag{7.62}$$

where the function of $f_{\Omega_2}(r, t)$ describes how the non-uniform external magnetic fields changes with time at the outer boundary of subdomain $\Omega_2$.

#### 7.6.1.2 Example of applications: levitation force simulation of HTSC over a monopole PMG with different cross-sectional configuration

In this section, the proposed numerical method will be applied to study the influence of the cross-sectional configuration of a monopole PMG on the levitation forces of a HTSC above the PMG. For the 3D modeling, we consider a cylindrical-shaped HTSC, which is levitated above a monopole PMG, as shown in Fig. 7.17. The computational region involves two subregions: R1 of the air domain and R2 of the HTSC domain.

Based on the proposed 3D method in this section, the levitation forces of a HTSC over a monopole PMG are simulated. The influence of the cross-sectional configuration of the PMG on the levitation forces is numerically estimated by a set of simulations. During the simulation, a cylindrical HTSC is used with a dimension of 30 mm in diameter and 15 mm in height. The calculation parameters are $M_0 = 9.6 \times 10^5$ A/m$^2$, $J_{c0} = 8 \times 10^7$ A/m$^2$, $E_0 = 1.0 \times 10^{-4}$ V/m, and $n = 21$. For the $\boldsymbol{E}$-$\boldsymbol{J}$ power law used in these simulations, the effect of the flux creep is related to the number $n$. The shape of the magnetization loop of the HTSC is determined by its flux creep effect. For HTS materials, $n$ usually varies between 5 and 25. According to Brandt's Assumption [34], when $n \to \infty$, the $\boldsymbol{E}$-$\boldsymbol{J}$ power evolves to the Bean model. For a HTS slab, the value of $n$ is dependent on the magnitude of magnetic field and its angle to the top face. For bulk YBCO, usually $n = 21$ is a typical case for the $\boldsymbol{E}$-$\boldsymbol{J}$ power law.

In order to study the influence of the cross-sectional configuration of the PMG on the levitation performance of the HTSC, two schemes are chosen for the monopole PMG. Table 7.2 shows the first scheme: keeping the height of the PM constant (TD = 40 mm), and changing the width of PM from 40 mm to 60 mm. Table 7.3 shows the second scheme: keeping the width of the PM constant (TB = 40 mm) and changing the height of PM from 40 to 60 mm. The magnetic forces with vertical movement of the

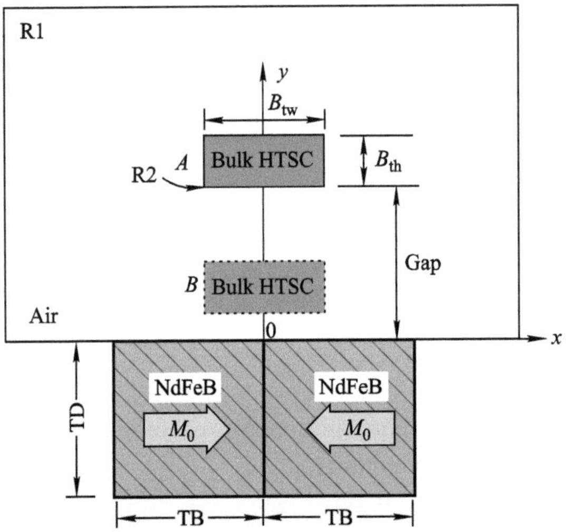


**Fig. 7.17:** Cross section of a HTSC levitating above a monopole PMG in 3D modeling. The monopole PMG is composed of two identical PMs. TD is the height of the PMG, whereas TB is the width of the PM.

HTSC are calculated separately. The MLFs are also recorded and tabulated in Tabs. 7.2 and 7.3.

In order to show the influence of the height and the width of the PM composing the PMG on the levitation performance of the HTSC, the MLF exerted on the HTSC with different values of height/width of the PM are shown in Fig. 7.18.

From Fig. 7.18, we can see that the MLF on the HTSC increases with the height/width of the PM while keeping the width/height constant. When TD changes from 40 to 60 mm, the width of the PM keeps a constant value of 40 mm. Similarly,

**Tab. 7.2:** Levitation force simulation of HTSC over a monopole PMG with different PM widths.

| Height TD (mm) | Width TB (mm) | Cross-section area (mm$^2$) | MLF (N) |
|---|---|---|---|
| 40 | 40 | 3200 | 78.14 |
| 40 | 42 | 3360 | 82.44 |
| 40 | 44 | 3520 | 84.27 |
| 40 | 46 | 3680 | 85.95 |
| 40 | 48 | 3840 | 87.5 |
| 40 | 50 | 4000 | 88.91 |
| 40 | 52 | 4160 | 90.21 |
| 40 | 54 | 4320 | 91.41 |
| 40 | 56 | 4480 | 92.51 |
| 40 | 58 | 4640 | 93.52 |
| 40 | 60 | 4800 | 94.46 |

**Tab. 7.3:** Levitation force simulation of HTSC over monopole PMG with different PM heights.

| Height TD (mm) | Width TB (mm) | Cross-section area (mm$^2$) | MLF(N) |
|---|---|---|---|
| 40 | 40 | 3200 | 78.14 |
| 42 | 40 | 3360 | 82.65 |
| 44 | 40 | 3520 | 84.72 |
| 46 | 40 | 3680 | 86.64 |
| 48 | 40 | 3840 | 88.43 |
| 50 | 40 | 4000 | 90.04 |
| 52 | 40 | 4160 | 91.64 |
| 54 | 40 | 4320 | 93.08 |
| 56 | 40 | 4480 | 94.42 |
| 58 | 40 | 4640 | 95.68 |
| 60 | 40 | 4800 | 96.85 |

when TB changes from 40 to 60 mm, the height of the PM keeps a constant value of 40 mm. The MLF of the HTSC increases with the cross-sectional area of the PMG. MLF increases faster when the height of the PM increases, compared with an increase of the width of the PM. Extremely, when the width extends to a very large value compared to its height, the MLF may tend to be constant for certain HTSC, which is similar to the case of changing the height.

One can see that the MLF increases rapidly while the cross-sectional area of the PMG increases in the initial stage. When the cross-sectional area gradually becomes larger, the increasing tendency of the MLF gradually becomes smaller. Regardless of whether it is the height or width that is increasing, the MLF always increases with the cross-sectional area.

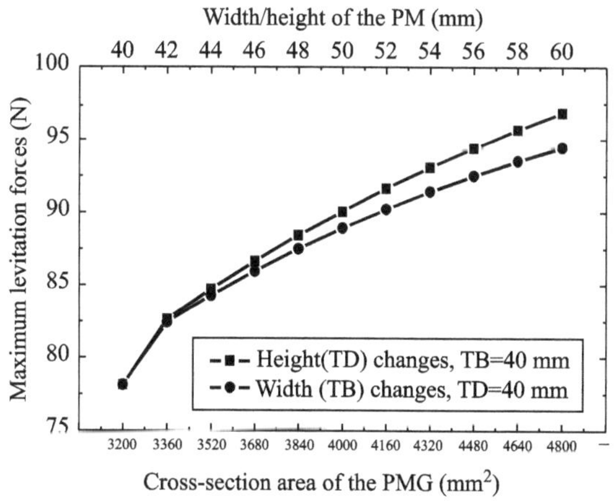


**Fig. 7.18:** Maximum levitation forces of the HTSC above the PMG with different cross-sectional areas. The width/height of the PM labels is shown on the top horizontal axis.

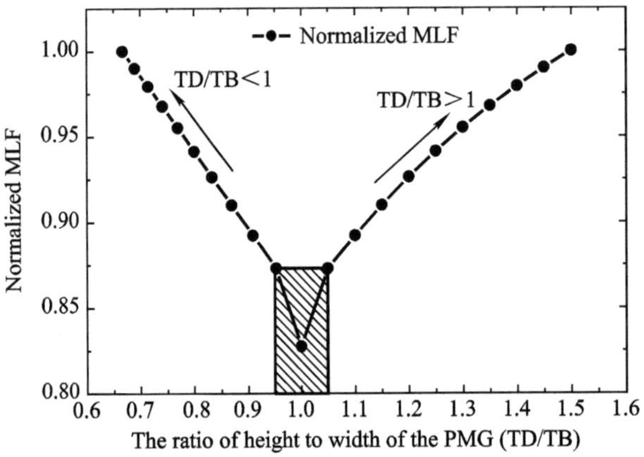


**Fig. 7.19:** Normalized maximum levitation forces with the ratio of height to width of the PMG.

In order to clearly show the relationship between the ratio of TD to TB of the PMG and the MLF of the HTSC, the curve of the normalized MLF with respect to the ratio of TD to TB is shown in Fig. 7.19. It is clear that when the ratio of TD/TB<1, which we call ultra-wide, the MLF will decrease with increasing TD/TB. When the ratio of TD/TB>1, which we call ultra-height, the MLF will increase with increasing TD/TB. It should be noticed that when increasing the cross section of the PMG while keeping the ratio of TD/TB constant, also increases the MLF simultaneously. This will, however, cause increase of the cost of the PMG.

From Fig. 7.19, we see that in the ultra-wide case, when the ratio of TD/TB changes from 1.0 to 0.95, the MLF changes faster compared to changing the ratio from 0.95 to 0.6. Similarly, prerequisite of ultra-height, when the ratio of TD/TB changes from 1.0 to 1.05, the MLF changes faster, compared with changing the ratio from 1.05 to 1.6. The shaded region in Fig. 7.19 may be effective in lowering the cost of the levitation parts.

On the basis of the discussions presented above, we can conclude that for HTS Maglev vehicle using a monopole PMG, low-levitation-cost optimization way can be achieved by utilizing a PMG whose ratio of height to width fulfils

$$\mathrm{TD}/(\mathrm{TB} + \mathrm{TB}) \subseteq (0.475, 0.525). \tag{7.63}$$

### 7.6.2 *T* method [10]

#### 7.6.2.1 Mathematical fundamentals

The *T* method referred here is an extended version of the traditional $T$-$\Omega$ method, which was proposed by Miya and coworkers [47] to calculate eddy-current problems and was adapted in recent years to electromagnetically model the HTS Maglev in 3D level. A prominent advantage of the $T$ method lies in the fact that the state variable $T$ is only defined in the conducting region, avoiding the mesh of the domain outside the conducting regions. Hence, this considerably reduces the number of degrees of freedom when the FEM technique is adopted to numerically discretize the mathematical equations.

From a macroscopic point of view, the Maxwell's equations are still valid to describe the electromagnetic behavior of the HTSC. According to Eq. (7.1) in the quasi-static approximation, the current density $\boldsymbol{J}$ is a divergence-free vector, which allows the introduction of a current vector potential $\boldsymbol{T}$ expressed as

$$\boldsymbol{J} = \nabla \times \boldsymbol{T}, \tag{7.64}$$

to which the Coulomb gauge is applied to guarantee the uniqueness of the solution, i.e. $\nabla \cdot \boldsymbol{T} = 0$.

Applying the Helmholtz's theorem to vector $\boldsymbol{T}$ yields

$$\begin{aligned} C(P)\boldsymbol{T}(P) = {} & \frac{1}{4\pi}\int\limits_V (\nabla' \cdot \boldsymbol{T}(P'))\nabla' \frac{1}{R(P,P')}\mathrm{d}V' - \frac{1}{4\pi}\int\limits_S (\boldsymbol{n}' \cdot \boldsymbol{T}(P'))\nabla' \frac{1}{R(P,P')}\mathrm{d}S' \\ & + \frac{1}{4\pi}\int\limits_V (\nabla' \times \boldsymbol{T}(P')) \times \nabla' \frac{1}{R(P,P')}\mathrm{d}V' \\ & - \frac{1}{4\pi}\int\limits_S (\boldsymbol{n}' \times \boldsymbol{T}(P')) \times \nabla' \frac{1}{R(P,P')}\mathrm{d}S', \end{aligned} \tag{7.65}$$

where $R(P, P')$ is the distance between the source point $P'$ and the field point $P$, the superscript $'$ refers to the quantity at the source point, $\boldsymbol{n}'$ is a unit vector out of the surface $S'$, and the coefficient $C(P)$ takes the following values [30]:

$$C(P) = \begin{cases} 1, & P \in V'(\text{excluding } S'), \\ 1/2, & P \in S', \\ 0, & \text{elsewhere.} \end{cases} \tag{7.66}$$

According to the physical fact that the normal component of $\boldsymbol{J}$ must be zero on the surface of the HTSC, i.e. $J_\mathrm{n} = 0$, $\boldsymbol{T}$ must fulfil the boundary condition that [30]

$$\boldsymbol{n}' \times \boldsymbol{T} = 0. \tag{7.67}$$

Therefore, only the normal component $T_\mathrm{n}$ exists on all surfaces of the HTSC.

Eq. (7.65) is reduced to the following form when the Coulomb gauge and boundary condition Eq. (7.67) are considered

$$C(P)\boldsymbol{T}(P) = \frac{1}{4\pi}\int_V (\nabla' \times \boldsymbol{T}(P')) \times \nabla' \frac{1}{R(P, P')} \mathrm{d}V' - \frac{1}{4\pi}\int_S (\boldsymbol{n}' \cdot \boldsymbol{T}(P'))\nabla' \frac{1}{R(P, P')} \mathrm{d}S'. \tag{7.68}$$

The *B-H* constitutive law of the HTSC can be assumed to be linear as that in vacuum to a good approximation because its applicable conditions [34] can be easily satisfied in a levitation system using bulk Y-Ba-Cu-O due to its small lower critical field $B_{c1}$ and large applied field as well as its geometry. Thus,

$$\boldsymbol{B} = \mu_0 \boldsymbol{H}. \tag{7.69}$$

The induced field $\boldsymbol{B}_s$ generated by the current in the HTSC can be expressed as a function of the vector $\boldsymbol{T}$ via combining Eq. (7.68) with Biot-Savart's law,

$$\boldsymbol{B}_s = \mu_0 C(P)\boldsymbol{T}(P) + \frac{\mu_0}{4\pi}\int_S (\boldsymbol{n} \cdot \boldsymbol{T}(P'))\nabla' \frac{1}{R(P, P')} \mathrm{d}S'. \tag{7.70}$$

When an equivalent conductivity $\sigma_s$, which is nonlinear and dependent on the local electrical field, is introduced, the traditional Ohm's law in the HTSC has the following form:

$$\boldsymbol{J} = \sigma_s(|\boldsymbol{E}|)\boldsymbol{E}. \tag{7.71}$$

By substituting Eqs. (7.64) and (7.71) into Eq. (7.65) and considering $\boldsymbol{B} = \boldsymbol{B}_e + \boldsymbol{B}_s$, where $\boldsymbol{E}_e$ is the applied field, the following equation is found:

$$\nabla \times \frac{1}{\sigma_s}(\nabla \times \boldsymbol{T}) + \frac{\partial(\boldsymbol{B}_e + \boldsymbol{B}_s)}{\partial t} = 0. \tag{7.72}$$

The governing equation of the HTSC based on the variable $T$ is finally derived from Eqs. (7.70) and (7.72) as follows,

$$\nabla \times \frac{1}{\sigma_s}(\nabla \times \boldsymbol{T}) + \mu_0 C(P)\frac{\partial \boldsymbol{T}}{\partial t} + \frac{\mu_0}{4\pi}\int_S \frac{\partial(\boldsymbol{n}' \cdot \boldsymbol{T}')}{\partial t}\nabla' \frac{1}{R(P, P')} \mathrm{d}S' + \frac{\partial \boldsymbol{B}_e}{\partial t} = 0. \tag{7.73}$$

When we replace $\sigma_s$ by a tensor resistivity $\overline{\overline{\rho_s}}$ to take the anisotropy of the HTSC into account in Eq. (7.73) and consider that

$$\overline{\overline{\rho_s}}(\nabla \times \boldsymbol{T}) = \begin{bmatrix} \rho_{ab} & 0 & 0 \\ 0 & \rho_{ab} & 0 \\ 0 & 0 & \alpha\rho_{ab} \end{bmatrix} \begin{bmatrix} \left(\frac{\partial T_z}{\partial y} - \frac{\partial T_y}{\partial z}\right)\hat{\boldsymbol{x}} \\ \left(\frac{\partial T_x}{\partial z} - \frac{\partial T_z}{\partial x}\right)\hat{\boldsymbol{y}} \\ \left(\frac{\partial T_y}{\partial x} - \frac{\partial T_x}{\partial y}\right)\hat{\boldsymbol{z}} \end{bmatrix}$$

$$= \rho_{ab}\left[\left(\frac{\partial T_z}{\partial y} - \frac{\partial T_y}{\partial z}\right)\hat{\boldsymbol{x}} + \left(\frac{\partial T_x}{\partial z} - \frac{\partial T_z}{\partial x}\right)\hat{\boldsymbol{y}} + \alpha\left(\frac{\partial T_y}{\partial x} - \frac{\partial T_x}{\partial y}\right)\hat{\boldsymbol{z}}\right]. \tag{7.74}$$

The following equality for the first term in the left side of Eq. (7.73) holds,

$$\nabla \times \overline{\overline{\rho_s}}(\nabla \times \boldsymbol{T}) = \rho_{ab}\left[\alpha\frac{\partial^2 T_y}{\partial x \partial y} - \alpha\frac{\partial^2 T_x}{\partial y^2} - \frac{\partial^2 T_x}{\partial z^2} + \frac{\partial^2 T_z}{\partial x \partial z}\right]\hat{\boldsymbol{x}}$$
$$+ \rho_{ab}\left[\frac{\partial^2 T_z}{\partial y \partial z} - \frac{\partial^2 T_y}{\partial z^2} - \alpha\frac{\partial^2 T_y}{\partial x^2} + \alpha\frac{\partial^2 T_x}{\partial y \partial x}\right]\hat{\boldsymbol{y}}$$
$$+ \rho_{ab}\left[\frac{\partial^2 T_x}{\partial z \partial x} - \frac{\partial^2 T_z}{\partial x^2} - \frac{\partial^2 T_z}{\partial y^2} + \frac{\partial^2 T_y}{\partial z \partial y}\right]\hat{\boldsymbol{z}}. \tag{7.75}$$

According to the Coulomb gauge, we have $\nabla(\nabla \cdot \boldsymbol{T}) = 0$, i.e.

$$\left(\frac{\partial^2 T_x}{\partial x^2} + \frac{\partial^2 T_y}{\partial y \partial x} + \frac{\partial^2 T_z}{\partial z \partial x}\right)\hat{\boldsymbol{x}} + \left(\frac{\partial^2 T_x}{\partial x \partial t} + \frac{\partial^2 T_y}{\partial y^2} + \frac{\partial^2 T_z}{\partial z \partial y}\right)\hat{\boldsymbol{y}}$$
$$+ \left(\frac{\partial^2 T_x}{\partial x \partial z} + \frac{\partial^2 T_y}{\partial y \partial z} + \frac{\partial^2 T_z}{\partial z^2}\right)\hat{\boldsymbol{z}} = 0. \tag{7.76}$$

The following identities can be derived from Eq. (7.76),

$$\frac{\partial^2 T_z}{\partial z \partial x} = -\frac{\partial^2 T_x}{\partial x^2} - \frac{\partial^2 T_y}{\partial y \partial x}, \quad \frac{\partial^2 T_z}{\partial z \partial y} = -\frac{\partial^2 T_x}{\partial x \partial y} - \frac{\partial^2 T_y}{\partial y^2}, \quad \frac{\partial^2 T_x}{\partial x \partial z} + \frac{\partial^2 T_y}{\partial y \partial z} = -\frac{\partial^2 T_z}{\partial z^2}. \tag{7.77}$$

Eq. (7.75) can be rewritten in the following form when Eq. (7.77) is taken into account:

$$\nabla \times \overline{\overline{\rho_s}}(\nabla \times \boldsymbol{T}) = \rho_{ab}\left[-\frac{\partial^2 T_x}{\partial x^2} - \alpha\frac{\partial^2 T_x}{\partial y^2} - \frac{\partial^2 T_x}{\partial z^2} + (\alpha-1)\frac{\partial^2 T_y}{\partial x \partial y}\right]\hat{\boldsymbol{x}}$$
$$+ \rho_{ab}\left[-\alpha\frac{\partial^2 T_y}{\partial x^2} - \frac{\partial^2 T_y}{\partial y^2} - \frac{\partial^2 T_y}{\partial z^2} + (\alpha-1)\frac{\partial^2 T_x}{\partial x \partial y}\right]\hat{\boldsymbol{y}}$$
$$+ \rho_{ab}\left(-\frac{\partial^2 T_z}{\partial x^2} - \frac{\partial^2 T_z}{\partial y^2} - \frac{\partial^2 T_z}{\partial z^2}\right)\hat{\boldsymbol{z}}. \quad (7.78)$$

Besides,

$$\nabla' = \frac{1}{R(P,P')} = \frac{\partial}{\partial x'}\left(\frac{1}{R(P,P')}\right)\hat{\boldsymbol{x}} + \frac{\partial}{\partial y'}\left(\frac{1}{R(P,P')}\right)\hat{\boldsymbol{y}} + \frac{\partial}{\partial z'}\left(\frac{1}{R(P,P')}\right)\hat{\boldsymbol{z}}, \quad (7.79)$$

and

$$\boldsymbol{B}_e = B_{ex}\hat{\boldsymbol{x}} + B_{ey}\hat{\boldsymbol{y}} + B_{ec}\hat{\boldsymbol{z}}. \quad (7.80)$$

Finally, the 3D PDEs for governing the electromagnetic behavior of HTSC were derived after Eqs. (7.78) to (7.80) are substituted into Eq. (7.73) and $\sigma_{ab}$ is replaced by its reciprocal $\sigma_{ab}$,

$$\frac{1}{\sigma_{ab}}\left(-\frac{\partial^2 T_y}{\partial x^2} - \alpha\frac{\partial^2 T_y}{\partial y^2} - \frac{\partial^2 T_y}{\partial z^2} + (\alpha-1)\frac{\partial^2 T_y}{\partial x \partial y}\right) + \mu_0 C(P)\frac{\partial T_x}{\partial t}$$
$$+ \frac{\mu_0}{4\pi}\int_S \frac{\partial(\boldsymbol{n}'\cdot\boldsymbol{T}')}{\partial t}\frac{\partial}{\partial x'}\left(\frac{1}{R(P,P')}\right)\mathrm{d}S' + \frac{\partial B_{ex}}{\partial t} = 0, \quad (7.81)$$

$$\frac{1}{\sigma_{ab}}\left(-\alpha - \frac{\partial^2 T_y}{\partial x^2} - \frac{\partial^2 T_y}{\partial y^2} - \frac{\partial^2 T_y}{\partial z^2} + (\alpha-1)\frac{\partial^2 T_x}{\partial x \partial y}\right) + \mu_0 C(P)$$
$$\frac{\partial T_y}{\partial t} + \frac{\mu_0}{4\pi}\int_S \frac{\partial(\boldsymbol{n}'\cdot\boldsymbol{T}')}{\partial t}\frac{\partial}{\partial y'}\left(\frac{1}{R(P,P')}\right)\mathrm{d}S' + \frac{\partial B_{ey}}{\partial t} = 0, \quad (7.82)$$

$$\frac{1}{\sigma_{ab}}\left(-\frac{\partial^2 T_z}{\partial x^2} - \frac{\partial^2 T_z}{\partial y^2} - \frac{\partial^2 T_z}{\partial z^2}\right) + \mu_0 C(P)\frac{\partial T_z}{\partial t}$$
$$+ \frac{\mu_0}{4\pi}\int_S \frac{\partial(\boldsymbol{n}'\cdot\boldsymbol{T}')}{\partial t}\frac{\partial}{\partial z'}\left(\frac{1}{R(P,P')}\right)\mathrm{d}S' + \frac{\partial B_{ez}}{\partial t} = 0, \quad (7.83)$$

where $\sigma_{ab}$ is the conductivity in the *a-b* plane. It is worth noting that compared with the traditional $T$-$\Omega$ method, the complexity of the governing equations is reduced due to the omission of the variable $\Omega$.

#### 7.6.2.2 Example of applications: optimization of the configuration of the PMG

PMG is a key component to provide the applied magnetic field for the present HTS Maglev system, and the cost needed in building the PMG comprises the most part of the entire investment because the PMG is required along the whole line. It is therefore meaningful to optimize the structure and geometric parameters of the PMG to obtain a PMG that has the required levitation capability while reducing the cost. Thus, on the basis of this 3D method, we have calculated the levitation forces and guidance forces of three HTSCs above two different PMGs to optimize the geometry of the PMG.

The geometric parameters of the PMG such as height and width are variable in the calculation, and then the dependence of levitation capability of the HTSC on those parameters is studied. Compared with the previous work, the merit of the present calculation is that the computed levitation/guidance forces are comparable to the real system with a reasonable value of the critical current density. Thus, the computed results can be used to conduct the practical design directly.

The geometric and material parameters of the HTSC bulk are shown in Tab. 7.4. The speed of the samples in both vertical and transverse direction is 1 mm/s and the magnetization $M_0$ of the PM employed to assemble the PMG is $8.9 \times 10^5$ A/m in all cases.

As for the structure of the PMG, a Halbach array is a better choice because this structure can concentrate the magnetic field above it where the HTSCs are placed, and thus it can improve the utilization of the magnetic field. From the basic structure of the Halbach array shown in Fig. 7.20, we can derive two different types of PMG, i.e. one has three PMs magnetized in the horizontal direction and two PMs magnetized in the vertical direction, the other has three PMs magnetized in the vertical direction

**Tab. 7.4:** Parameters of HTS used in the simulation for obtaining an optimized PMG.

| $w_{sc}$(mm) | $l_{sc}$(mm) | $t_{sc}$(mm) | $J_c^{ab}$(A/m$^2$) | $E_c$(V/m) | $U_0$ (ev) | $\alpha$ |
|---|---|---|---|---|---|---|
| 42 | 21 | 9 | $2.5 \times 10^8$ | $1 \times 10^{-4}$ | 0.1 | 3 |

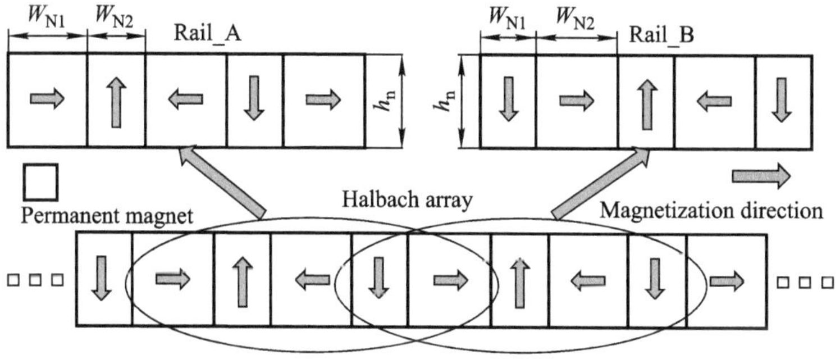


**Fig. 7.20:** Two different PMGs with five PMs deriving from the Halbach array.

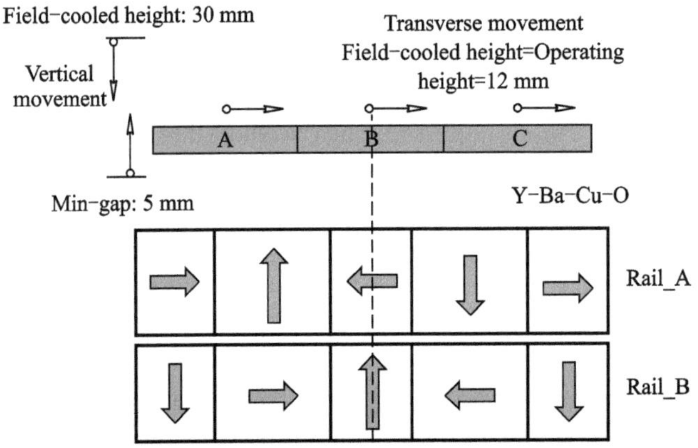


**Fig. 7.21:** Chart of the Maglev system with three HTSCs undergoing the vertical or transverse movement above the PMG.

and two PMs magnetized in the horizontal direction. These two PMGs, like those in Fig. 7.8, derived from the Halbach array, are presented in Fig. 7.20.

In the calculation, we ignore the possible interaction among the HTSCs for simplicity, and calculate only two HTSCs because of the symmetry of the levitation system shown in Fig. 7.21. In the default case, the HTSCs are field-cooled at a position of 30 mm above the surface of the PMG for the calculation of the levitation forces and the guidance forces, the transverse movement occurs at the same height as the FC position, that is, 12 mm above the surface of the PMG. The main parameters that should be optimized in the two structures, i.e. Rail_A and Rail_B shown in Fig. 7.20, are the ratio between the width of two different magnetized PMs and the width and height of the PMG. The following section shows the computed results of the optimization by altering those parameters.

In this part, the levitation forces and guidance forces on the HTSCs with the variation of the width ratio, i.e. $w_{A1}/w_{A2}$ for Rail_A or $w_{B1}/w_{B2}$ for Rail_B, are calculated. In this calculation, the total width and height of both PMGs are assumed to be 130 and 30 mm, respectively, and invariant with the change of the width ratio. For simplicity in drawing the following figures, the width ratio is replaced by an order and the corresponding relationship between the width ratio and the order is given in Tab. 7.5.

**Tab. 7.5:** Corresponding relationship between the width ratio and the order in optimizing the width ratio of the PMGs.

| **Order** | **1** | **2** | **3** | **4** | **5** | **6** | **7** | **8** | **9** | **10** | **11** |
|---|---|---|---|---|---|---|---|---|---|---|---|
| | 12 | 13 | 14 | 15 | 16 | 17 | 18 | 19 | 20 | 21 | – |
| Width | 0 | 0.01 | 0.1 | 0.16667 | 0.25 | 0.333333 | 0.4 | 0.5 | 0.66667 | 0.83 | 1 |
| ratio | 1.22 | 1.5 | 2 | 2.5 | 3 | 4 | 6 | 10 | 100 | $\infty$ | – |

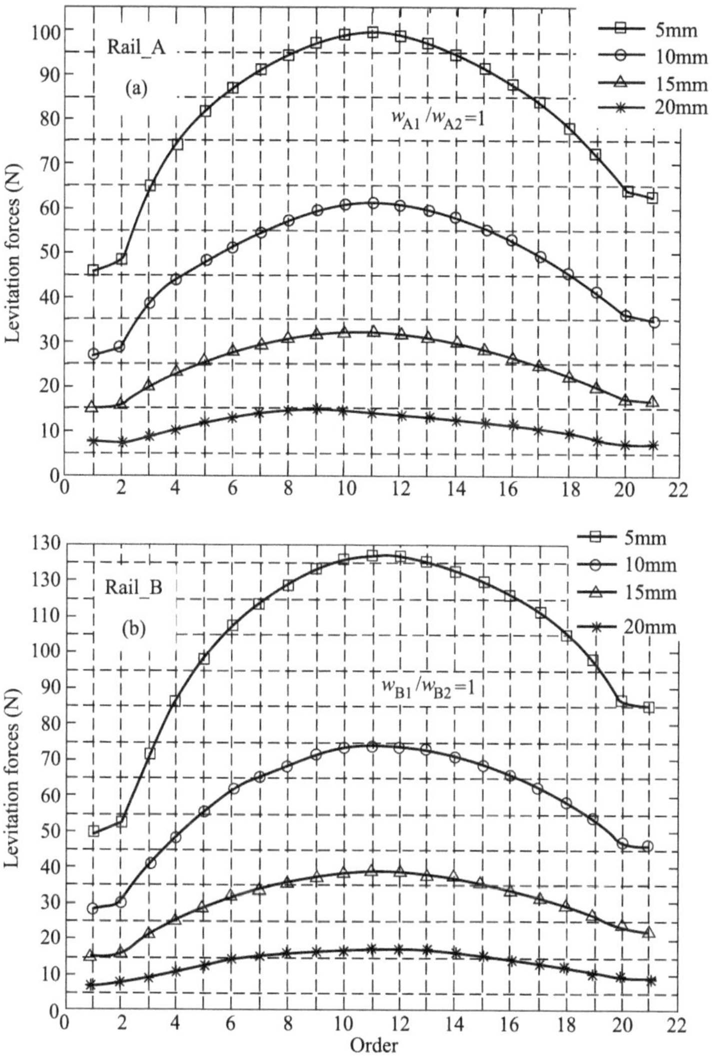


**Fig. 7.22:** Changing curves of the levitation forces on the HTSCs at position of 5, 10, 15, and 20 mm above Rail_A (a) and Rail_B (b) with the growth of the width ratio. The width ratio is replaced by an order and the corresponding relationship between the width ratio and the order (abscissa) is given in Tab. 7.5.

For the case of vertical movement, 21 different width ratios from zero to infinity are considered and the changing curves of the levitation forces along with the width ratio at position of 5, 10, 15, and 20 mm above Rail_A and Rail_B are shown in Figs. 7.22a and b, respectively. Note that the extreme case with a width ratio of zero or infinity denotes that Halbach structure disappears and PMs employed in the PMG only have one magnetized direction (vertical or horizontal). Both Figs. 7.22a and b display that

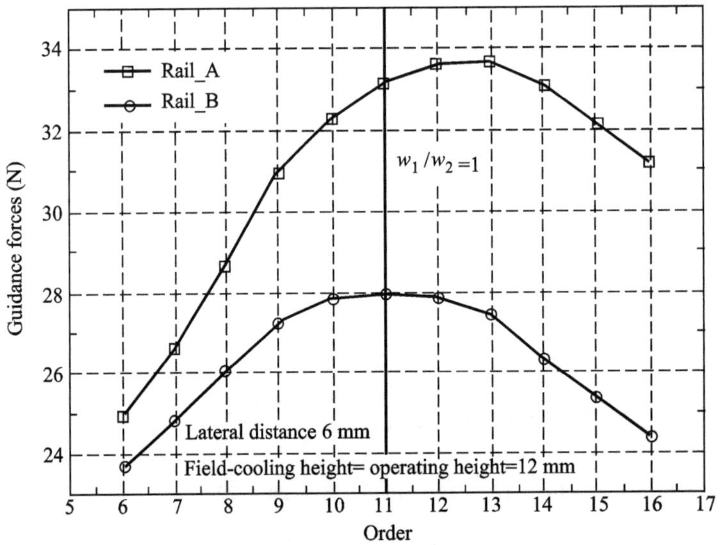


**Fig. 7.23:** Changing curve of the guidance forces on HTSCs at a lateral displacement of 6 mm with the growth of the width ratio. The corresponding relationship between the width ratio and the order (abscissa) can be found in Tab. 7.5. The total number of width ratios considered in this calculation is reduced and thus the order starts at 6 and ends at 16.

the levitation forces at all positions, increases with the growth of the width ratio, and reaches a maximum value when the width ratio is ~0.83 for Rail_A and is ~1 for Rail_B and then drops with continuous growth of the width ratio. This finding indicates that the Halbach array has a better performance because the levitation forces in the extreme case are always the smallest as has been verified by experiment. On the other hand, the optimized structure for Rail_A is that the width of the PMs magnetized in the horizontal direction is slightly larger than those magnetized in vertical direction, but for Rail_B, the two types of PMs have an identical width ratio.

For the case of transverse movement, only 11 different width ratios around 1 are considered because a larger levitation force can be obtained in this range according to the above computed results of levitation forces. The HTSCs have an FC height of 12 mm and then are driven transversely at the same height with a maximum lateral displacement of 6 mm. Fig. 7.23 presents the guidance forces with respect to the width ratio for Rail_A and Rail_B. We can find from this figure that the guidance forces exhibit the same tendency with the growth of the width ratio as that found in the case of the levitation forces, but the width ratio where the largest guidance force occurs is found to be 1.22–1.5 for Rail_A and is still ~1 for Rail_B.

As a whole, this suggests that for the Halbach array as the PMG to the HTS Maglev vehicle, the optimized width ratio between the vertical and transverse magnetized PMs is ~1.

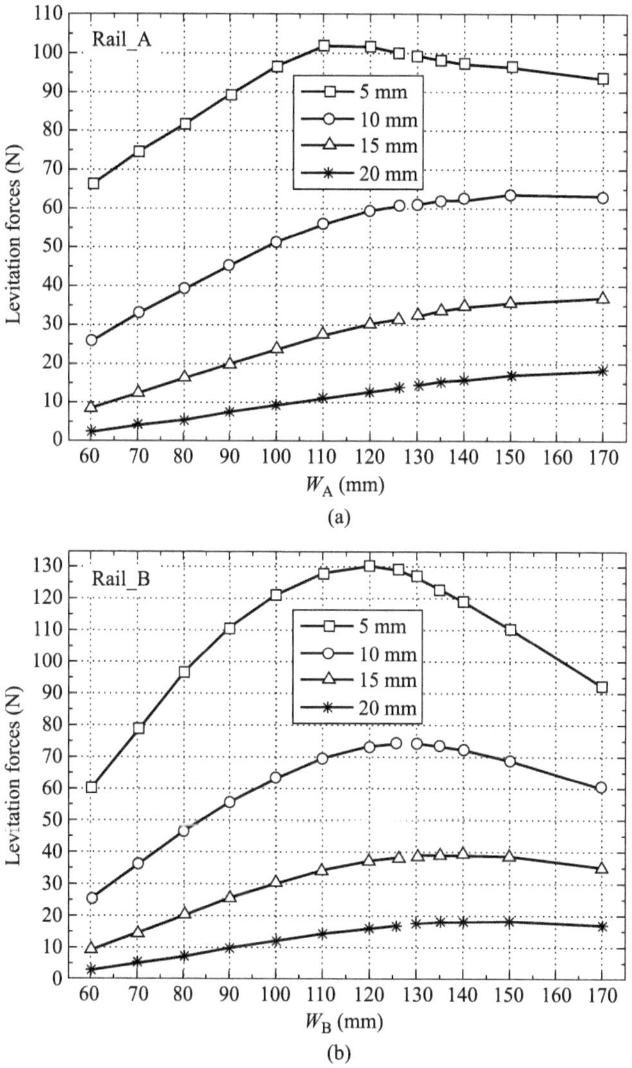


**Fig. 7.24:** Changing curve of the levitation forces on the HTSCs at positions of 5, 10, 15, and 20 mm above Rail_A (a) and Rail_B (b) with the growth of the total width. The total width is varied from 60 mm to 170 mm.

In this calculation, the total width of both PMGs is a variable parameter with a fixed height of 30 mm and a fixed width ratio of 1. Fig. 7.24 shows the changing curve of the levitation forces at positions of 5, 10, 15, and 20 mm above Rail_A and Rail_B with the growth of the total width from 60 to 170 mm. These figures show that with the increase of the position, the total width where the MLF appears is shifted from small to large value, e.g. the levitation forces increase almost linearly with the total width

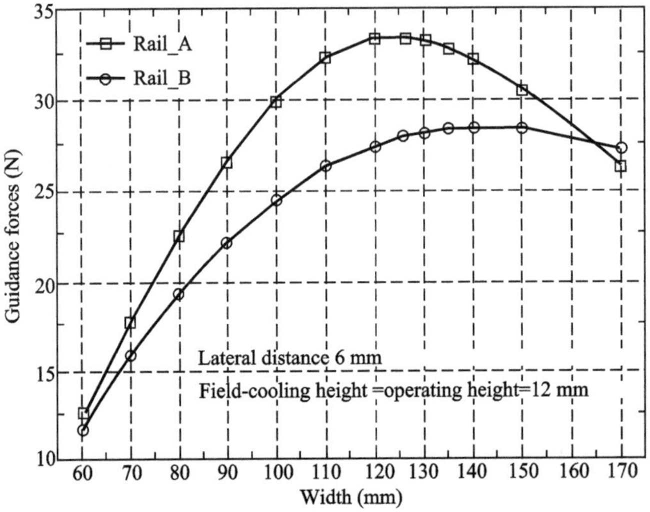


**Fig. 7.25:** Changing curve of the guidance forces on the HTSCs at a lateral displacement of 6 mm with the growth of the total width. The total width is varied from 60 to 170 mm.

at the position of 20 mm for both PMGs. The MLF even does not appear within our selected range. This figure also indicates that the increase of the total width does not always bring an enhancement of the levitation forces, especially for the case at a low position. Of course, if a larger levitation capability is required at a higher position, it is viable to lengthen the total width of the PMG. For the typical position of 15 mm in the Maglev vehicle, the optimized value of the total width appears around 130 mm for both PMGs because the slope of the curve begins to decrease at this value.

Fig. 7.25 shows the changing curve of the guidance forces at a lateral displacement of 6 mm with the growth of the total width from 60 mm to 170 mm. From this figure, it can be found that the guidance forces also increase firstly and then decreases with the growth of the total width, and the maximum value of the guidance forces occurs around 120 mm for Rail_A and around 135 mm for Rail_B. Although the total width where the maximum value of the guidance forces occurs is scattered above Rail_A and Rail_B, the variation of the guidance forces above both PMGs in the range of 120–140 mm is not evident. Therefore this range can be considered as the optimized range, which is comparable to the total width of the HTSC (126 mm), in determining the total width of the PMG for a practical design.

The height of the PMG is another factor influencing the levitation capability of the HTSC. Here, we fixed the width ratio to be 1 and total width to be 130 mm, and varied the height of the PMG from 15 to 100 mm. The levitation forces above Rail_A and Rail_B are plotted as a function of the height in Fig. 7.26. This figure clearly displays that no matter the position where the HTSCs are, the levitation forces increase drastically at first and then gradually saturates with increase in height of the PMG. This indicates

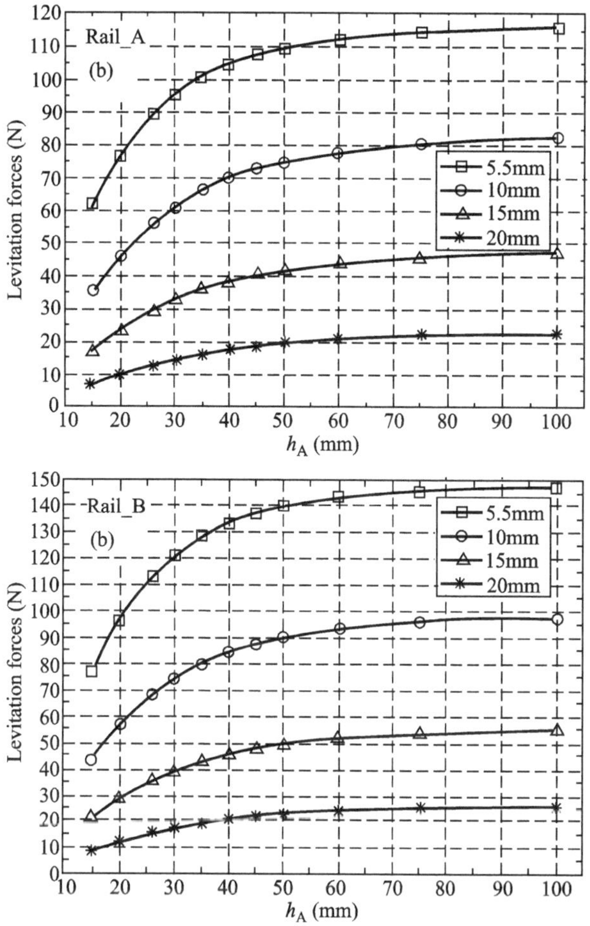


**Fig. 7.26:** Levitation forces versus height of Rail_A (a) and Rail_B (b). The height of both PMGs varied from 15 to 100 mm.

that considering the cost of the PMG, it is not reasonable to get better levitation by increasing the height of the PMG further.

## References

[1] Davis LC, Logothetis EM, Soltis RE. Stability of magnets levitated above superconductors. J Appl Phys. 1988;64(8):4212–4218.

[2] Kordyuk AA. Magnetic levitation for hard superconductors. J Appl Phys. 1998;83(1):610–612.

[3] Hull JR, Cansiz A. Vertical and lateral forces between a permanent magnet and a high-temperature superconductor. J Appl Phys. 1999;86(11):6396–6404.

[4] Yang Y, Zheng XJ. Method for solution of the interaction between superconductor and permanent magnet. J Appl Phys. 2007;101:113922.
[5] Wang XR, Ren ZY, Song HH, Wang XZ, Zheng J, Wang SY, Wang JS, Zhao Y. Guidance force in an infinitely long superconductor and permanent magnetic guideway system. Supercond Sci Technol. 2005;18(2):S99.
[6] Song HH, Wang JS, Wang SY, Ren ZY, Wang XR, Hass O, Fuchs G, Schultz L. Martin BP, editors. New topics in superconductivity research. Nova Science Publishers; 2006. pp. 107–156.
[7] Sanchez A, Valle ND, Pardo E, Chen DX, Navau C. Magnetic levitation of superconducting bars. J Appl Phys. 2006;99:113904.
[8] Zhang JH, Zeng YW, Cheng J, Tang X. Optimization of permanent magnet guideway for HTS Maglev vehicle with numerical methods. IEEE Trans on Appl Supercond. 2008;18(3): 1681–1686.
[9] Lu YY, Wang JS, Wang SY, Zheng J. 3D-Modeling numerical solutions of electromagnetic behavior of HTSC bulk above permanent magnetic guideway. J Supercond Nov Magn. 2008;21(8):467–472.
[10] Ma GT, Wang JS, Wang SY. 3-D Modeling of High-$T_c$ Superconductor for magnetic levitation/suspension application—Part I: Introduction to the method. IEEE Trans on Appl Supercond. 2010;20(4):2219–2227.
[11] Dias DHN, Motta ES, Sotelo GG, de Andrade R, Jr. Experimental validation of field cooling simulations for linear superconducting magnetic bearings. Supercond Sci Technol. 2010;23:075013.
[12] Ma GT. Considerations on the finite-element simulation of high-temperature superconductors for magnetic levitation purposes. IEEE Trans on Appl Supercond. 2013;23(5):3601609.
[13] Navau C, Del-Valle N, Sanchez A. Macroscopic modeling of magnetization and levitation of hard type-II superconductors:The critical-state model. IEEE Trans on Appl Supercond. 2013;23(1):8201023.
[14] Ma GT, Liu H, Li XT, Zhang H, Xu YY. Numerical simulations of the mutual effect among the superconducting constituents in a levitation system with translational symmetry. J Appl Phys. 2014;115:083908.
[15] Del-Valle N, Sanchez A, Navau C, Chen DX. A theoretical study of the influence of superconductor and magnet dimensions on the levitation force and stability of maglev systems. Supercond Sci Technol. 2008;21:125008.
[16] Ma GT, Wang JS, Wang SY. 3-D Modeling of High-$T_c$ Superconductor for magnetic levitation/suspension application—Part II: Validation with experiment. IEEE Trans on Appl Supercond. 2010;20(4):2228–2234.
[17] Del-Valle N, Sanchez A, Navau C, Chen DX. Lateral-displacement influence on the levitation force in a superconducting system with translational symmetry. Appl Phys Lett. 2008; 92 (4) :1554.
[18] Ma GT, Wang JS, Wang SY. Numerical investigation of the lateral movement influence on the levitation force of the bulk HTS based on a 3-D model. IEEE Trans on Appl Supercond. 2010;20(3):924–928.
[19] Del-Valle N, Sanchez A, Navau C, Chen DX. Magnet guideways for superconducting Maglevs: comparison between halbach-type and conventional arrangements of permanent magnets. J Low Temp Phys. 2011;162(1–2):62–71.
[20] Lu Y, Zhuang S. Magnetic forces simulation of bulk hts over permanent magnetic railway with numerical method. Low J. Temp Phys. 2012;169(1–2):111–121.
[21] Del-Valle N, Sanchez A, Pardo E, Chen DX, Navau C. Optimizing levitation force and stability in superconducting levitation with translational symmetry. Appl Phys Lett. 2007;90:042503.

[22] Del-Valle N, Sanchez A, Pardo E, Navau C, Chen DX. Enhanced stability by field cooling in superconducting levitation with translational symmetry. Appl Phys Lett. 2007;91:112507.
[23] Del-Valle N, Sanchez A, Navau C, Chen DX. Theoretical hints for optimizing force and stability in actual Maglev devices. IEEE Trans on Appl Supercond. 2009;19(3):2070–2073.
[24] Del-Valle N, Agramunt-Puig S, Navau C, Sanchez A. Shaping magnetic fields with soft ferromagnets: Application to levitation of superconductors. J Appl Phys. 2012;111:013921.
[25] Agramunt-Puig S, Del-Valle N, Navau C, Sanchez A. Optimization of a superconducting linear levitation system using a soft ferromagnet. Physica C. 2013;487(4):11–15.
[26] Jackson JD. Classical Electrodynamics, 3rd ed. John Wiley & Sons, Inc.; 1998.
[27] Moon FC. Magneto-solid mechanics. John Wiley & Sons, Inc.; 1984.
[28] Iwasa Y. Case studies in superconducting magnets: design and operational issues. Springer; 2009.
[29] Bean CP. Magnetization of high-field superconductors. Rev Mod Phys. 1964;36(1):886–901.
[30] Hashizume H, Sugiura T, Miya K, Ando Y, Akita S, Torii S, Kubota Y, Ogasawara T. Numerical analysis of a.c. losses in superconductors. Cryogenics. 1991;31(7):601–606.
[31] Luo Y, Takagi T, Miya K. Reduction of levitation decay in high $T_c$ superconducting magnetic bearings. Cryogenics. 1999;39(4):331–338.
[32] Gou XF, Zheng XJ, Zhou YH. Drift of levitated/suspended body in high-$T_c$ superconducting levitation systems under vibration—Part II: Drift velocity for gap varying with time. IEEE Trans on Appl Supercond. 2007;17(3):3803–3808.
[33] Rhyner J. Magnetic properties and AC-losses of superconductors with power law current—voltage characteristics. Physica C. 1993;212(3–4):292–300.
[34] Brandt EH. Superconductors of finite thickness in a perpendicular magnetic field: Strips and slabs. Phys Rev B. 1996;54(6):4246–4264.
[35] Anderson PW. Theory of flux creep in hard superconductors. Phys Rev Lett. 1962;9(7): 309–311.
[36] Ma GT, Rauh H. Thermo-electromagnetic properties of a magnetically shielded superconductor strip: theoretical foundations and numerical simulations. Supercond Sci Technol. 2013;26:105001.
[37] Kim YB, Hempstead CF, Strnad AR. Critical persistent currents in hard superconductors. Phys Rev Lett. 1962;9(7):306–309.
[38] Dinger TR, Worthington TK, Gallagher WJ, Sandstrom RL. Direct observation of electronic anisotropy in single-crystal Y1Ba2Cu3O7-x. Phys Rev Lett. 1987;58(25):2687–2690.
[39] Mikitik GP, Brandt EH. Critical state in thin anisotropic superconductors of arbitrary shape. Phys Rev B. 2000;62(10):6800–6811.
[40] Matsushita T. Flux pinning in superconductors. Springer-Verlag; 2007.
[41] Furlani E. Permanent magnet and electromechanical devices. Academic Press; 2001. p. 61.
[42] Ma GT, Liu HF, Wang JS, Wang SY, Li XC. 3D Modeling permanent magnet guideway for high temperature superconducting Maglev vehicle application. J Supercond Nov Magn. 2009;22(8):841–847.
[43] Prigozhin L. On the Bean critical-state model in superconductivity. Europ J Appl Math. 1996;7(3):237–247.
[44] Prigozhin L. The bean model in superconductivity: Variational formulation and numerical solution. J Comput Phys. 1996;129(1):190–200.
[45] Prigozhin L. Analysis of critical-state problems in type-II superconductivity. IEEE Trans on Appl Supercond. 1997;7(4):3866–3873.
[46] Murakami M, Oyama T, Fujimoto H, Gotoh S, Yamaguchi K, Shiohara Y, Koshizuaka N, Tanaka S. Melt processing of bulk high $T_c$ superconductors and their application. IEEE Trans on Magnetics. 1991;27(2):1479–1486.

[47] Miya K, Hashizume H. Application of T-method to AC problem based on boundary element method. IEEE Trans on Magnetics. 1988;24(1):134–137.

## Further readings

Detailed introduction of the finite-element method for electromagnetics:

[1] Jin JM. The finite element method in electromagnetics. Wiley; 2002.

Three-dimensional model the HTS bulk using magnetic vector potential:

[2] Ueda H, Azumaya S, Tsuchiya S, Ishiyama A. 3D electromagnetic analysis of levitating transporter using bulk superconductor. IEEE Trans on Appl Supercond. 2006;16(2):1092–1095.

[3] Alloui L, Bouillault F, Mimoune SM. Numerical study of the influence of flux creep and of thermal effect on dynamic behaviour of magnetic levitation systems with a high-$T_c$ superconductor using control volume method. Eur Phys J Appl Phys. 2009;45:020801.

Numerical techniques to solve the nonlinear problems and the large algebraic equations:

[4] Alloui L, Ben Alia K, Bouillault F, Mimoune SM, Bernard L, Lévêque J. Numerical study of the relation between the thermal effect and the stability of the levitation system excited by an external source. Physica C. 2013;487(4):1–10.

[5] Knoll DA, Keyes DE. Jacobian-free Newton-Krylov methods: a survey of approaches and applications. J Comput Phys. 2004;193(2):357–397.

[6] Saad Y, Schultz MH. GMRES: A generalised minimal residual algorithm for solving nonsymmetric linear systems. SIAM J Sci Stat Comput. 1986;7(3):856–869.

[7] Saad Y. Iterative methods for sparse linear systems. Philadelphia: Society for industrial and applied mathematics; 2003.

[8] Kelley CT. Iterative methods for linear and nonlinear equations. Philadelphia: Society for industrial and applied mathematics; 1995.

Jun Zheng, Zi-Gang Deng, Jia-Su Wang, and Su-Yu Wang

# 8 New progress of HTS Maglev vehicle

## 8.1 Introduction

Since there is no mechanical contact, Maglev trains can easily move faster, more smoothly, and quietly than any conventional wheeled rail transport system. Up to now, the superconducting Maglev is still the holder of the world speed record for the whole rail transit, which is 581 km/h by the Japanese experimental low-temperature superconducting (LTS) Maglev vehicle, MLX-01, in 2003 [1]. On May 27, 2011, the Japanese government announced that the Chou Shinkansen Maglev [2], the fruit of half a century of tireless research, was granted 9 trillion JPY to start constructing a first real commercial superconducting Maglev line in the world. The most dramatic technology innovation is the remarkable designed running speed of 505 km/h. This event became a significant landmark in the field of superconducting Maglev technology, which stated that "the prospects of the establishment of technology for commercial operation as an ultra high-speed mass transportation system have become clear". Regarding the newly born high-temperature superconducting (HTS) Maglev using high-temperature superconductor (HTSC) bulks, which is also referred to as a flux-trapping type HTS Maglev, it is predicted that it will be ready to use in 2020 by 35% of the respondents of the survey of Bentō Strategy's Superconductor Market Research in 2010 [3], although the R&D history of HTS Maglev vehicle is only 16 years old.

Due to the inherent flux-pinning characteristics, HTSC bulks can realize passive stable levitation equilibrium in an applied magnetic field without any control. It is amazing that this large-scale passive levitation exists or is elastic over a continuous range of stable equilibrium positions or orientations to allow a HTS bulk to levitate or suspend motionless [4]. From the viewpoint of technology application, it promises simple levitation and guidance implementations as well as a light vehicle body. These prominent advantages based on this levitation principle bring wide potential for application, including the Maglev vehicle, in the field of rail transit. Scientists and engineers have focused on the first-stage research on the feasibility of HTS Maglev vehicles since the late 1990s. On December 31, 2000, the first manned HTS Maglev test vehicle "Century" was tested successfully carrying up to five people at a net levitation gap larger than 20 mm above the 15.5-m-long double permanent magnetic guideway (PMG) in China [5]. (see Chapter 6). Afterward, two other manned HTS Maglev demonstration vehicles were developed in Germany [6, 7] and Russia [8]. A small-scaled vehicle model in Japan experimentally reached a 42-km/h running speed over a circular test PMG [9]. More experimental HTS Maglev vehicles were launched from different research and application views, including a traffic exhibition model, a new PMG for better Maglev performance, a high-speed launch model [10–15].

https://doi.org/10.1515/9783110538434-008

These preliminary explorations not only proved the feasibility, but also demonstrated the following potential technological merits: simple structure, no controller, lighter vehicle, environmental-friendly with liquid nitrogen or cryo-coolers, energy-efficient without any electrical power supporting levitation or guidance in any operation conditions, comfort based on the inherent elastic and damping HTS levitation [16], low requirements on infrastructure, and low construction and maintenance costs [17].

Table 8.1 sums up the main R&D events of HTS Maglev vehicle systems in the world. It is obvious that the first-stage research on the feasibility for the rail transit applications has been widely accepted based on past intensive studies on the quasi-static levitation tests and related manufacturing schemes.

In the second-stage research after 2008, the focus was on differentreal operation conditions, dynamic Maglev behaviors, long-distance HTS Maglev test lines, and even the future transportation forms of the HTS Maglev vehicle. In early 2006, the construction cost of a 1.0-km full-scale HTS Maglev line was evaluated to be cheaper than a light rail vehicle, mainly because of the simple structure and corresponding low infrastructure cost [17]. Based on this, a 200 m real-scale HTS Maglev test line "MagLev-Cobra" in Brazil was studied as a substitute design for an urban light track in 2014 [10, 11, 18]. In fact, all worldwide efforts, including the Chinese work, are focused on speeding up the engineering and commercialization of a real full-scale HTS Maglev vehicle system. Thus, in this chapter, the present second-stage research and corresponding conclusions of an HTS Maglev vehicle, especially those from ASCLab,

**Tab. 8.1:** R&D history of HTS Maglev vehicle systems.

| Years | Progress |
|---|---|
| 1934 | Patent of the EMS-type Maglev train in Germany |
| 1966 | Patent of the LTS Maglev train in USA |
| 1988 | Discovery of the first HTS suspension in USA |
| 1996 | Levitation of a 220-kg sumo wrestler by a HTS disk in Japan |
| 1997 | A HTS Maglev model with a 20-kg weight and 7 mm levitation gap as a China-Germany joint project |
| 1997 | HTS Maglev test vehicle project at SWJTU launched by the "National 863 Program" in China |
| 2000 | The first manned HTS Maglev test vehicle "Century" in the world successfully developed by ASCLab, SWJTU, China |
| 2004 | A manned HTS Maglev test vehicle successfully developed by Germany (SupraTrans I) and Russia |
| 2009 | A manned HTS Maglev test vehicle "Cobra I" in Brazil |
| 2010 | A manned HTS Maglev test vehicle "SupraTrans II" in Germany |
| 2011 | A full-scale HTS Maglev test vehicle "Cobra II" and related new transportation concept |
| 2013 | A 45-m-long HTS Maglev ring test line and vehicle "Super-Maglev" in China |
| 2014 | The first ETT HTS Maglev ring test line "Super-Maglev" in the world in China |
| 2014 | A 200-m-long full-scale HTS Maglev test line and vehicle "Maglev-Cobra" in Brazil |

SWJTU, China, will be systemically summarized and reviewed in the next sections to illustrate the continuous R&D on HTS Maglev vehicles in the field for rail transit.

## 8.2 Dynamic characteristics

### 8.2.1 Vibration performance

As a significant experimental dynamic testing method, the vibration test is performed to determine the physical nature, vibration parameters and possible motion modes of a HTS Maglev vehicle system as the preliminary to the experimental design. A scaled HTS Maglev vehicle model (see Fig. 8.1) with 4:1 ratio to the first manned HTS Maglev test vehicle [5] was made and measured with different field cooling heights (FCHs) and working heights (WHs). In this model, the on-board liquid nitrogen vessels on each side of the vehicle model were those employed in the first manned HTS Maglev test vehicle [5] which guarantees that the vibration parameters were the same value for both.

The vehicle model was 0.8 m in length, 1.2 m in width, and 0.35 m in height. The total weight of the model (including the vessels with liquid nitrogen) was 32 kg. There were 43 cylindrical melt-textured YBCO HTS bulks in each vessel. Each YBCO bulk was of 30 mm in diameter and 17 mm in height.

An 11-ch vibration measurement (3560C, Bruel & Kjear S&V) was applied together with six piezoelectric accelerometers (4507B-004, Bruel & Kjear S&V). Cartesian coordinate was set-up to analyze the equivalent dynamic model of the HTS Maglev vehicle. Seven accelerometers were used to sample the vibration signals at seven keypoints on the upper surface of the model, represented by A, B, C, D, E, and F, respectively, in Fig. 8.2. They were divided into two groups A–C and E–F, which were fixed on each side of the vehicle model levitated over the PMG. All vibration signals were analyzed

**Fig. 8.1:** Photo of the vibration experimental set-up. Points A, B, C, D, E, and F stand for the accelerometers.

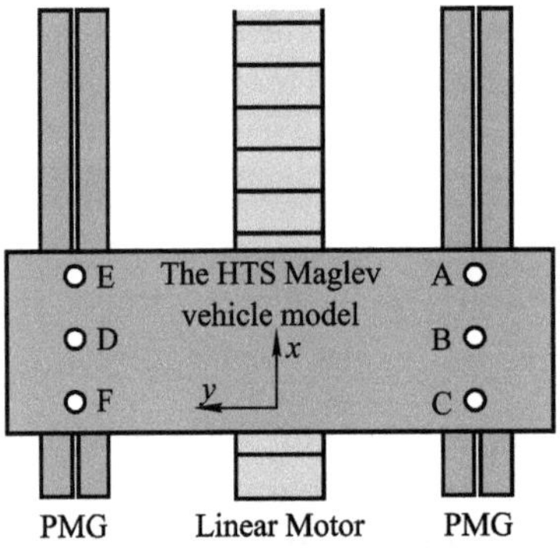


**Fig. 8.2:** Top view of the vibration experimental set-up.

by the vibration measurement system which was put on the center of the vehicle model, as shown in Fig. 8.1.

Figure 8.3 gives the 2D dynamic model of the freely levitated HTS Maglev vehicle combined with a mass, two springs, and two dampers. The corresponding dynamics characteristics in the vertical direction are described by

$$m\ddot{z} + c\dot{z} + kz = f, \tag{8.1}$$

where $z$ is the vertical displacement, $m$ is the mass, $c$ is an equivalent damping coefficient, $k$ is the equivalent stiffness and $f$ is the external impulse force.

Eq. (8.1) is further transformed into

$$\ddot{z} + 2\gamma\omega_n\dot{z} + \omega_n^2 z = \frac{f}{m}, \tag{8.2}$$

where $\gamma = c/(2\sqrt{mk})$ is the damping ratio and the resonant frequency $\omega_n = \sqrt{k/m}$.

The resonance frequency, dynamic stiffness, damping coefficient, and even the possible dynamic motion modes were the unsolved significant dynamic parameters

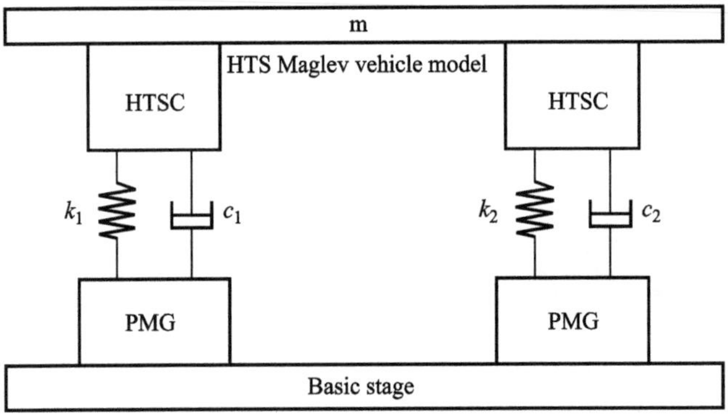


**Fig. 8.3:** Simplified 2D dynamic model of the HTS Maglev vehicle mode.

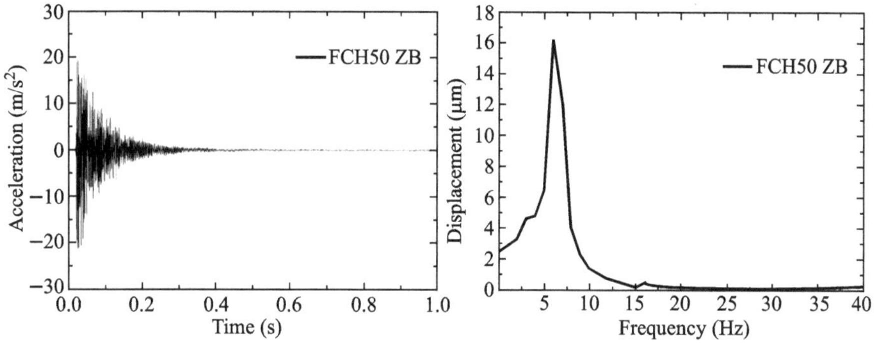


**Fig. 8.4:** Typical vertical dynamic response of the HTS Maglev vehicle model at the B point after the impulse force acting on the B point (FCH: 50 mm): (a) time and (b) frequency domains.

based on this two-dimensional dynamic model. Figure 8.4a and b shows the typical dynamic response in the time domain and frequency domain analyzed by the free-vibration experiments after a single impulse force excitation. The sampling time was set to 1 s and the sampling frequency was up to 800 Hz. The FCH of the HTS Maglev vehicle model was set as 60, 50, 45, 40, 35, 30, 25, 20, 15, and 10 mm, respectively. Each field cooling (FC) duration time was 30 minutes.

The free vibration curve in Fig. 8.4a belongs to the typical damped free vibration curve which decreases exponentially. The amplitude of the vibration curve decreases quickly to zero after only 0.4 s. Figure 8.4b further verifies the distinct low-frequency resonance characteristics of this HTS Maglev vehicle system since the resonant frequency was 6 Hz and the amplitude of the vibration was very small with the maximum of 16.2 $\mu$m. In fact, all the resonant frequencies were low: 6, 6, 7, 8, 7, 8, 9, 10, 11, and the highest 13 Hz, respectively, for every FCH from the highest 60 mm to the lowest 10 mm. Moreover, similar to the conclusion of some reports on HTS bearings, etc., the dynamic characteristics of the HTS Maglev vehicle exhibited vibration isolation at a wide high frequency range because of the strong flux-pinning nature [19, 20]. More time-domain vibration experiments with the HTS Maglev system have comprehensively validated the inherent dynamic characteristics of the low-frequency resonance and high-frequency isolation from the viewpoint of both vertical levitation and horizontal guidance [21, 22].

The experimental data were used with Eqs. (8.1) and (8.2) to determine the dynamic stiffness and damping coefficient of the HTS Maglev vehicle. The relationships with FCH are plotted in Figs. 8.5 and 8.6. The dynamic stiffness tends to decrease with the increasing FCH. This can be explained by the flux-pinning forces between the on-board HTSC and the PMG which increases with decreasing FCH. The lower the FCH is, the more stable the vehicle system is. The maximum dynamic stiffness, 5408 N/m, occurs at the lowest, 10 mm, FCH.

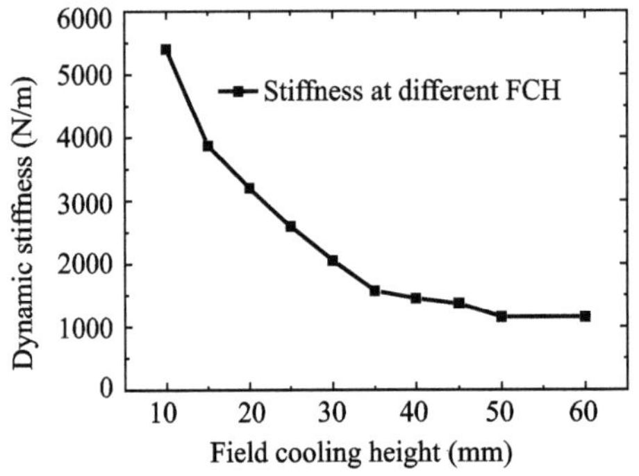


**Fig. 8.5:** Dynamic stiffness at different FCHs of the HTS Maglev vehicle model.

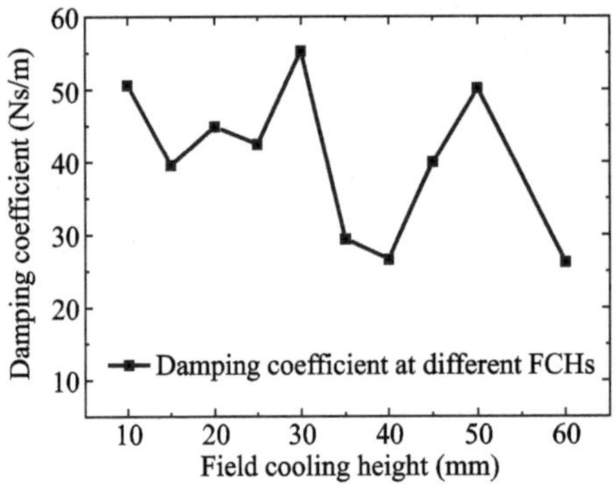


**Fig. 8.6:** Damping coefficient at different FCHs of the HTS Maglev vehicle model.

The relationship between the damping coefficient and the FCH in Fig. 8.6 appears to be more complicated than that of stiffness. The damping coefficient curves are irregular and concavo-convex. It should be noted that there are two big peaks at the 30- and 50-mm FCH, whose damping coefficients are 55.24 and 50.15 Ns/m, respectively. At the same time, the maximum 55.24 Ns/m occurs at the 30-mm FCH. As an important dynamic parameter, the damping coefficient seems to be correlated to many factors at a particular FCH.

These experimental results show that the FCH is an important factor which plays an important role among the dynamic characteristics of the HTS Maglev vehicle model. The stiffness and damping coefficient of the vehicle model can be adjusted by changing the FCH. It is a fact that the damping coefficient is one of the most effective methods to suppress the resonant amplitude. For the running vehicle model, the bigger the damping coefficient is, the better the anti-vibration ability and stability are. Therefore, considering the stiffness and damping coefficient curves together, the optimal FCH is 30 mm to obtain the best dynamic performance, and the corresponding stiffness and damping coefficient are 2048 N/m and 55.24 Ns/m.

As for the dynamic movement, the HTS Maglev vehicle acts like a "plane" flying over the PMG because the magnetic field gradient is almost zero along the forward direction. The degree of freedom is obvious in the forward direction. Although the vehicle system is designed to equilibrate with gravity and HTS bulk has strong flux-pinning property, it may leave the equilibrium position for a new position when disturbed. Six degrees of freedom as well as the coupled freedom may be the source of a dynamic problem and there has few reports of dynamic stability of HTSC vehicle.

Figure 8.7 shows the vibration curves of points A, B, C, E, D, and F in the three different directions. The forward speed is 56.6 mm/s and the FCH is 30 mm for the initial position. It can be seen that the vehicle system has sufficient damping to ensure free smooth running over the PMG. However, a better FCH still does not avoid existence of slight dynamic motion.

Figure 8.7a shows the accelerometer signals in the forward direction. Points A, B, and C have the same waveforms, and points E, D, and F have the same waveforms. However, the initial phase difference of two groups is about 180°. From the viewpoint of structure, the signal group with points A, B, and C means the right part of the vehicle model while the other group means the left part. Obviously, the vehicle prefers moving forward together with a swing motion. Moreover, the vibration curves are more like random signals. This is because the present PMG is composed of many rectangular magnets. The magnetism in homogeneity and air gap between each two magnets are the main factors for friction. Once the left part and right part do not move forward at the same time, the resistances have different effects for the two sides of the vehicle. In the practical assembly of a PMG, the uniformity of the PMG should be guaranteed as much as possible since it enhances frictionless movement of HTS Maglev vehicle. Then, the swing motion can be very weak.

On the other hand, the coupled lateral vibration cannot be ignored because both dynamic motions are in the same horizontal plane. Figure 8.7a and c indicates that the vehicle model will produce a lateral-yaw dynamic. As shown in Fig. 8.7a, the waveforms of points B and C or the waveforms of points D and F are the same but with inverse phase. It is a typical lateral motion phenomenon. However, the back part (points A and E) of the vehicle model have stronger lateral vibration because of the damping effect of the vehicle model body.

As for the vertical dynamic motion in Fig. 8.7b, point A is almost the same as points E and C is close to point F. The waveforms of the front part (points A and E) are almost the inverse of the back part (points C and F) of the vehicle model. Vibrations from the front and the back parts couple and counteract at points B and D, which implies the existence of a pitch motion. When the speed increases, vibration also happens periodically in the middle part of the vehicle. More investigation indicates the trend of the pitch-heave dynamic that happens in the HTS Maglev vehicle system.

By combining all of the results from the vertical dynamics, it appears that the levitation forces change with the slight lateral displacement when the vehicle model moves forward with a lateral-yaw motion. The present position adjusts to another new equilibrium. At a higher WH, the HTS Maglev vehicle model is a natural motion because of the smaller effect of the magnetic flux density and its gradient. Compared with Fig. 8.7a–c, the pitch-heave motion is stronger than the lateral-yaw motion,

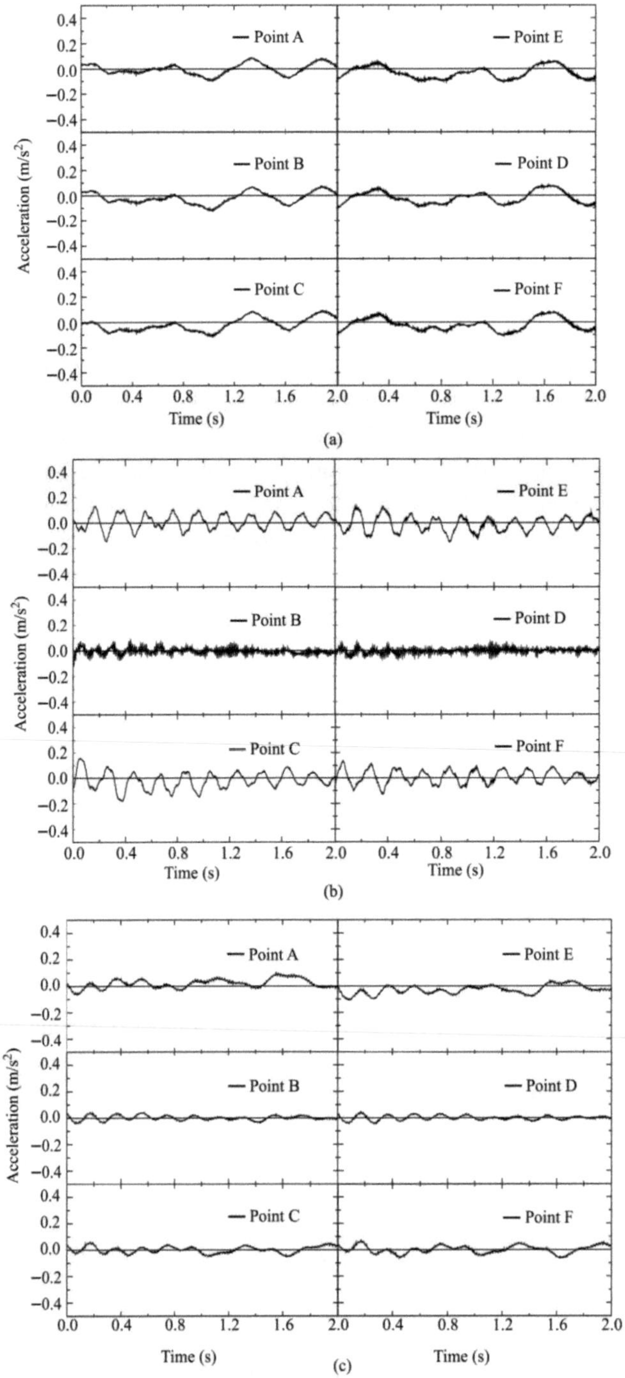


**Fig. 8.7:** Six vibration curves at the speed of 56.6 mm/s at 30 mm FCH: (a) forward direction, (b) vertical direction, (c) lateral direction.

which can be attributed to the fact that the force stiffness in the vertical direction is larger than that in the axial direction. Magnetic flux density $B_z$ and its derivative $dB_z/dz$ around the on-board HTSC, which contribute to the levitation forces, are larger than the fields and field gradients in the lateral and forward directions due to the PMG's joint disturbances. Thus, PMG's optimization and high-precise assembly are both effective in suppressing those dynamic motions as well as improving the other quasi-static levitation properties.

When the HTS Maglev vehicle model suffers some undesirable disturbances, different from other kinds of Maglev, the nonlinear repulsive forces from the on-board bulk YBCO pull the vehicle back due to the inherent HTS flux-pinning property. The more flux penetrate into the HTS bulk, the better anti-disturbance capability the vehicle will have. According to the distribution of the PMG magnetic field density, a lower FCH means a larger vertical magnetic flux density and its gradient. That is why the maximum amplitude decreases with increasing FCH as shown in Fig. 8.8. The maximum amplitude, 414 $\mu$m, happens when the bulks are cooled at the 60 mm height and the vehicle moves at the speed of 90 mm/s. The dynamic phenomenon is still slight because of the low speed.

However, Fig. 8.8 implied that the effect of the running speed, a conventional factor resulting in the instability, on the dynamic motion exists. The vibration intensity increases with the speed at the higher FCH. However, the effect is not so strong with a decreased FCH. When the forward speed of the vehicle is considered, the vertical magnetic field gradient of the PMG with time should be formulated as $(dB_z(x, y, z)/dz)dt$. The improvement of speed brings a much faster change of the magnetic flux, then a derivative larger restoring force will be produced to pull the vehicle back. Thus, the higher speed may not bring more danger in the operation of a HTS Maglev vehicle over a PMG unlike other kinds of Maglev vehicles. However, this phenomenon needs more experiments for verification.

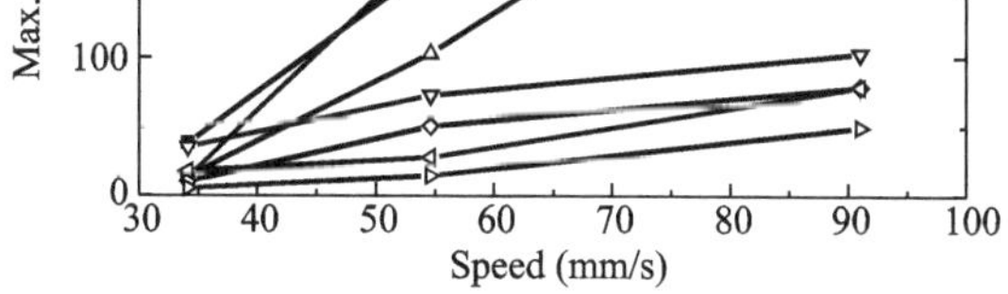


**Fig. 8.8:** Maximum amplitude of point E under the forward speed at different FCHs.

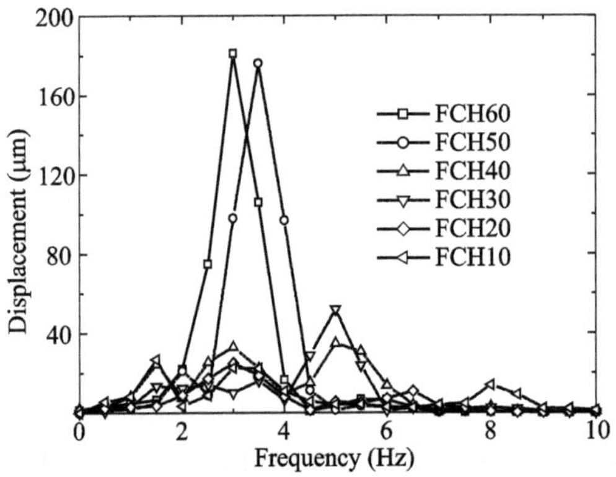


**Fig. 8.9:** Autospectra in the vertical direction at different FCHs.

As shown in Fig. 8.9, the autospectrum curves in the frequency domain indicate that the natural frequency is about 3 Hz. The autospectrum curves become flatter at lower FCHs and have more peaks. Moreover, it is the other experimental evidence that the effective FCH should be 30 mm to obtain a more dynamically stable HTS Maglev vehicle. Under the condition of 30 mm FCH, the WH of the HTS Maglev vehicle is around 25 mm which is also a perfect levitation height for HTS Maglev vehicle applications.

### 8.2.2 Dynamic Maglev characteristics under moving applied magnetic field

As is well known, the series of quasi-static "force-displacement" or "force-time" measurements have been accepted as an effective laboratory procedure to validate the application feasibility of the first HTS Maglev test vehicle (see Chapter 6) and help the successful development of HTS Maglev vehicle demonstrations or short-distance test line models. Based on those quasi-static results, the dynamic behavior characteristics of the HTS Maglev appear to be a significant science problem, especially the running situation under normal rail speeds. To understand the dynamic running performance, the HTS Maglev dynamic measurement system, SCML-03 [23] (see Section 5.6) at ASCLab of SWJTU in China, was developed in-house to simulate the HTS Maglev vehicle's translational motion above a circular PMG of 1.5 m diameter under different running speeds. The rotation of the circular PMG could be controlled to realize the translational movement of HTS bulks with respect to the PMG. The rotational speed could be set to simulate the different running speed operations of a HTS Maglev vehicle. The highest experimental linear equivalent speed was 238 km/h.

Previous designs and studies on quasi-static levitation performance were under the assumption that the magnetic field generated by the PMG is uniform along the longitudinal direction. However, the magnetic field generated by a real long-distance PMG is inevitably inhomogeneous due to the magnetism differences between adjacent permanent magnets (PMs), assembly errors such as PM cracks, junctions between PMs, etc. Consequently, the on-board YBCO bulks experience a spatially varying external magnetic field when traveling along the PMG, this moving field depends on the running speed, and it may be cyclical. Figure 8.10 plots the vertical component of the central magnetic field density of the PMG circumferential field at different heights. The smallest field position was assigned to zero degrees and the highest circumferential field appeared at 221.52 degree which was chosen as the cooling position of the onboard HTS bulk over the PMG. During the running experiment, the dynamic levitation forces at gaps of 10 and 15 mm were of most concern.

Figure 8.11 shows the schematic diagram of the experimental YBCO bulks array above the PMG in SCML-03 (see Section 5.6). It is seen that the circular PMG assembly in SCML-03 belonged to the "Century" PMG type and that there were four pieces of three-seeded melt-textured YBCO bulks (64 mm in length, 32 mm in width and 13 mm in thickness) employed as the HTSC array.

With the above experimental set-up for SCML-03, the dynamic levitation forces of the YBCO bulks were systematically studied under different linear equivalent speeds from 0 to 238 km/h, in acceleration and constant speed conditions. First, Fig. 8.12 describes the actual levitation transition during the experimental process as the PMG was accelerated from 0 km/h to 238 km/h and then decelerated back to 0 km/h in steps. The steady speeds at every step were 1.5 km/h, 60 km/h, 119 km/h, 159 km/h, 179 km/h, 208 km/h and 238 km/h. The staying time at each speed step was 4 minutes. In Fig. 8.12, the dark zone corresponds to the oscillation range of the levitation forces while the light grey average line indicates the general tendency of the levitation forces. This dark zone in Fig. 8.12 shows the sharp rise and decline behavior of the levitation force with speed change which most likely was caused by interference. The average line in Fig. 8.13 shows the true attenuation tendency of the levitation forces with respect to speed. The average levitation forces decreased from 179 to 170 N after 65 minutes in operation. The static levitation forces decreased from 201 N before the PMG rotated to 191 N after the PMG rotated. Meanwhile, there was no significant force

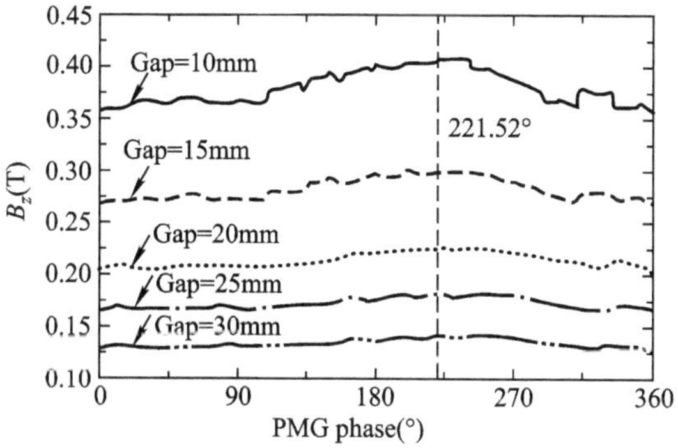


**Fig. 8.10:** Vertical central magnetic field component among the circular PMG.

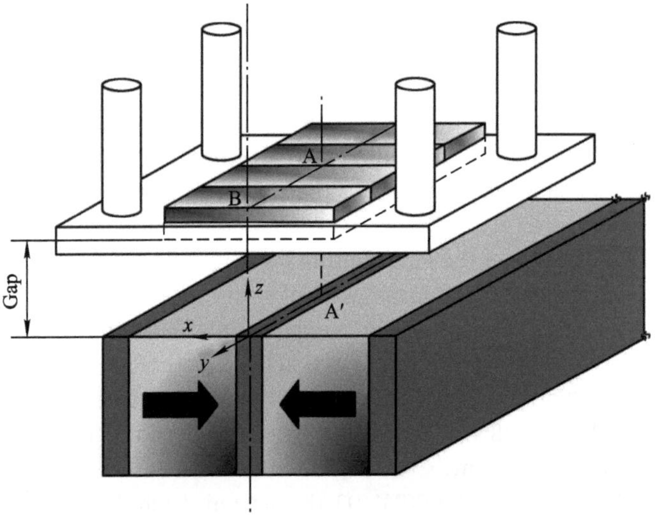


Fig. 8.11: Schematic diagram of the YBCO bulks array above the PMG in SCML-03.

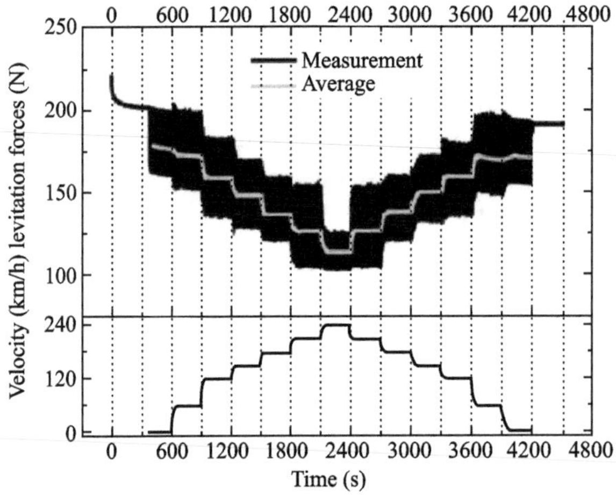


Fig. 8.12: Levitation forces and their average result during the step-by-step acceleration and deceleration.

attenuation over 60 km/h, the transition among all velocity parts of the curve was smooth. The force attenuation mainly appeared at speeds less than 60 km/h. It has been reported that the diffusion of the shielding current caused by an AC field might be restricted at high frequency and then rarely penetrate deeper [24]. Therefore, the reasonable explanation is that the total shielding current attenuation due to the field perturbation caused by the motion above PMG might also be restrained at high velocities so that the force attenuation saturated with speeding up. Thus, the force attenuation

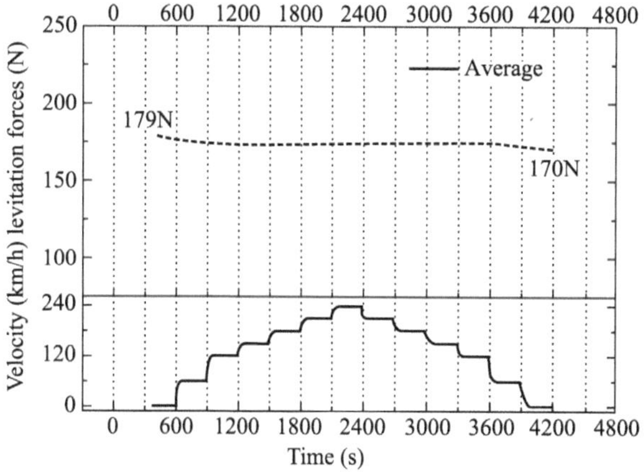


**Fig. 8.13:** Corrected levitation force average in each steady velocity.

caused by energy loss due to field perturbations was only about 5% and only happens at speeds below 60 km/h. Between 60 and 238 km/h during speed-up or speed-down, the whole levitation performance of the HTS Maglev seemed not to be affected.

To compare the speedup history effect, more dynamic levitation measurements under direct speedup were carried out to understand how the levitation forces performed without the force relaxation duration. Because the shielding current in the inner region gradually attenuates in an applied field perturbation according to the *E-J* characteristic of the bulk, it causes an irreversible energy loss so that the varying field would intensify the attenuation of the levitation forces (as shown in Fig. 8.14). The four force averaged curves located below the static force relaxation curve can be classified into two groups according to their test velocity. At the same velocity, the force attenuation showed a similar saturation tendency, no matter which

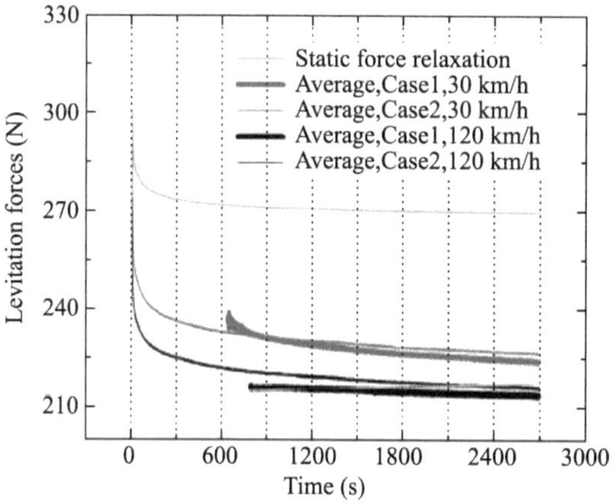


**Fig. 8.14:** Averaged levitation force curves in different measurement cases.

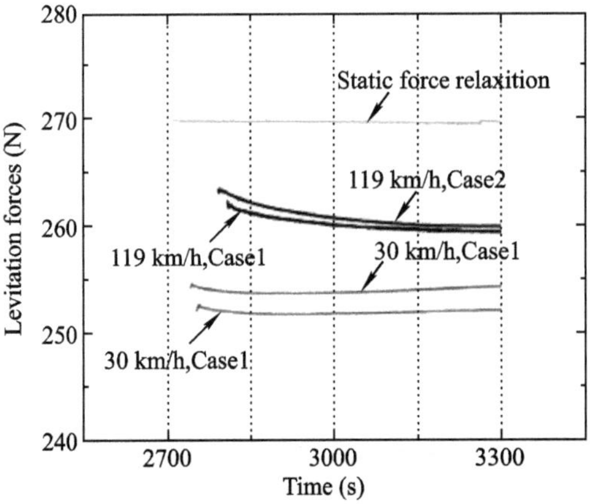


**Fig. 8.15:** Levitation force relaxation after different dynamic measurement cases.

speedup history it had been through. The force decay at 30 km/h seemed to be less than that at 120 km/h. Although the correction curves were not adopted in this part for these complex cases, as shown in Fig. 8.15, the levitation force relaxation after the two different dynamic measurement cases indicated that the forces decreased more for 30 km/h, and the forces decayed more at low velocities as discussed above. Meanwhile, the levitation force difference caused by magnetic field history between the former step-by-step speed-up (indirect, case 1) and the direct speedup (case 2) was not significant. The levitation force difference between the two velocity groups became closer with time, less than 10 N at 3300 s.

The levitation force behavior versus levitation gap after the bulk superconductors were subject to different running cases is displayed in Fig. 8.16. The approaching part in all curves nearly overlapped with the "directly return" (the quasi-static

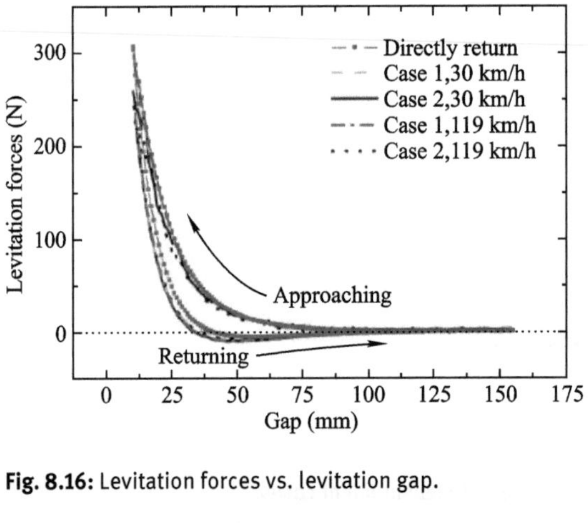


**Fig. 8.16:** Levitation forces vs. levitation gap.

"force-displacement" measurement), except for case 2 because the forces in case 2 oscillated with the rotating PMG. Case 1 and 2 curves experienced a bigger hysteresis area with a smaller force in the returning part when compared to the "directly return" curve. This further validates that the shielding current decrease was caused by irreversible energy loss.

Further exploration will improve the preliminary understanding of the dynamic equilibrium of an HTS Maglev vehicle. It is believed that these observed phenomena are of great importance in the future running and design considerations for the HTS Maglev system.

### 8.2.3 Levitation performance at different working temperatures

The relationship between critical current density ($J_c$) and working temperature ($T$) is given by the following equation:

$$J_c \propto [1 - (T/T_c)^2]^{3/2}. \tag{8.3}$$

According to the Eq. (8.3), $J_c$ of the YBCO bulk will be increased with decreasing temperature. Eq. (8.4) below gives the common solution of the levitation force produced by $J_c$ inside a HTSC bulk in magnetic field,

$$F_{Lev} = \int J_s dV = \int [J_c(B_x(\delta_0) - B_x(0))/\delta_0] dV, \tag{8.4}$$

where $J_c$ is the critical current density at a certain temperature, and $V$ represents the volume where the induced supercurrent flows in the superconductor. $B_x$ is the horizontal component of the original magnetic field, and $\delta_0$ is the penetration depth of the magnetic field which is given by the following relation:

$$\delta_0 = (B_z - B_{FC})/\mu_0 \lambda J_c, \tag{8.5}$$

where $B_z$ is the vertical component of magnetic field at the smallest levitation gap during the levitation measument, $B_{FC}$ is the vertical component of magnetic field at FCH, $\lambda$ is the coefficient determined by the configuration of the HTSC sample, and $\mu_0$ is the permeability of vacuum.

Eqs. (8.3) and (8.4) indicate that the levitation performance of HTS bulk will become better at lower working temperature in an applied magnetic field. In this section, the specific effects of temperature on some significant levitation performances are reviewed for the HTS Maglev vehicle application. Then, a ring-shaped YBCO bulk above a two-pole Halbach array's PMG [25]. (Halbach PMG, Section 8.4.1) was set up as the basic model for an HTS Maglev vehicle. The ring-shaped YBCO superconductor

**Fig. 8.17:** Ring-shaped YBCO bulk sample fixed on the second-stage cooling head of the cryocooler.

bulk was 30 mm in outside diameter, 18 mm in inside diameter, and 15 mm in thickness. It was fixed on the second-stage cooling head of the cryocooler (Fig. 8.17).

Figure 8.18 shows the photograph of the deep cryogenic measurement system assembled in the HTS Maglev measurement system, SCML-01 [26] (see Section 5.3), a model K535 Stirling cryocooler, a model 331 temperature controller, and some low temperature sensors. The experiment temperature was controlled by the deep cryogenic measurement system, ranging from 77 to 60 K. For the force measurements, the YBCO bulk was cooled in different FCHs above the Halbach PMG. The quasi-static

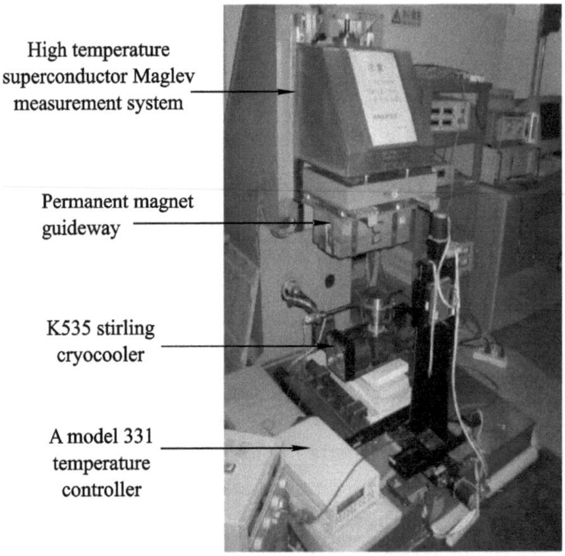


**Fig. 8.18:** Deep cryogenic measurement system.

"force-displacement" experiments were performed with a levitation gap range from 80 to 10 mm.

Figure 8.19 shows the levitation forces of YBCO bulk at 77, 73, 68, 63, and 60 K in the 20-, 25-, 30-, and 35-mm FCHs while Fig. 8.20 summarizes the maximum levitation forces of YBCO bulk as a function of the temperature at 10 mm levitation gap under the different FCHs. It is interesting to note that at 35, 30, and 25 mm FCHs, the forces drastically increased with temperature from 77 to 60 K. However, a much different phenomenon was observed for 20-mm FCH, The temperature effect on the levitation forces was negligibly small at the 20-mm FCH.

That was because the variation of the penetration depth, $\delta_0$ [see Eq. (8.5)], was different when the FCH of YBCO bulk was different. For the 20-mm FCH, the vertical component of magnetic field at FCH was strong. Thus, when YBCO was moved from $B_{FC}$ to $B_z$, $\delta_0$ of YBCO at the 20-mm FCH was small. The variation of $\delta_0$ at the 20-mm FCH was also small when the temperature of YBCO was decreased from 77 to 60 K. Thus, according to Eq. (8.4), the variation of the levitation forces was negative with decreasing temperature. That was, the levitation forces

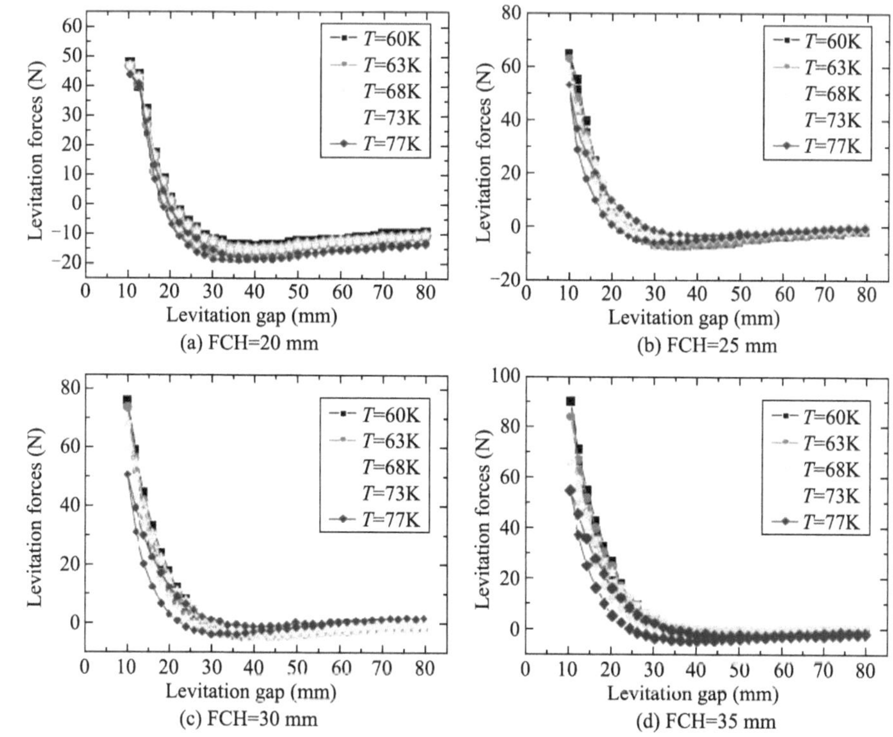


**Fig. 8.19:** Levitation forces between YBCO bulk and the Halbach PMG at 60–77 K under different FCHs are shown in (a), (b), (c), and (d).

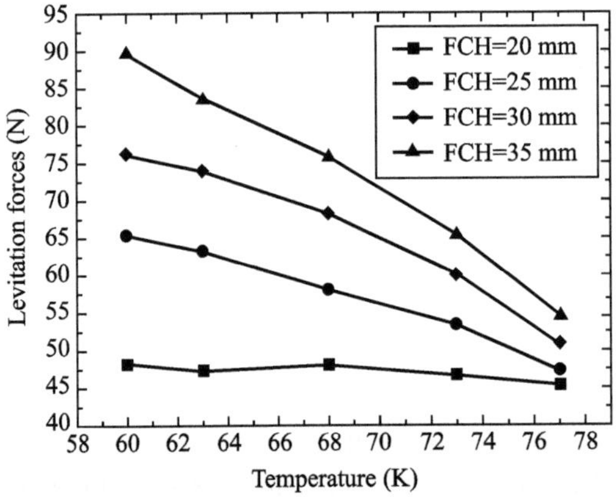


**Fig. 8.20:** Maximum levitation forces dependence on the working temperature under different FCHs.

were dependent on not only the temperature of the HTS but also the original FCH. When the FCH was less than 20 mm, the vertical component of the magnetic field at FCH was strong. The $\delta_0$ of the HTSC was large enough to disturb the invasion of the magnetic field, so the effect of the working temperature on the levitation forces was ultra-weak at the low FCH. This implied that the original FCH of an HTS Maglev system was the key parameter for the enhancement of the levitation forces. When the levitation gap was more than 40 mm, the PMG magnetic field was less than 0.1 T. The $J_c$ of the HTSC bulk was high enough to disturb the invasion of the PMG magnetic field, so the potential of the working temperature to increase the levitation forces was very small in the low magnetic field. It implied the need to have a relatively high magnetic field for the levitation enhancement of an HTS Maglev vehicle in addition to considering the improvement effect of a low working temperature.

In addition, multi-time measurements of the working temperature effect on the levitation forces were explored and an estimate of the effect of passengers leaving or boarding the HTS Maglev vehicle, resulting in an up-and-down (vertical) round trip movement (change of working position) of the on-board superconductor bulk, was considered. The experimental procedure was designed based on the deep cryogenic measurement system (see Fig. 8.18) as follows. First, the deep cryogenic measurement system was initialized with the midline of the ring-shaped YBCO bulk in accordance with the peak $B_x$ line, $x = 0$ mm, of the Halbch PMG. Then the guideway was moved down and stopped at a height of 35 mm over the upper surface of the ring-shaped YBCO bulk in the K535 Stirling cryocooler. When the temperature of the YBCO bulk reached the pre-set value (63–77 K), the Halbach PMG was moved at a pre-speed. At the same time, levitation forces were measured continuously three times under the control of the SCML-01. Then the next set of levitation experiments were performed when the ring-shaped YBCO sample was aligned with the peak line of $B_z$, $x = -25$ mm, of the Halbach PMG.

Figure 8.21 shows the experimental curves. It is interesting to note that the levitation forces decreased visibly from the first movement to the third movement at the relative higher temperature of 77 or 73 K at $x = -25$ mm. However, the decrease rates gradually reduced although levitation forces continuously decreased when increasing the round-trip movement times. Moreover, the influence of the round-trip movement times of the bulk on levitation forces was not obvious at lower temperatures, for example, at 63 or 68 K, three groups of curves were almost the same during different cycling movements at $x = -25$ mm. This indicated that the influence of YBCO bulk's round-trip movement times on the levitation forces was larger at higher temperature and the influence became smaller at lower temperature. This was because the strength of $B_z$ was extremely small at $x = 0$ mm, 20 times smaller than at the −25 mm position. The influence of the bulk's round-trip movement times on the levitation force was very small at $x = 0$ mm position at any temperature. Based on Maxwell's equations, the induced current was related to the magnetic field difference between the applied field and the trapped magnetic field in the bulk HTSC. In addition, according to the critical state model [27], it was well known that lower hysteresis of the levitation forces happens with decreasing temperature. Therefore, the influence of the YBCO

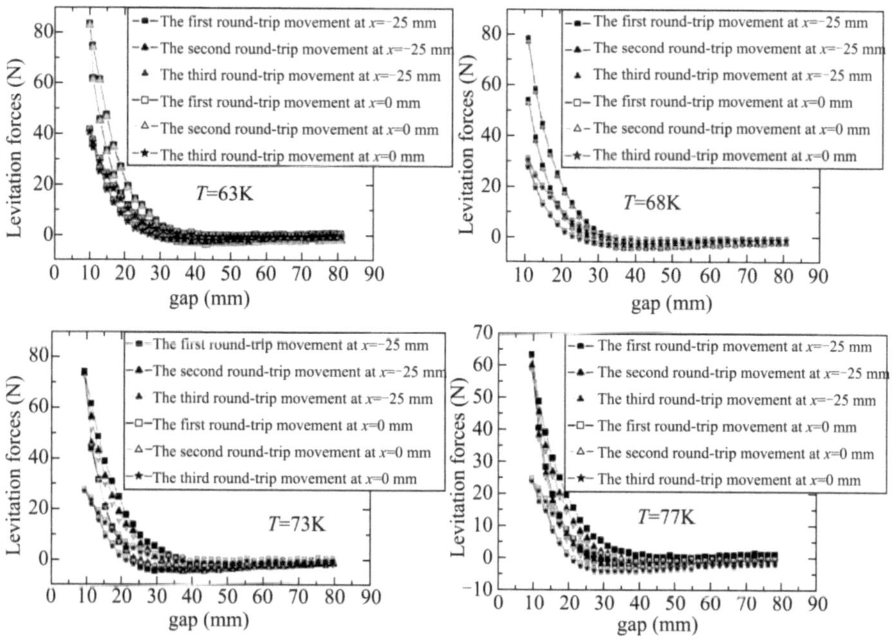


**Fig. 8.21:** Levitation forces curves under three different round-trip movements at 35-mm FCH and different positions of $x = -25$ mm and $x = 0$ mm.

bulk's round-trip movement times on levitation forces was larger at higher temperature while it was smaller at lower temperature. On the other hand, when the bulk lay at $x = 0$ mm, the variation of $B_z$ became small during the vertical movements which led to a smaller induced current in the bulk. Therefore, reduction of the variation of $B_z$ during movement was the main reason why the influence of the bulk's round-trip movement times on the levitation forces was so small at any temperature.

Considering the practical running situation of the HTS Maglev vehicle, we chose four gap points in the experiment curves, i.e. 11, 12, 15, and 16 mm, respectively. Figure 8.22 shows levitation forces under different movement speeds for the above gaps for four working temperature conditions.

It is interesting to note that the levitation forces decreased with increasing movement speeds from 1 mm/s to 6 mm/s for the gap at $x = -25$ mm when the temperature of the bulk was held at 77 or 73 K. For example, when the gap was 11 mm and the temperature was at 77 K, the levitation force of the bulk was 67 N at 1-mm/s movement speed, while it decreased to 61 N at 6-mm/s movement speed. However, at the lowest 63 K, the forces did not decrease with increasing movement speed for the gap at $x = -25$ mm. The influence of movement speed on the levitation forces was negligibly small at 63 K or lower. These conclusions were also obtained for a large variation of $B_z$ field. Once the variation of $B_z$ becomes small, the influence of the bulk's round-trip movement times and speed on the levitation forces were very small at any temperature. With $x = 0$ mm position, for example, the strength of $B_z$ was extremely

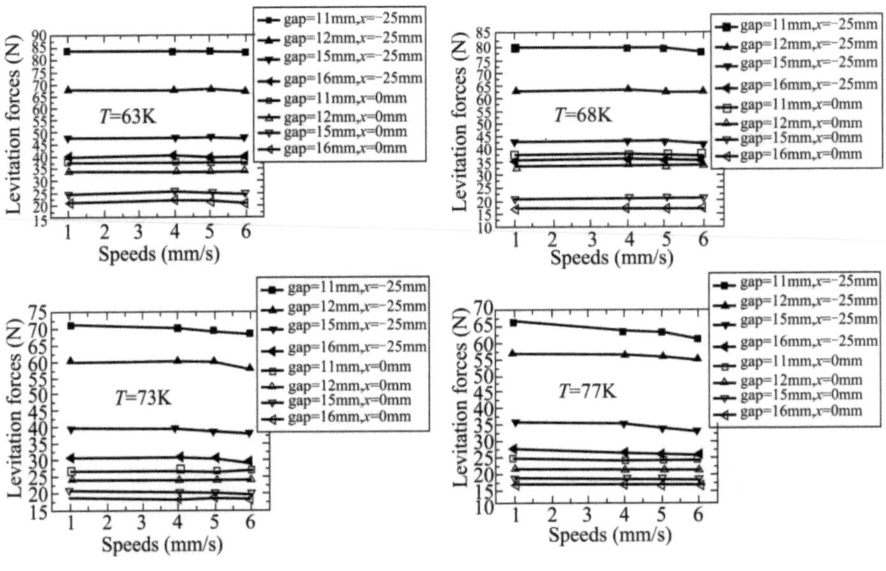


**Fig. 8.22:** Levitation forces under different movement speeds at the 11-, 12-, 15-, and 16-mm gap at four working temperatures and at 35-mm FCH.

small, 20 times smaller than that at 25 mm, and the influence of the bulk's movement speed on the levitation forces was very small at any temperature.

Hence, if more magnetic energy with strong field and a lower temperature condition are utilized, the enhanced of levitation performance will directly promote the development of HTS Maglev vehicle applications.

## 8.3 Methods to improve Maglev performances

### 8.3.1 Pre-loading method

The pre-load method is an effective improvement to suppress the levitation height and force decay and enhance the stability of the HTS Maglev vehicle system. The typical pre-load execution steps include:

1. Lower the levitation height of the on-board HTS bulks after the FC initialization by adding additional load or other ways.
2. Keep the levitation system, including the on-board HTS bulk, staying at this above lower levitation position for some time.
3. Release the levitation system to the working height before running.

Finally, it will be found that better Maglev performance could be obtained for HTS Maglev vehicle application. As to the pre-load method, because more magnetic flux is trapped at a lower height (larger applied magnetic field) than at the working height before any movement, the magnetism of the HTS bulk is enhanced during vertical movement and the probability of the HTS bulk to trap more magnetic flux during lateral movement is reduced. That is why the pre-load method can suppress not only the levitation forces decay but also the guidance force decay.

The following implementation of the pre-load method was applied to the scaled HTS Maglev vehicle model (see Fig. 8.1 and Section 8.2.1) before running to verify the improvement effect on the levitation stability of the HTS Maglev vehicle system. Figure 8.23 shows the schematic procedures for the pre-load method carried on HTS Maglev vehicle.

The pre-load was added to the vehicle step by step, with 10 kg every time. The new WH was measured after each loading as shown in Fig. 8.24. The maximal pre-load was set at 130 kg due to an ideal acceptable levitation height from 10 mm to 15 mm for the HTS Maglev vehicle. Then unloaded the vehicle model, so that it returned to a new equilibrium position. Figure 8.24 shows that the levitation height linearly decreased with the increase of the load. The slope of the curve after the pre-load was smaller than that before the pre-load. It indicated that the HTS Maglev vehicle was an elastic system, so that an elastic restoring force would act when the vehicle is pulled down or up. However, the restoring force was limited, and the vehicle would not return to the original position if the load was too big. The levitation height reduced by 8 mm

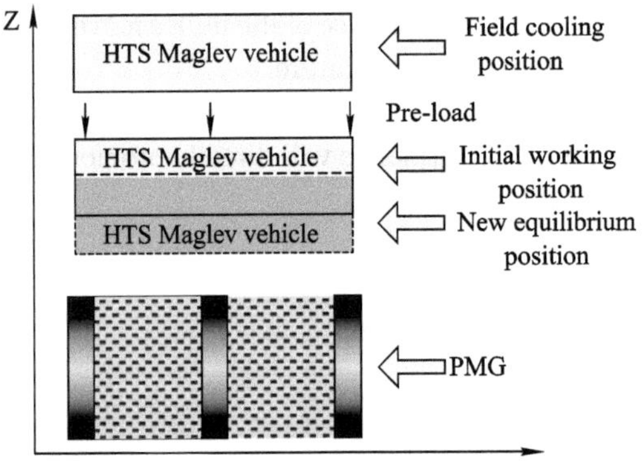


**Fig. 8.23:** Schematic diagram of the pre-load method.

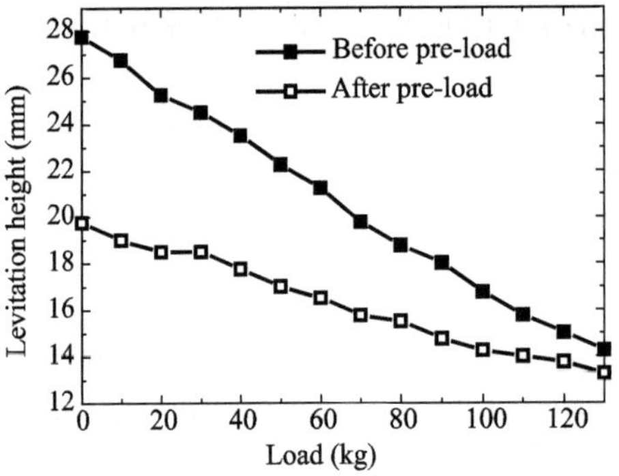


**Fig. 8.24:** Levitation height dependence on load.

without the pre-load process. But there was only a 1-mm decrement when the pre-load was applied. Thus, the pre-load can suppress the levitation height decay effectively. The following data will further prove this conclusion.

Table 8.2 compares the measured dynamic results of the pre-load method applied to the HTS Maglev vehicle model. After the pre-load process, the HTS Maglev vehicle became more stable indicated by the higher stiffness, damping coefficient and better anti-vibration performance. It was also found that the levitation height decay was suppressed by the pre-load method. For example, using the pre-load method, the levitation decay rate was reduced from 14% to 2.5%.

**Tab. 8.2:** Enhancement of dynamic stiffness and damping coefficients after pre-load.

| Pre-load | *k*(N/m) | *c*(Ns/m) |
|---|---|---|
| No | 53,740 | 144 |
| Yes | 87,050 | 214 |

To exhibit the advantages of the pre-load method on lateral movements, Figs. 8.25a and b shows the continuous multi-time levitation measurements when two HTS bulks (S1 is 48 mm in diameter and 12 mm in thickness, and S2 is 68 mm long, 33 mm wide, and 13 mm thick) went across the center position above the Halbach PMG during the lateral movement, respectively. As shown in Fig. 8.25a, for the no pre-load case, the levitation forces were decreased from 29.697 to 15.386 N when S1 initially and finally crossed the center position. Thus, about half of the levitation forces were reduced due to lateral movement in the experimental time. For the pre-load case, the levitation force decay of S1 was also existent, but was obviously larger obviously than that in the no pre-load case after the fourth time across the center position; moreover, the decay rate was a little smaller than that in the no pre-load case. Therefore, the pre-load method was applicable to reduce the levitation forces decay in this case. Figure 8.25b

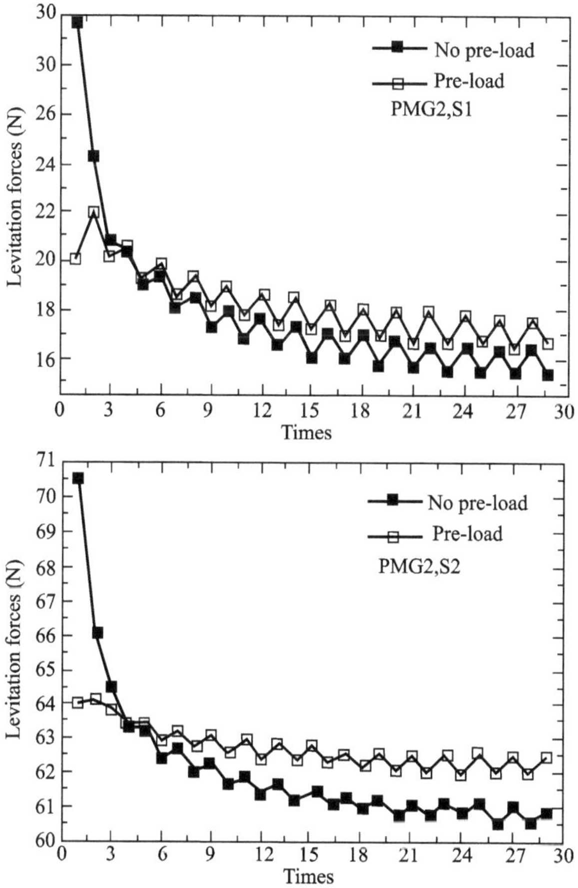


**Fig. 8.25:** Levitation forces dependence on the lateral movement times in the continuous multi-time no-pre-load and pre-load procedures above the Halbach PMG. (a) S1, (b) S2.

shows that, for the no pre-load case, the levitation forces decreased from 70.471 to 60.808 N when S2 initially and finally crossed the center position. Thus, 14% of the levitation forces were reduced due to lateral movement in the experimental time in this case. For the pre-load case, the levitation forces also decreased with the increase of the lateral movement, but the levitation forces were obviously larger than that in the no pre-load case after the fourth time across the center position. The levitation forces started from 63.997 N and decreased to 62.423 N when the last time it went across the center position, i.e., only 2.5% of the initial levitation forces was reduced due to lateral movement, so the levitation force decay was also successfully suppressed in this continuous multi-times lateral pre-load implementation.

Figure 8.26a shows the guidance forces variation for the S1 sample moved across the Halbach PMG. The second value jumped to 8.896 from 7.771 N for no pre-load case. In succession, the value decreased to 8.567 N and the last guidance force was 8.321 N.

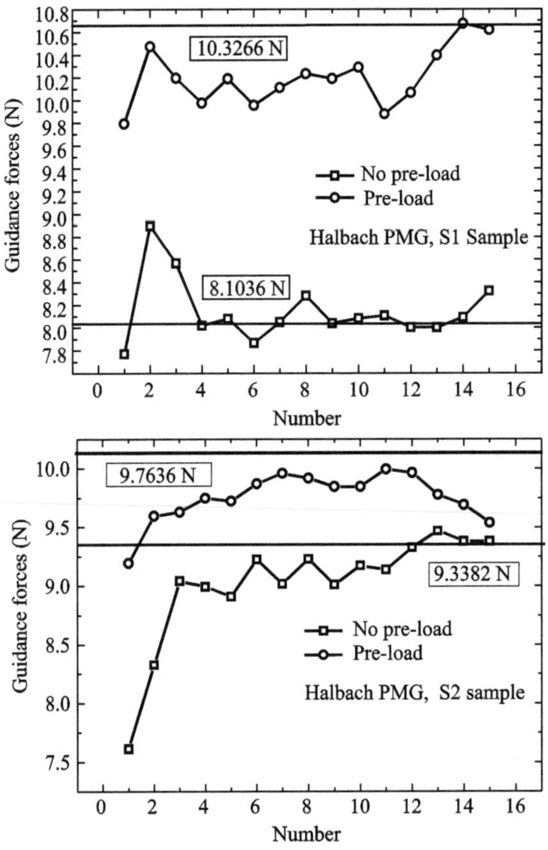


**Fig. 8.26:** Guidance forces dependence on the lateral movement times in the continuous multi-time no pre-load and pre-load procedure above the Halbach PMG. (a) S1, (b) S2.

For the pre-load case, the change of the guidance forces was fluctuated. The initial value was 9.794 N, and the final value was 0.825 N, larger than the initial value. At the same time, Fig. 8.26b shows the guidance forces variation for the S2 sample moved across the PMG2. In the no pre-load case, the first and third values were 7.613 and 9.042 N, respectively. Then the guidance forces changed circuitously. The final force was 9.379 N, 23.2% larger than the initial value. In the pre-load case, the curve could be divided into three phases: increasing phase, middle phase, and decreasing phase. The last guidance force was just 0.346 N larger than the initial one.

By comparison, the guidance forces became larger after the pre-load procedure, too. The increase ratio was 127.43% for sample S1 and 104.56% for sample S2. At the same measure height, the guidance forces increased with the decrease of FCH due to the increase of the trapping magnetic field. After the implementation of the pre-load method, the HTS bulk trapped more magnetic flux, so the guidance forces increased, too.

### 8.3.2 Magnetization process

#### 8.3.2.1 Pre-magnetization

It is noted that the present HTS Maglev vehicle and related optimization studies 28–34 are conducted under well-recognized field-cooling magnetization (FCM) conditions, which is a convenient way to realize the stable superconducting levitation. During the FCM process, the HTS bulk stays in the same magnetic field as the magnetization field. That is, the applied magnetic field is the magnetization field for the initial levitation stability. Compared with the best trapped field capability of about 3.5 T at the working temperature (liquid nitrogen temperature, 77 K), the actual utilization level of the HTSC material is low and inefficient because the FCM field source is generally from PM materials. Taking the first manned HTS Maglev test vehicle [5] (see Chapter 6) as an example, the on-board bulk YBCO trapped only 0.174 T (the maximum magnetic field density) at the applied 0.243 T PMG surface magnetic field at the 40 mm FCH. On the other hand, it has been found that a higher trapped field brings a better levitation performance of the HTS Maglev system (the HTSC-PMG system). Thus, it is potential to introduce the HTS bulk magnet (HTSCM) with a higher trapped field into the HTS Maglev system for better levitation. That means the HTSCM should be first magnetized in a higher magnetic field rather than in the relative low applied field and then put into the actual applied field. One more advantage of the pre-magnetization method that can be predicted is that more magnetic field energy from the outside field will be imported to the HTS Maglev system without changing its working conditions.

Usually, a static magnetic field was regarded as the most efficient magnetization condition for a HTSCM. Here, the static magnetization field was produced by an electromagnet pair (Model EM4-CV, Lakeshore) which was charged by a magnet

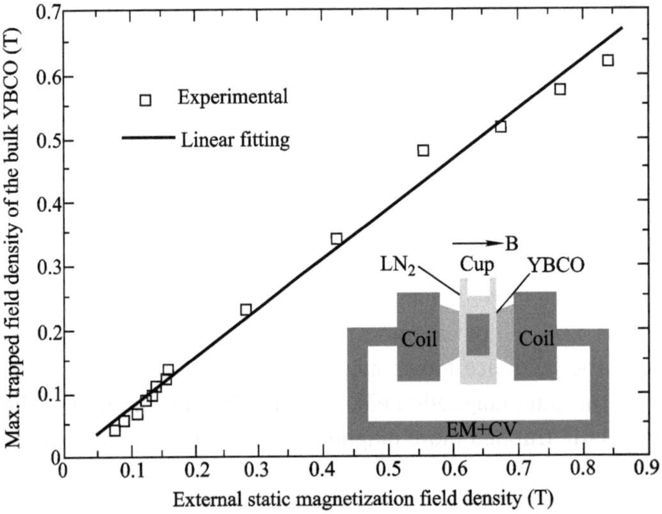


**Fig. 8.27:** Maximum trapped flux density dependence on the external static magnetization field density of the Lakeshore electromagnet.

power supply (Model 647, Lakeshore) to magnetize a cylindrical melt-processed YBCO superconductor with *c*-axis oriented grain with 30 mm in diameter and 18 mm height.

In the inset of Fig. 8.27, the arrow shows the magnetization direction which was parallel to the *c*-axis direction of the YBCO sample for obtaining a better HTS bulk magnet based on the $J_c$ anisotropy of the YBCO superconductor. The maximum trapped field at the surface of the bulk was measured by a Lakeshore low-temperature Hall probe and a Lakeshore model 450 gauss meter. Every magnetic measurement was conducted after 15 minutes when the magnetization was completed according to IEC 61788–9: 2005. In such static magnetization, the maximum external static magnetization field could be up to 0.836 T, and the maximum surface trapped field of the YBCO HTSCM was 0.620 T when the distance between the two electromagnet poles was 55 mm with a small LN2 cryostat just clamped between the poles. Figure 8.27 shows that the maximum trapped field of the YBCO HTSCM increased approximately linearly with the external magnetization field, and that the magnetization ratio was around 77.8% for this YBCO superconductor in the static magnetization field environment. That was higher than after a pulsed-field magnetization.

More specific levitation performance of the HTS bulk magnet was further investigated over the Halbach PMG [25] (see Section 8.4.1). Both levitation and guidance forces, two important levitation parameters of the HTS Maglev vehicle system, were measured in the SCML-02 [35] (see Section 5.5). During the experiments, the levitation force measurement was conducted after the guidance force measurement. It aimed to avoid the PMG magnetic field affecting the original trapped field in the HTS bulk

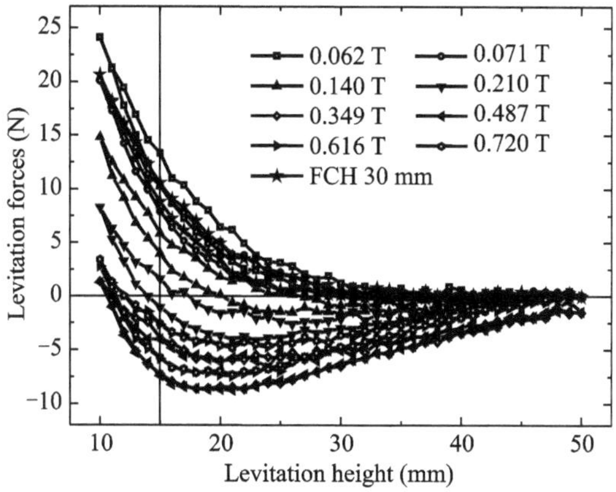


**Fig. 8.28:** Levitation forces vs. height on the different trapped-field conditions.

magnet during the levitation force measurement loops because the guidance forces mainly depended on the trapped flux inside the HTS bulk.

Figure 8.28 shows that the higher the trapped field of the HTS bulk magnet was, the smaller the repulsion levitation force became and the bigger the attractive force got. The reason was that the magnetized YBCO bulk behaved more like a "magnet" with the "S pole" surface close to the nearby concentration flux pole, as an "N pole" of the PMG. Then the attractive forces were generated between the HTS bulk magnet and the PMG as the negative part of the eventual levitation forces. Moreover, the bigger external magnetization field caused a stronger HTSC "magnet", which further produced a larger attractive force. Thus, the levitation forces decreased with the stronger HTS bulk magnet. In contrast, the levitation forces would increase by changing the negative effect into positive effect only if the "S pole" orientation of the HTSCM was placed in reverse over the PMG, which was feasible and easy to put into effect in practice for load enhancement.

On the other hand, the levitation forces did not continue the above negative trend of the HTS bulk magnet with field higher than 0.349 T, like the 0.487-, 0.616-, and 0.720-T curves in Fig. 8.28. This was attributed to the limitation of the maximum flux-trapping capacity of the YBCO sample. Similar to the guidance force trend as shown in Fig. 8.29, it will be of little help to improve the Maglev performance with a stronger HTS bulk magnet. At 15-mm WH, the interaction vertical forces began to change from a repulsive force to an attractive force when the trapped field was more than 0.210 T. The attractive forces achieved a maximum 7.4 N at the 0.349-T trapped field. Thus, the HTS bulk magnet will be more suitable to realize stable suspension in the future.

On the other hand, Fig. 8.29 shows all the guidance forces at the 15-mm WH and the guidance stiffness was negative like the conventional stable HTSC-PMG levitation. The guidance forces increased with the increase of the trapped field of the bulk magnet above the PMG due to the more flux in the bulk magnet from the external static

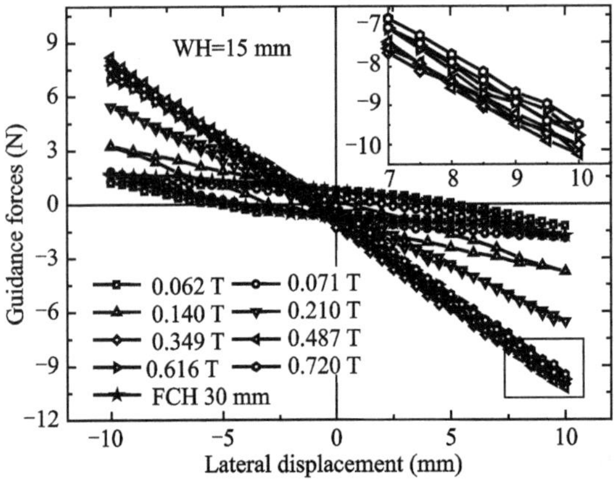


**Fig. 8.29:** Guidance forces vs. lateral displacement on the different trapped-field conditions.

magnetization field. As shown in the upper right inset of Fig. 8.29, the guidance forces saturated from the 0.349-T trapped field because of the trapped flux saturation in the YBCO sample.

Combined with the levitation force curves in Fig. 8.28, both the levitation and guidance forces on the 0.071-T trapped-field condition could almost achieve 20.6 N at a 10-mm levitation height and 1.3 N at a –10-mm lateral displacement on the conventional 30-mm FCH condition. Moreover, larger levitation and guidance forces can also be accomplished at other trapped-field conditions.

It is very interesting to find that the HTS bulk magnet with the 0.349-T trapped field achieved a 9.8-N maximum guidance force which was more than 7 times that of a conventional FCM condition for the 30-mm FCH. These experimental data further confirmed that the HTS bulk magnet can be of benefit to some HTS Maglev curve applications which needed the larger guidance forces and stiffness.

Furthermore, it can be a feasible and efficient way to improve the HTS Maglev performance by introducing the HTS bulk magnet, but without changing the original design considerations because the HTS bulk magnet works at the same temperature condition of the HTSC with the conventional FCM. In fact, because the trapped flux density of the HTS bulk magnet can be easily artificially controlled according to the fixed magnetization ratio of the superconductor material, the HTS bulk magnet will be easily magnetized to the required magnetic flux density, and then the required levitation performance and stability in an applied magnetic field according to practical design parameters.

#### 8.3.2.2 Zero field-cooling re-magnetization

ZFC and FC are two popular magnetic activation ways for HTS Maglev vehicle application. The HTS bulks with ZFC processes can have a bigger levitation forces (load capability) but poor stability when interacting with an applied field. While with FC processes, the HTS bulk can have a good guidance forces (stability) due to the trapped flux, but a reduced levitation force. Hence, the FC way seems more suitable for practical applications. In FC cases, the bigger levitation force can be obtained by increasing the FCH between the HTS bulks and the applied field at the price of some stability, and vice versa [36–38]. The two forces in a levitation system act more like a contradictory couple. Under this rule, it is very difficult to improve the levitation forces (load capability) and guidance forces (stability) at the same time with the given HTS bulks and applied field in practice. Referring to the recent reports about the re-magnetization ability of HTS bulks [39, 40], the HTS bulks after the first ZFC process are considered to be able to trap some flux in the following re-magnetization process, thus creating stable levitation. This characteristic implies a possible way to improve both the levitation and guidance forces. Thus, the optimization method with an additional re-magnetization process after ZFC is proposed and examined in this section.

As shown in Fig. 8.30, the levitation unit was composed of four three-seeded bulk samples. The magnetic forces between the levitation unit and the Halbach PMG under different excitation ways were measured by the SCML-01 [26] (see Section 5.3). For the re-magnetization process after ZFC, the typical pre-load execution steps include

1. The HTS bulk levitation unit was first cooled by liquid nitrogen far away the PMG in the ZFC process.
2. After about 20 minutes when the bulks have been cooled completely, the bulk levitation unit was lowered and stopped above the PMG with a smaller gap to experience the PMG field thus being magnetized again.
3. This re-magnetization process also lasted about 20 minutes to make the bulks trap as much flux as they can in the PMG field.

After that, the force measurements were performed to verify the Maglev performance obtained for HTS Maglev vehicle application.

Figure 8.31a and b shows the levitation and guidance force curves of the bulk levitation unit above the PMG at different re-magnetization heights after ZFC, respectively. All the levitation and guidance force curves have the similar profiles to the normal FC

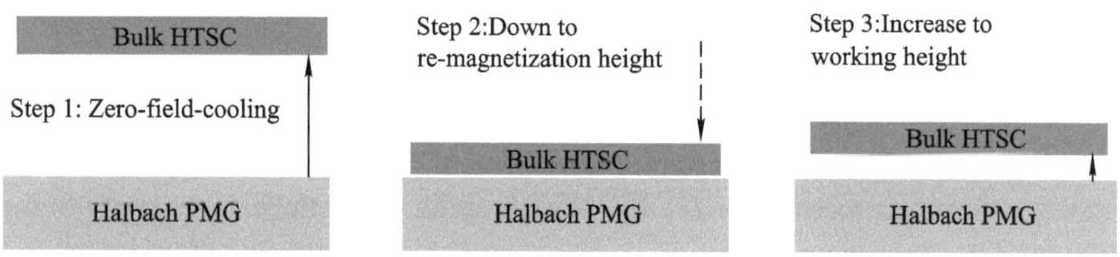


**Fig. 8.30:** Schematic illustration of working process of the bulk levitation unit with a re-magnetization process after zero-field cooling.

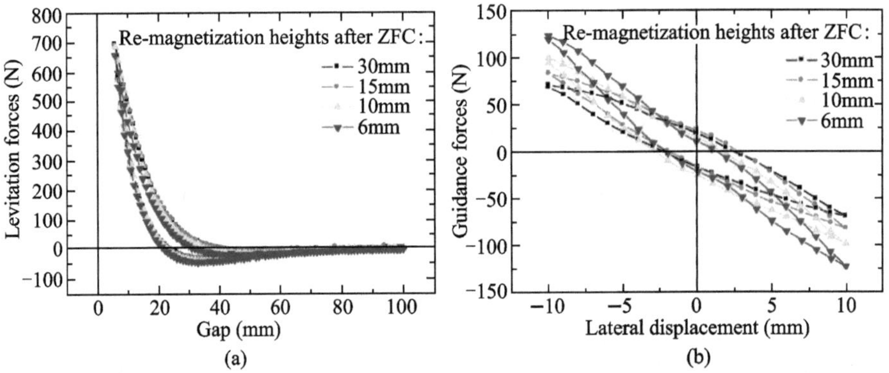


**Fig. 8.31:** Levitation (a) and guidance forces (b) of the bulk levitation unit above the PMG at different re-magnetization heights after ZFC.

cases. By adding the re-magnetization process after ZFC, the levitation force curves are very close to each other and show a slight decrease with the decrease of the re-magnetization height between bulks and the PMG. The guidance performance of the HTS bulk levitation unit shows an obvious improvement with the decrease of the re-magnetization height. It is worth noting that such stability partly benefited from the double-pole magnetic field distribution of the experimental Halbach PMG [25].

With the decrease of the re-magnetization height from 30 to 6 mm, the maximum levitation forces decreased from 695 to 658 N, corresponding to a decrease ratio of 5.3%. At the case of 6 mm re-magnetization height, the maximum levitation force of 658 N was still bigger than the biggest levitation force of 627 N in the direct 30-mm FCH case. It indicated that by employing the re-magnetization process after ZFC, bulks had inherited the advantage of the ZFC activation with a big levitation force. Meanwhile, the maximum guidance forces increased from 70.8 to 123.2 N with the decrease of the re-magnetization height from 30 to 6 mm. Although the guidance forces had increased greatly with an increase ratio of 74%, the maximum guidance force of 123.2 N was only about two thirds of the largest one of 181.0 N in the 6-mm FCH case. On one hand, the increased guidance forces indicated HTS bulks can indeed re-trap certain flux in the re-magnetization process after the first ZFC activation. On the other hand, the smaller guidance forces indicated the re-trap ability of HTS bulk is limited in the present PMG field condition. In other words, the re-trap ability is not as effective as the one in the direct FC case.

As indicated in Section 8.2.1, 30-mm FCH is regarded as an interesting FCH for the present HTS Maglev vehicle system. It was chosen to verify the effectiveness of this re-magnetization process after ZFC by comparing the magnetic force results with the 30-mm FCH case. Figure 8.32a and b show comparisons of the levitation forces and guidance forces of the bulk levitation unit above the PMG. The two levitation force curves are very close to each other, and the inset shows an enlargement from the gap

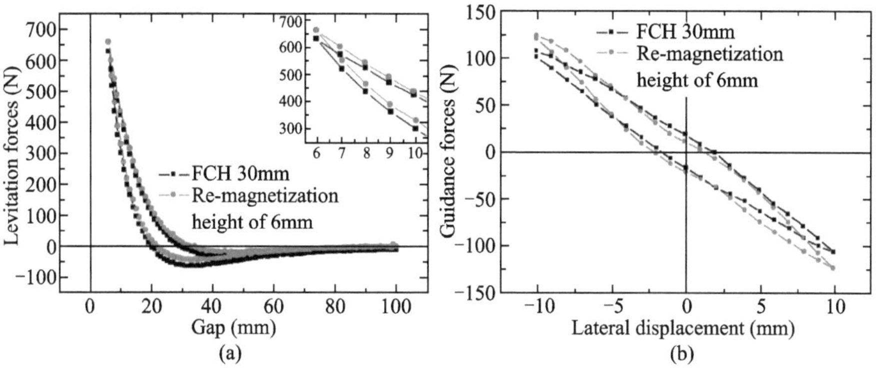


**Fig. 8.32:** Levitation (a) and guidance force (b) comparisons of the bulk levitation unit above the PMG between the cases of direct 30- and 6-mm FCH re-magnetization height after ZFC. The inset in (a) is the enlargement of the two force curves at gaps of 6–10 mm.

of 6 to 10 mm. The maximum levitation forces were 627 and 658 N, respectively, for the case of the direct 30- and 6-mm FCH re-magnetization height. The corresponding maximum guidance forces had a larger difference, 106.4 and 123.2 N, respectively. That is to say, by a re-magnetization process at a gap of 6 mm above the PMG after ZFC, the levitation capability and stability of the HTS bulk levitation unit had been improved simultaneously with respect to the direct 30-mm FCH case. Although the increase ratio was not very large, it had broken through the present defect in performance improvement with the sacrifice of one important force. Furthermore, the re-magnetization method after ZFC presents a possible working way for applications of HTS bulks to levitation systems.

### 8.3.3 An on-board double-layered HTSC array

At present, all HTS Maglev test vehicles employ a single-layered HTS bulk layout inside the cryostat [5–15]. Previously, there were many reports on the dependence of the magnetic levitation forces on the thickness or even layer number of the HTS bulks [41–43]. It was reported that the levitation forces kept increasing with the thickness of a HTS bulk and did not saturate until 12 mm [44]. This may be the reason why the bulk samples in common use are around 12–15 mm in thickness which is also dependent on the bulk processing technologies. If two layers of HTS bulk are applied over a PMG, the question is whether there is still some magnetic field available above the upper surface of the single-layered bulk samples to interact with the new upper bulk layer. Because the levitation force is derived from the electromagnetic force, where $J$ is the induced current inside the HTS bulk, $B_{ex}$, the external magnetic field, and $V$ is the volume of the bulk body. If the answer is positive, some levitation forces could be further obtained by

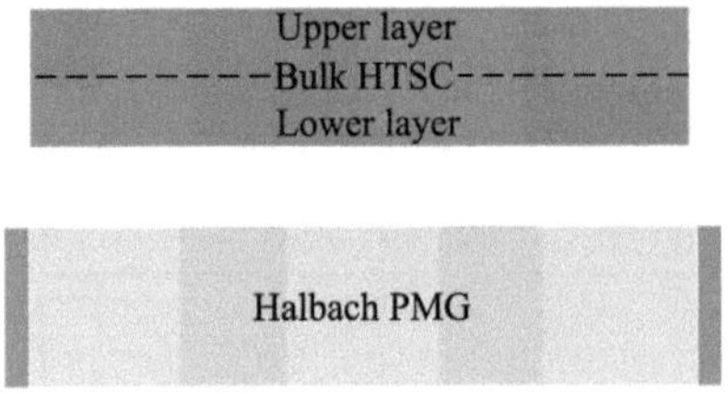


Fig. 8.33: Schematic diagram of the double-layer HTS bulk array with a width of 32 mm × 4 and a height of 13 mm × 2 working above the applied Halbach PMG.

adding a new bulk layer. In this section, the performance improvement from adding a new layer of HTS bulk above the original on-board HTS bulk is examined. It is reasonable to believe that some flux can go through and be trapped by pinning centers of the new upper bulk layer based on the following exploration.

As shown in Fig. 8.33, eight HTS bulk samples were used to investigate the Maglev performance of the double-layered HTS bulk array. Either layer of HTS bulk levitation unit was composed of four bulk samples, denoted as the upper and lower layers, respectively, where the upper layer of four HTS bulks were directly put at the upper surface of the lower layer. Both ZFC and FC conditions were performed on the double-layer HTS bulk in the experiments. In the FC condition, a representative FCH of 30 mm was chosen. After about 20 minutes, when the bulks were cooled completely, levitation and guidance forces between the HTS bulk levitation unit and the PMG were measured by the SCML-01 [26]. It should be noticed that the smallest measurement gap was 6 mm for the lower-layer HTS bulk, while it was 19 mm for the upper-layer HTS bulk due to the thickness of the lower-layer HTS bulk.

In the ZFC condition (Fig. 8.34), the levitation force curves of the individual and double-layer of HTS bulk levitation unit above the PMG were a big hysteresis loop composed of two branches, descending and ascending. The levitation forces showed an approximate exponential increase with the decrease of the levitation gap, and reached a maximum value at the lowest measurement gap. It was observed that the levitation capability of the single-layer HTS bulk levitation unit had been improved

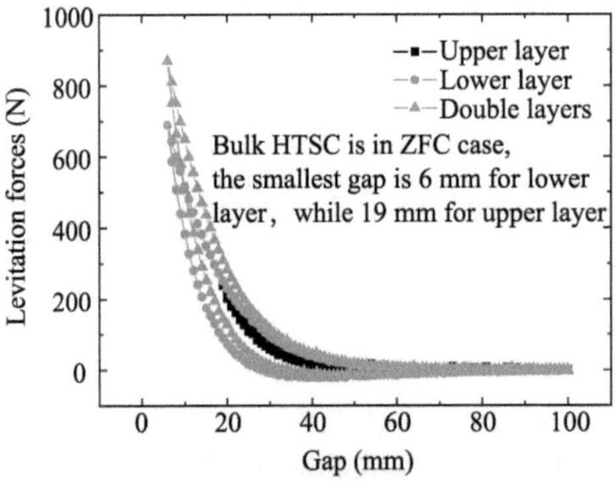


Fig. 8.34: Levitation forces of the individual and double-layer of the HTS bulk levitation unit above the PMG in ZFC conditions.

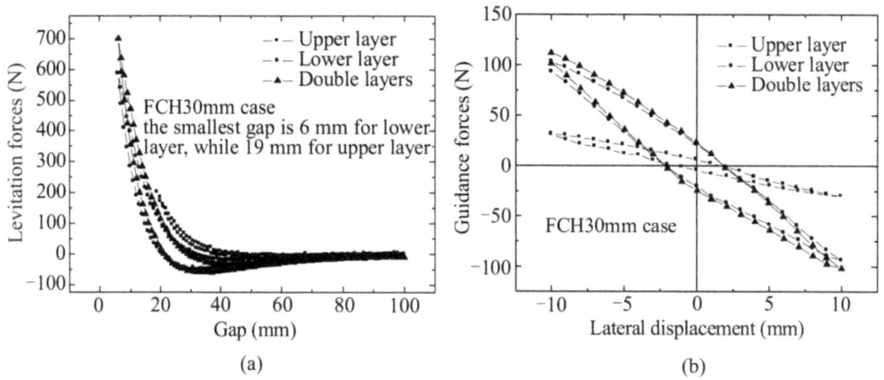


**Fig. 8.35:** Levitation (a) and guidance force curves (b) of the individual and double-layer of HTS bulk levitation unit above the PMG in 30-mm FCH conditions.

by adding the upper-layer HTS bulk so that the double-layer HTS bulk levitation unit obtained the biggest levitation force at the smallest gap.

In the 30-mm FCH condition, the profiles of levitation force curves in Fig. 8.35a are similar to those of the ZFC case in Fig. 8.34. Only the maximum levitation forces at the smallest gap showed some decrease due to the trapped flux in the FC case. The double-layer HTS bulk levitation unit still obtained the biggest levitation force at the smallest gap in the FC case, followed by the lower layer and then the upper layer. The guidance force curves in 30-mm FCH cases, shown in Fig. 8.35b, exhibited the stable levitation feature, i.e. a restoring force would be generated against any displacements to either side [4]. The maximum guidance force was found at the maximum lateral displacement of 10 mm. In the figure, the guidance force curves of the lower-layer and double-layer HTS bulk were very close to each other and showed some improvement in the latter. As the measurement height of the upper-layer HTS bulk was 28 mm due to the thickness of the lower-layer HTS bulk, the guidance forces of the upper levitation unit were obviously smaller than the other two.

To compare quantitatively, Tab. 8.3 shows the maximum levitation forces and guidance forces of the individual and double-layer of HTS bulk levitation unit above the

**Tab. 8.3:** Maximum levitation forces of the individual and double-layer of HTS bulk levitation unit above the PMG at the smallest gap of 6 mm.

| Conditions | | Upper layer | Lower layer | Sum.[a] | Double-layer | Double increase.[b] | Upper efficiency[c] |
|---|---|---|---|---|---|---|---|
| ZFC | $F_{Lev}$[d] (N) | 239 | 690 | 929 | 870 | 26.1% | 75.3% |
| FC | $F_{Lev}$ (N) | 200 | 587 | 787 | 701 | 19.4% | 57.0% |

[a] Sum = Upper layer + Lower layer; [b] Double increase = (Double-layer − Lower layer)/Lower layer; [c] Upper efficiency = (Double-layer − Lower layer)/Upper layer; [d] $F_{Lev}$ is levitation force.

PMG at the smallest gap of 6 mm. In the ZFC case, the maximum levitation forces were 239 N, 690 N, and 870 N, respectively for the upper, lower, and double-layer HTS bulk. By adding the upper-layer HTS bulk, the levitation forces of the double-layer HTS bulk increased by 26.1% compared to the lower layer. However, the maximum levitation force of the double-layer of 870 N was smaller than the sum of the upper and lower HTS bulk's levitation force of 929 N, 75.3% of levitation capability of the upper-layer HTS bulk was excited. This decrease of levitation forces between the sum of the upper and lower layer and the double-layer HTS bulk can be explained by the shielding of the HTS bulks in the lower layer [45]. In the FC case, the maximum levitation force of the double-layer HTS bulk increased by 19.4% compared to the lower layer and 57% of levitation capability of the upper-layer HTS bulk was excited. It was implied that adding the upper-layer HTS bulk is not as effective in FC case as in ZFC case.

In practice, the levitation performance of the on-board HTS bulk at the working gap is more useful than the maximum forces at the smallest gap. Table 8.4 shows the levitation forces and guidance forces of the individual and double-layer of HTS bulk levitation unit above the PMG at a typical working gap of 15 mm. The levitation forces of the double-layer HTS bulk were still smaller than the sum of that of the upper and lower layer in both ZFC and FC cases. Moreover, the increased ratio of the double-layer to the lower layer and the efficiency of the upper layer at the working gap of 15 mm were close to the corresponding values at the smallest gap, shown in Tab. 8.3. An increase of 24.9% in levitation force was obtained by the double-layer HTS bulk in the ZFC case, which means 79.8% of levitation capability of the upper-layer HTS bulk was excited at the working gap, while, in the FC case, the increase in the ratio of the double-layer was only 16.9% and the efficiency of the upper layer was 53.5%. The improvement of performance by adding the upper HTS bulk layer was still not as effective in the FC case as it was in the ZFC case. As for the guidance forces, the efficiency of adding the upper-layer HTS bulk was lower than that of the levitation forces. Only an increase of 8.8% was obtained by the double-layer HTS bulk, which means 27.3% of guidance performance of the upper-layer HTS bulk was excited. Hence, the performance improvement in guidance forces was not increased much by adding the upper-layer HTS bulk.

**Tab. 8.4:** Levitation forces and guidance forces of the individual and double-layer of HTS bulk levitation unit above the PMG at the working gap of 15 mm.

| Conditions | | Upper layer | Lower layer | Sum | Double-layer | Double increase | Upper efficiency |
|---|---|---|---|---|---|---|---|
| ZFC | $F_{Lev}$ (N) | 109 | 349 | 458 | 436 | 24.9% | 79.8% |
| FC | $F_{Lev}$ (N) | 68 | 215 | 283 | 252 | 16.9% | 53.5% |
| | $F_{Gui}$[e] (N) | 30.0 | 93.6 | 123.6 | 101.8 | 8.8% | 27.3% |

[e] $F_{Gui}$ is the guidance force.

Thus, from these Maglev experiments of individual and double-layer HTS bulk above the PMG, it was verified that the levitation forces and guidance forces can be indeed improved by adding the upper-layer HTS bulk, but the final Maglev performance of the double-layer HTS bulk cannot be calculated by simple addition. Due to the shielding and interaction of the two adjacent HTS bulk layers, only part of the Maglev performance of the upper-layer HTS bulk had been excited. In the typical FCH, an improvement of 16.9% in levitation forces and 8.8% in guidance forces was obtained. The added upper-layer HTS bulk was more effective to improve the Maglev performance in the ZFC case than the FC case and more effective to improve the levitation forces than the guidance forces. The increase of a layer of HTS bulk meant a double increase in the cost of superconductor. However, compared with the huge PMG cost, the cost of the upper-layer HTS bulk was not so significant. Thus, this optimization method was able to enhance the load capability and stability of the levitation system simultaneously without high cost.

### 8.3.4 A laying mode using the *c*-axis orientation of bulk HTSC

In this section, a laying mode of the HTS bulk above the PMG will be introduced based on the anisotropy properties of HTS bulk materials and the magnetic field feature of the applied double-pole Halbach PMG so that they can match to each other efficiently as shown in Fig. 8.36.

Due to the layered structure of the Cu-O plane in the YBCO crystal, the HTS bulk exhibits a high anisotropy between the *c*-axis and the *a*-*b* plane. The typical critical current density, $J_c$, in the *a*-*b* plane is found to be about three times that along the *c*-axis [28]. This property indicates that it will be more efficient to excite the induced current in the *a*-*b* plane compared with along the *c*-axis. At the field-pole position, it reaches the maximum vertical component of magnetic field, while, at the middle of the Halbach PMG, it was the maximum horizontal component of the magnetic field, where most of the magnetic flux lines are parallel to the horizontal direction. In an HTS levitation system, the levitation or guidance force (electromagnetic force)

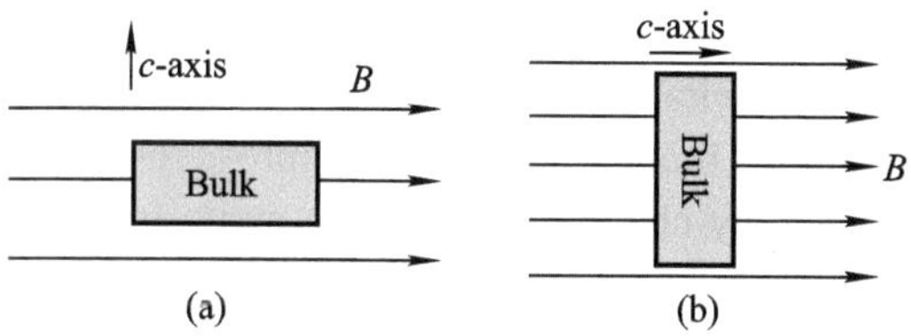


**Fig. 8.36:** Schematic illustrations of the different laying modes of a HTS bulk working in the horizontal-applied magnetic field, where the bulk keeps its *c*-axis vertical to the applied field (a) or parallel to the field (b).

is closely related to the intensity and the gradient of the applied magnetic field, as follows,

$$\mu_0 J = dB_{ex}/dr, \tag{8.6}$$

$$F_{Lev} = \int J \times B_x dv, \tag{8.7}$$

$$F_{Gui} = \int J \times B_z dv, \tag{8.8}$$

where the induced current $J$ is dependent on the gradient of the external magnetic field $B_{ex}$ and levitation force $F_{Lev}$ and guidance force $F_{Gui}$ correspond to the Lorentz force between the induced current and the horizontal component of the magnetic field $B_x$ or vertical component of the magnetic field $B_z$, respectively. This relationship implies that we can use the magnetic field configuration to design a reasonable laying mode of HTS bulks. As shown in Fig. 8.36b, the laying mode of a HTS bulk with its *c*-axis parallel to the horizontal applied field is more efficient to excite a larger induced current in the *a*-*b* plane than the counterpart in Fig. 8.36a. At present, the HTS bulk is put inside the cryostat with its *a*-*b* plane parallel to the surface of the PMG in HTS Maglev prototypes [5–15]. In this layout, the bulk gains the maximum levitation force at the maximum vertical field position and the maximum guidance force at the maximum horizontal field position [46]. It means that it will be feasible to improve the levitation forces or guidance forces by changing the *c*-axis laying direction of the HTS bulk. Thus, the on-board HTS bulk appears in two different laying modes. One is a traditional horizontal laying mode in which the *c*-axis of the bulk directs to the vertical direction, the other one is a vertical laying mode in which the *c*-axis of the bulk is along the horizontal direction.

Figure 8.37a and b depicts the levitation force curves of the bulk sample with different laying modes measured at the PMG middle and field-pole positions in the cases of ZFC and 30-mm FCH, respectively. The bulk shows a similar levitation force curve in the two cooling conditions. When the bulk was put in a horizontal mode, the levitation forces in the field-pole position were larger than that at the middle of the PMG. In contrast, when the bulk was laid in a vertical mode, the levitation forces in the PMG middle were larger than that at the field-pole position. Furthermore, at the middle of the PMG, the bulk with a vertical laying mode got a larger levitation force than that with a horizontal laying mode. These results verified again that the bulk with a horizontal mode can obtain the best levitation force performance at the field-pole position. At the same time, it also implied that it is possible to improve the bulk's levitation forces at the middle of the PMG by changing the laying mode from the horizontal to the vertical one. By laying the bulk as a vertical way, the width of the bulk facing the PMG surface was 13 mm rather than the former 32 mm (see Fig. 8.36). At the same available space in width for the horizontal bulk, at least two bulks can be laid in the vertical way. The two bulk's levitation performance can be estimated by an approximate superposition method [47].

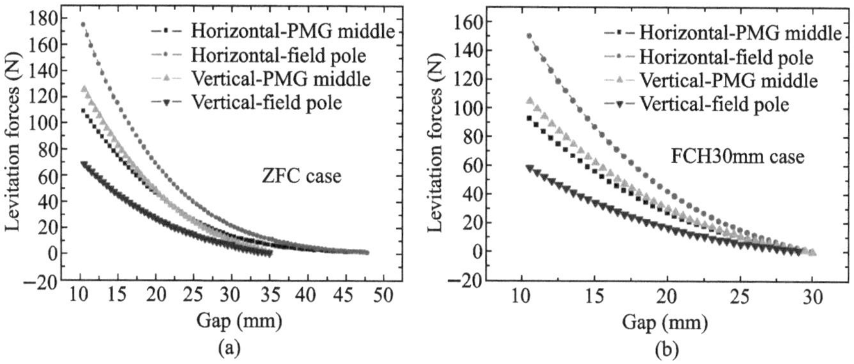


**Fig. 8.37:** Levitation force curves of the bulk sample with horizontal or vertical laying modes measured at the PMG middle and field-pole positions in the cases of ZFC (a) and 30 mm FCH (b).

In the ZFC case, the levitation force of two bulks with a vertical laying mode was approximately 125.5 N × 2 = 251.0 N at a measurement gap of 10 mm, which was larger than 108.2 N for the horizontal laying mode. In the case of 30-mm FCH, the levitation force of the vertical mode was 104.6 N × 2 = 209.2 N, also larger than 92.1 N in the horizontal mode. Moreover, the estimated levitation force by a vertical mode was also larger than that when the bulk is put horizontally above the field-pole of the PMG. These results indicated that by adopting a vertical bulk-laying mode at the middle of the PMG, the levitation forces of the bulk could be improved significantly, because in such mode, the induced current in the *a*-*b* plane could interact with the horizontal component of the applied magnetic field efficiently, as shown in Fig. 8.36b.

Next, the stability (guidance force) of the bulk sample with different laying modes at the two measurement positions was compared. Figure 8.38 shows the guidance force curves of the bulk sample with different laying modes measured at the PMG middle and field-pole positions at the 30-mm FCH. All guidance force curves exhibited the stable feature of a levitation system. It notes that when the bulk was laid at the field-pole position, the magnetic field at the two sides of the measurement position was asymmetric. Correspondingly, the guidance force curves measured above the field-pole position were not symmetric either. When the bulk was put horizontally, the maximum guidance force was obtained at the middle of the PMG, and the minimum at the field-pole position. When the bulk was laid vertically, it was just reversed. The maximum guidance force at the field-pole position was larger than that at the middle of the PMG. These results were consistent with the former conclusions, i.e. the horizontal bulk-laying mode can generate the maximum guidance force at the middle of the PMG. Meantime, it was implied that the stability of the levitation system can be improved by changing the horizontal bulk-laying mode to the vertical mode at the field-pole position. At the maximum lateral displacement of 5 mm, the maximum guidance force increased from −3.4 to −17.3 N when changing the horizontal bulk

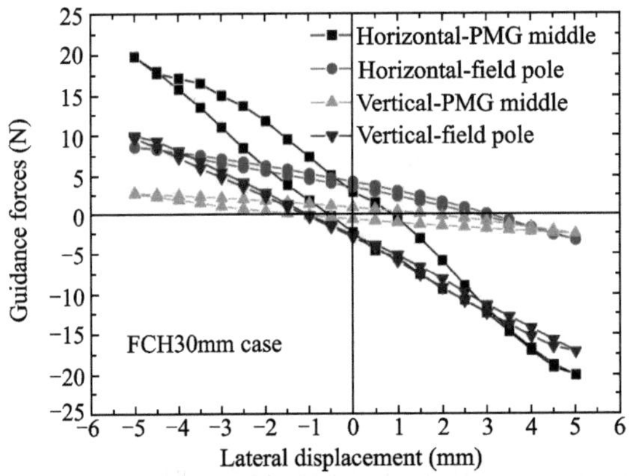


**Fig. 8.38:** Guidance forces of the bulk sample with horizontal and vertical laying modes measured at the PMG middle and field-pole positions in the case of the 30-mm FCH and the 15-mm WH.

mode to the vertical mode at the field-pole position. If considering the same space in width for bulks, an approximate factor of 2 can be multiplied, indicating a further improvement of the guidance force.

Table 8.5 further indicates that the levitation or guidance forces can be alternatively improved by changing the bulk-laying mode. At the middle of the PMG, the maximum levitation force was approximately improved from 92.1 to 209.2 N, corresponding to an increasing factor of 2.27, when the bulk-laying mode was changed from horizontal to the vertical one. While the maximum guidance force at the maximum lateral displacement of 5 mm decreased approximately from −20.3 to −5.2 N, indicating a decrease factor of 3.9. At the other measurement position of the field pole, 10.2 times of improvement in the guidance forces and 1.28 times' decrease in the guidance forces were obtained when changing the bulk-laying mode from the horizontal to the vertical one. Hence, in practical applications, the *c*-axis direction of the bulk could be designed according to the applied field configuration to meet the requirements on the load capability or stability of the levitation system.

**Tab. 8.5:** Maximum levitation forces and guidance forces of the bulk sample with different laying modes in the case of 30-mm FCH.

| Experimental conditions | | Field-pole position | PMG middle |
|---|---|---|---|
| Max. $F_{Lev}$ (N) | Horizontal mode | 149.5 | 92.1 |
| | Vertical mode | 58.4 (×2) | 104.6 (×2) |
| Max. $F_{Gui}$ (N) | Horizontal mode | −3.4 | −20.3 |
| | Vertical mode | −17.3 (×2) | −2.6 (×2) |

Note: In the vertical mode, (×2) means an approximate coefficient of 2 considering the same laying width facing the PMG surface as in the horizontal mode.

### 8.3.5 Introduction of ferromagnetic materials

As described in Eqs. (8.6)–(8.8), the electromagnetic forces (levitation forces and guidance forces) obtained by HTS bulks is closely dependent on the magnetic field at the bulk position. If the magnetic field at the bulk position is enhanced, correspondingly, the levitation forces of the bulk can be improved [48]. A ferromagnetic material like iron with a high permeability is used to guide the flux distribution of the PMG in order to enhance the magnetic field at the bulk-laying position. The principle of this method is similar to the PMG design, in which the ferromagnetic materials are usually employed as flux collectors to increase the intensity and gradient of the generated magnetic field of the PMG. The key question is how and where to add the iron plates. Figure 8.39 presents an example in which an optimized 1-mm-thick iron plate was added on the surface of the bulk unit. It was observed that after adding the iron plate, the magnetic field distribution was obviously changed at the bulk position.

In the confirmatory experiments, a single iron plate was employed just covering the upper surface of the bulk unit. Since there would be an obvious attractive force between the iron plate and the PMG, the added iron plate should be thin. After magnetic field simulations by FEMM software, a 1-mm-thick iron plate was chosen by optimization. To check the effectiveness, the influence of the added iron plate on the magnetic forces in the typical FC condition was examined with the SCML-01 [26] (see Section 5.3).

Figure 8.40a and b shows the levitation force and guidance force curves of the bulk levitation unit with and without adding the iron plate on its surface, respectively. It is found that the added iron plate will weaken the levitation forces and that the levitation forces decrease more with increased size of the iron plate (see Fig. 8.40a). The maximum levitation force of the bulk unit is 474, 396, and 433 N, respectively, for the cases of a full iron plate, no iron plate and partial iron plate on the bulk surface. This result is thought to result from the attractive forces between the iron plate and

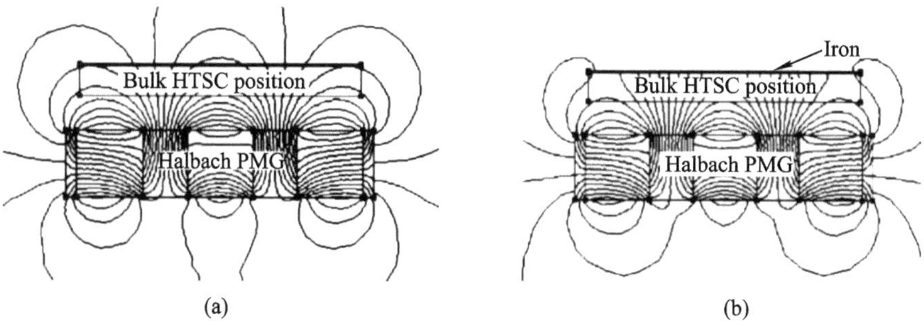


**Fig. 8.39:** Magnetic field distribution of the Halbach PMG at the HTS bulk position. (a) Without iron, (b) with a 1-mm-thick iron plate added on the HTS bulk surface.

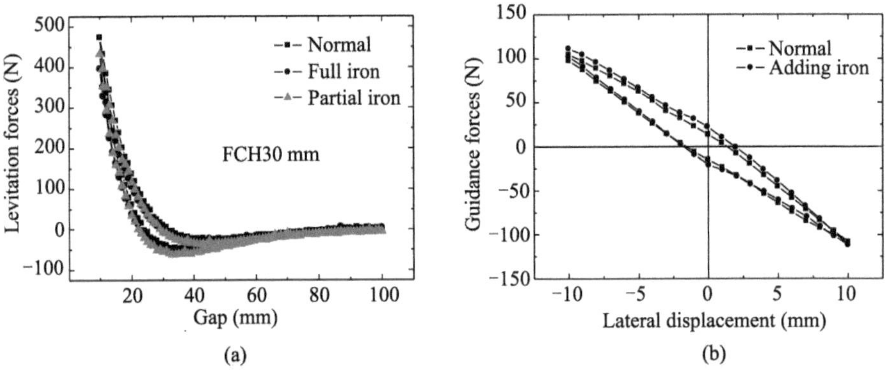


**Fig. 8.40:** Levitation (a) and guidance force (b) curves of the bulk levitation unit with and without introducing iron plates on its surface.

the PMG. For the guidance forces (see Fig. 8.40b), there is some increase when adding the iron plate. The maximum guidance force increases from 109.0 to 112.0 N when a full iron plate is added on the bulk surface. However, it is not feasible to add a full iron plate on the bulk surface considering the obvious decrease of the levitation forces.

To find a suitable position to add the iron plate, the force behavior was investigated with an iron plate of the same size as one bulk sample at the two typical working positions, i.e., above the field pole and in the middle of the PMG. It was found that the attractive forces always exist at the two measured positions, but a stable guidance force can be obtained only at the middle of the PMG. This stable guidance force will be useful for practical applications, so in the following measurements, the influence of an added iron plate on a bulk sample laid above the middle of the PMG was investigated in the typical FC condition with the SCML-02 [35] (see Section 5.5).

When adding the iron plate on the bulk sample above the middle of the PMG, the magnetic field distribution would be changed at the bulk position (see Fig. 8.41a). The horizontal component of the magnetic field decreased obviously at the middle of the PMG, while, the vertical component of the magnetic field increased at the two field-pole positions (see Fig. 8.41b). As a result, the levitation forces and guidance forces of the bulk also changed. As shown in Fig. 8.42a, there was also some decrease of the levitation force when adding the iron plate. The maximum levitation force at the minimum gap of 10 mm decreased from 92.1 to 78.7 N, decreasing by 14%. However, for the guidance forces in Fig. 8.42b, the maximum guidance force increased from −20.3 to −30.4 N, increasing by 50% at the maximum lateral displacement of 5 mm. These data meant that with a reasonable introduction position, the positive benefit in the guidance forces brought by the ferromagnetic materials would surpass the negative influence on the levitation forces. This method of introducing the

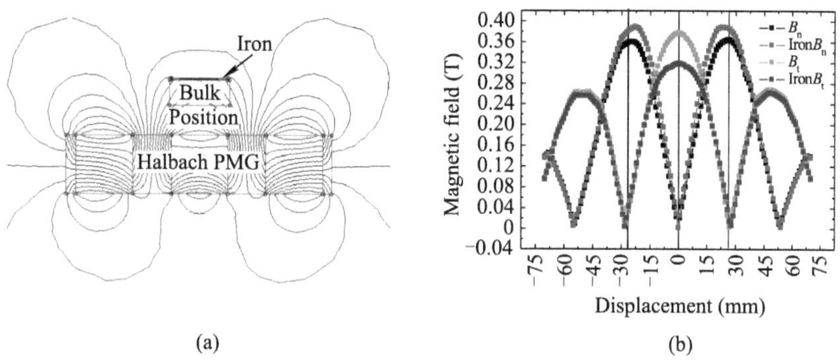


**Fig. 8.41:** Magnetic field distribution of the Halbach PMG. (a) The profile at a gap of 15 mm. (b) When a 1-mm-thick iron plate is introduced to the bulk surface above the middle of the PMG. $B_n$ is the vertical component and $B_t$ is the horizontal component of the magnetic field.

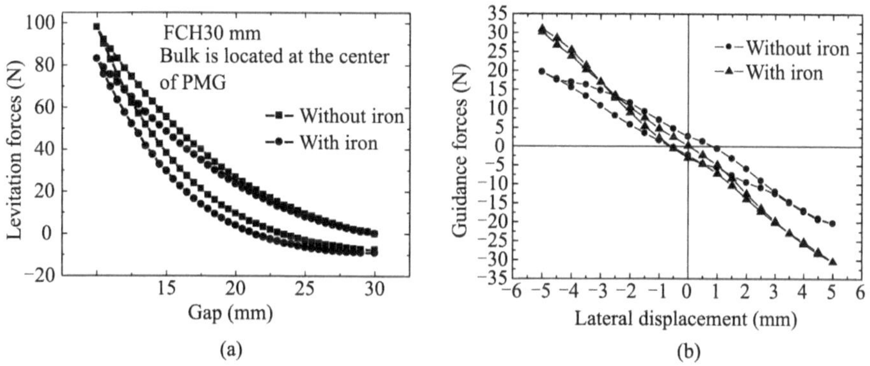


**Fig. 8.42:** Levitation (a) and guidance force (b) curves of the bulk sample measured above the middle of the PMG with and without introduced iron plates.

ferromagnetic material on the bulk levitation unit can be used for cases where a large guidance force was needed. At the same time, the experimental results also confirmed the conclusion that it was very difficult to improve the levitation forces (load capability) and guidance forces (stability) simultaneously with the HTS bulks and applied magnetic fields in practice. Usually, we improve one of them according to actual requirements. For instance, in the case of a curved-route running line, a large guidance force was required to overcome the centrifugal force. A new HTS Maglev design making use of the results of this chapter is schematically as shown in Fig. 8.43. With the reasonable choice of the *c*-axis direction of the HTS bulk (see Section 8.3.4) and introduction of iron plates, the guidance performance (stability) can be enhanced greatly compared with the common levitation system, so that it is able to meet some special requirements like curved-route cases.

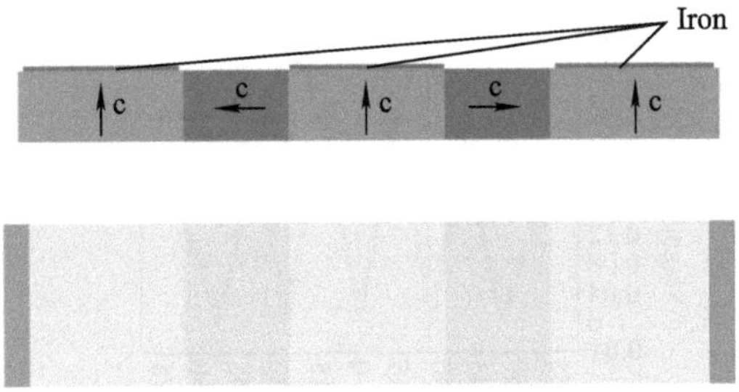


Fig. 8.43: Schematic diagram of a new HTS Maglev design using the *c*-axis direction of bulk superconductors and iron introduction for better guidance performance.

## 8.4 Some developed designs of the HTS Maglev vehicle system

### 8.4.1 Multi-pole PM guideway structure

The monopole PMG is most popular in HTS Maglev vehicle systems [5–15]. The most popular structure, similar to the "Century" PMG, is that of two PMs with opposite horizontal magnetization direction connected by a flux-concentration iron with a single peak in the vertical component of magnetic field. However, an axially symmetric HTS magnetic bearing with only a 200-mm diameter realized about 1-ton load in two representative applications [49, 50]. It is reasonable to believe that high load capability of the axially symmetric HTS bearing is mainly due to the PM rotor structure [49–51]. In the general HTS-bearing design, the PM rotor is composed of multi-pole PMs for larger force stiffness because one pole can create one higher magnetic field region. More poles are effective in enhancing the stability [52]. Application of the multi-pole magnetic field to the HTS Maglev PMG design, the expected to be improved is efficiency under the same PM material cost through the reasonable design and arrangement of HTS bulk.

Figure 8.44 illustrates a double-pole PMG with Halbach array that is designed and fabricated to verify its improved capability for use in a HTS Maglev vehicle system. Two vertical magnetic field peaks appear at the two poles. In comparison, the double-pole Halbach PMG has nearly the same PM cross-section area of a traditional monopole PMG, 3900 mm$^2$ and 4000 mm$^2$, respectively. More details about the two PMGs can be found in Ref. 25.

Figure 8.45 shows the levitation force comparison of this unit above two PMGs at the 30-mm FCH. The bulk unit with the double-pole Halbach PMG had a bigger levitation force and hysteresis loop than the traditional monopole PMG. At the smallest gap of 7 mm, the maximal levitation force was 231.5 N for the former, which was 1.73 times larger than that of the latter, 133.7 N. The bigger force and hysteresis indicated that the double-pole Halbach PMG produces a better magnetic field distribution at the interacting range with the bulk. In Fig. 8.44, it shows that the average magnetic

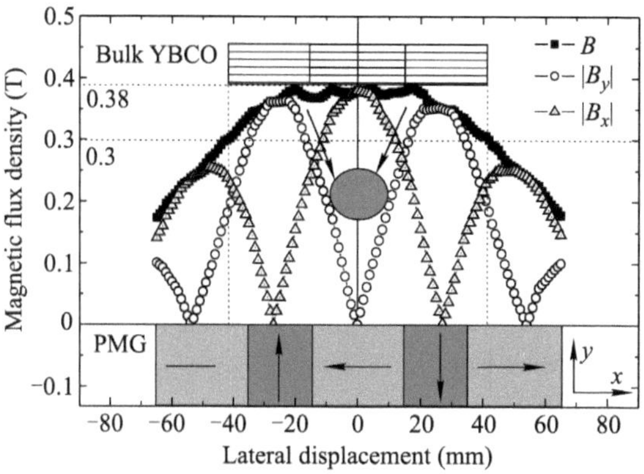


**Fig. 8.44:** Cross-section structure and magnetic field distribution at the 15-mm gap of the double-pole Halbach PMG.

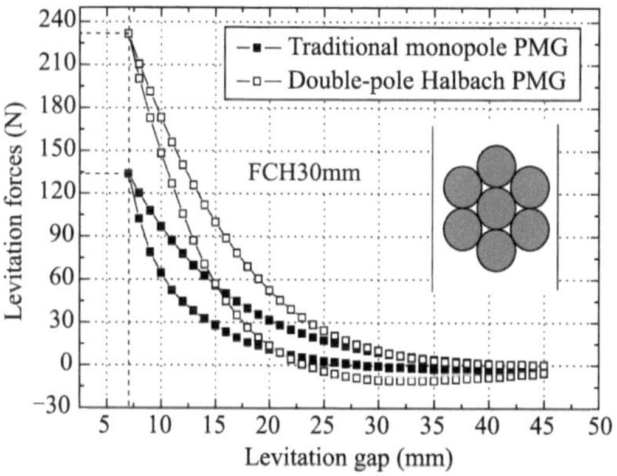


**Fig. 8.45:** Levitation force comparison of the levitation unit above two PMGs under the 30-mm FCH.

flux density at gap of 15 mm was 0.328 T above that of the traditional monopole PMG while it was 0.36 T for the double-pole Halbach PMG at the 82 mm width area of the levitation unit. The average field rather than the peak magnetic field factor was the main factor of levitation forces and hysteresis. The average magnetic field embodies most of the flux motion tendency in HTSC. The double-pole Halbach PMG increased its average magnetic field by broadening the strong magnetic field region with two vertical magnetic field peaks (see Fig. 8.44).

The corresponding levitation stiffness are shown in Fig. 8.46. Similar to the levitation force curve, the levitation stiffness curve shows an approximate exponential increase with decrease of the levitation gap. The levitation stiffness of the bulk unit above the double-pole Halbach PMG was bigger than that of the monopole PMG from a gap of 40 to 7 mm. At the smallest gap of 7 mm, the levitation stiffness was 10.2 and 19.4 N/mm, respectively, whereas at large gap over 40 mm, the two curves are

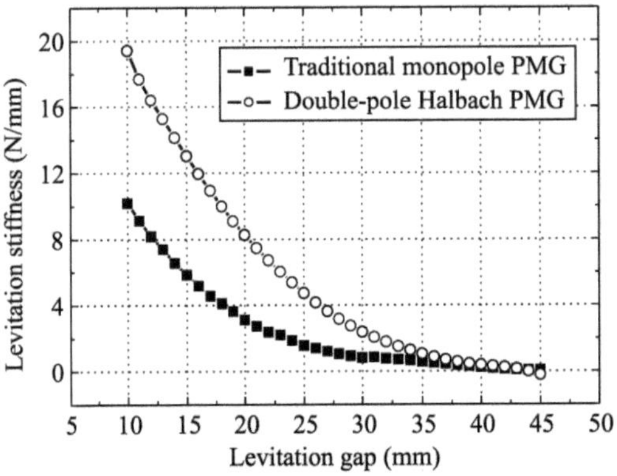


**Fig. 8.46:** Levitation stiffness comparison of the levitation unit above two PMGs under the 30-mm FCH.

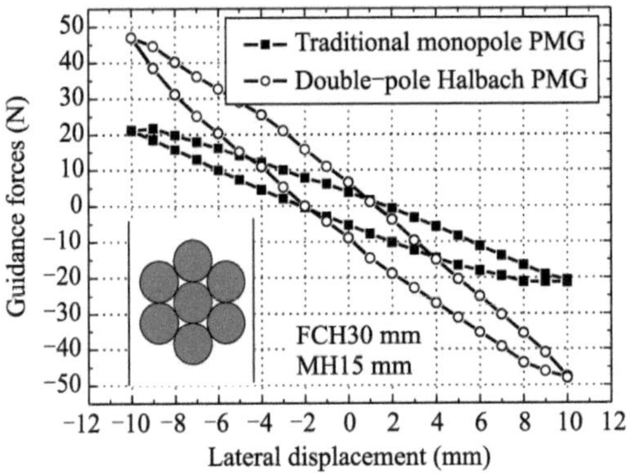


**Fig. 8.47:** Guidance force comparison of the levitation unit above two PMGs under the 30-mm FCH and 15-mm WH.

close and even cross each other. That is, the double-pole Halbach PMG did not have an advantage over the monopole PMG at large levitation gap.

For HTS levitation system, the guidance forces are used to evaluate whether the system is stable in the lateral direction and determine its restoring capability. Figure 8.47 shows the guidance force comparison of a seven-bulk levitation unit above two PMGs at the 30-mm FCH and 15-mm MH. Due to the field cooling condition, the two guidance force curves both imply a stable Maglev system where a restoring force was generated against lateral displacement. The bulk unit above the double-pole Halbach PMG had a better guidance performance and bigger hysteresis loop than that above the monopole PMG. At the maximum lateral displacement of −10 mm, the maximum guidance force was 47.0 and 21.1 N, respectively, for the two PMGs. The guidance force ratio was 2.2, which was larger than the levitation force ratio. Figure 8.48 shows

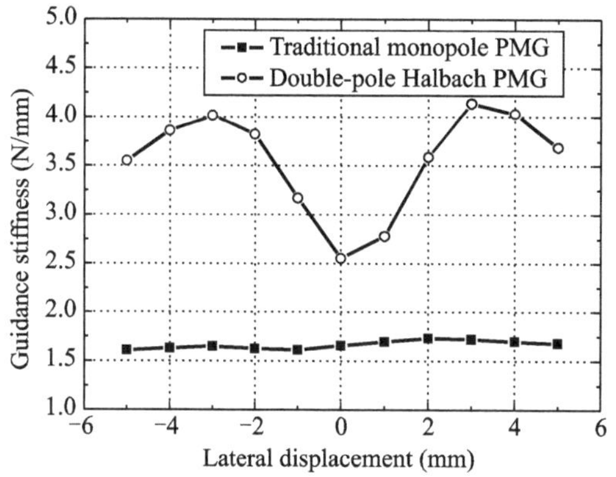


**Fig. 8.48:** Guidance stiffness comparison of the levitation unit above two PMGs under the 30-mm FCH and 15-mm WH.

the guidance stiffness comparison of the bulk unit above two PMGs at the same measurement conditions. The two guidance stiffness curves appear symmetric about their origin in the lateral displacement range of −5 to 5 mm. It was clear that the guidance stiffness of the double-pole Halbach PMG was about 2–3 times larger than that of traditional monopole PMG even when the guidance stiffness was at its smallest value. It is well known that guidance force performance depends on the quantity of trapped flux in the bulks. On the one hand, the experimental trapped-field saturation of a bulk was a little low, only about 0.26 T [53] at liquid nitrogen temperature; thus, a much smaller or bigger applied magnetic field was not very effective to enhance its levitation and guidance capability. Moreover, the magnetic field distribution from 0.3 to 0.38 T of the double-pole Halbach PMG was thought to be more reasonable than that of the traditional monopole PMG from 0.25 to 0.4 T at the measuring position. Thus, it is important to note that the PMG configuration should be optimized according to the performance of the HTS bulk.

On the other hand, it is interesting to find that the double-pole Halbach PMG has an additional effect on the guidance performance of levitation bulks. Even under ZFC, the seven-bulk unit can realize stable levitation above a double-pole Halbach PMG, which is contrary to the case above the monopole PMG. The stable guidance force curves at ZFC with a double-pole Halbach PMG are shown in Fig. 8.49. It implies that the stability is strong. At the maximum lateral displacement of −10 mm, the maximum guidance force of 34.26 N was obtained, which was 1.62 times larger than that of the bulk unit in 30-mm FCH 15-mm WH above monopole PMG. Moreover, the value was 72.9% of the maximum guidance force (47.0 N) in 30-mm FC 15-mm WH above the same double-pole PMG. Different from stable levitation in FC by flux-pinning in the HTS bulk, the stable levitation in ZFC results from the electromagnetic interaction between the HTS bulk and the double-pole magnetic field. At the center of the double-pole Halbach PMG, an obvious magnetic potential-well, marked by a dark circle in

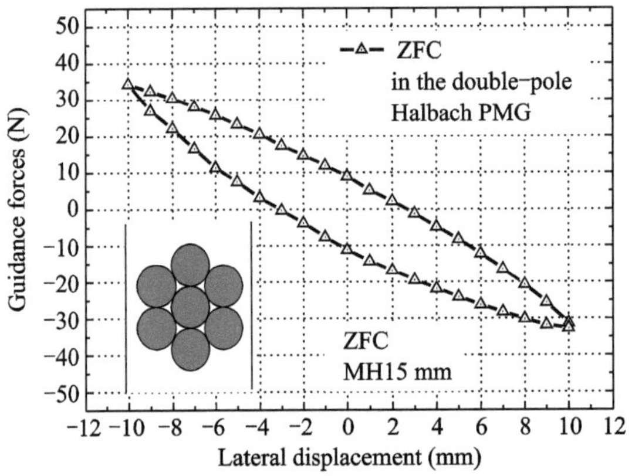


**Fig. 8.49:** Guidance forces of the seven-bulk levitation unit above the double-pole Halbach PMG under ZFC and the 15-mm WH.

Fig. 8.44, was formed between the two peaks of the vertical magnetic field component. At the same time, the three middle bulks just fell into the potential-well area. When a lateral displacement on any side happened to the bulks in the potential-well, the vertical magnetic field at the bulk position would be larger. According to Faraday's law of electromagnetic induction, the shielding current would be induced in the bulk material, and a corresponding repulsive Lorentz force would result from the interaction of induced current and the applied PMG field to resist the lateral displacement. Thus, the levitation was stable. This was also the reason why more poles were effective in enhancing he levitation stability [52]. This experiment indicated another way to realize stable superconducting levitation beyond the FC condition by PMG design. It is very attractive to use the potential-well field configuration to enhance the guidance performance in the multi-pole PMG design.

As listed in Tabs. 8.6 and 8.7, the values of the dynamic parameters measured in the vibration experiment procedure (see Section 8.2.1) are consistent with the above quasi-static force research. The dynamic stiffness of the bulk unit is thought to correlate with the quantity of trapped flux inside it. At lower FCH, higher trapped flux can bring a bigger dynamic stiffness, which is a benefit to the HTS Maglev system. At the same FCH, the dynamic stiffness of the bulk unit above the double-pole PMG

**Tab. 8.6:** Vertical and lateral dynamic stiffness above two PMGs at different FCHs.

| FCH (mm) | Dynamic stiffness in vertical $k_z$(N/mm) | | Dynamic stiffness in lateral $k_x$(N/mm) | |
|---|---|---|---|---|
| | Monopole PMG | Double-pole PMG | Monopole PMG | Double-pole PMG |
| 40 | 5.54 | 7.44 | 3.96 | 7.44 |
| 30 | 6.42 | 11.03 | 6.42 | 13.62 |
| 20 | 10.62 | 20.97 | 13.11 | 32.25 |

**Tab. 8.7:** Vertical and lateral damping coefficient above two PMGs at different FCHs.

| FCH (mm) | Damping coefficient in vertical $c_z$ (Ns/m) | | Damping coefficient in lateral $c_x$ (Ns/m) | |
|---|---|---|---|---|
| | Monopole PMG | Double-pole PMG | Monopole PMG | Double-pole PMG |
| 40 | 11.63 | 8.84 | 13.96 | 15.47 |
| 30 | 16.58 | 12.09 | 9.37 | 14.96 |
| 20 | 15.36 | 17.62 | 15.27 | 23.05 |

is always larger than that of the monopole PMG, which implies that better dynamic stability can be achieved with the double-pole Halbach PMG.

Unlike the dynamic stiffness, another dynamic parameter, the damping coefficient, does not show monotonic relationship with FCH or trapped flux, as shown in Tab. 8.7, which implies that other factors, like vibration amplitude, can affect the damping coefficient. With almost the same magnitude of pulse force excitation, trapped flux is still considered as the main factor for the rough analysis of the HTS Maglev dynamic system. This is because, for a well-known low-damping HTS Maglev system, the main energy loss associated with damping is the hysteresis loss of the bulk, which is related to its trapped flux. More trapped flux can cause more hysteresis loss leading to a bigger damping coefficient. Generally, the damping coefficient in Tab. 8.7 still tends to increase with a decrease of FCH due to higher trapped flux. At a lower FCH, such as 20 mm, the vertical damping coefficient with a double-pole Halbach PMG is larger than that of monopole PMG. However, the double-pole PMG does not keep the advantage at higher FCH from the viewpoint of damping efficiency. It is because that the main magnetic field is concentrated and larger than that of the monopole PMG at the low height of the double-pole Halbach PMG. Similarly, the lateral damping coefficient of the double-pole PMG is larger than that of the monopole PMG due to the same magnetic field comparison situation. Thus, the same bulk unit with the double-pole Halbach PMG has a better dynamic stability because a larger damping coefficient implies a better anti-vibration ability.

From the point of practical application, both the static and dynamic experiments show that the seven-bulk levitation unit with the double-pole Halbach PMG has a better load capability, guidance performance, dynamic stability, and cost performance. Based on this study, several PMG design guidelines are given:

1. The average magnetic field at the applied position will be more important than the peak field.
2. The optimum PMG configuration is correlated to the performance of the HTS bulks. They should be matched to each other. Otherwise, the performance of good material cannot be excited and will be wasted.
3. The optimal PMG configuration is a function of the gap, at which the vehicle is operated.
4. The Halbach PMG has a notable effect to concentrate the magnetic field into its upper surface in order to increase its efficiency and reduce the cost of PMG.

5. The multi-pole PMG configuration can play an important additive role for guidance performance of a levitation system due to the potential-well field configuration.
6. The homogeneity of magnetic field along the PMG is very important for the practical application. Since the Halbach PMG does not produce a homogenous magnetic field, much attention should be paid to the precision and assembling precision of each small PM.
7. For practical application, the pole number should be chosen according to the shape, distribution, and working gap of the on-board bulks. If the HTS Maglev vehicle is designed to run at a large levitation height (over 15 mm) and the precisions of each small PM can be guaranteed, the highly efficient Halbach PMG is recommended.

### 8.4.2 A T-shaped HTS Maglev monorail system

A T-shaped monorail HTS Maglev system was also proposed to achieve the design targets of low-cost and high-performance. This T-shaped monorail produced not only a strong periodic magnetic profile in the upper space of the array, but also a strong symmetric magnetic field in the lateral spaces of the array.

The 2D view of the T-shaped HTS Maglev monorail system is shown in Fig. 8.50. The system's PMG with T-shaped cross section was notlike the present typical PMGs with a rectangular cross section. Conversely, the HTSCs could be cooled in the upper (#1), left (#2), and right (#3) liquid nitrogen cryostats which were fixed in the bed plate of the vehicle, corresponding to the up, left, and right sides of the T-shaped PMG. It also can be observed that the structure of the T-shaped PMG took advantage of the Halbach PMG in the horizontal direction and the "Century" PMG in the vertical direction. Figure 8.50 also shows the magnetization directions of PMs and the magnetic field distribution of the T-shaped PMG.

Figure 8.51a and b shows the simulation results of the magnetic field distributions above two kinds of PMG under the same calculation conditions. It is obvious that the T-shaped PMG had a wider high-field region than the present popular PMG with equal PM cross-sectional area. At the height of 15 mm, the magnetic field of the T-shaped PMG was still up to 0.45 T which was a better applied magnetic field for a Maglev system. In addition, Fig. 8.51b showed that at the 15-, 10-, and 5-mm lateral gaps of PMG, the magnetic fields of the T-shaped PMG were about 3, 2, and 1.5 times stronger than the present popular "Century" PMG [25], respectively.

Figure 8.52a further shows the calculated results of levitation forces at 45 and 40 mm FCHs of the T-shaped monorail HTS Maglev system. At 45-mm FCH, the levitation force density at the #1 liquid nitrogen cryostat, i.e. the levitation forces due to the per unit volume PM, could be about $55.9 \times 103\ \text{kg/m}^3$ when the levitation gap was 5 mm. At 20-mm levitation gap, the levitation force density still was $13.9 \times 103\ \text{kg/m}^3$. It

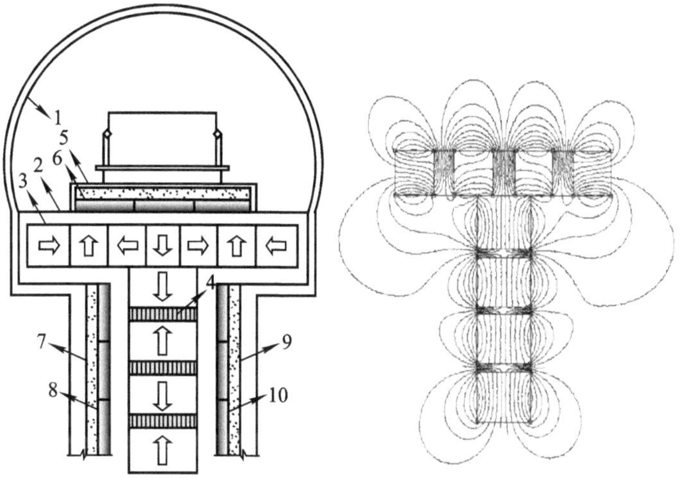


(a) (b)

**Fig. 8.50:** (a) Schematic drawing of the overall HTS Maglev system. (b) Calculated magnetic field profile of the T-shaped PMG. (1) Vehicle body. (2) Bed plate. (3) T-shaped PMG (arrow shows the magnetization direction). (4) Iron. (5) #1 liquid nitrogen cryostat. (6) HTSCs. (7) #2 liquid nitrogen cryostat. (8) HTSCs. (9) #3 liquid nitrogen cryostat. (10) HTSCs.

was exciting that there was still a large levitation force density at higher positions. On the other hand, the calculated values for the guidance force density at different lateral gaps are shown in Fig. 8.52b. It can be seen that at the same FCH, the guidance force density of the system increased greatly with decreasing lateral gap. For example, the guidance force density of the #2 liquid nitrogen cryostat at 5-mm lateral gap was much larger than that at 15-mm lateral gap. This was because, for this system, the guidance forces depended not only on the pinning forces of the HTSCs in the #1 cryostat but also on the lateral restoring forces. The lateral restoring forces came from two parts. One part resulted from the interaction of the left magnetic field of the T-shaped PMG and the induced currents of the HTSCs in the #2 liquid nitrogen cryostat. The other part resulted from the interaction of the right magnetic field of the T-shaped PMG and the induced currents of the HTSCs in the #3 liquid nitrogen cryostat. The sum of the two parts was much larger at smaller lateral gap since the gradient of magnetic field increased greatly with decrease of the lateral gap. Therefore, the system possesses a larger lateral restoring force and better stable levitation for small lateral gaps.

For supplement, Tab. 8.8 compares the calculated levitation force density for two kinds of PMG with different levitation gaps at 45 mm FCH while Tab. 8.9 shows the guidance force density for these two kinds of PMG at different lateral gaps. It is seen at 15-mm levitation gap, the levitation force density of the T-shaped PMG was 23,216 kg/m$^3$, and the present popular PMG was 12,584 kg/m$^3$, which was reduced by 46%. Thus, with almost the same amount of PM, the T-shaped PMG had a larger

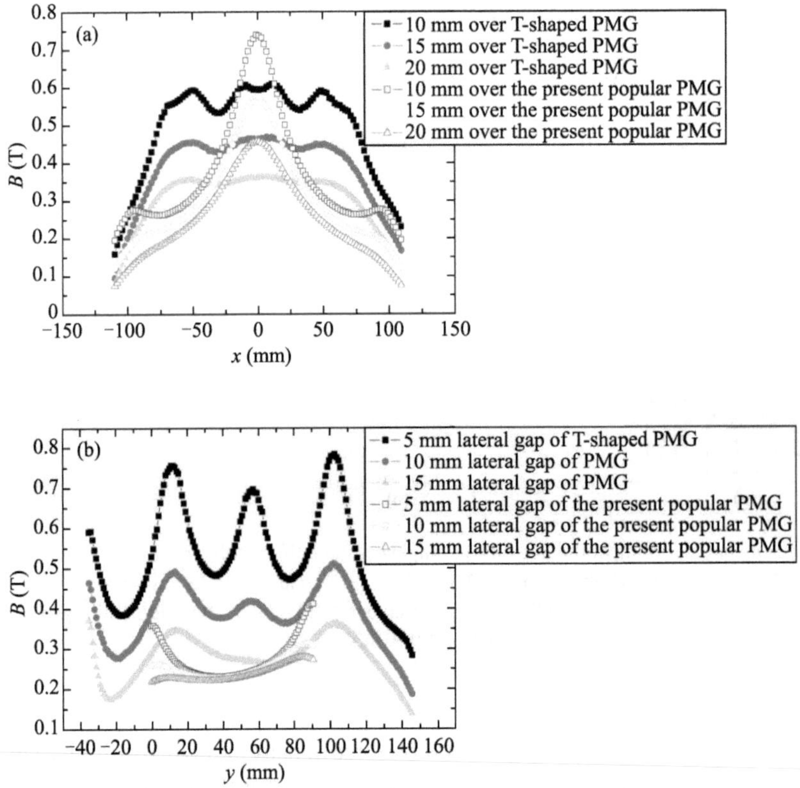


**Fig. 8.51:** Comparison of the calculated magnetic fields above the T-shaped PMG and the present popular PMG. (a) At three different heights: 10, 15, and 20 mm; (b) for different lateral gaps: 15, 10, and 5 mm.

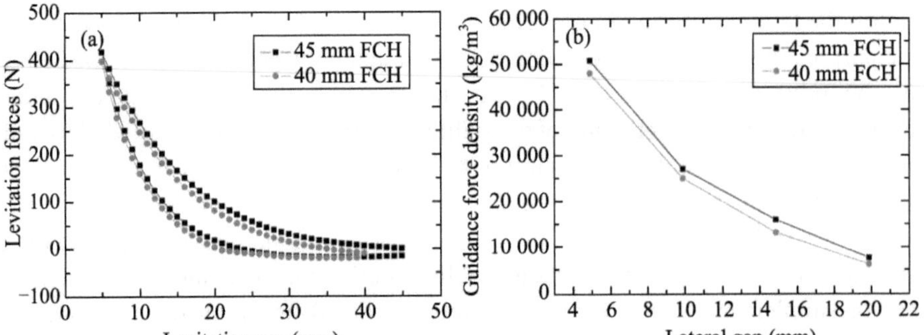


**Fig. 8.52:** (a) Levitation forces vs. gap between the T-shaped PMG and HTSCs. (b) Guidance force density vs. lateral gaps between the T-shaped PMG and HTSCs (FCH: 40 and 45 mm).

**Tab. 8.8:** Calculated results of levitation force density.

| Levitation gap (mm) | 5 | 10 | 15 | 15 |
|---|---|---|---|---|
| The T-shaped PMG($kg/m^3$) | 55,944 | 37,342 | 23,216 | 23,216 |
| The present PMG ($kg/m^3$) | 27,972 | 18,181 | 12,584 | 12,584 |

**Tab. 8.9:** Calculated results of guidance force density.

| Lateral gap (mm) | 5 | 10 | 15 | 15 |
|---|---|---|---|---|
| The T-shaped PMG($kg/m^3$) | 51,748 | 27,972 | 16,783 | 8391 |
| The present PMG ($kg/m^3$) | 5874 | 3776 | 2797 | 2097 |

lift force compared to the present popular PMG. As listed in Tab. 8.9, the guidance force density of the T-shaped PMG was much larger than that of the popular PMG. Especially, at 5-mm lateral gap, the guidance force of T-shaped PMG was about 8.8 times larger than that of the present popular PMG. The reason was mainly that the T-shaped PMG had a strong periodic magnetic field in its upper space and strong symmetric magnetic field in its lateral spaces.

### 8.4.3 An asymmetric HTS Maglev curve design

Considering the attractive symmetric double-pole Halbach PMG (see Fig. 8.44 and Section 8.4.1) for example, the effective magnetic circuit idea of the field potential well was introduced with an asymmetric PMG design which produced the additional guidance improvement especially for the HTS Maglev curve application.

From the view of the conventional considerations of the HTS Maglev system, the double-pole Halbach PMG is regarded as a symmetric magnetic field source with the line of symmetry at $x = 0$ mm, as shown in Fig. 8.53a. This center line, $x = 0$ mm, is also the working line of symmetry of the on-board superconductor and the relative HTS Maglev system. When the working center line is at $x = -25$ mm ($x' = 0$ mm, the peak position of $B_z$), the applied magnetic field is asymmetric from the view of the working center line, but a larger guidance force was obtained with the same 10-mm lateral displacement of the HTS Maglev system. This is due to a effective magnetic circuit design of the multi-pole PMG. Figure 8.53b shows a percentage increase of 84.5% when $x = -25$ mm working as the center line compared with the case of $x = 0$ mm.

Figure 8.54 presents four asymmetric PMG designs and their magnetic field distributions which are evolved from two original optimized symmetric PMGs of PMG I and PMG II. PMG I is the “Century” PMG type and PMG II is the Halbach PMG type. Furthermore, combined with the asymmetric design of the PMG II (b) in Fig. 8.54, the

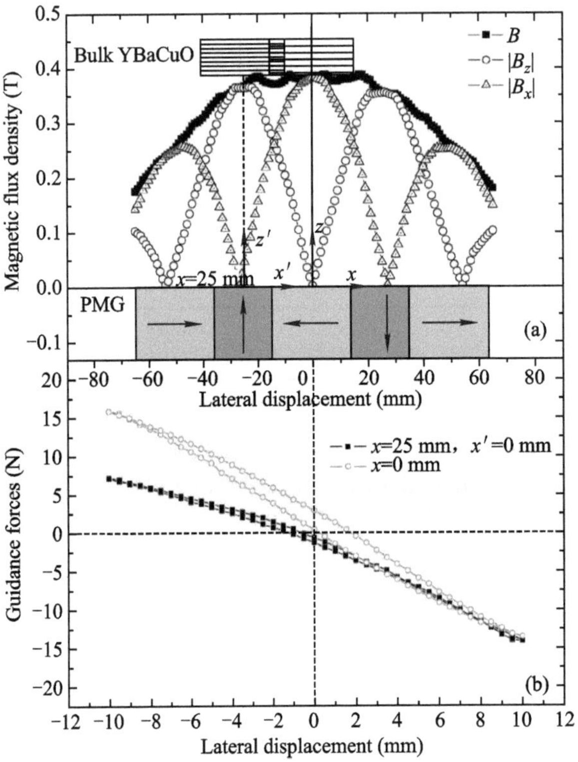


**Fig. 8.53:** (a) Magnetic field distribution ($z = 15$ mm) of the double-pole Halbach PMG. (b) Guidance forces at different working center lines of an on-board superconductor above the Halbach PMG. The superconductor sample was a bulk YBaCuO, 50 mm in diameter and 12 mm in thickness (FCH = 30 mm and WH = 15 mm).

range of $x > 0$ mm is set as the outside PMG part for the future HTS Maglev with a curve a guide in order to produce a faster increase of the guidance forces and better curve negotiation ability to avoid the vehicle's lateral displacement along the curve.

As shown in Fig. 8.54, first from the view of the magnetic field distribution, the additional outside PMs widen the range of the strong horizontal magnetic field density, $B_x$. The guidance forces correspondingly become much bigger so that the lateral eccentricity of the levitated superconductor or vehicle due to a centrifugal effect is greatly inhibited along the curve. Moreover, as shown in the magnetic field distribution profiles of PMG I (b) and PMG II (b), with more PMs assembled to the outside part of the existing PMG, the magnetic circuit becomes a multi-pole structure. The potential-well effect of the vertical magnetic field density, $B_z$, is produced to further inhibit the lateral eccentricity over PMG I (b) or PMG II (b). These important changes of the magnetic circuit of the PMG will result in a larger guidance force and better curve negotiation ability.

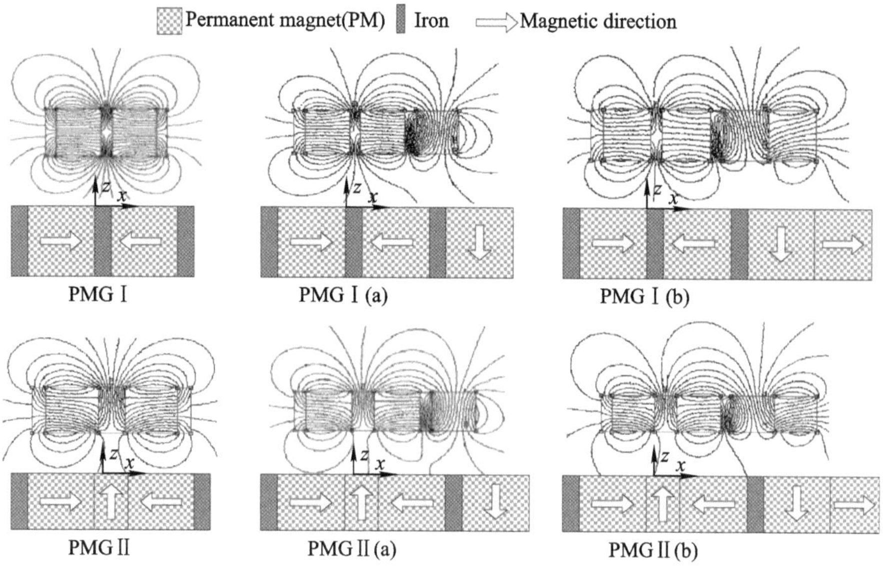


**Fig. 8.54:** PMG I (a), PMG I (b) and PMG II (a) and PMG II (b) are asymmetric PMG designs. PMG I and PMG II represent the original symmetric PMG designs.

**Tab. 8.10:** Guidance force increase of the bulk YBCO above the four asymmetric PMGs compared to the two symmetrical PMGs.

| | PMG I | PMG I (a) | PMG I (b) | PMG II | PMG II (a) | PMG II (b) |
|---|---|---|---|---|---|---|
| Increase magnitude (N) | 0 | 1.5 | 1.5 | 0 | 2.4 | 3.1 |
| Increase percent (%) | 0 | 6.5% | 22.4% | 0 | 10.1% | 12.9% |

To evaluate the improvement effect on the guidance forces by the asymmetrical PMG designs, the finite element method was used to calculate all the magnetic field distributions from those PMGs in Fig. 8.54. The 2D simulation program [54] was used to calculate the guidance forces with the lateral displacement of $x = 0$–$25$ mm above the six PMGs depicted in Fig. 8.54. The calculation parameters used were following: a rectangular bulk YBCO with dimensions 30 mm width, 36 mm length, and 15 mm thickness, FCH = 30 mm, and WH = 15 mm. Table 8.10 shows results from the calculations. Thus, with the assumption of an asymmetric curved PMG, the HTS Maglev vehicle system will allow the vehicle to overcome larger centrifugal forces and run with the small or even zero lateral displacement when traveling on a curved path.

From the viewpoint of the practical engineering assembly, the additional PMs will be directly assembled out-board of the original symmetric PMG of the straight line when the guideways need to offer a larger guidance force or have a curvature. The assembly of the outside magnets will be similar to the original assembly of the original PMG. Thus, it is simple, feasible, low-cost, and widely applicable with any kind of

**Tab. 8.11:** Main parameters of the HTS Maglev vehicle.

| | |
|---|---|
| PMG distance | 8 m diameter |
| Levitation height | 12–15 mm |
| Load capability | 75–50 kg |
| Propulsion mode | Airscrew propulsion |
| Max. acceleration | 1 m/s$^2$ |
| Running speed | Maximum 15 km/h |

PMG. Consequently, this asymmetric PMG design can help to realize the "straight line to curve to straight line" running environment for HTS Maglev vehicle application.

## 8.5 New developments in HTS Maglev vehicle system

### 8.5.1 An 8-m-diameter PM guideway test line

In the practical application, the homogeneity of an infinite PMG paved along its route is a key problem. A homogeneous magnetic field distribution along the PMG's extended direction determines the fundamental frictionless operation of HTS Maglev vehicles. Otherwise, there will be some magnetic drag forces produced during running at any speed. This disturbance phenomenon is not expected and will violate the freely stable levitation advantage of a HTS Maglev vehicle. Hence, before using any PMGs, the homogeneity should be checked carefully in advance. Thus, in 2007, under the support of the National 863 Plan, an 8-m-diameter PMG test line for the HTS Maglev vehicle system was built to validate the assembly feasibility and accuracy of a long-distance PMG. Table 8.11 gives the main parameters of this HTS Maglev vehicle system based on Fig. 8.55.

As shown in Fig. 8.55, the levitation load was measured from 600 to 1034 N since the 10–15-mm WH range is the most important. It is obvious that the practical FCH should be higher than 30 mm; otherwise, the design levitation load target cannot be obtained. It should be noted that the highest central magnetic field density of the PMG was 1.557 T (see Fig. 8.56) by the optimization [55] and careful consideration of the ferromagnetic material as a highly efficient flux collector during the PMG design.

From the viewpoint of the construction procedure for this HTS Maglev vehicle system, Fig. 8.57 shows the schematic drawing of the 8-m-diameter PMG test line and the inset is a photo of two jointed PMG unit segments. The PMG test line was composed of 240 PMG segments whose corresponding central angle was 3°. Among those 240 PMG segments, the production pass percentage was 95.8%. It was necessary that careful attention be paid to the homogeneity of the infinite PMG for HTS Maglev vehicle application.

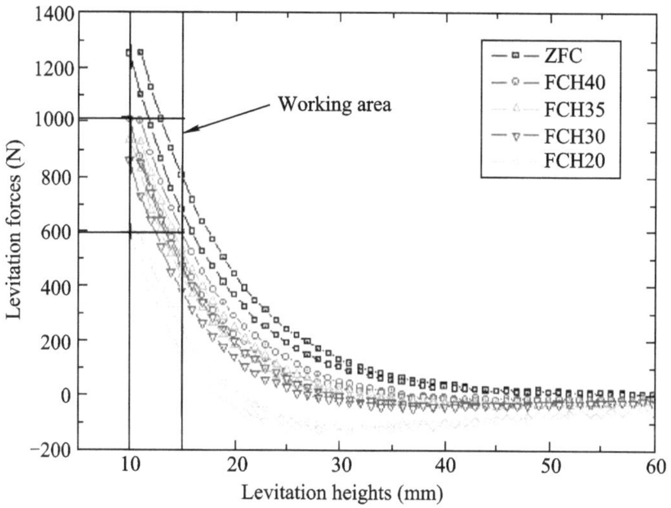


**Fig. 8.55:** Total levitation force curves under ZFC and different FCHs.

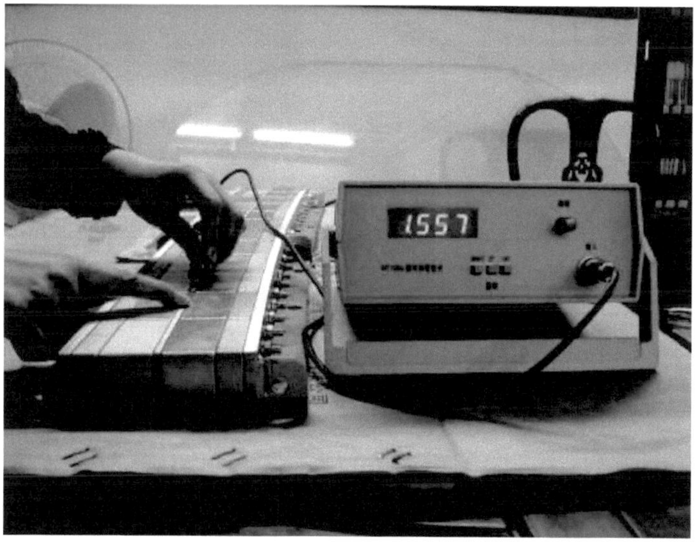


**Fig. 8.56:** Central magnetic field density of the PMG of 1.557 T.

Actually, assembly errors, like a material crack, junction gap, misalignment, etc., occur with any PMG lines whatever its distance is. To guarantee reasonable homogeneity of the whole PMG test line, an authorized patent [56] was proposed and applied to the assembly process. Finally, as shown in Fig. 8.58, a variation error of the magnetic field along the test line at the 10-mm WH of about 5% was obtained and could be accepted in this HTS Maglev vehicle system because the magnetic resistance forces

Fig. 8.57: Schematic drawing of the 8-m-diameter PMG test line. The inset is the photo of two jointed PMG unit segments.

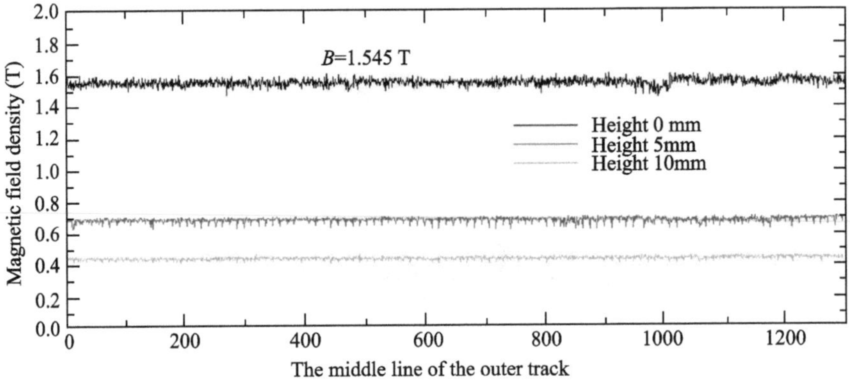


Fig. 8.58: Center magnetic field density curves along the circular PMG at 0-, 5-, and 10-mm heights.

were predicted to be smaller than 1 N. Thus, the average central magnetic field density along the circular 8-m-diameter PMG was still as high as 1.545 T compared with those of all the other PMGs [5–15].

### 8.5.2 A 45-m-long HTS Maglev ring test line "Super-Maglev"

Based on the first man-loading HTS Maglev test vehicle "Century" in 2000 (see Chapter 6), a 45-m-long HTS Maglev ring test line "Super-Maglev" [57] which was planned to run in an evacuated tube, as shown in Fig. 4.13, had been successfully developed in Chengdu, China, in March 2013.

After 12 years of continuous study, the HTS Maglev vehicle "Super-Maglev" (2.2 m length, 1.1 m width) was designed for one passenger with a levitation height of 15–20 mm. Table 8.12 lists the design parameters. It should be noted that one remarkable research target of this HTS Maglev test line was to build the first ETT HTS Maglev system in the world. The maximum running speed was designed as 50 km /h under 0.1 atm without passengers.

From the viewpoint of the vehicle system, "Super-Maglev" integrates a highly efficient bogie (see Fig. 8.59a), a wireless real-time controller (see Fig. 8.59b) of the linear motor propulsion system, and an on-board tablet computer (see Fig. 8.59c) which can monitor, display, and collect real-time all the running parameters of levitation weight, levitation height, running speed, acceleration, lateral offset, and total running distance of the HTS Maglev vehicle. Thus, this second-generation HTS Maglev vehicle system can be expanded easily to a car of a full-scale HTS Maglev vehicle system.

The technology highlight of this second-generation HTS Maglev vehicle system is that it successfully achieved more than 1000 kg levitation load capability at 10 mm WH with only 3000 mm$^2$. cross-sectional area of the PM in an optimized PMG (see Fig. 8.60). As is well known, the previously smallest cross-section area of PM was 5980 mm$^2$ to obtain the same 1000-kg levitation load at the 10-mm WH with the same on-board bulks YBCO arrangement [10]. The 49.8% decrease of the total PM used will not only greatly reduce the construction cost of the HTS Maglev vehicle system, but also lighten the gravitational ground load from the PMG so as to lower the design requirements and thus the infrastructure cost. Therefore, the total cost of a practical HTS Maglev vehicle system will be further reduced even if only limited funding for the longer HTS Maglev running line is available.

**Tab. 8.12:** Main parameters of the HTS Maglev ring test line "Super-Maglev".

| | |
|---|---|
| PMG distance | 45 m combined with 2 × 3.6 m straight lines and 6-m-radius curve lines |
| Levitation height | 15 mm |
| Load capability | 300 kg (one passenger) |
| Propulsion mode | Long stator linear induction motor |
| Max. acceleration | 0.5 m/s$^2$ |
| Running speed | Max. 25 km/h (with one passenger); 50 km/h (non-load) |
| Vacuum degree | 0.1 atm (the diameter of the evacuation tube is 2 m.) |

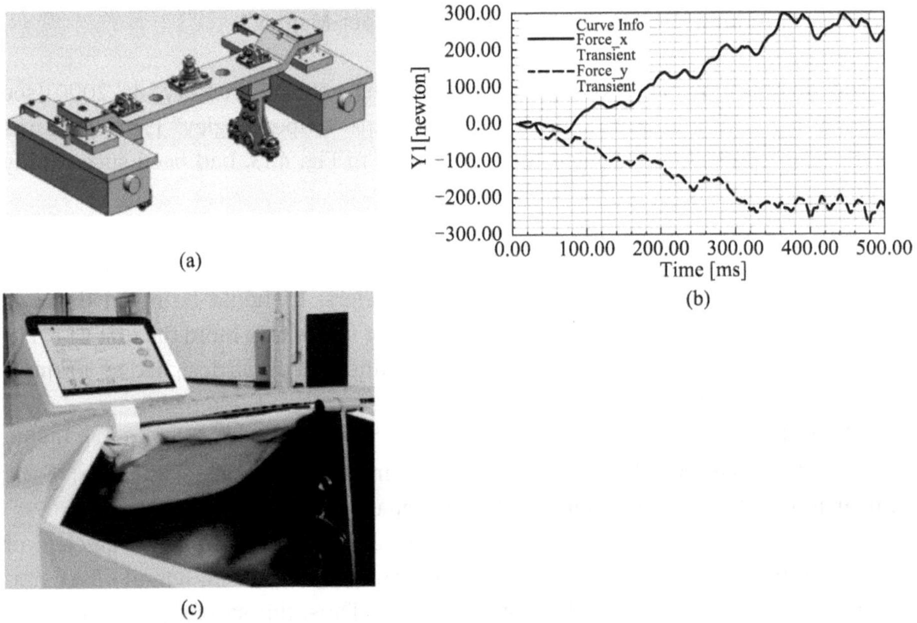


**Fig. 8.59:** Subsystems of the HTS Maglev vehicle system. (a) The Bogie, (b) the linear motor control curve and the wireless controller, (c) the on-board tablet computer with the functions of control, communication, and running data processing.

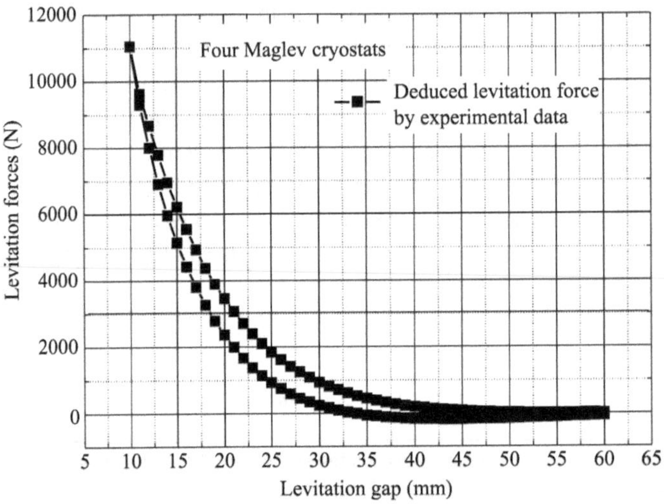


**Fig. 8.60:** Levitation force sum curve of the HTS Maglev vehicle, "Super-Maglev".

Table 8.13 lists the experimental levitation load and lateral guidance capabilities under several FCHs which are of most concern for the real running tests. All the experimental data met the design parameters of the HTS Maglev vehicle "Super-Maglev"

**Tab. 8.13:** Levitation load and lateral guidance capabilities of "Super-Maglev".

| | ZFC | FCH40 mm | FCH30 mm | FCH20 mm |
|---|---|---|---|---|
| Levitation load capability | | | | |
| WH20mm | 325 kg | 225 kg | 162 kg | |
| WH15mm | 612 kg | 473 kg | 444 kg | 237 kg |
| WH10mm | 1013 kg | 848 kg | 779 kg | 622 kg |
| Lateral guidance capability | | | | |
| LD20mm | 128 kg | 246 kg | 352 kg | 555 kg |

thanks to the rational utilization of the guidelines mentioned in Sections 8.3 and 8.4. Up to now, a maximum speed of 25 km/h has been reached with one passenger sitting in it. The maximum 50-km/h target is still being attempted in low air pressure in the vacuum tube now. Further experimental data and information on the running performance will be reported in the near future.

### 8.5.3 ETT HTS Maglev

Unhampered by wheels, Maglev vehicles undoubtedly can run faster than 350 km/h. However, with the speeds above 350 km/h, air friction becomes a serious resistance drag, and the aerodynamic noise also becomes a huge pollution issue. Eqs. (8.9) and (8.10) show the relationships of the air resistance force, $F_{\text{air}}$, the aerodynamic noise, *Noise* ($v$), with running speed, $v$.

$$F_{\text{air}} \propto v^2, \tag{8.9}$$

$$Noise\ (v) \propto \begin{cases} v^7, v < 120\ \text{km/h}, \\ v^8, v > 120\ \text{km/h}. \end{cases} \tag{8.10}$$

If the running speed exceeds 400 km/h, more than 83% of the traction energy will be dissipated against air resistance, and the aerodynamic noise will exceed 90 dB (the corresponding environmental standard is 75 dB for the noise issue) [58]. Therefore, even if the HTS Maglev vehicle has the potential for higher running speed, to reduce air pressure around the vehicle is the only way to achieve speeds from 400 up to 1000 km/h.

Some systems (notably the Swissmetro system [59]) have proposed of rail trains, even Maglev train technology used in evacuated (airless) tubes, which eliminates air drag and allows higher speed. Analogically, evacuation tube transportation (ETT) systems might allow HTS Maglev trains to attain speeds of a new order of magnitude, like a super-high 3000 km/h which could be applied to some military or space launch systems. On January 12, 2004, the project meeting of a super-high speed ETT HTS

(a)

(b)

**Fig. 8.61:** Photos of the ETT HTS Maglev vehicle system. (a) Panorama, (b) "Super-Maglev" running in the evacuation tube.

Maglev train was held at SWJTU in China. Jia-Su Wang presented an ETT HTS Maglev train proposal with a proposed speed of more than 600 km/h [60]. Then in early 2005, design consideration of the super-high-speed HTS Maglev vehicle system in a low-pressure tube (LPT) or ETT was proposed based on the 5-year continuous studies on the first HTS Maglev vehicle [61] (see Sections 6.17, 6.19, 6.20, and Fig. 4.17). Figure 4.17 presents the preliminary structure design of the LPT/ETT HTS Maglev vehicle system.

Up to now, neither a real-scale nor an experimental HTS Maglev vehicle system has been built to operate in an ETT or LPT. In May 2014, an evacuated tube was successfully assembled to cover the HTS Maglev ring line "Super-Maglev", which can now be considered to be a real ETT HTS Maglev vehicle system. As the new transportation concept, "Super-Maglev", as shown in Fig. 8.61, has successfully run in the evacuated tube since May 2013. A speed of 40 km/h was obtained under the low air pressure of near 0.1 atm, which is only 10 km/h below the maximum design speed of 50 km/h without passengers.

The significance of the "Super-Maglev" is that it will be the first ETT HTS Maglev prototype to realize this new transportation concept for future ETT. At this moment, more running tests of the HTS Maglev vehicle prototype in the evacuation tube are being conducted under around 0.1 atm of air pressure. It is believed that more valuable results and achievements will be public in the future.

On the other hand, it also suggests that cars of the future HTS Maglev vehicle should be miniaturized and aim at home users considering its levitation capability and feasible turning radius. Miniaturization, super-high speed, and incorporation with ETT technology may reasonably to be regarded as its dominant potential characteristics during future HTS Maglev development.

## References

[1] Igarashi M, Nemoto K, Kutomi T, Hirano S, Kuwano K, Tominaga T, et al. The project overview of the HTS magnet for superconducting maglev. IEEE Trans Appl Supercond. 2007;17(2): 2111–2116.

[2] Shirakuni N, Terai M, Watanabe K, Takahashi K. The status of development and running tests of superconducting maglev. International Conference Magnetically Levitated System, Dresden, Germany; 2006.

[3] Strategy B. Superconductor Market Research. 2010.

[4] Brandt EH. Levitation in Physics. Science. 1989;243:349–355.

[5] Wang JS, Wang SY, Zeng YW, Huang HY, Luo F, Xu ZP, et al. The first mad-loading high temperature superconducting maglev test vehicle in the world. Physica C. 2002;378–381: 809–814.

[6] Schultz L, De Haas O, Verges P, Beyer C, Rohlig S, Olsen H, et al. Superconductively levitated transport system – the SupraTrans Project. IEEE Trans Appl Supercond. 2005;15(2):2301–2305.

[7] http://www.supratrans.de/en/home/.

[8] Kovalev KL, Koneev SMA, Poltavec VN. Magnetically levitated high-speed carriages on the basis of bulk HTS elements. Proceeding of the 8th International Symposium on Magnetic Suspend Technology (ISMST'8), 2005;51:45–49.

[9] Okano M, Iwamoto T, Furuse M, Fuchino S, Ishii I. Running performance of a pinning-type superconducting magnetic levitation guide. J Phys Conf Ser. 2006;43:999–1002.

[10] Sotelo GG, Dias DHN, Machado OJ, David ED, de Andrade R, Jr., Stephan RM, Costa GC. Experiments in a real scale maglev vehicle prototype. J Phys Conf Ser. 2010;234:032054.

[11] Sotelo GG, Dias DHN, De Oliveira RAH, Ferreira AC, De Andrade R, Stephan RM. Maglev Cobra: test facilities and operational experiments. J Phys Conf Ser. 2014;507:032017.

[12] http://www.maglevcobra.coppe.ufrj.br/.

[13] Okano M, Iwamoto T, Fuchino S, Tamada N. Feasibility of a good transportation system with a superconducting magnetic levitation guide–load characteristics of a magnetic levitation guide using a bulk high-$T_c$ superconductor. Physica C. 2003;386:500–505.

[14] D'Ovidio G, Crisi F, Lanzara G. A "V" shaped superconducting levitation module for lift and guidance of a magnetic transportation system. Physica C. 2008;468:1036–1040.

[15] Yang WJ, Min Q, Liu Y, Wen Z, Duan Y, Chen XD. Levitation characteristics in an HTS maglev launch assist test vehicle. Supercond Sci Technol. 2007;20:281–286.

[16] Zheng J. PhD thesis, Southwest Jiaotong University, Chengdu, China; 2007.

[17] David EG, Stephan RM, Costa GC, Ferreira AC, Andrade R, Jr., Nicolsky R, Nicolsky R, Moreira M. Feasibility study of an HTS-Maglev line at the Federal University of Rio de Janeiro. Proceedings

of the 19th International Conference Magnetically Levitated Systems and Linear Drives (Maglev'2006), Dresden, Germany; 2006.

[18] Sotelo GG, de Oliveira RAH, Costa FS, Dias DHN, de Andrade R, Stephan RM. A full scale superconducting magnetic levitation (MagLev) vehicle operational line. IEEE Trans Appl Supercond. 2015;25(3):3601005.

[19] Komori M, Matsushita T, Fukata S, Tsuruta A, et al. Dynamic characteristics of a High-$T_c$ superconducting bearing with a set of alternating-polarity magnets. J Appl Phys. 1995;77(2):899–904.

[20] Yu JH, Postrekhin E, Ma KB, Chu WK. Vibration isolation for space structures using HTS-Magnet interaction. IEEE Trans Appl Supercond. 1999;9(2):908–910.

[21] Pan ST, Wang SY, Jiang DH, Liu W, Li F. Influence of vertical vibrations on array of bulk HTSC above the permanent magnet guideway. J Supercond Nov Magn. 2010;23:475–480.

[22] Pan ST, Wang SY, Jiang DH, Wang JS. Influence of horizontal vibrations on the lateral stability of bulk high temperature superconductors. IEEE Trans Appl Supercond. 2010;20(3):911–914.

[23] Wang JS, Wang SY, Deng CY, Zeng YW, Zhang LC, Deng ZG, et al. A high-temperature superconducting maglev dynamic measurement system. IEEE Trans Appl Supercond. 2008;18(2):791–794.

[24] Ueda H, Itoh M, Ishiyama A. Trapped field characteristic of HTS bulk in AC external magnetic field. IEEE Trans Appl Supercond. 2003;13(2):2283–2286.

[25] Jing H, Wang JS, Wang SY, Liu W, Deng CY, Jiang M, et al. A two-pole halbach permanent guideway for high temperature superconducting maglev vehicle. Physica C. 2007; 463–465:426–430.

[26] Wang JS, Wang SY, Lin GB, Huang HY, Zhang CF, Zeng YW, et al. High $T_c$ superconducting magnetic levitation measurement system. High technol lett. 2000,10:56–58. (In Chinese).

[27] Xu M, Shi D, Fox RF. Generalized critical-state model for hard superconductors. Physical Review B. 1990;42(16):10773–10776.

[28] Murakami M. Melt processed high temperature superconductors. World Scientific Publisher, Singapore; 1992.

[29] Krabbes G, Fuchs G, Canders W, May H, Palka R. High temperature superconductor bulk materials. Wiley-VCH, Weinheim, Germany; 2006.

[30] Tomita M, Murakami M. High-temperature superconductor bulk magnets that can trap magnetic fields of over 17 tesla at 29K. Nature. 2003;421:517–520.

[31] Song HH, Zheng J, Liu MX, Zhang LC, Lu YY, Huang YG, et al. Optimization and design of the permanent magnet guideway with the high temperature superconductor. IEEE Trans Appl Supercond. 2006;16(2):1023–1026.

[32] Valle ND, Sanchez A, Pardo E, Chen DX, Navau C. Optimizing levitation force and stability in superconducting levitation with translational symmetry. Appl Phys Lett. 2007;90:042503.

[33] Ma KB, Postrekhin YV, Chu WK. Superconductor and magnet levitation devices. Rev Sci Instrum. 2003;74(12):4989–5017.

[34] Kamijo H, Fujimoto H. Repeated pulsed-field magnetization with temperature control in a High-$T_c$ bulk superconductor. IEEE Trans Appl Supercond. 2001;11(1):1816–1819.

[35] Wang SY, Wang JS, Deng CY, Lu YY, Zeng YW, Song HH, et al. An update high-temperature superconducting maglev measurement system. IEEE Trans Appl Supercond. 2007;17(2): 2067–2070.

[36] Yang WM, Zhou L, Feng Y, et al. The effect of different field cooling processes on the levitation force and attractive force of single-domain YBa2Cu3O7-x bulk. Supercond Sci Technol. 2002;15:1410–1414.

[37] Del-Valle N, Sanchez A, Navau C, Chen DX. A theoretical study of the influence of superconductor and magnet dimensions on the levitation force and stability of maglev systems. Supercond Sci Technol. 2008;21:125008.
[38] Hennig W, Parks D, Weinstein R, Sawh RP. Enhanced levitation forces with field cooled $YBa_2Cu_3O_{7-\delta}$. Appl Phys Lett. 1998;72(23):3059–3061.
[39] Vanderbemden P, Hong Z, Coombs TA, et al. Remagnetization of bulk high-temperature superconductors subjected to crossed and rotating magnetic fields. Supercond Sci Technol. 2007;20(9):S174–S183.
[40] Coombs TA, Hong Z, Zhu X, Krabbes G. A novel heat engine for magnetizing superconductors. Supercond Sci Technol. 2008;21:034001.
[41] Yang WM, Zhou L, Nicolsky R. The relationship of levitation force between individual discs and double-layer disc $YBa_2Cu_3O_{7-x}$ superconductors. Supercond Sci Technol. 2003;16(4):451–454.
[42] Rudnev IA, Ermolaev YS. Non-additivity of magnetic levitation force. J Phys Conf Ser. 2006;43:983–986.
[43] Rudnev IA, Ermolaev YS. Effect of the characteristics of a superconductor on the levitation properties of the magnet-superconductor system. J Exp Theor Phys. 2007;105(1):256–258.
[44] Shi DL, Lahiri K, Qu D, Sagar S. Surface nucleation, domain growth mechanisms, and factors dominating superconducting properties in seeded melt grown $YBa_2Cu_3O_x$. J Mater Res Technol. 1997;12(11):3036–3045.
[45] Wang XR, Wang JS, Wang SY, Ren ZY, Song HH, Wang XZ, et al. The relationship of guidance force between single and multiple cylindrical Y–Ba–Cu–O superconductors. Physica C. 2003;390:113–119.
[46] Liu W, Wang SY, Jing H, Zheng J, Jiang M, Wang JS. Levitation performance of YBCO bulk in different applied magnetic fields. Physica C. 2008;468:974–977.
[47] Ren ZY, de Haas O, Wang XR, Song HH, Wang XZ, Jun Z, et al. Study of relationship of levitation force between single and multiple YBCO bulk above a permanent magnetic guideway. China J Low Temp Phys. 2003;25:182–186.
[48] Ueda H, Tsuchiya M, Ishiyama A. Development of pipeline transport system using pinning effect of high-temperature superconducting bulks. IEEE Trans Appl Supercond. 2007;17(2):2323–2326.
[49] Walter H, Bock J, Frohne C, Schippl K, May H, Canders WR, et al. First Heavy Load Bearing for Industrial Application With Shaft Loads Up To 10 kN. J Phys Conf Ser. 2006: 995–998.
[50] Floegel-Delor U, Rothfeld R, Wippich D, Goebel B, Riedel T, Werfel FN. Fabrication of HTS bearings with ton load performance. IEEE Trans Appl Supercond. 2007;17(2):2142–2145.
[51] Hull JR. Topical review: Superconducting bearings. Supercond Sci Technol. 2000;13(2):R1–R15.
[52] Nagashima K, Otani T, Murakami M. Magnetic interaction between permanent magnets and bulk superconductors. Physica C. 1999;328:137–144.
[53] Deng Z, Zheng J, Li J, Ma G, Lu Y, Zhang Y, Wang S, Wang J. Superconducting bulk magnet for maglev vehicle: Stable levitation performance above permanent magnet guideway. Materials Science and Engineering B. 2008;151:117–121.
[54] Wang XR, Ren ZY, Song HH, Wang XZ, Zheng J, Wang SY, et al. Guidance force in an infinitely long superconductor and permanent magnetic guideway system. Supercond Sci Technol. 2005;18(2):S99-S104.
[55] Wang J, Wang S, Zheng J. Recent development of high temperature superconducting maglev system in China. IEEE Trans Appl Supercond. 2009;19(3):2142–2147.
[56] Wang JS, Wang SY, Zheng J, et al. Magnetic rail construction methods of high temperature superconducting maglev rail. Patent application 200710048306. 4, Aug. 24, 2011.
[57] Deng ZG, Zhang WH, Zheng J, et al. A high-temperature superconducting Maglev ring test line developed in Chengdu, China. IEEE Trans Appl Supercond. 2016;26(6):3602408.

[58] Shen Z. On developing high-speed evacuated tube transportation in China. Journal of Southwest Jiaotong University. 2005;40:133–137.

[59] Mossi M, Rossel P. Swissmetro: a revolution in the high-speed passenger transport system. 1st Swiss Transport Research Conference, Monte Ascona, Switzerland; 2001.

[60] Wang JS, Wang SY, Deng CY, Zeng YW, Song HH, Huang HY. A Superhigh speed HTS maglev vehicle. Int J Mod Phys B. 2005;19:399–401.

[61] Wang JS, Wang SY, Deng CY, et al. Design consideration of a high temperature superconductor maglev vehicle system. IEEE Trans Appl Supercond. 2005;15(2):2273–2276.

Zi-Gang Deng, Qun-Xu Lin, Wei Liu, Jia-Su Wang, and Su-Yu Wang

# 9 HTS Maglev bearing and flywheel energy storage system

## 9.1 Introduction

A bearing is a machine element which offers the desired movement or rotation with low friction, and constrains other relative motions [1]. Generally, the bearing is typically classified as mechanical, permanent magnet (PM), gas suspension, electromagnetic, and high-temperature superconducting bearing (HTSB).

The mechanical bearing is one of the most important machine elements, which is widely applied in industry, agriculture, traffic, military, and so on. There are many kinds for particular applications. For example, a high-speed precision ball bearing enables a high rotational speed of 150,000 rpm [2]. A large-load thrust bearing with elastic metal-plastic pads can support the magnitude of 6000 tons [3], a huge-size turnplate bearing for wind power generators has an outer diameter of more than 7.3 m [4]. However, due to the inherent solid-solid or solid-liquid contact between moving parts (rotor and stator), the mechanical bearing has some unavoidable drawbacks, like vibration and noise. Furthermore, mechanical abrasion and related temperature rise also limit the ultimate rotational speed and the service life. At the same time, special lubricants have to be used in almost all applications of mechanical bearings, even including the food processing and pharmaceutical industries which do not welcome any lubricants [5]. At low temperature, lubricants cannot be used in the cryogenic pump system since they will freeze. Thus, the service life of the mechanical bearing is greatly reduced. For the flywheel energy storage system (FESS), rotational loss from mechanical bearings is the main technological obstacle for long time high-efficient energy storage.

To solve the above problems, PM bearings [6], gas suspension Bearings [7], electromagnetic bearings [8], and HTSBs have been developed based on the consideration of physically separating the moving parts. Strong points of these bearings are non-contact, no abrasion, low rotational loss and long service life. PM bearings have the advantages of low cost, simple structure, and low rotational loss. Yet, they cannot achieve complete free and stable levitation in both axial and radial directions, so other auxiliary bearings are combined to use. Gas suspension bearings support the load by a thin layer of gas [9]. They operate at very high speed (of $8\times10^5$ rpm [10]), have a large working temperature range, and are non-polluted [7], but their load capacity and suspension gap are small and rotational loss is higher than that of PM and electromagnetic bearings. Moreover, gas suspension bearings cannot be used in a vacuum environment. Electromagnetic bearings work under the principle of controlling the position of the rotor with an electromagnet assembly [11]. The rotational loss is small

https://doi.org/10.1515/9783110538434-009

and the stiffness is large (50 N/cm$^2$) [12]. The feedback control of electromagnetic bearings is complicated and requires continuous power input, which implies danger in the case of sudden power failures.

HTSBs have the lowest rotational loss among all kinds of bearings (coefficient of friction (COF) is as low as $10^{-7}$) [13], and they can reach very high speed (up to 520,000 rpm) [14]. Compared with mechanical bearings, HTSBs have low noise, long service life (~ 20 years), no abrasion, and no need of lubricant. Unlike PM bearing, HTSBs are stable in all directions. Furthermore, HTSBs do not need any active control system like that necessary in electromagnetic bearings. Even in the case of power failures, HTSBs have enough time to stop the rotor safely before the superconducting state disappears. Recognized as a green technology, HTSB technology has good environmental adaptation and can be used in a vacuum environment, which is impossible for gas-suspending bearings. Table 9.1 shows the performance comparison of HTSB, electromagnetic bearings, and mechanical bearings.

Because of these outstanding characteristics, HTSBs attract much research interest all over the world. Both Moon from Cornell University [15] and Hull from Argonne National Laboratory [13] had conducted comprehensive and systematic fundamental work on the interaction forces between HTSC and PM, relaxation and drift of forces, low-temperature performance, vibration performance, stiffness, damping, dynamic stability, magnetic resistance, rotational loss, long-term operation performance, structure optimization, etc. In 1990, the rotational speed of a HTSB prototype reached 120,000 rpm, and the maximum rotational speed of 520,000 rpm was reported 2 years later [14]. With the development of HTSC material (trapped field, $B_{trap} \approx 3.7$ T at 77 K) and PM material (NdFeB with magnetic energy product of 52 MGOe), the HTSB technology has a gradually mature. The research has entered the competitive phase of the full-scale prototype since the end of the last century. USA, Germany, Japan, and Korea have already developed full-scale prototypes of HTSB for different applications. Among these applications, FESS is one of the most promising applications. HTS Maglev FESS is the perfect technology match of HTSB and FESS because the high rotational speed and low rotational loss of HTSB significantly improve energy density and prolong storage time of FESS, as well as give a longer working life. As mentioned

**Tab. 9.1:** Performance comparison of HTSB, electromagnetic bearing, and mechanical bearing.

| | HTSB | Electromagnetic bearing | Mechanical bearing |
|---|---|---|---|
| COF (minimum) | $10^{-7}$ | $10^{-4}$ | $10^{-3}$ |
| Abrasion | No | No | Yes |
| Control system | No | Yes | No |
| Accessory | Cooling system | Sensor and electronic component | No |
| Speed limit | No | No | Yes |
| Carrying capacity | Low | High | Very high |
| Stiffness | Low | High | Very high |

before, HTSB can work well in vacuum even with a short-time failure of the cooling system. Thus, there is still enough time to stop the flywheel safely before the HTSB failed.

## 9.2 Characteristics of HTSB

The stable HTS levitation comes from the unique magnetic flux pinning property of HTS bulk (see Chapters 1 and 3). It is interesting that the stable levitation can be realized when the PM is located below or beside the bulk HTSC. If the PM is a cylinder with an axial symmetric magnetic field, it can levitate above the HTS bulk and rotate along its axis smoothly regardless of the direction of its axis. This is the basic principle of HTSB.

The smooth rotation of the cylindrical PM is not affected by its initial position or the direction of its axis. Researchers have developed an axial (ASB) and a radial high-temperature superconducting magnetic bearing (RSB). ASB consists of a HTS stator (with disc or ring shaped) and a PM rotor. The PM rotor just levitates above the HTS stator (Fig. 9.1a). Before and after field-cooling activation, there is a gap difference between the original cooling position and levitation position of the PM rotor. The ratio of weight and gap difference of the PM rotor is defined as its average axial stiffness. The RSB consists of a HTS stator (pipe-shaped) and a PM rotor. The PM rotor levitates at the center of the HTS stator (Fig. 9.1b). After field-cooling activation, there is also a height difference of the PM rotor. The definition of its average axial stiffness is the same as for the ASB. When a radial force is exerted, the PM rotor will move along the force to a new equilibrium position. The ratio of radial forces and radial displacement of the PM is defined as the average radial stiffness for both the ASB and the RSB. The PM rotor for both the ASB and the RSB is usually composed of PM rings and iron rings. The PM rings are the source of magnetic field, and their polarity directions are usually opposite, as shown by the arrows in the PM rings in Fig. 9.1. Iron rings are employed to concentrate the magnetic field. The direction of magnetic field in the working gap

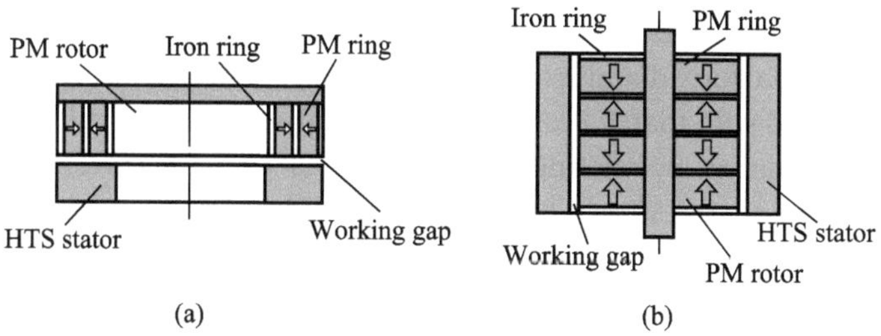


**Fig. 9.1:** Schematic diagrams of two types of HTSB: (a) ASB and (b) RSB. The arrows in the PM rings show the magnetization directions.

between HTS stator and PM rotor is mainly axial for the ASB and radial for the RSB, respectively. This is the basic distinction for these two types of HTSB. The HTS stator is cooled down by cryogenic liquid or a cryocooler. The usual material for the HTS stator is REBaCuO (RE denotes the rare earth elements), such as YBaCuO and GdBaCuO, because REBaCuO has a high magnetic irreversibility field and can be grown as a large-grain bulks [13].

ASB can reach 520,000 rpm [14] due to its simple structure and easy manufacture, but its stiffness is low. The Boeing Company's ASB prototype has a PM rotor with diameter of about 220 mm, and its axial and radial stiffness is 144 and 69 N/mm, respectively [16, 17]. The RSB stands out because of its compact structure, larger carrying capacity, and higher stiffness, which is more suitable for heavy-load applications. In 2007, ATZ Company in Germany manufactured a RSB achieving an axial force of 10.08 kN and stiffness of 4.5 kN/mm [18]. The diameter and length of the PM ring are 200 and 120 mm, respectively. Nexans SuperConductors GmbH in Germany developed a RSB with a radial force of 6.9 kN and radial stiffness of 5.1 kN/mm [19]. The inner cylindrical surface of the HTS stator is 325 mm in diameter and 305 mm in length. During the actual design process, RSB can be classified into two types: RSB with inner PM rotor (as shown in Fig. 9.1b) and RSB with outer PM rotor. A RSB with an inner PM rotor has been chosen by ATZ and Nexans SuperConductors GmbH. Compared to the outer PM rotor structure, a RSB with an inner PM rotor has advantages, such as easy location of the PM rotor during field cooling, simple design of touch-down bearing, high homogeneity of rotor fields, low rotational loss, and easy assembly.

The good operation performance of the HTSB is fundamental to engineering applications, so researchers have carefully investigated the primary characteristics like stiffness, damping, rotor dynamics, stabilization, force creep, temperature dependence, and rotational loss. The stiffness, rotational loss, and stability are particularly important for the practical application of HTSB.

1. Stiffness. Stiffness means the resistance ability to exterior force and torque for a HTSB, which can be improved by lowering the working temperature [18] to improve the material performance of HTS bulks and PMs. For example, ATZ found that the radial stiffness of the RSB increases from 1.4 kN/mm at 79 K to 1.8 kN/mm at 72 K and that the axial stiffness increases from 3 kN/mm at 78.5 K to 4.5 kN/mm at 72 K. Moreover, the axial force doubles to 10,080 N with an axial displacement of 3.3 mm at 72 K compared with that at 82.3 K [18]. On the other hand, the stiffness also changes during rotation. By rotating the HTSB to 1500 rpm, the corresponding force decreases by almost 50% at 80 K. This is considered to be correlated to the dependence of the critical current density on the temperature of HTS bulks [20].

Cross stiffness is an interesting parameter for HTSB. This is just the influence of axial displacement on radial stiffness or the influence of radial displacement on axial stiffness. Experimental results demonstrated that a radial displacement of 1.2 mm leads to a decrease of the axial force by 25%, whereas an axial displacement of 2 mm causes an increase of radial force by 20% [18].

As for the stiffness calculations, Takashi Hikihara introduced a deflection angle stiffness equation for ASB, but the dimension parameters were simplified [21]. Korea Electric Power Research Institute (KEPRI, Korea) developed a calculation method for radial stiffness with the error of 12.14% [22].

2. Rotational loss. Masaie assumed that rotation loss comes from the Lorentz force between the irregular magnetization of HTS bulks and the current induced by the PM rotor [23–26]. Researchers of International Superconductivity Technology Center (ISTEC, Japan) pointed out that rotation loss can be reduced by improving the magnetic field distribution of the HTS stator and adding insulators into the multi-layer PM rotor [27]. Rotational loss is also related to other factors. Boeing Company explored the rotational loss of ASB of a 5-kWh FESS with a PM rotor accelerated to 14,500 rpm [28]. They found that the rotation loss decreases with decreasing temperature and the loss with conduction cooling is not significantly higher than that with direct liquid nitrogen cooling. They also measured the loss of a 15.4-mm-diameter, 150.3-mm-long rotor with two ASB stators at a maximum speed of 100,860 rpm [29]. It was found that the loss increased to the maximum value at about 16,440 rpm when the speed was higher than 4800 rpm.
3. Stability. Stability is the guarantee of the normal and reliable operation for a HTSB. In ISTEC, the decay of the levitation forces due to flux creep was reported to be suppressed by pre-loading and super-cooling methods, and electromagnetic bearings were applied to increase the rotational stability [27]. The axial displacement of the rotor was less than 0.06 mm, and the radial amplitude was less than 0.03 mm in normal operation and as small as 0.15 mm at the critical rotational speed [27]. ATZ developed an optimal design for the dynamic structure of rotor (stiffer) and added two dynamic damping systems at both ends of the rotor shaft to increase the stability [30]. KEPRI found that the thermal insulating bolts can increase the damping of RSB and discussed two damping mechanisms in details [31]. The damping coefficient of RSB as optimized by KEPRI was 12 times larger than the estimated value [32]. Boeing found that inserting a secondary damper between the HTS stator and the foundation ground increased the stability [33].

In Sections 9.2.1–9.2.3, the basic stiffness characteristics of HTSB are significantly discussed.

### 9.2.1 Axial stiffness characteristics of HTSB

The first-generation RSB prototype in our group was designed and developed as shown in Fig. 9.2. It is made up of a PM rotor with 52 mm in diameter and 66 mm in length (Fig. 9.3) and a HTS stator with six HTS bulks. Each HTS bulk had the dimension

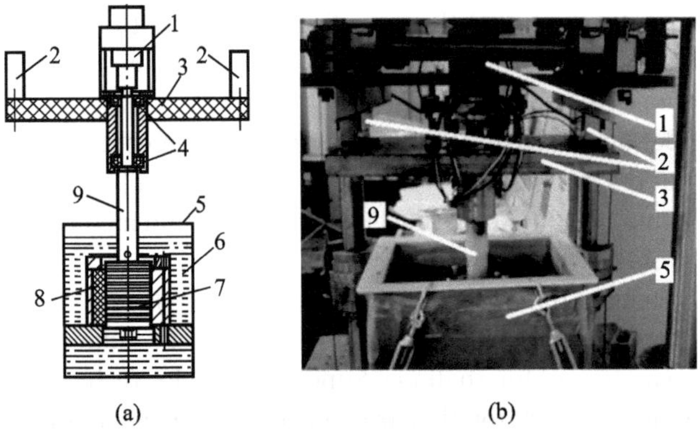


**Fig. 9.2:** Axial force measurement device for RSB prototype: (a) schematic view and (b) photograph. (1) DC motor. (2) Axial force sensor of the SCML-02 measurement system. (3) Epoxy resin board. (4) Mechanical bearing. (5) Dewar. (6) Liquid nitrogen. (7) PM rotor. (8) HTS stator. (9) Spindle.

of 13 × 32 × 64 mm. The working gap of HTSB was 1.5 mm. The PM rotor was composed of PM rings and iron rings with five magnetic poles (Fig. 9.3a). The magnetic field of the central pole at the surface of the PM rotor was 1.34 T (Fig. 9.3b) and the magnetic field decreases with the increasing distance away from the PM rotor. At the position of 11 mm away from the PM rotor, magnetic field was less than 0.2 T. This RSB was manufactured to study the axial stiffness at different rotational speeds. As shown in Fig. 9.2b, the axial force measurement device included a RSB, a Dewar, a DC motor, a DC power supply, a tachometer and a transmission mechanism. This measurement device was installed in the HTS Maglev measurement system SCML-02 [34] (see Section 5.5). The Dewar with a HTS stator inside was fixed on an *X-Y* motion stage, and the

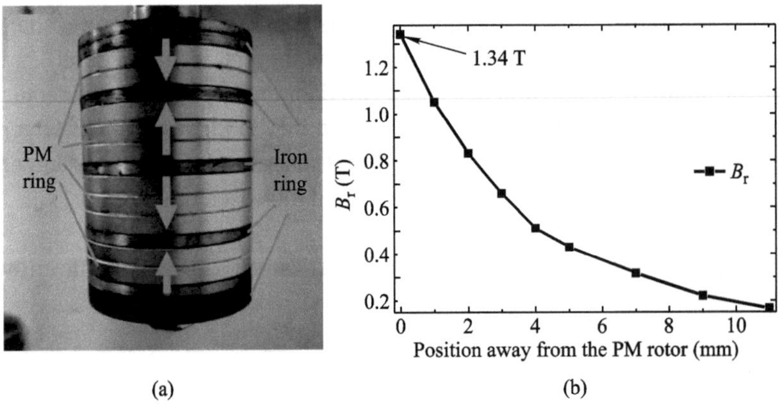


**Fig. 9.3:** PM rotor: (a) photograph of PM rotor, the arrows in the PM rings show the magnetization directions Schematically, and (b) magnetic field at the center pole of the PM rotor at different distances from the PM rotor.

PM rotor was fixed to the axial force sensors through a transmission mechanism and an epoxy resin board. In this case, the HTS stator could move horizontally and PM rotor could move axially with a precision of 0.05 mm. Force sensors could measure the axial forces with a precision of 0.02 N. The PM rotor was located at the center of the HTS stator with liquid nitrogen for cooling. After field cooling, the DC power was switched on to drive the PM rotor, through a 280-mm-long stainless steel spindle. The rotational speed was measured by a non-contact tachometer and is controlled by voltage of DC power. The long spindle was used for reducing heat loss from the motor to liquid nitrogen and eliminating influence of motor field on the PM rotor. When the PM rotor moves axially, the axial forces of the RSB at different rotational speeds were measured by SCML-02.

As is well-known, the experimental results of axial stiffness of a HTSB at different rotational speeds are helpful for practical applications. Figure 9.4 shows these results of the axial forces. The maximum axial force was 178.7 N (at 0 rpm) at an axial displacement of 1.05 mm, and the relevant axial stiffness was 170.2 N/mm. When speeds of the PM rotor were 645, 1827, and 3801 rpm, the maximum axial forces were 174.5, 176.2, and 164.4 N with an axial stiffness of 166.2, 167.8, and 156.6 N/mm, respectively. The maximum decrease of axial force was 14.3 N from 178.7 N (0 rpm) to 164.4 N (3801 rpm), and the maximum decrease rate of axial stiffness was about 8%. This result indicated that the rotational performance of this RSB was good compared with former reports in Ref. 19. It was found that the radial force was reduced by about 50% at 1500 rpm and 80 K, which was thought to be correlated to the critical current density of HTS bulk and its temperature dependence [20].

The large decline of force and stiffness influences the normal operation of HTSB. Siemens found that the radial force of the RSB was reduced from 3700 N (at 0 rpm)

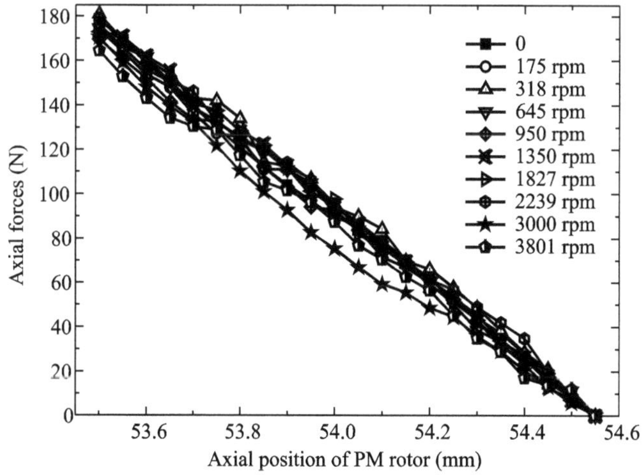


**Fig. 9.4:** Axial forces of RSB at different speeds.

to 2700 N (1500 rpm) when the temperature of the HTS bulk was increased from 68 to 72 K [35]. Synthesizing the reported research results, it is clear that the inhomogeneity of the PM rotor field causes loss and heat inside the HTS bulks. The heat increases the temperature and decreases the critical current density within HTS bulks. This will reduce the stiffness of the RSB. In the RSBs of ATZ and Siemens, HTS bulks are cooled by copper conduction cooling. The cooling efficiency is lower than the case of direct liquid nitrogen cooling shown in Fig. 9.3. HTS bulks of Fig. 9.3 are immersed in liquid nitrogen, then heat inside the bulks can be easily dissipated. Finally, the decrease of the axial stiffness of RSB is only 8%.

It has been proved that the liquid nitrogen cooling method has better cooling efficiency, which assures the stiffness stability of the RSB at different speeds. On the other hand, although the copper cooling method has a lower cooling efficiency, introduction of a cryocooler to the copper cooling method will make the cooling operation convenient because no Dewar cover is needed between PM rotor and HTS stator (while the liquid nitrogen cooling method needs a Dewar which covers the whole HTS stator).

Overall, the temperature rise problem can be solved by the following methods: reducing the working temperature of HTS bulk, optimizing the cooling system, increasing cooling efficiency, improving the manufacture technique to increase the homogeneity of the PM rotor, etc. All these methods can reduce the influence of rotation of the PM rotor on the stiffness of RSB. In practice, the selection criterion should be based on cost, technical requirements, and difficulty of liquid nitrogen cooling method vs. the copper conduction cooling method.

### 9.2.2 Calculation method of radial stiffness

Radial stiffness of the RSB is usually obtained by experiment or calculation (Q. X. Lin, Z. G. Deng, D. H. Jiang, G. T. Ma, J. Zheng, J. S. Wang, and S. Y. Wang, Calculation method for radial stiffness and deflection angle stiffness of radial high temperature superconducting magnetic bearing, unpublished data). The experimental result can provide a precise value, but it costs much time and effort. The calculation is simple, but its precision and effectiveness are sensitive to the selected parameters.

KEPRI developed a method to obtain the radial stiffness of a RSB. It applies a calculation equation based on the experimental test to obtain the radial stiffness. The error of this testing and calculation method was reported to be 12.14% [22]. This calculation method may be able to be improved for higher precision.

Considering that the influence between HTS bulks is small [22], the radial stiffness of a RSB can be equal to the sum of a radial stiffness between the PM rotor and each HTS bulk. In Fig. 9.5a, the center circle represents a PM rotor, and the arrow represents the radial movement direction of the PM rotor. The surrounded tile shaped objects are HTS bulks in different positions. When the PM rotor moves, the distance between

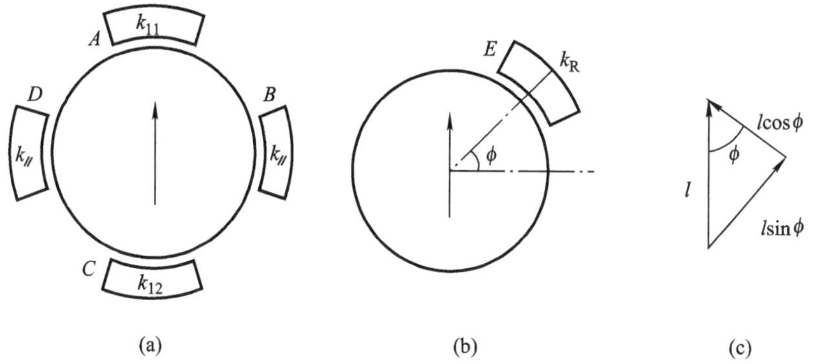


**Fig. 9.5:** Schematic diagrams of the interaction between the PM rotor and HTS bulks. (a) The PM rotor and four HTS bulks. (b) The HTS bulk with a position angle of $\phi$. (c) Decomposition of radial displacement of the PM rotor in (b).

HTS bulk $A$ and the PM rotor is reduced and their mutual radial stiffness is marked as $k_{\perp 1}$. For the HTS bulk $C$, the distance is increased and the radial stiffness is marked as $k_{\perp 2}$. For HTS bulks $B$ or $D$, the distance to the PM rotor almost doesn't change and the radial stiffness is marked as $k_{//}$. In Fig. 9.5b, the position angle of a HTS bulk $E$ is $\phi$. As shown in Fig. 9.5c, the radial displacement of PM rotor, $l$, can be decomposed into a vertical component and a parallel component relative to the HTS bulk $E$: $l \sin \phi$ and $l \cos \phi$. Thus, the radial stiffness between the HTS bulk $E$ and PM rotor is

$$k_{R,\phi} = k_{//} \cos^2 \phi + k_{\perp 1} \sin^2 \phi \quad (\text{if } 0 \leqslant \phi < \pi). \tag{9.1}$$

If the position angle $\phi$ is larger than $\pi$, the radial stiffness is

$$k_{R,\phi} = k_{//} \cos^2 \phi + k_{\perp 2} \sin^2 \phi \quad (\text{if } \pi \leqslant \phi < 2\pi). \tag{9.2}$$

The calculation precision of Eqs. (9.1) and (9.2) are discussed basing on the experimental data from Ref. 21. The HTS stator is made up of 8 HTS bulks $A \sim H$. The position angle of the HTS bulk $A$ is $\frac{\pi}{2}$ and its radial stiffness $k_{\perp 1}$ is 32.5 N/mm. The position angle of the HTS bulk $E$ is $\frac{3\pi}{2}$ and its radial stiffness $k_{\perp 2}$ is also 32.5 N/mm. For the HTS bulk $C$ or $G$, the radial stiffness $k_{//}$ is 3.5 N/mm. The position angle of the HTS bulk $B$ is $\frac{\pi}{4}$ and its radial stiffness can be calculated by Eq. (9.1):

$$k_{R,\phi} = k_{//} \cos^2 \phi + k_{\perp 1} \sin^2 \phi = 3.5 \times \cos^2 \frac{\pi}{4} + 32.5 \times \sin^2 \frac{2}{4} = 18(\mathrm{N} \cdot \mathrm{m}). \tag{9.3}$$

The practical value for HTS bulk $B$ is 17.5 N/mm, while the calculated result of the former method is 23 N/mm [22]. The calculated result of 18 N/mm obtained with Eqs. (9.1) and (9.2) indicate that the calculation precision has been improved.

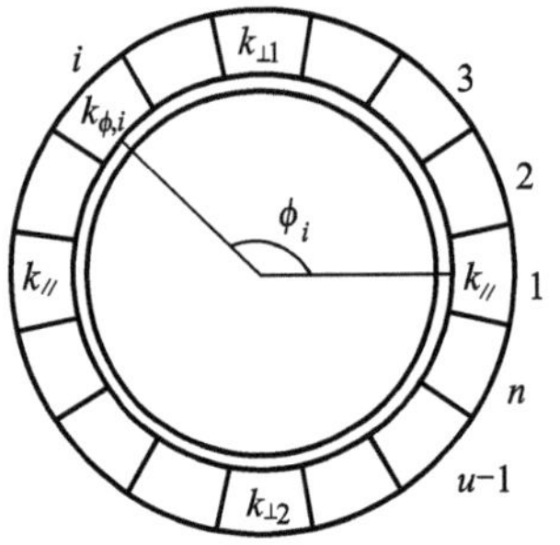


**Fig. 9.6:** Schematic diagram of PM rotor and HTS stator with $n$ pieces of HTS bulks.

If the number of the HTS bulks in the stator is $n$ (Fig. 9.6), each bulk is marked with number 1 to $n$. The position angle of No. 1 HTS bulk is 0 and the position angle of No. $i$ $(1 \leqslant i \leqslant n)$ HTS bulk is

$$\phi_i = \frac{2\pi(i-1)}{n}. \tag{9.4}$$

The radial stiffness of No. $i$ HTS bulk is

$$k_{R,\phi_i} = k_{//} \cos^2 \phi_i + k_{\perp 1} \sin^2 \phi_i \quad (\text{if } 0 \leqslant \phi_i < \pi) \tag{9.5}$$

or

$$k_{R,\phi_i} = k_{//} \cos^2 \phi_i + k_{\perp 2} \sin^2 \phi_i \quad (\text{if } \pi \leqslant \phi_i < 2\pi). \tag{9.6}$$

The radial stiffness of RSB can be calculated as

$$k_R = \sum_1^n k_{R,\phi_i}. \tag{9.7}$$

Because the center angle of each HTS bulk is $2\pi/n$, so the angle density of $k_{//}$, $k_{\perp 1}$ and $k_{\perp 2}$ is $nk_{//}/2\pi$, $nk_{\perp 1}/2\pi$, and $nk_{\perp 2}/2\pi$. Based on Eqs. (9.5), (9.6), and (9.7), the radial stiffness of a RSB can be calculated as

$$\begin{aligned} k_R &= \int_0^{\pi} \left( \frac{nk_{//}}{2\pi} \cos^2\theta + \frac{nk_{\perp 1}}{2\pi} \sin^2\theta \right) \mathrm{d}\theta + \int_{\pi}^{2\pi} \left( \frac{nk_{//}}{2\pi} \cos^2\theta + \frac{nk_{\perp 2}}{2\pi} \sin^2\theta \right) \mathrm{d}\theta \\ &= \frac{n(k_{\perp 1} + k_{\perp 2})}{4} + \frac{nk_{//}}{2}. \end{aligned} \tag{9.8}$$

With the data in Ref. 21, the values of $n$, $k_{//}$, $k_{\perp 1}$, and $k_{\perp 2}$ are 8, 3.5 N/mm, 32.5 N/mm, and 32.5 N/mm, respectively. The radial stiffness calculated with Eq. (9.8) is

$$k_R = \frac{8 \times (32.5 + 32.5)}{4} + \frac{8 \times 3.5}{2} = 144(\text{N/mm}). \tag{9.9}$$

In comparison, the experimental value of radial stiffness of RSB is 140 N/mm while the former calculated value is 157 N/mm (with an error of 12.14%) [22]. Thus, the new calculation method gives an error of only 2.86% [Eq. (9.8)].

### 9.2.3 Deflection angle stiffness

Deflection angle stiffness represents the capacity to resist an external torque, which is an important parameter for dynamical operation of a HTSB.

When a torque is exerted on a PM rotor as shown in Fig. 9.7, the axis of the PM rotor will deflect along the horizontal line which passes through the center point $A$. If the torque is $M$, the deflection angle is $\theta$. The deflection angle stiffness is expressed as

$$k_\theta = \frac{M}{\theta}. \tag{9.10}$$

Because the effective gap between stator and rotor of RSB is small (usually 1–3 mm), the deflection angle is small too. In the calculation, only the radial displacement of each part of the PM is involved, and the axis of each part is considered to be vertical. In Fig. 9.7, the axial distance between points $B$ and $A$ is $l$, and the radial displacement of $B$ is

$$D_\mathrm{r} = \sin\theta l \approx \theta l. \tag{9.11}$$

When $l > 0$, point $B$ is located above point $A$, and its radial displacement is to the right. When $l < 0$, point $B$ is located below point $A$, and its radial displacement is to the left. At point $B$, there is a fraction of the PM rotor whose length is $\mathrm{d}l$, with a radial force:

$$\mathrm{d}F_R = \mathrm{d}l \times \frac{k_R}{L} \times D_\mathrm{r} = \frac{\theta l k_R}{L}\mathrm{d}l, \tag{9.12}$$

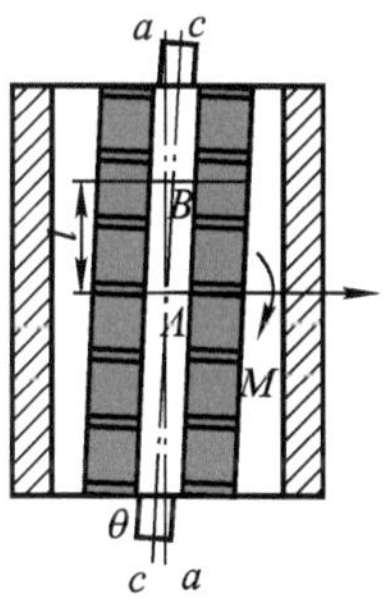


**Fig. 9.7:** Schematic force diagram of the PM rotor when a torque is exerted on.

where $L$ is the length of PM rotor and $k_R$ is the radial stiffness of RSB. If $l > 0$, the direction of $dF_R$ is left. If $l < 0$, the direction of $dF_R$ is right. The corresponding moment of $dF_R$ relative to the horizontal line passing through point $A$ is

$$dM = dF_R \times l = \frac{\theta l^2 k_R}{L} dl. \tag{9.13}$$

Based on the principle of balance of torque, the following equation can be obtained:

$$M = \int_{-0.5L}^{0.5L} dM = \int_{-0.5L}^{0.5L} \frac{\theta l^2 k_R}{L} dl = \frac{\theta k_R L^2}{12}. \tag{9.14}$$

Thus, the angular deflection stiffness is

$$k_\theta = \frac{M}{\theta} = \frac{k_R L^2}{12}. \tag{9.15}$$

A new parameter, the linear density of radial stiffness, $\varepsilon$, is introduced. This is the radial stiffness per unit length of the PM rotor:

$$\varepsilon = \frac{k_R}{L}. \tag{9.16}$$

When all the material characteristics of the HTS bulk and the PM ring, the dimensional parameters of the HTSB (except the length of PM rotor and HTS stator) and the relative position between the HTS bulk and the PM ring, are the same, then $\varepsilon$ is a constant value. From Eqs. (9.15) and (9.16), it is

$$k_\theta = \frac{k_R}{L}\frac{L^3}{12} = \frac{\varepsilon L^3}{12}. \tag{9.17}$$

It is obvious that deflection angular stiffness is proportional to the cube of the length of the PM rotor.

## 9.3 Application of HTSB

It is noticed that FESS is one of the most important and widespread applications of HTSB (also see Section 9.4.2). Some new applications of HTSB are also significant.

### 9.3.1 Typical applications

1. Rotor support for machine [19]: Utilization of a HTSB for rotor support can reduce rotational loss and increase total efficiency of a motor/generator, with no need to

worry about wear and generated heat present with conventional bearings. A RSB was designed to support the 1 ton shaft of a 4-MW HTS generator of Siemens. This application was claimed as the first industrial application of HTSB [19]. The HTS stator was composed of 270 YBCO bulks which were arranged in nine rings and was subdivided into an upper half shell and a lower half shell. Its inside diameter was 325 mm and length was 305 mm. PM rotor had an outer diameter of 319 mm, so the working gap of RSB was 3 mm. During activation, the upper half shell of the HTS stator was fixed and the lower half shell moved upward. By this method, radial force of the RSB reached 6900 N at 28 K and 5000 N at 65 K, and the radial stiffness reached 5.1 kN/mm at 28 K. Its maximum speed was 4400 rpm and the shaft oscillation was less than 0.01 mm at 3600 rpm.

2. Laser beam deflection polygon scanner [36]: A light beam scanner can be used to deflect a laser not beam in xerographic printers or laser display units. Fluid and gas suspension bearing can not be accelerated to above 120,000 rpm for a 60-mm diameter rotor due to friction, so a RSB has chosen for the scanner. The main advantages of introduction of the RSB are its high speed with good optical resolution, low power consumption due to elimination of wear and heat, no need for electronic control and safety during power failure. In the scanner, a 0.2-kg rotor was suspended by two RSBs with 36-mm-outer diameter, 20-mm-inter-diameter, 15-mm-thick YBCO rings and 18-mm-outer diameter, 10-mm-inter-diameter, 5-mm-thick PM rings. The maximum speed of the rotor was 174,000 rpm. When the magnetic inhomogeneity of the PM ring was less than 2%, rotational loss per rotation is $3.8 \times 10^{-4}$ J.
3. Centrifuge [37]: A centrifuge can separate different materials quickly by the centrifugal force. It has been used in chemical engineering, oil industry, food processing, pharmaceuticals, etc. In a centrifuge, the motor generates torque and transfers it to the centrifuge rotor through shafts, couplers and spindles. This structure often causes breakdowns and a whirling up of the sediments. RSB can solve these problems, because it can handle much higher imbalance mass compared to conventional bearings. During resonance, an extreme eccentricity of 2 mm is safely accommodated by introduction of the RSB. A centrifuge with 1 kg titanium rotor levitated by two RSBs with inner PM rotors was developed [37]. PM rings were selected with a magnetic inhomogeneity of the PM rings less than 2%, and the magnetic field was 0.5–0.8 T in the gap center. HTS stators were cooled by a 15-W/80-K Stirling cryocooler, and the working temperature of the upper and lower RSB was typically 68 and 65 K, respectively. A 250-W motor between two RSBs drove the rotor to 30,000 rpm. A radial emergency bearing was placed under the upper RSB and an axial emergency bearing was placed in the bottom of the rotor.
4. Liquid hydrogen tank [38–40]: Hydrogen can provide environmentally friendly energy. It is usually stored as liquid hydrogen in a tank. Heat loss through convection and radiation is reduced by vacuum and radiation shields between

inner tank and outer tank. While support between the inner and outer tanks is a problem, a HTSB can be used in this case to reduce the conductive heat. A RSB is applicable, because the exterior tank is more compact than the planar design of a superconducting bearing. When two RSBs with inner PM rotors were designed to support the inner tank of a liquid hydrogen tank, the heat leakage was reduced from 2.7 W with a conventional support to 0.3 W with a RSB. The PM rotors were fixed to the outer tank. Outer HTS stators with YBCO rings and BSCCO rings were fixed to the inner tank and cooled by 20 K liquid hydrogen. The advantage of this design was that there was no need for extra cooling device.

5. Non-contact mixer [41–43]: Most pharmaceutical solutions on an industrial scale require highly controlled, thorough mixing to achieve a satisfactory yield and a uniform distribution of ingredients in the final product. Agitator tanks are frequently used to complete the mixing process, and a large proportion of conventional mixing technologies are normally achieved by using a mechanical stirrer. However, a significant shortcoming of such an arrangement is the danger of contamination or leakage during mixing, and the contamination is unavoidable. For superconducting mixers, the proposal for stirring fluids under sterile conditions is to use a magnetic coupling with a combination of PMs and HTS bulks. At the top of the agitator vessel, a rotating PM bar is levitated by magnetic coupling between PMs and HTS bulks coupled with flux pinning. It is rotated by a driving motor positioned external to the vessel. In the agitator vessel, another rotating PM bar attached to the mixer wing is suspended by magnetic coupling between PMs and HTS bulks based on similar principles. The two magnetic bars are coupled by HTS bulks. At the bottom of the agitator vessel, a radial PM bearing is applied for the compensation of lateral stiffness and to enhance the stability of the mixer wing. The use of such an externally driven magnet bar avoids the need for a dynamic bearing in the vessel to transfer the rotational forces from the driving magnet to the stirring magnet. Therefore, a completely enclosed system is provided which is able to prevent leakage and possible contamination from mechanical bearings. The cleanup and sterilization are also easy. Experimentally, this superconducting mixer has provided a rotational force strong enough to stir a 30-L solution [43].

### 9.3.2 Liquid nitrogen pump

In this section, a liquid nitrogen pump and a related prototype are introduced as interesting cryogenic applications beyond the above-mentioned hot applications of HTSB (Q. X. Lin, D. H. Jiang, Z. G. Deng, X. Chen, H. L. Jing, Y. Y. Xu, R. B. Wang, G. T. Ma, J. S. Wang, J. Zheng, and S. Y. Wang. Operation of liquid nitrogen pump with radial high temperature superconducting magnetic bearing, unpublished data) [44].

As one kind of cryogenic pump, a liquid helium circulation pump can be used for cooling of the superconducting magnets in a Tokamak reactor. First, the cryogenic pump is employed to speed up the liquid flow rate and enhance the cooling efficiency. Second, this pump can improve stability of superconducting magnets and increase the pressure at the inlet to ensure the supercritical helium at the outlet. Moreover, with this pump the flow rate of liquid helium can be controlled conveniently.

At present, the cryogenic pump usually works with cryogenic mechanical bearings. However, lubricating oil cannot be used in the cryogenic mechanical bearing because it freezes and would also pollute to the cryogenic liquid. With these limitations, the cryogenic mechanical bearing has a short service life. In addition, the cryogenic pump has to be shut down in order to replace the cryogenic mechanical bearing. To solve this problem, RSB may replace the conventional cryogenic mechanical bearing. The obvious advantages of this RSB application include no abrasion to affect the working life and no need for an extra cooling device for the HTS stator of RSB.

#### 9.3.2.1 Liquid nitrogen pump with HTSB

Structure of a liquid nitrogen pump with a RSB is shown in Fig. 9.8a. The pump is laid vertically and contains six components: a motor, a PM coupling, a RSB, a pump body, a blade wheel, and a radial PM bearing.

The motor is a DC motor with rated voltage of 24 V and rated speed of 1200 rpm. The PM coupling is made up of upper and lower PMs. Driving torque is transmitted

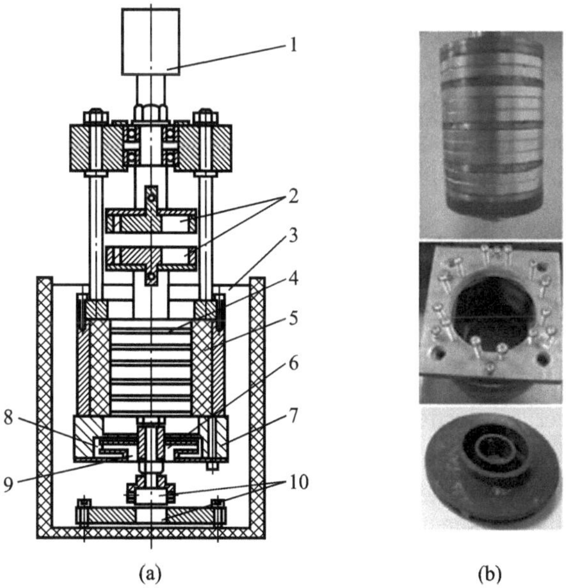


**Fig. 9.8:** (a) Schematic diagram of the structure of the liquid nitrogen pump. (b) Photos of PM rotor, HTS stator and blade wheel. (1) DC motor. (2) PM coupling. (3) Liquid nitrogen. (4) PM rotor. (5) HTS stator. (6) Blade wheel. (7) Pump body. (8) Outlet of pump. (9) Inlet of pump. (10) Radial PM bearing.

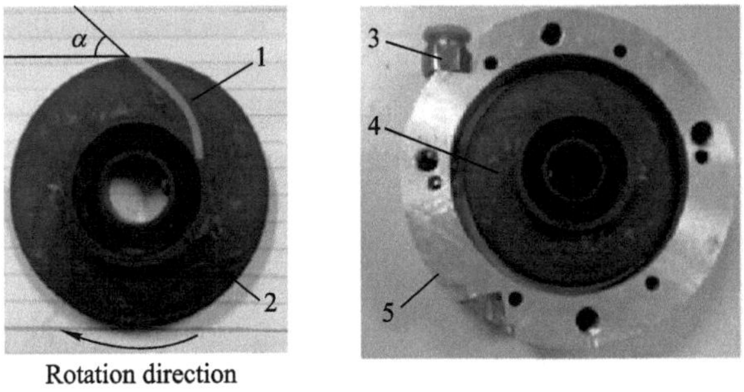


**Fig. 9.9:** (a) Blade wheel. (b) Blade wheel inside the pumping chamber of pump body. (1) Shape of blades inside the blade wheel. (2) Inlet of blade wheel. (3) Outlet of pump. (4) Blade wheel. (5) Pump body.

from the upper PM to the lower PM by magnetic forces. Due to non-contact between the upper and lower PMs, heat leakage can be reduced as much as possible. The RSB shown in Fig. 9.8 is installed in the middle of the pump. The HTS stator includes six pieces of HTS bulks and each bulk has the dimension of 13 × 32 × 64 mm while the PM rotor has a diameter of 52 mm and a length of 66 mm (Fig. 9.8b). The working gap of RSB is 1.5 mm. As discussed in Fig. 9.4, the maximum decrease of the axial stiffness of the RSB at different speeds is less than 8%. This is the minimum requirement for the application of the RSB to liquid nitrogen pump. The blade wheel with six blades is made of plastic, with a diameter of 65 mm and a blade discharge angle $\alpha$ of about 40° (Fig. 9.9a). This type of blade is called a back curved blade, which can reduce the flow loss and increase the efficiency of the pump. Fig. 9.9b shows the blade wheel placed inside the pumping chamber of the pump body. It is clear that the section of channel between the blade wheel and pumping chamber is variable. The radial PM bearing enforces the radial stiffness and stability of the rotor during the operation of the pump.

#### 9.3.2.2 Operation of liquid nitrogen pump

Figure 9.10 shows the testing device for the liquid nitrogen pump with a RSB, which is used to test the flow rate and head of the pump at different rotational speeds. The testing device includes the liquid nitrogen pump with a RSB, a Dewar, a DC power supply, a non-contact tachometer, an electronic scale, and a liquid nitrogen vessel. When the DC motor is operated, liquid nitrogen is pumped from the Dewar into the liquid nitrogen vessel. The speed of the pump is controlled by the DC voltage of the motor and measured with the tachometer. Flow of the pump is determined by measuring the mass of liquid nitrogen pumped into the vessel during one minute. The head of the pump means the increase of potential energy of 1 kg liquid nitrogen pumped from the output tube. In Fig. 9.10b, the inner diameter of output tube is 12 mm, and the height difference of center line of outlet and surface of liquid nitrogen in the Dewar is 40 mm.

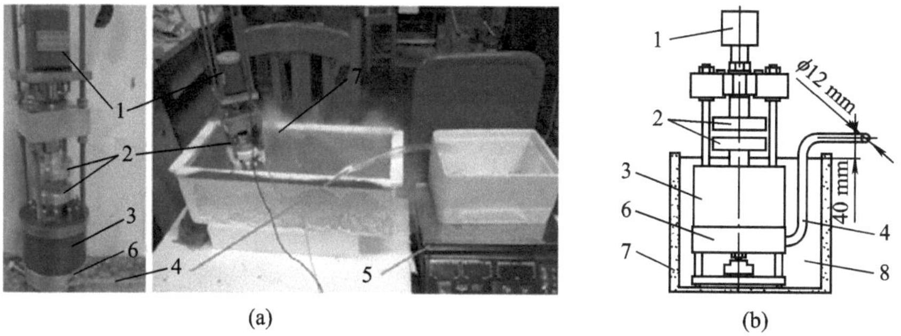


**Fig. 9.10:** Experimental testing device for a liquid nitrogen pump with a RSB. (a) Photograph of testing device. (b) Schematic view of liquid nitrogen pump in testing. (1) DC motor. (2) PM coupling. (3) RSB. (4) Output tube. (5) Electronic scale and liquid nitrogen vessel. (6) Main body of pump. (7) Dewar. (8) Liquid nitrogen.

If the flow rate of the pump is $Q$ (L/s), the speed of the liquid nitrogen in the output tube, $v$ (m/s), is

$$v = \frac{\frac{Q}{1000}}{2\pi \times \left(\frac{0.012}{2}\right)^2} = 8.84Q. \tag{9.18}$$

The head of the pump, $H_{\text{pump}}$, and the speed of the liquid nitrogen in the output tube have the following relation:

$$H_{\text{pump}} a_g = \frac{1}{2} v^2 + 0.04 a_g, \tag{9.19}$$

where $a_g$(=9.8 m/s$^2$) is acceleration of gravity and 0.04 (m) is height difference between the center line of outlet and surface of liquid nitrogen in the Dewar. The head of the pump can be calculated as

$$H_{\text{pump}} = \frac{1}{19.6} v^2 + 0.04 = 3.987Q^2 + 0.04. \tag{9.20}$$

Figure 9.11 and Tab. 9.2 show the flow rate and the head of the liquid nitrogen pump with the RSB operation at a rotational speed range between 756 rpm and 2245 rpm. It was observed that operation of the pump was stable and that the flow rate and the head increased with the increasing rotational speeds. During more tests at 2061 rpm, the pump worked very well for 1200 s without any abnormal vibration or noise.

There are still some problems that must be solved before commercial application of the RSB to a cryogenic pump. First, the total efficiency of the pump is not high. At

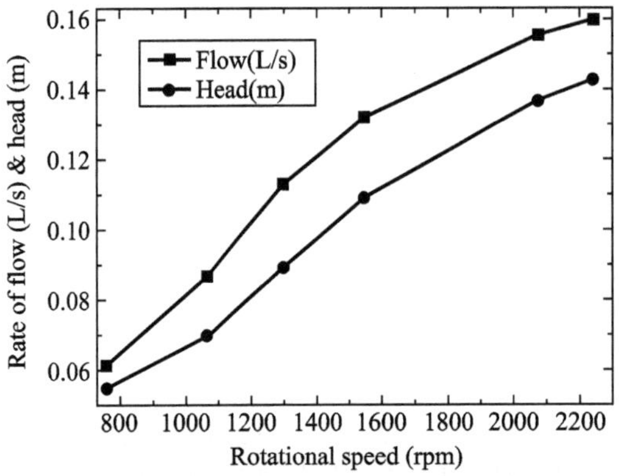


**Fig. 9.11:** Flow rate and head of liquid nitrogen pump with RSB at different rotational speeds.

**Tab. 9.2:** Relation of rotational speed, flow rate, and head of liquid nitrogen pump with RSB.

| **Rotation speed (rpm)** | **756** | **1065** | **1298** | **1547** | **2075** | **2245** |
|---|---|---|---|---|---|---|
| Flow rate (L/s) | 0.0611 | 0.0866 | 0.1130 | 0.1319 | 0.1558 | 0.1603 |
| Head (m) | 0.0549 | 0.0699 | 0.0894 | 0.1094 | 0.1368 | 0.1425 |

2245 rpm, the current and the voltage of the DC power is 2.8 A and 18.5 V, so the power is 51.8 W. Increase of mechanical energy of liquid nitrogen in 1 s can be calculated:

$$\Delta E = \frac{Q}{1000} \times \rho a_g H_{\text{pump}} = \frac{0.1603}{1000} \times 808 \times 9.8 \times 0.1425 = 0.181(\text{J}), \tag{9.21}$$

where, $\rho$ (=808 kg/m$^3$) is the density of liquid nitrogen, $a_g$ the acceleration of gravity, and 0.1425 m is the head of the pump at 2245 rpm.

The total efficiency of the pump is

$$\lambda = \frac{0.181\,\text{J}/1\,\text{s}}{51.8\,\text{W}} = \frac{0.181\,\text{W}}{51.8\,\text{W}} = 0.35\%. \tag{9.22}$$

As a cryogenic pump prototype, the efficiency is low, although some mechanical parts are not specially designed for the liquid nitrogen pump. The gap between some parts should be redesigned to be smaller in order to reduce the leakage of liquid nitrogen.

Second, for practical application, the displacement of the rotor due to force creep of the RSB should be reconsidered, too. The critical current density of the HTS bulk increases with decreased temperature (such as in liquid helium), so a lower working temperature benefits to the performance of the RSB. More research on characteristics of the RSB at lower temperature is needed for the application of the RSB to liquid hydrogen and helium pumps.

### 9.3.3 Future development of HTSB

What is the future development of HTSB? It is certain that researchers will improve its performance, i.e. larger load capacity and force density. Because force density is proportional to the square of magnetic field [45], three kinds of new HTSB designs have been developed to increase the magnetic field. Patel [46] from University of Cambridge put forward a RSB concept that uses superconductors for both rotor and stator, so that the magnetic field of the superconducting rotor was greatly enlarged. Central Japan Railway Company designed superconducting magnetic bearings with a superconducting coil and an iron core in the rotor [47, 48]. Japan Railway Technical Research Institute developed a superconducting bearing with a HTS bulk rotor and a superconducting coil stator that finally obtained a stable 20 kN axial force [49]. ATZ company also simulated this type of HTSB, and found that the force density was 5 times the value (10–15 N/cm$^2$) obtained with a PM rotor [45].

Second, rotational loss must be reduced for practical applications of the HTSB. In a small prototype, very low rotational loss is easy to realize [13]. However, for large-scale HTSBs whose stator and rotor are made up of numerous HTS bulks and PM rings, respectively, homogeneity of the trapped field in the HTS stator and the magnetic field of the PM rotor are hard to guarantee. Rotational loss also comes from gas friction, eddy current loss, etc. More focus should be systematically made to eliminate rotational loss in practical applications of HTSB.

The last direction is to broaden the application fields of HTSB, i.e. the liquid nitrogen pump [44], the centrifuge [37], the superconducting non-contact mixer [42], etc.

## 9.4 HTS FESS principle model

### 9.4.1 Introduction of FESS

FESS stores kinetic energy which is proportional to the rotational inertia and the square of the rotational speed of its flywheel. A complete FESS system is composed of a flywheel, bearing, motor/generator, vacuum chamber and safety protection device, control system and electronic convertor. The three working stages for FESS are (1) the energy storage stage, when the electric energy is converted into kinetic energy by the motor while the speed of flywheel increases; (2) the standing stage, when the motor/generator stops working (if standing losses from motor/generator, bearing, and air drag are low enough, the energy storage time can be greatly extended); (3) the electricity generation stage, when the kinetic energy is converted into electric energy by the generator while the speed of the flywheel decreases. Speed of flywheel is controlled over the range between the maximum speed and the minimum speed. At the maximum speed, the stored energy of FESS reaches the peak. The maximum speed is mainly limited by the ratio of strength and density of the material of flywheel as well as its rotational stability. The minimum speed depends on the rotational stability of flywheel and the efficiency of electricity generation. To reduce standing losses, optimal design of the motor/generator, suitable selection of the bearings and high vacuum in the chamber (lower than $10^{-2}$ Pa) are required. The higher ratio of strength and density the materials are, the higher limit of the energy storage density will be as the basic technology demands. For instance, the ratio of strength and density for an aluminum alloy is $2.1 \times 10^5\ m^2/s^2$, and its energy density limit is 23.8 Wh/kg. Ratio of strength and density for T700 carbon fiber is $2.6 \times 10^6\ m^2/s^2$ with the corresponding energy density limit of 218.7 Wh/kg [50]. In comparison, the carbon fiber material is a good candidate for large-energy storage flywheels. In addition, stability can be improved by optimal design of the dynamic structure of flywheel and the speed range should be kept away from resonance speeds. Safety can be guaranteed by many methods, such as introducing the carbon fiber flywheel whose broken fragments are less dangerous than those from a metal flywheel and further strengthening the vacuum chamber. Response time depends on the control system and the electronic convertor.

FESS has high energy efficiency of the range of 90%–95% during short periods. It allows a very large number of charge/discharge cycles and has a long lifetime. Monitoring the state of charge is simple and reliable because it is easy to measure the rotational speed. It is an environmentally friendly technology because all of the components can be recycled [51]. FESS can also be applied to vehicles, spaceflights, and uninterrupted power supplies (UPSs). Twelve units of FESS for an American Flywheel Systems Inc. with a total mass of 237 kg have been used in the IMPACT car of GM Company, which reached a cruising range of 480 km [52]. Hongkong Citybus uses the UPT tr100 FESS to successfully save 11% of the electric energy of an electrically driven trolley bus [53]. Research shows that 30% of fuel can be saved and 75% of automobile exhaust can be reduced by employing FESS [54]. For spaceflight, NASA, Fare Company, and University of Maryland developed a 50-Wh FESS with an open core composite flywheel for satellites. The FESS replaced the original chemical battery [55] and also had the function of attitude control [56]. FESS can be also applied as UPS which prevent important systems from shutting down in the case of power failure, i.e. for the computer networks of banks, cloud computing centers, and communication base stations. Active Power Company launched FESS UPS 150i, which can supply AC power at 380 V, 228 A, and has a standard ride-through time of 24 s [57].

Table 9.3 summarizes the technical, economic, and environmental state of all the energy storage technologies including FESS [51].

**Tab. 9.3:** Comparison of technical characteristics of different energy storage systems [51].

| | Efficiency (%) | Capacity (MW) | Energy density (Wh/kg) | Response time | Life time (years) | Maturity | Environmental impact | Capital ($/kW and $/kWh) |
|---|---|---|---|---|---|---|---|---|
| FESS | 93–95 | 0.25 | 10–30 | Very fas (<ms) | ~15 | Demonstration | Almost none | 350 and 5000 |
| Li-ion battery | 85–90 | 0.1 | 75–200 | Fast | 515 | Demonstration | Negative | 4000 and 2500 |
| Fuel cells | 20–50 | 0–50 | 800–10,000 | Good (<1s) | 5–15 | Developing | Small | 500–1500 and 10–20 |
| Super-capacitors | 90–95 | 0.3 | 2.5–15 | Very fast | 20+ | Developed | Small | 300 and 2000 |
| SMES* | 95–98 | 0.1–10 | 0.5–5 | Very fast | 20+ | Demonstration | Benign | 300 and 10,000 |

*SMES = Superconducting magnetic energy storage.

### 9.4.2 HTS Maglev FESS

As discussed above, HTSB is very suitable for the application to FESS because of these inherent advantages. China [58, 59], Germany [60–62], USA [17], Japan [23–27, 47, 48], England [63], Korea [64–68], Brazil [69], Italy [70], have carried out research on HTS Maglev FESS. The HTS Maglev FESSs from Boeing Company [16, 17], ATZ [60, 61], ISTEC [23, 24, 26, 27, 47], and Korea Advanced Institute of Science and Technology (KAIST) [64–68] are the most typical ones. Table 9.4 shows the parameters of the HTSBs in these typical FESSs. The parameters of RSB from Nexans SuperConductors GmbH are also listed for comparison.

In Boeing's FESS, the upper PM bearing supports most of the rotor mass of 132 kg and ASB plays the function of stability. The maximum storage energy is 5 kWh and the FESS can supply 208 V, 3 kW AC power. The ASB is composed of a YBCO HTS stator and a ~220-mm-diameter PM rotor. Axial and radial stiffness are 144 and 69 N/mm, respectively, when the working gap is 3.2 mm. Boeing tested its rotational loss under the maximum speed of 14,500 rpm.

The FESS from ATZ can store 10-kWh energies when the flywheel speeds up to 10,000 rpm and supply a power of 250 kW. The 600-kg rotor is levitated by the upper RSB and the lower PM bearing. At 72 K, the maximum axial carrying capacity and axial stiffness of the RSB are 10.08 kN and 4.5 kN/mm (3.3 mm) and 4.7 kN and 1.8 kN/mm (3.2 mm), respectively. Diameter and length of the PM rotor are 200 and about 120 mm, respectively. The inner diameter of the HTS stator is 205 mm and working gap is 2 mm. Force density at 70 K is 13 N/cm$^2$ axially and 6.5 N/cm$^2$ radially. The HTS stator is cooled by a cryocooler and the G-10 support structure can support the 2-ton load with a heat loss less than 0.3 W.

ISTEC FESS can store 10 kWh at 15,860 rpm. The 425-kg rotor is suspended by the RSB and the top axial electromagnetic bearing. Rotational stability is increased by introducing two radial electromagnetic bearings. The RSB has an inner HTS stator and an outer PM rotor. The PM rotor has an outer diameter of about 170 mm and length of about 300 mm. The maximum axial force of the RSB reaches 8.7 kN.

Compared with the above large-scale FESSs, KAIST has developed a small-sized FESS with ASB. This FESS is used for the energy storage and the attitude control of nanosatellites. The HTS stator is made up of seven 10-mm-diameter, 1.8-mm-depth HTS bulks. A 50-mm-diameter and 63.2-g flywheel is levitated above the HTS stator with a gap of 1–7 mm. At 0.13 Pa, the rotor can rotate at 51,000 rpm with stored energy of 337 J. The whole rotation lasts 3 h 20 min at 7-mm working gap.

Presently, KEPRI is developing a 35-kWh FESS with a 1588-kg rotor which is levitated by two RSBs, a PM bearing, and a thrust electromagnetic bearing. The RSB is composed of 12 YBCO bulks and the working gap is 2.5 mm. The radial stiffness and damping coefficient of RSB are 346.6 N/mm and 1255 Ns/m, respectively. The maximum design speed is 12,000 rpm, and it has achieved an experimental speed of 6500 rpm successfully [31, 73].

**Tab. 9.4:** Parameters of HTSBs of Boeing, ATZ, ISTEC, KAIST and Nexans.

| | **Boeing** | **ATZ** | **ISTEC** | **KAIST** | **Nexans** |
|---|---|---|---|---|---|
| HTSB Type | ASB | RSB (inner PM rotor) | RSB (outer PM rotor) | ASB | RSB (inner PM rotor) |
| HTS stator | Regular hexagon single-grain YBCO bulks, 36 mm × 36 mm × 4.5 mm | 55 pieces of three- seeds rectangular YBCO bulks, 65 mm × 35 mm × 13 mm, OD 230 mm, ID 205 mm × 120 mm [60] | 40 pieces of roof tile shaped YBCO bulks with epoxy resin protection, OD 123.2 mm, ID 93.2 mm, length 60 mm × 5 stories [24] | 7 pieces of disc type YBCO bulks, $\phi$10 mm × 1.8 mm [64] | 270 pieces of square single-grain YBCO bulks, 35 mm × 35 mm × 10 mm, ID 325 mm × 305 mm [19] |
| PM rotor | ~ $\phi$220 mm, PM ring (radial polarity) + iron ring, 4 magnetic poles [17] | $\phi$200 mm × ~ 120 mm, PM ring (axial polarity, OD 200 mm × ID 150 mm × 8 mm) + iron ring [18], about 8 magnetic poles [30] | OD ~ $\phi$170 mm × ~ 300 mm [71], PM ring (axial polarity) + iron ring, 16 magnetic poles [24] | ~ $\phi$25 mm, Single PM ring (axial polarity) [64] | OD 319 mm, PM ring (axial polarity) + iron ring, 20 magnetic poles [19] |
| Working gap | Several mm [17] | 2 mm [60] | 1.8 mm [71] | 1 ~ 7 mm [64] | 1 mm [19] |
| Cooling system | Stirling-cycle cryocooler (15 W @ 77 K), G-10 supportor [17] | Liquid nitrogen, 72 ~ 79 K, G-10 cryostat [18] (Now by 35 W/80 K G/M cryocooler [30]) | Liquid nitrogen [24] | Stirling cryocooler [64] | "Coolpower 140T" (Leybold Vakuum GmbH), stainless steel Dewar [19] |
| Experimental speed | 14510 rpm [28] | 3200 rpm [60] | 11250 rpm [27] | 51000 rpm [64] | 4400 rpm [19] |
| Main performance | Working gap is 3.2 mm, axial stiffness of 144 N/mm, radial stiffness of 69 N/mm [72] | At 72 K, radial force and stiffness of 4.7 kN (3.2 mm) and 1.8 kN/mm, axial force and stiffness of 10.08 kN (3.3 mm) and 4.5 kN/mm [18] | Maximum axial force of 8.7 kN [23] | Mass of the flywheel is 63.2 g [64] | Radial carrying capacity of 6900 N, radial stiffness of 5.1 kN/mm@ 28 K [19] |
| Loss | ~ −4 rpm/min (working gap = 3.9 mm, 5000 rpm, 50 K) [28] | COF of $2.8 \times 10^{-6}$, $-5 \times 10^{-4}$ Nm [18] | Simulated energy loss of 0.55 mW/N at 6000 rpm for 100 kWh HTSB model [27] | COF of 0.001 ~ 0.002 (working gap of 7 mm) [64] | |
| Application | 5 kWh/ 3 kW-208 V AC FESS [17], design speed of 20000 rpm (10 kWh) [16] | 5 kWh/250 kW FESS with 600 kg rotor [30], design speed of 10000 rpm (10 kWh) [60] | 10 kWh/15 kW FESS with 425 kg rotor, design speed of 15860 rpm (10 kWh) [23] | 337 J FESS for energy storage [64] and attitude control of nano-satellites [66] | 4 MW HTS generator, supporting the 1000 kg shaft [19] |

In China, four HTSB prototypes have been developed. The Institute of Electrical Engineering of the Chinese Academy of Sciences constructed a prototype with ASB, PM bearing, and electromagnetic bearing, whose maximum speed goes up to 9600 rpm [74]. Beijing Jiaotong University also developed a prototype with ASB, PM bearing, and electromagnetic bearing [75]. In ASCLab at SWJTU, J. S. Wang et al. manufactured a FESS principle model with two ASBs[76] and conducted basic studies on the characteristics [77, 78]. They then developed a magnetic motor based on a RSB with a maximum speed of 9901 rpm [79].

### 9.4.3 Double ASBs system

The common problems of HTS Maglev technology in large-scale applications, especially for FESS, are the relatively low load capability and stiffness compared with active magnetic levitation systems. With continuous improvements of the HTS bulk materials, it is believed that the load capability, stiffness, and dynamic stability of HTS Maglev system can be largely improved so that it can meet the requirements of industrial applications. With further efforts, the whole bearing system will be greatly simplified [62]. From this point of view, a double ASBs system (full HTSB system) was fabricated for a FESS model.

As shown in Fig. 9.12, the double ASBs system is a vertical structure and its total height is 610 mm. The system is mainly composed of three parts: a support platform, two ASBs and a driving module. The support platform is a circular base

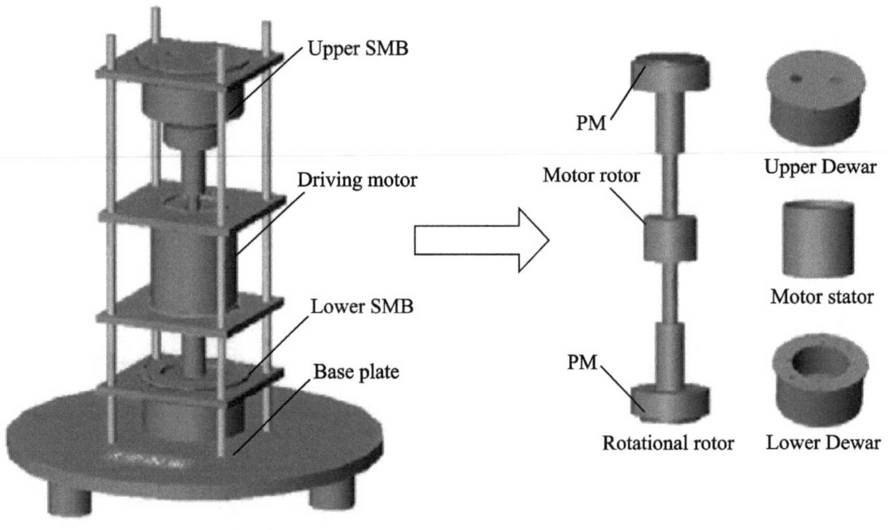


**Fig. 9.12:** Schematic overall view of the double ASBs system.

plate made of epoxy resin with diameter of 500 mm. The upper and lower ASBs are fixed at both ends of a vertical shaft to suspend and levitate the shaft. At the middle of the shaft, an inductive motor is mounted to drive the rotational rotor. The working gap of this motor is 2.5 mm, which can realize the fully stable levitation of ASBs. A transformer is used to control the motor by changing the voltage of the power supply, as shown in Fig. 9.13. The maximum rotational speed of the motor is about 3000 rpm.

The two axial ASBs are the core part of the system. One of them is used to levitate and the other is used to suspend the rotor. Both HTS stators are independently composed of seven single-domain melt-textured YBCO bulks with diameter of 30 mm and height of 18 mm (as shown in Fig. 9.14) and the PM rotors are both made of three cylindrical axial-magnetized NdFeB PMs by superposition (as shown in Fig. 9.15). Each PM is 63 mm in diameter and 9 mm in height and its maximum surface magnetic field

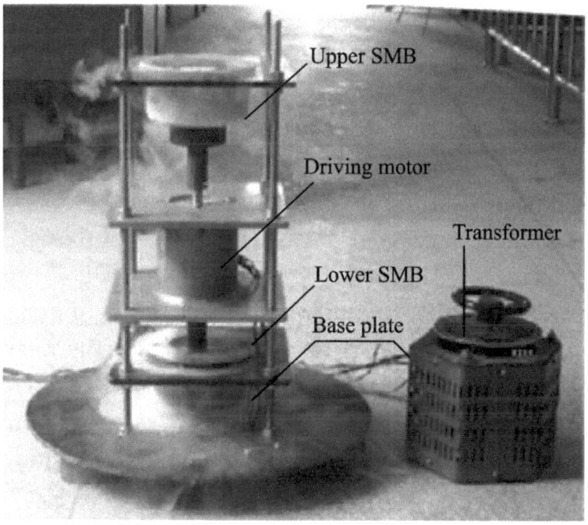


**Fig. 9.13:** Photo of the double ASBs system model in free rotation with a transformer.

**Fig. 9.14:** Photo of HTS (YBCO) stator of ASB.

**Fig. 9.15:** Photo of PM rotor of ASB.

is 0.4 T. Two in-house epoxy fiberglass Dewars are used as the holder and cryostat for the YBCO bulks. Those two Dewars are with outer diameter of 140 mm and height of 75 mm and the thickness of wall is only 2 mm. Eventually, the double ASBs can provide a levitation force of 79.2 N and a lateral force of 9.7 N at the lateral displacement of 3 mm under a certain working condition [58].

### 9.4.4 FESS model with double ASBs

The HTS FESS model is mainly composed of the double ASBs, flywheel rotor, PM motor, and a set of control circuit. Table 9.5 gives the detailed parameters of this HTS FESS model.

The motor employed in this system is a PM motor which is brushless and can work as a generator. The rated three-phase voltage of the motor is 380 V, the rated efficiency 95.8%, the maximum speed 13,000 rpm, and the air gap is 1.5 mm. The PM motor is mounted at the middle of the FESS model and controlled by a frequency converter to study the rotational characteristics of the rotor under different acceleration models.

A steel disk with diameter of 200 mm, weight of 1.4 kg, and rotary inertia of 0.01 kg/m$^2$ is used as the flywheel of the HTS FESS model. The flywheel is installed between the PM motor and the upper SMB, as shown in Fig. 9.16. As the energy carrier

**Tab. 9.5:** Parameters of the FESS model with double ASBs.

| | |
|---|---|
| Stored energy | 9 kJ |
| Maximum speed | 13,000 rpm |
| Bearings | Double ASBs |
| Motor | PM motor |
| Flywheel | A steel disk with a diameter of 200 mm, weight of 1.4 kg |
| Load | A lamp of 12 V/10 W |

**Fig. 9.16:** Upper part of HTS FESS model.

**Fig. 9.17:** Power generation demonstration of the HTS FESS model.

of the FESS, the flywheel rotates with high speed and stores the electrical energy as mechanical energy. The stored kinetic energy can be calculated as $E = 0.5Iw^2$, where $I$ is the rotary inertia of the rotor and $w$ is the rotational angular velocity. When the speed is up to the maximum value of 13,000 rpm, the stored kinetic energy of the flywheel model is about 9 kJ.

A set of control circuits, AC-DC converter, and a lamp load were used for the power charge and discharge processes of the HTS FESS model. After the shaft levitates, acceleration and deceleration of the flywheel rotor are realized by closing or opening the switch of the circuit, thus realizing the charge and discharge of electrical energy.

Figure 9.17 demonstrates the electricity generation after the speed of flywheel rotor is up to 2000 rpm. When the acceleration of the FESS is finished, by switch from the driving circuit of the motor to the load circuit, the motor will run as a generator, and then the stored energy of the flywheel rotor will be released to the lamp load. This 12-V/10-W lamp is lit for ten of seconds, as shown in Fig. 9.17. The lighting of the lamp illustrates the whole energy storage and conversion of the FESS system from electricity energy to mechanical energy and then mechanical energy back to electricity energy intuitively.

## 9.5 Development of a 5-kWh HTS FESS prototype

### 9.5.1 Design and constitution

A HTS FESS prototype has been developed by the authors with the primary goal of energy savings in a subway system. Total energy storage capacity (ESC) is the key parameter for a FESS. For tests and application in subway system, the ESC of the HTS FESS prototype is designed for 5 kWh. Figure 9.18 shows the feedback kinetic energy of a subway train with different weights and speeds. For the metropolis like Beijing or Shanghai, the weight of a subway train under full load is usually more than 320 tons with an average operation speed of more than 60 km/h. From Fig. 9.18, it is clear that the ESC of 20 MJ (5.56 kWh) can satisfy this requirement. Furthermore, the depth of discharge (DOD) should be considered because not all of the storage energy can be used. With the consideration of voltage attenuation, the lowest revolving speed of a HTS FESS during operation is usually half of the highest speed, which implies a DOD of 75%.

The input/output (I/O) power of HTS FESS is also important to the application in a subway system, which is determined by the power requirement of the subway train during starting, acceleration, and braking. A conventional subway train has about 16 onboard drive motors and the peak power demand is more than 2 MW. The I/O power of HTS FESS depends on the combined electrical machine and it is difficult to design and manufacture a single FESS unit with a 2-MW power level. It is better to have the FESS array consisting of several uniform FESS units with lower power level to satisfy the large power demand.

To simplify the design, manufacture, and fabrication of the FESS unit, we planned to install at least three HTS FESS units each with ESC of 5 kWh and I/O power of

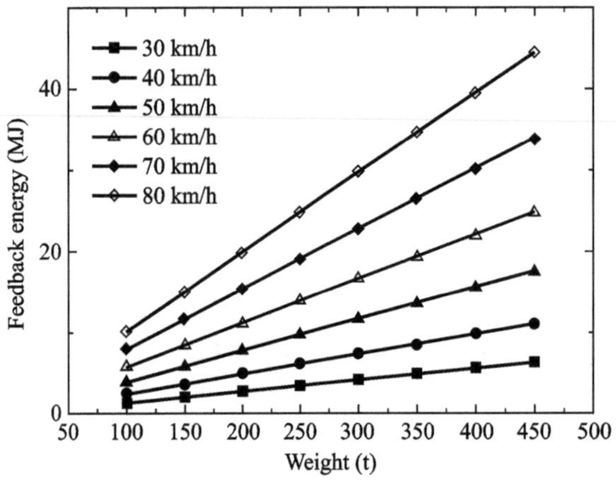


**Fig. 9.18:** Energy storage capacity for a subway train with different weight and speed. The feedback energy equals 40% of the total kinetic energy.

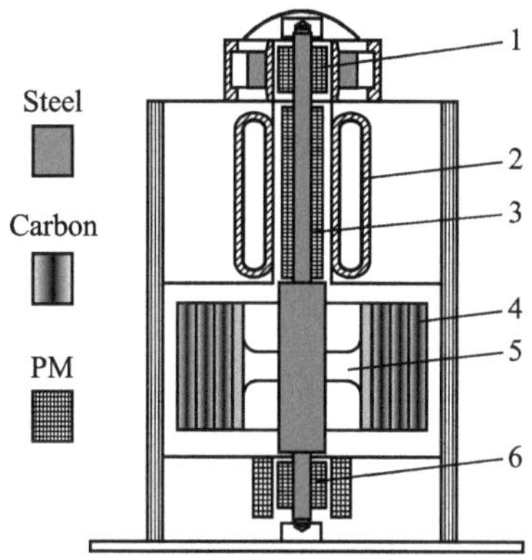


**Fig. 9.19:** Design concept of a 5-kWh HTS FESS. (1) PM assistant bearing. (2) Stator of motor. (3) Rotor of motor. (4) Carbon fiber composite flywheel. (5) Wheel hub. (6) HTS bearing.

400 kW on one subway station. When the SC units are used in series, the HTS FESS array can supply more than 11 kWh with 75% DOD and average I/O power more than 1.2 MW, which satisfies most energy and power demands of a subway regenerative braking system. The conceptual design of a 5-kWh HTS FESS is shown in Fig. 9.19.

### 9.5.2 HTS Maglev bearing

The stator of HTS Maglev bearing with inner diameter of 151 mm is composed by YBCO bulk. The length, width, and height of the YBCO bulk are 64, 32, and 12 mm, respectively. The rotor of HTS Maglev bearing is composed of PM rings and iron hoops just like the typical axial assembly mode (RSB) as shown in Refs. 11 and 17. The diameter of the PM rotor is 147 mm. All of the PM rings have a magnetic energy product of 50 MGOe (398 kJ/m$^3$). The total length of the PM rotor of the HTS bearing is about 240 mm in order to supply enough axial forces to suspend the flywheel. A cryogenic refrigerator is introduced to supply the 40–50-K working temperature for the YBCO bulks. It can supply a cooling capacity of 100 W at 40 K. The designed load capacity of this HTS Maglev bearing is 600 kg. The assembled PM rotor of the HTS Maglev bearing is shown in Fig. 9.20. The assembled PM rotor has been tested for its dynamic balance characteristic which is adjusted by the de-weighting method with the drilled holes on the flange shown in Fig. 9.20. After this adjustment, the dynamic balance performance index of the PM rotor is about 0.1 g.

### 9.5.3 Flywheel rotor

The composite material of flywheel rotor was chosen as T-700 carbon fiber, which had a high tensile strength, more than 1600 MPa. For safety considerations, the upper limit of strength was set to 800 MPa in the analog calculation and design, only the half

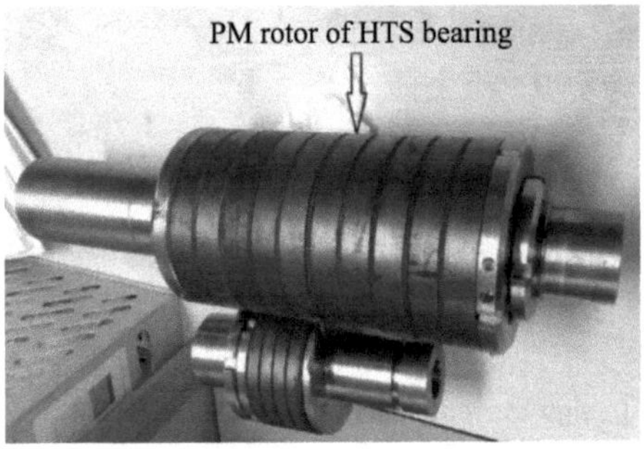


**Fig. 9.20:** PM rotor of HTS Maglev bearing.

of maximum strength of T-700. With the requirement of 5 kWh ESC, the calculated outer diameter, inner diameter and height of the carbon flywheel rotor were 770 mm, 540 mm and 380 mm, respectively.

The shape of the flywheel hub was optimized and the comparison of stress is shown in Fig. 9.21. From the results of Fig. 9.21, it is clear that the optimization can reduce the maximal radial stress from 320 to 240 MPa, which is safe for most aluminum alloys.

For an integral flywheel rotor, the eigen-frequencies are very important because there may be serious vibration or deformation of the rotor. The first and second calculated eigen-frequencies of the rotor are about $3.5 \times 10^{-4}$ and 3.4 Hz, respectively. No serious distortion occurs at those two eigen-frequencies. The third eigen-frequency is about 433.3 Hz, and it causes serious deformation, as shown in Fig. 9.22. Fortunately, the eigen-frequency of 433.3 Hz, which equals a rotation speed of 25,998 rpm, is much above the designed operation speed of 15,000 rpm for the flywheel rotor. However, enough attention should be paid to the first and second eigen-frequency because of

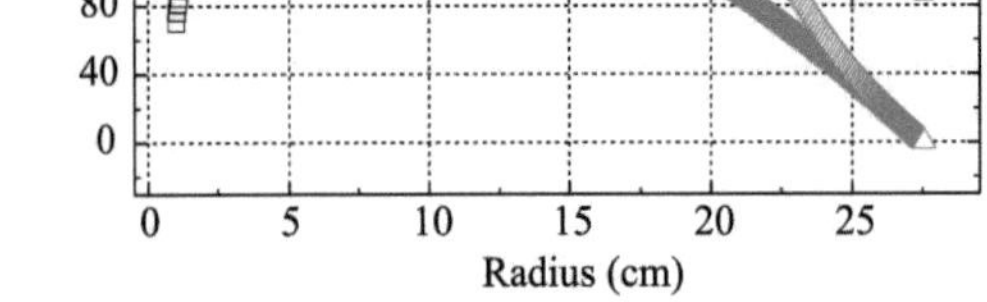


**Fig. 9.21:** Optimal results for the flywheel hub. Stress-phi and stress-r represents the azimuthal and axial stress, respectively.

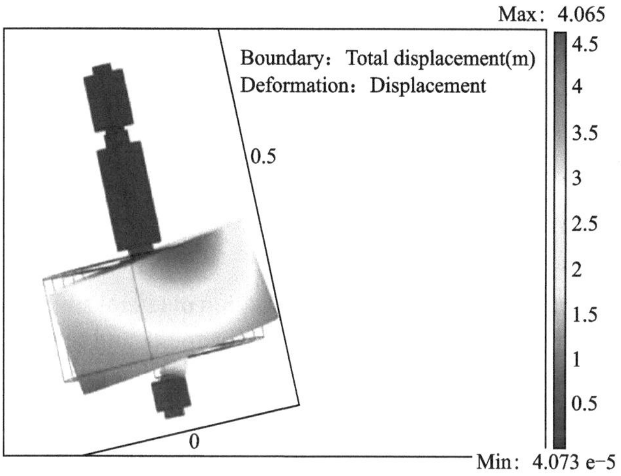


**Fig. 9.22:** Serious deformation of flywheel rotor at the third calculated eigen-frequency of 433.3 Hz.

the probable vibration. It is better to pass these eigen-frequencies quickly in the actual operation.

The weight of the flywheel rotor is nearly 433 kg and its axial moment of inertia is about 22.38 $m^2$·kg. The ESC calculated from the moment of inertia at 15,000 rpm is about 7.66 kWh, which is more than the required 5 kWh. The excess contribution to ESC comes from the flywheel hub, the PM rotors, and the steel spindle. With a DOD of 75%, the available ESC of an individual HTS FESS unit is about 5.75 kWh, which is suited to the practical application. At the first stage, the flywheel rotor will be tested without the carbon composite. After the preliminary experiments and adjustment of dynamic balance, the carbon composite will be convolved onto the flywheel hub by the pre-stressing method. At the first stage, the flywheel rotor has the weight of 340 kg and axial moment of inertia of 4.1 kg·$m^2$ and an ESC of 1.5 kWh at 15,000 rpm.

### 9.5.4 Motor/generator

The PM synchronous motor can switch between the motor function and generator function quickly. Thus, it is a suitable choice for the HTS FESS system. Because of a lack of design and fabrication experience of PM synchronous motor at 400-kW and 15,000-rpm level, a PM synchronous motor with I/O power 100 kW has been designed and manufactured with outer diameter of 350 mm and height of 370 mm. Because of the compact configuration and size, the PM synchronous motor can be integrated into the FESS easily.

The output voltage of this PM synchronous motor is about 600 V DC, which is lower than the standard voltage of a power supply network for a subway system, 750 or 1500 V DC. To satisfy the voltage, the HTS FESS units will be arrayed in series. Furthermore, a DC-DC voltage converter must be introduced for each HTS FESS unit to stabilize the output voltage because the output voltage of a PM synchronous motor will decrease with the decrease of the rotation speed during the discharge process.

### 9.5.5 PM assistant bearing

A PM assistant Maglev bearing was designed for the compensation of lateral stiffness. The outer and inner diameters of the PM ring of bearing stator were 148 and 128 mm, respectively. The outer and inner diameters of the PM ring of the bearing rotor were 120 and 100 mm, respectively. All of the PM rings used had a magnetic energy product of 50 MGOe (398 kJ/$m^3$). The stator and rotor of the PM bearing were both composed of eight pieces of PM ring. The calculated maximal axial and radial stiffness of the PM bearing were about 1300 and 640 N/mm, respectively.

### 9.5.6 Customized machining equipment

The YBCO bulks must have an arc surface to make up the entire stator of a HTS Maglev bearing. For the YBCO bulk at 293 K, the Vickers hardness and the fracture toughness is about $5.2 \pm 0.5$ GPa and 1.3 MPa·$m^{1/2}$, respectively, which means the YBCO bulk is a hard and brittle material [80]. J. S. Wang et al. have designed and manufactured customized equipment to machine the arc surface on the YBCO bulk. This customized equipment is composed of three servo motors, which are controlled by programmed logic controller (PLC) modules, as shown in Figs. 9.23 and 9.24. A grinding wheel is fixed on the vertical servo motor, which supplies the grinding forces for the surface machining of the HTS bulk. The HTS bulk will be fixed on a specially designed fixture and be driven by the horizontal servo motors. The horizontal servo motors can move in two directions separately.

The routes of the servo motors can be set by the operation staff through the PLC modules. Thus, the YBCO bulk can be machined automatically. During the machining

**Fig. 9.23:** Customized equipment to machine the surface of HTS bulk.

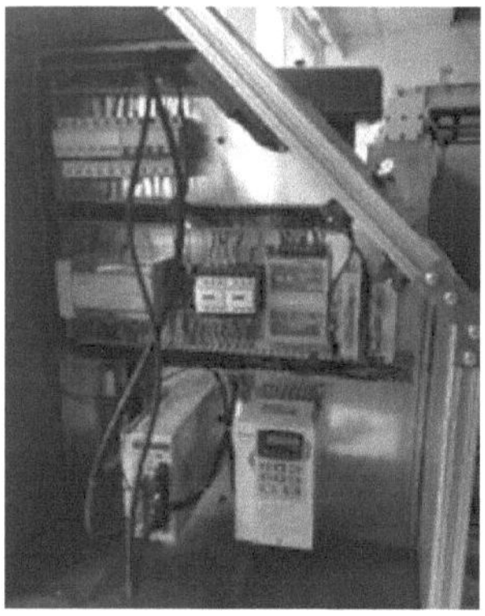

**Fig. 9.24:** PLC modules of customized machining equipment for HTS bulk.

**Fig. 9.25:** Processed bulk YBCO with an arc surface.

process, the grinding wheel is set to descend vertically in multiple cycles. After every machining cycle, the grinding wheel will hold in a non-contact position and stay for a short time. It lets the revolving grinding wheel cool in the air. Thus, there is no need of any cooling fluid that may have a bad effect on the material properties of the HTS bulk. The minimum displacement of the servo motor is about 0.01 mm so that the machining precision of the HTS bulk is about 0.1 mm. Figure 9.25 shows a YBCO bulk processed with the customized equipment.

### 9.5.7 Complete HTS FESS test system

The complete HTS FESS system is shown in Fig. 9.26. The refrigerator (1) and the vacuum pump (4) are used to supply the cryogenic operation conditions of the HTS bulks. Furthermore, the vacuum environment could reduce the air friction loss of the flywheel rotor under high rotation speed. The designed vacuum is 0.1 Pa.

The power cabinet (2) transmits the electrical energy from the power grid to the HTS FESS unit and the accessory equipment, which can supply a maximum power of 140 kW. Between the power cabinet and the HTS FESS unit, the frequency converter (6) is necessary regulate the speed of the PM synchronous motor. This frequency converter has a power level of 110 kW with the maximum current modulation frequency of 300 Hz, which is equivalent to 20,000 rpm rotation speed for the PM synchronous

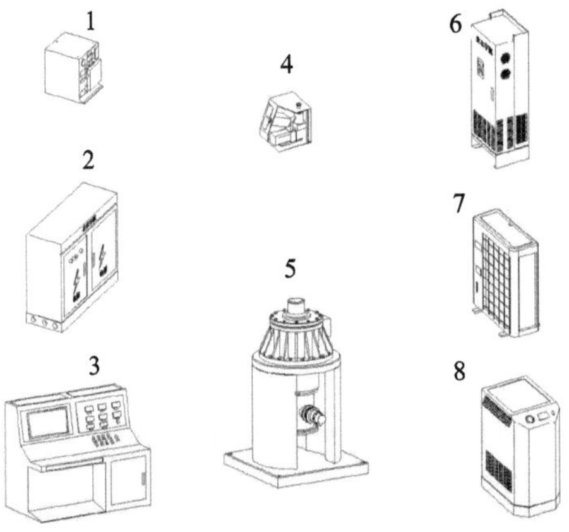


**Fig. 9.26:** Complete test system of HTS FESS. (1) Cryogenic refrigerator. (2) Power cabinet. (3) Control and test system. (4) Vacuum pump. (5) HTS FESS unit. (6) Frequency converter. (7) Radiator of the water chilling system. (8) Circulating water cooling machine.

motor. The radiator (7) and water cooling machine (8) constitute the circulating water chilling system, which is used to eliminate heat produced by the PM synchronous motor and the refrigerator. The HTS FESS unit (5) is connected to the refrigerator, the vacuum pump and the frequency converter cryogenic cooling, vacuum and speed modulation, respectively, as shown in Fig. 9.27.

The control and test system (3) was developed by J. S. Wang et al. It is used not only to operate the HTS FESS, but also to test its performance. The levitation forces, guidance forces, and force relaxation can be measured by this system. Furthermore, the vibration and temperature characteristics are also measured and supervised by this system, if any vibration or temperature signals exceed the threshold value, the control system will stop the operation of the HTS FESS promptly.

**Fig. 9.27:** Connection status of HTS FESS unit.

## 9.6 Application of HTS FESS

### 9.6.1 Subway regenerative braking

The huge electric energy consumption of a subway system is a problem chronically troubling the modern metropolis. Regenerative braking technology is a good candidate to solve this problem. Nearly 30%–40% of the kinetic energy of a subway vehicle can be fed back to the supply network of the subway system through regenerative braking technology [81]. Unfortunately, the feedback energy from a braking subway vehicle cannot be utilized by other subway vehicles or electric equipment synchronously since it is mostly consumed by the brake resistor in the form of heat. It is better to introduce an energy storage unit to receive the feedback energy and then supply the feedback energy to other subway vehicles or electrical equipment.

Compared with other energy storage models, the FESS has higher power density, long service life, and many more charge cycles, so it is very suitable for the application in a regenerative subway braking system. In fact, several FESS units have been successfully applied in a subway or light railway system. They have given excellent performances in energy savings and power quality improvements.[82, 83] However, the bearings used in those FESS units are conventional mechanical type, which give frictional loss and inconvenience in maintenance for the subway application.

The HTS Maglev bearing has low rotating friction and requires no active control, which means higher rotate speed, larger energy storage capacity, smaller size, and lower standby loss. Furthermore, the HTS Maglev bearing has good capacity of resisting disturbances and excellent self-recovery ability, which come from the flux pinning characteristics of HTS materials.

Besides the ESC and the I/O power, the installation and operation model is also important for application of HTS FESS in subway systems. If the HTS FESS operates on-board the subway vehicle, there is a strict limitation in volume and weight. Then, the HTS FESS must work at a very high speed, usually more than 30,000 rpm. However, if the HTS FESS is installed and operates on the station of subway system, it has no volume and weight limitation, which is beneficial to the design and test, especially for the dynamic balance characteristics of the flywheel. Furthermore, the storage energy of FESS in the station model can be used conveniently by other auxiliary equipment of the subway system, such as air-conditioning and lighting facilities of the subway stations.

The array of three HTS FESS prototypes with an available ESC of 5.75 kWh and maximal I/O power of 400 kW can satisfy the basic operation requirement of a subway system. Before the practical application, this HTS FESS prototype system will be tested to simulate a subway system in the next 1–2 years. When used in the subway system, the HTS FESS units can bring obvious energy savings. For example, under the average departure frequency of the Beijing subway system, the operation interval of a HTS FESS unit is about 3 minutes. Assuming 50% of the 5.75-kWh ESC can be utilized ever during ever operation processes, the HTS FESS unit can save 345,000 kWh of electric energy over 6000 working hours per year, which is an excellent result. Moreover, there are other benefits from the usage of HTS FESS units. One benefit comes from the improvement of the power quality: namely, the relaxation of serious fluctuation of voltage and current in the subway supply network can improve the service life and operation performance of the electric equipment. Another benefit is the reduction of power demand, which reduces the huge cost associated with expansion of the supply network under the need for ever-increasing transport capacity of the subway system.

### 9.6.2 Renewable energy system

The renewable energy systems like solar power generation and wind power generation need energy storage systems to improve the quality and stability of generated electrical energy. The principal technical point of the running test of this 5-kWh HTS FESS prototype in renewable energy system is power capacity. For example, the power of an individual wind turbine generator is about 2–5 MW, which is much higher than the 100-kW power of our HTS FESS prototype. Application in renewable energy system requires a high power level, so the maximal I/O power should be improved to 400 or 500 kW for a single HTS FESS unit. This is the next upgrade for this HTS FESS prototype. For a renewable energy station with hundreds of megawatts power level, there would be thousands of HTS FESS units in operation.

### 9.6.3 Power grid system

Their application in power grid is the optimal and most beneficial usage of FESS units. The FESS units can supply protection or improvements on the power grid system, such as the power angle stability, the grid voltage stability, the grid frequency excursion suppression, and the regulation control. Those protective applications in a power grid require large power capacity, as much as hundreds of megawatts, but with a short working time, such as several minutes. The single HTS FESS unit used for the protective applications should have enough ESC to support the maximal I/O power at least 1–5 minutes. This HTS FESS prototype has an available ESC of 5.75 kWh of 75% DOD, which can support the I/O power of 100 kW for 3.45 minutes. Thus, it can be tested for the protective applications. Because the 100-kW power of this HTS FESS prototype is much smaller than the actual power of the grid, its I/O power needs to improve to 300–500 kW, and at least 10–20 prototypes are needed to run the test. Another application of FESS units in the power grid is for long-term load shifting (LS), such as 3 or 7 hours LS, which is used to adjust the power demand of the gird or store electrical energy at valley times to sell later [84]. The long-term LS applications require the HTS FESS have enough ESC to support the maximal I/O power for several hours.

For example, this HTS FESS prototype needs an ESC of 100 kWh to support the I/O power of 100 kW for 1 hour. The design and manufacture of a 100-kWh FESS unit is very difficult, so the HTS FESS unit used for the long-term LS applications will have a smaller I/O power level, to reduce the individual ESC, such as a HTS FESS unit with the I/O power of 10 kW and ESC of 10 kWh, respectively.

### 9.6.4 Uninterrupted power supply

The most commercial FESS products are used in the UPS application. Because of the numerous and different operation conditions, it is difficult to design a single FESS module to satisfy most of the requirements for UPS. It is better to choose a modularization scheme to improve the flexibility of application and reduce the cost of research and development. For example, four 1.6-MVA FESS UPS units combined with a diesel generator could easily supply voltage protection and power relay to a 20-kV distribution network [85]. The simulation results display the feasibility of this FESS UPS and the improvement in power quality. By the introduction of FESS unit, the voltage dips are divided by a factor of 3, even in the worst case. If the I/O power of our HTS FESS prototype could be improved to 400 kW, using 4 HTS FESS units can provide similar operation performance to the test FESS of 1.6 MVA. Furthermore, the individual I/O power of most types of the commercial FESS UPS products is 100–250 kW, with a discharge time of 10–30 seconds, which can provide useful reference for the design of a HTS FESS for the application of UPS [84].

### 9.6.5 Electromagnetic launch and weapon

The HTS FESS is a suitable power source for the electromagnetic launch (EML) system. The electric machine of HTS FESS unit should be redesigned to the compulsator type to produce a large pulsed power supply. The compulsator is a special synchronous machine with a damping winding to minimize the interior impedance of the machine, which can output peak power or peak current maximally during a very short time. A 50-MW/650-MJ FESS system consisted of 25 FESS units was investigated for the operation performance of an EML system [82]. This FESS system can output the peak power of 50 MW within 13 s, with maximal overall efficiency of 95%. The FESS units were arrayed in the parallel connection style to output a 1200-V DC voltage. The requirements of FESS system were quite different for various EM launch and weapon systems. For example, the ESC and peak power of the electromagnetic aircraft launch system was 122 MJ and 326 MW, respectively [86]. The available ESC of our HTS FESS prototype is 5.75 kWh, namely, about 21 MJ, so it needs six HTS FESS units to satisfy the requirement of 122 MJ ESC. The US Army has contracted a hypervelocity railgun of 2-MJ muzzle kinetic energy, based on the electrothermal chemical technology and a hybrid electric drive ground combat vehicle [87]. The US Navy has contracted a ship-based hypervelocity railgun of 64-MJ muzzle kinetic energy [88]. The ESC of our HTS FESS units can satisfy those electromagnetic weapon systems well. The key point is the design and manufacture of the compulsator to the HTS FESS unit to satisfy the pulsed power demand of EM launch or weapon system. Furthermore, the possible applications of HTS FESS as pulse power supply also includes plasma experiments, laser experiments, ion accelerator, pre-acceleration of spacecraft, etc.

### 9.6.6 Electric vehicle

The HTS FESS prototype can be used as the energy supply for an electric vehicle (EV) or hybrid electric vehicle (HEV). The applications in EV/HEV require small size and light weight of the HTS FESS unit. Compared with the application in an EM launch system or UPS system, the ESC is more important for an EV/HEV system. For example, the commercial product, Model S EV of Tesla Motors Inc., has an ESC of 60 kWh and 85 kWh, with maximal output power of 222–310 kW [89]. Under the normal driving mode, the maximal range per charge can be achieved is 390 and 500 km for an ESC of 60 and 85 kWh, respectively. An I/O power of 200–300 kW can be conveniently achieved with a single HTS FESS unit, but the ESC is a problematic issue. Because of the limitation of size and structure of a vehicle, it is better to choose a distributed design of HEV FESS, i.e. an array of HTS FESS units. For example, a HTS FESS array for an EV can consist of 20 HTS FESS units, each with ESC of 3 kWh and I/O power of 15 kW, which can achieve the similar operation performance as the Model S EV of 60 kWh ESC. Besides the ESC and power, the resistance to vibration and the need for cryogenic working conditions are more essential problems for the HTS FESS units in EV/HEV.

## 9.7 Summary

As one of the most important applications of HTS Maglev technologies, HTSB exhibits outstanding advantages, like the lowest rotational loss in all kind of bearings (COF is as low as $10^{-7}$) [13] and very high speed up to 520,000 rpm [14]. It has low noise, long lifetime, no abrasion, and no need for lubricant. It is stable at all directions, and does not need an active control system. Even in the case of power failure, it has enough time to stop the rotor safely before the HTS stator quenches. It has good environment adaptation and can be applied in a vacuum environment which is impossible for gas-suspended bearing.

Due to the excellent performances, HTSB attracts research interests all over the world. HTSB characteristics, such as magnetic force, stiffness, damping, rotor dynamics, stabilization, force creep, temperature dependence, and rotational loss, have been comprehensively studied. The research status of stiffness, rotational loss, and stability characteristics is reviewed in this chapter. The stiffness work in our group has also been described. It is found that axial stiffness of a RSB is reduced by no more than 8% at different speeds, which is thought to result from the high cooling efficiency in liquid nitrogen. An efficient calculation method of radial stiffness has been proposed, and the calculation error is just 2.86% compared with experimental results.

HTSB applications to FESS, rotor support for machine, laser beam deflection polygon scanner, centrifuge, liquid hydrogen tank, and non-contact mixer have been introduced. The first industrial application of HTSB was realized by Nexans SuperConductors GmbH to support the rotor of a 4-MW HTS generator in 2006. A liquid nitrogen pump is one of the new applications of HTSB, while other applications like a high-speed motor, wind power generation and an electric aircraft engine, are also promising. A liquid nitrogen pump prototype with HTSB was developed to verify the application feasibility. Experiment shows that operation of this pump is very stable without abnormal vibration or noise and that the flow rate and head increase with rotational speeds.

Among the wide applications of HTSB, FESS is the most popular and most promising one. Because of the passive stability, low rotational loss, and potential high rotational speed of HTSB, HTS FESS owns distinct advantages of high energy density, high efficiency, longevity of service, simple control, low maintenance, etc., which provides a new approach to the global energy crisis. Several large-scale (over 5 kWh) HTS FESS prototypes have been developed by Boeing, ATZ and ISTEC to demonstrate the technical advantages. As a high*quality power source, HTS FESS is very promising for EVs, spaceflight, and UPS.

A FESS model with two axial ASBs is made to demonstrate the working principle of the storage system. With a 1.4-kg levitated steel plate as spinning flywheel, a lamp could be lighted by the stored energy in the FESS model, which is useful to demonstrate the mutual conversion processes of electrical energy to mechanical energy. In order to develop a large-scale FESS prototype, a measurement set-up for HTSB was first designed. As a basic platform, it was able to measure the force and stiffness characteristics of HTSB under conditions of different cooling methods and working temperatures. Based on the research results of the HTSB, a 5-kWh HTS FESS prototype was designed and manufactured. It is planned to apply it in a regenerative braking system subway for energy savings. Furthermore, the 5-kWh HTS FESS prototype also has good application prospects in renewable energy, power grid, UPS, electromagnetic launch, and EVs.

For the practical applications of HTSB and FESS, continuous work is being conducted on the load capacity, force density, rotational loss, stability, and new designs. Once the reasonable compromise between performance and cost is achieved, HTSB and FESS will immediately become widely commercial, because there are great demands for highly efficient energy storage technologies, from many fields of energy, transportation, military, etc.

## References


[1] http://en.wikipedia.org/wiki/Bearing_(mechanical).

[2] Zhou FZ, Liu BC, Tang QH, Wang K, Shen JX. Design and test of high-speed permanent magnet brushless DC motor. Electric Machines & Control Application. 2010;37(1):13–15. (In Chinese).

[3] Wu ZD, Zhang H, Liang GT, Wu JL, Zhang RJ. Thrust bearings for Three Gorges generators. China Three Gorges Construction. 2003;9:7–8. (In Chinese).

[4] http://www.lyxql.com/cn/newsshow.aspx?id=403.

[5] Safety of machinery – Lubricants with incidental product contact – Hygiene requirements, ISO 21469:2006, IDT.

[6] Zhang G, Yin QZ, Ruan J, Jiang DD, Gao G. Analysis on stiffness of permanent magnetic bearing. Bearing. 2011;4:4–8. (In Chinese).

[7] Ma WQ, Yu HC, Sun A. Research situation of gas-bearing/rotor system. Lubrication Engineering. 2010;06:121–125. (In Chinese).

[8] Tang SY. Magnetic bearing and its application in industry. Micromotors Servo Technique. 1994;27(4):28–31. (In Chinese).

[9] Chen XM, Li WL. Gas bearing technology and its application. Lubrication Engineering. 2000;4:61–63. (In Chinese).

[10] Chen RG, Hou Y, Yuan XL, Chen CZ. Application of gas bearings on high speed turbo-machinery. Lubrication Engineering. 2007;35(4):28–32. (In Chinese).

[11] Li CG, Qi R, Lin H. The application and developing trend of magnetic bearings. Industrial Instrumentation & Automation. 2006;3:75–77. (In Chinese).

[12] Werfel FN, Floegel-Delor U, Riedel T, Rothfeld R, Wippich D, Goebel B. HTS magnetic bearings in prototype application. IEEE Trans Appl Supercond. 2010;20(3):874–879.

[13] Hull JR. Topical review: Superconducting bearings. Supercond Sci Technol. 2000;13(2):R1–R15.

[14] Ma KB, Postrekhin YV, Chu WK. Superconductor and magnet levitation devices. Rev Sci Instrum. 2003;74:4989–5017.

[15] Moon FC. Superconducting levitation. John Wiley & Sons Inc.; 1994.

[16] Day AC, Strasik M, McCrary KE, Johnson PE, Gabrys JW, Schindler JR, et al. Design and testing of the HTS bearing for a 10 kWh flywheel system. Supercond Sci Technol. 2002;15:838–841.

[17] Strasik M, Hull JR, Mittleider JA, Gonder JF, Johnson PE, McCrary KE, McIver CR. An overview of Boeing flywheel energy storage systems with high-temperature superconducting bearings. Supercond Sci Technol. 2010;23:034021.

[18] Floegel-Delor U, Rothfeld R, Wippich D, Goebel B, Riedel T, Werfel FN. Fabrication of HTS bearings with ton load performance. IEEE Trans Appl Supercond. 2007;17(2):2142–2145.

[19] Walter H, Bock J, Frohne Ch, Schippl K, May H, Canders WR, Kummeth P, Nick W, Neumueller HW. First heavy load bearing for industrial application with shaft loads up to 10 kN. J Phys Conf Ser. 2006;43:995–998.

[20] Werfel FN, Floegel-Delor U, Riedel T, Rothfeld R, Wippich D, Goebel B, et al. Progress toward 500 kg HTS bearings. IEEE Trans Appl Supercond. 2003;13(2):2173–2178.

[21] Hikihara T, Adachi H, Moon FC. Dynamical behavior of flywheel rotor suspended by hysteretic force of HTSC magnetic bearing. J Sound Vib. 1999;228(4):871–887.

[22] Park BJ, Han YH, Jung SY, Kim CH, Han SC, Lee JP, Park BC, Sung TH. Static properties of high temperature superconductor bearings for a 10 kW h class superconductor flywheel energy storage system. Physica C. 2010;470:1772–1776.

[23] Ichihara T, Matsunaga K, Kita M, Hirabayashi I, Isono M, Hirose M, Yoshii K, Kurihara K, Saito O, Saito S, Murakami M, Takabayashi H, Natsumeda M, Koshizuka N. Fabrication and evaluation of superconducting magnetic bearing for 10kWh-class flywheel energy storage system. Physica C. 2005;426–431:752–758.

[24] Ichihara T, Matsunaga K, Kita M, Hirabayashi I, Isono M, Hirose M, Yoshii K, Kurihara K, Saito O, Saito S, Murakami M, Takabayashi H, Natsumeda M, Koshizuka N. Application of superconducting magnetic bearings to a 10 kWh-class flywheel energy storage system. IEEE Trans on Appl Supercond. 2005;15:2245–2248.

[25] Matsunaga K, Tomita M, Yamachi N, Iida K, Yoshioka J, Murakami M. YBCO bulk of the superconducting bearing for a10kWh flywheel. Supercond Sci Technol. 2002;15(5):842–845.

[26] Koshizuka N, Matsunaga K, Yamachi N, Kawaji A, Hirabayashi H, Murakami M, Tomita M, Unee S, Saito S, Isono M, Nasu H, Maeda T, Ishikawa F. Construction of the stator installed in the superconducting magnetic bearing for a 10 kWh flywheel. Physica C. 2004;412–414: 756–760.

[27] Koshizuka N. R&D of superconducting bearing technologies for flywheel energy storage systems. Physica C. 2006;445–448:1103–1108.

[28] Strasik M, Hull JR, Johnson PE, Mittleider J, McCrary KE, McIver CR, Day AC. Performance of a conduction-cooled high-temperature superconducting bearing. Materials Science and Engineering B. 2008;151:195–198.

[29] Hull JR, Strasik M, Mittleider JA, Gonder JF, Johnson PE, McCrary KE, McIver CR. High rotational-rate rotor with high-temperature superconducting bearings. IEEE Trans Appl Supercond. 2009;19(3):2078–2082.

[30] Werfel FN, Floegel-Delor U, Riedel T, Rothfeld R, Wippich D, Goebel B, Reiner G, Wehlau N. Towards high-capacity HTS flywheel systems. IEEE Trans Appl Supercond. 2010;20(4): 2272–2275.

[31] Han YH, Park BJ, Jung SY, Han SC, Lee WR, Bae YC. The improved damping of superconductor bearings for 35 kWh superconductor flywheel energy storage system. Physica C. 2013;485:102–106.

[32] Han YH, Park BJ, Jung SY, Han SC. Study of superconductor bearings for a 35 kWh superconductor flywheel energy storage system. Physica C. 2012;483:156–161.

[33] Hull JR, Strasik M, Mittleider J, McIver C, McCrary K, Gonder J, Johnson P. Damping of sub-synchronous whirl in rotors with high-temperature superconducting bearings. IEEE Trans Appl Supercond. 2011;21(3):1453–1459.

[34] Wang S, Wang J, Deng C, Lu Y, Zeng Y, Song H, Huang H, Jing H, Huang Y, Zheng J, Wang X, Zhang Y. An update high-temperature superconducting Maglev measurement system. IEEE Trans Appl Supercond. 2007;17(2):2067–2070.

[35] Kummeth P, Ries G, Nick W, Neumuller HW. Development and characterization of magnetic HTS bearings for a 400 kW synchronous HTS motor. Supercond Sci Technol. 2004;17(5): S259–S263.

[36] Werfel FN, Floegel-Delor U, Rothfeld R, Wippich D, Riedel T. Laser beam deflection polygon scanner using HTS bearings. IEEE Trans Appl Supercond. 2001;11(1):1737–1740.

[37] Werfel FN, Flogel-Delor U, Rothfeld R, Wippich D, Riedel T. Centrifuge advances using HTS magnetic bearings. Physica C. 2001;354:13–17.

[38] Walter H, Arsac S, Bock J, Siems SO, Canders WR, Leenders A, Freyhardt HC, Fieseler H, Kesten M. Liquid hydrogen tank with cylindrical superconducting bearing for automotive application. IEEE Trans Appl Supercond. 2003;13(2):2150–2153.

[39] Leenders A, Ullrich M, Freyhardt HC, Kesten M, Fieseler H, Canders WR, May H, Weh H, Gauss S, Bock J. Fabrication of HTS monoliths for a bearing system in a cryogenic vessel. IEEE Trans Appl Supercond. 1999;9(2):992–995.

[40] Gauss S, Albering JH, Bock J, Kesten M, Fieseler H, Canders WR, May H, Freyhardt HC, Ullrich M. Cryotank with superconducting, magnetic suspension of the interior tank. IEEE Trans Appl Supercond. 1999;9(2):1004–1007.

[41] Murakami M. Processing and Applications of Bulk RE–Ba–Cu–O Superconductors. Int J Appl Ceram Tec. 2007;4(3):225–241.
[42] Koyama F, Akiyama S, Murakami M. Developments of superconducting mixers for medical applications. Supercond Sci Technol. 2006;19(7):S572–S574.
[43] Wongsatanawarid A, Suzuki A, Seki H, Murakami M. Magnetic torque transferring study for bulk High-$T_c$ superconductors and permanent magnets. J Phys Conf Ser. 2009;153:012035.
[44] Lin QX, Jiang DH, Ma GT, Wang JS, Deng ZG, Zheng J, Wang SY. Research of radial high temperature superconducting magnetic bearings for cryogenic liquid pumps. IEEE Trans Appl Supercond. 2012;22(3):5201604.
[45] Werfel FN, Floegel-Delor U, Rothfeld R, Riedel T, Goebel B, Wippich D, Schirrmeister P, Koenig R. Next generation of HTS magnetic application: HTS bulk and coil interaction. J Phys Conf Ser. 2014;507:032055.
[46] Patel A, Palka R, Glowacki BA. New bulk – bulk superconducting bearing concept using additional permanent magnets. Przeglad Elektrotechniczny. 2012;88(6):108–110.
[47] Koshizuka N, Ishikawa F, Nasu H, Murakami M, Matsunaga K, Saito S, Saito O, Nakamura Y, Yamamoto H, Takahata R, Oka T, Ikezawa H, Tomita M. Present status of R&D on superconducting magnetic bearing technologies for flywheel energy storage system. Physica C. 2002;378–381:11–17.
[48] Matsunaga K, Tomita M, Yamachi N, Iida K, Yoshioka J, Isono M, Hirose M, Nasu H, Kameno H, Kubo A, Takahata R, Kitai N, Yamamoto H, Nakamura Y, Koshizuka N, Murakami M. Fabrication and evaluation of superconducting bearing module for 10 kWh flywheel. Physica C. 2002; 378–381:883–887.
[49] Seino H, Nagashima K, Tanaka Y, Nakauchi M. Study of superconducting magnetic bearing applicable to the flywheel energy storage system that consist of HTS-bulks and superconducting-coils. J Phys Conf Ser. 2010;234:032052.
[50] Dai XJ, Li YL, Yu H. Design of high specific energy density flywheel. Journal of Tsinghua Universigy (Science and Technology). 2008;48(3):379–382. (In Chinese).
[51] Kousksou T, Bruel P, Jamil A, ElRhafiki T, Zeraouli Y. Energy storage: Applications and challenges. Solar Energy Materials & Solar Cells. 2014;120:59–80.
[52] Tang SQ, Liao DX, Wu ZJ. Key technology and prospects on the electric vehicles. Mechanical Science and Technology. 2003;22:189–244. (In Chinese).
[53] Colin Tarrant. Hong Kong Citybus-UPT tr100 trial success. (http://202.193.70.166/mirror/Materials0322115.pdf).
[54] Wen SB, Jiang SY. Application study of flywheel energy storage system in automobile. Machinery Design and Manufacture. 2010;12:82–84. (In Chinese)
[55] Kirk JA, Walsh GC, Hronmada LP, Zmood RB, Sullivan GE. Open core composite flywheel. Proceedings of the 32nd intersociety energy conversion engineering conference (IECEC-97), 1997;3:1748–1752.
[56] Fausz JL, Richie DJ. Flywheel simultaneous attitude control and energy storage using a VSCMG configuration. Proceedings of the 2000 IEEE International Conference on Control Applications, 2000:991–995.
[57] www.activepower.com.
[58] Deng ZG, Lin QX, Wang JS, Zheng J, Jiang DH, Ma GT, et al. Prototype of high temperature superconducting magnetic levitation flywheel energy storage system. Chinese J Low Temp Phys. 2009;31(4):311–314. (In Chinese).
[59] Deng Z, Lin Q, Ma G, Zheng J, Zhang Y, Wang S, et al. A double-superconducting axial bearing system for an energy storage flywheel model. J Phys Conf Ser. 2008;97:012283.
[60] Werfel FN, Floegel-Delor U, Riedel T, Rothfeld R, Wippich D, Goebel B, et al. 250 kW flywheel with HTS magnetic bearing for industrial use. J Phys Conf Ser. 2008;97:012206.

[61] Werfel FN, Floegel-Delor U, Riedel T, Wippich D, Goebel B, Rothfeld R. HTS flywheel from R&D to pilot energy storage system. J Phys Conf Ser. 2010;234:032062.
[62] Siems SO, Canders WR, Walter H, Bock J. Superconducting magnetic bearings for a 2 MW/10 kW h class energy storage flywheel system. Supercond Sci Technol. 2004;17:S229–S233.
[63] Coombs TA, Samad I, Ruiz-Alonso D, Tadinada K. Superconducting micro-bearings. IEEE Transaction on Applied Superconductivity. 2005;15(2):2312–2315.
[64] Lee K, Kim B, Ko J, Jeong S, Lee SS. Advanced design and experiment of a small-size flywheel energy storage system using a high-temperature superconductor bearing. Supercond Sci Technol. 2007;20:634–639.
[65] Kim B, Ko J, Jeong S, Lee SS. Experiment and analysis for a small-sized flywheel energy storage system with a high-temperature superconductor bearing. Supercond Sci Technol. 2006;19:217–222.
[66] Lee E, Kim B, Ko J, Song CY, Kim SJ, Jeong S, Lee SS. An integrated micro HTS system for energy storage and attitude control for three-axis stabilized nanosatellites. IEEE Trans Appl Supercond. 2005;15(2):2324–2327.
[67] Lee E. A micro HTS renewable energy/attitude control system for micro/nano satellites. IEEE Trans Appl Supercond. 2003;13(2):2263–2266.
[68] Lee E. A high-temperature superconductor – magnet energy storage and attitude control system for space MEMS. 2002 IEEE Aerospace conference proceedings. 2002;5:2365–2371.
[69] Sotelo GG, de Andrade R, Ferreira AC. Magnetic bearing sets for a flywheel system. IEEE Trans Appl Supercond. 2007;17(2):2150–2153.
[70] Perini E, Giunchi G. Field cooling of a MgB2 cylinder around a permanent magnet stack: prototype for superconductive magnetic bearing. Supercond Sci Technol. 2009;22:045021.
[71] Koshizuka N, Ishikawa F, Nasu H, Murakami M, Matsunaga K, Saito S, Saito O, et al. Progress of superconducting bearing technologies for flywheel energy storage systems. Physica C. 2003;386:444–450.
[72] Day AC, Hull JR, Strasik M, Johnson PE, McCrary KE, Edwards J, et al. Temperature and frequency effects in a high-performance superconducting bearing. IEEE Trans Appl Supercond. 2003;13(2):2179–2184.
[73] Han YH, Park KJ, Park BJ, Jung SY, Han SC, Lee WR, Bae YC. Results and analysis of an accident in 35-kWh SFES. IEEE Trans Appl Supercond. 2013;23(6):5701806.
[74] Fang JR, Lin LZ, Yan LG, Xiao LY. A new flywheel energy storage system using hybrid superconducting magnetic bearings. IEEE Trans Appl Supercond. 2001;11(1):1657–1660.
[75] Li YL, Fang J, Guo MZ, Xiao L, Zheng MH, Jiao YL. Analysis on a new-style HTS magnetic bearing's characteristics of levitation force. Rare Metal Materials and Engineering. 2008;37(S4):221–225. (In Chinese).
[76] Deng Z, Lin Q, Wang J, Zheng J, Ma G, Liu L, Zhang Y, Wang S. Basic design and characteristics study of a double-axial superconducting magnetic bearing system. Cryogenics. 2009;49: 259–262.
[77] Deng ZG, Lin QX, Wang JS, Zheng J, Zhang Y, Wang SY. Dynamic test platform for high temperature superconducting magnetic bearings. Cryogenics and Superconductivity. 2009;37(3):1–4. (In Chinese).
[78] Deng ZG, Lin QX, Wang JS, Zheng J, Jiang DH, Zhang Y, Wang SY. Study on the coefficient of friction of high temperature superconducting magnetic bearings. Chinese J Low Temp Phys. 2009;31(5):227–230. (In Chinese).
[79] Lin QX. Research on characteristics of radial high temperature superconducting bearing, PhD thesis, Southwest Jiaotong University; 2013. (In Chinese).
[80] Grigorashvili Y. Superconductors – properties, technology, and applications. InTech; 2012.
[81] Gyuk IP. EPRI Technical Report, December, 2003.

[82] Reiner G, Wehlau N. Concept of a 50 MW/650 MJ power source based on industry-established MDS, 28th IEEE International Conference on Plasma Science and 13th IEEE International Pulsed Power Conference, 2001: 248.
[83] Richardson MB. Flywheel energy storage system for traction applications. International Conference on Power Electronics, Machines and Drives, 2002; 275–279.
[84] Gyuk IP. EPRI-DOE handbook of energy storage for transmission and distribution applications. EPRI Technical Report, 2003.
[85] Richard T, Belhomme R, Buchheit N, Gorgette F. Power quality improvement case study of the connection of four 1.6 MVA flywheel dynamic UPS systems to a medium voltage distribution network. IEEE/PES Transmission Distribution Conference Exposition, Atlanta GA, USA; 2001.
[86] Doyle MR, Samuel DJ, Conway T, Klimowski RR. Electromagnetic aircraft launch system – EMALS. IEEE Trans Magn. 1995;31(1):528–533.
[87] Goodell B. Electro-thermal chemical (ETC) armament system integration into a combat vehicle. IEEE Trans on Magnetics. 2007;43(1):456–459.
[88] Fair HD. Progress in electromagnetic launch science and technology. IEEE Trans Magn. 2007;43(1):93–98.
[89] http://www.teslamotors.com/.

Wei Liu, Jing Li, Jia-Su Wang, and Su-Yu Wang

# 10 HTS Maglev launch technology

## 10.1 Introduction

The electromagnetic launch system can be divided into three typical modes depending on the type of propulsion, such as rail gun, coil gun, and linear motor modes.

The rail gun mode is composed of two rails and a projectile to construct a closed loop of current. A very high current pulse can be excited in the loop by a pulsed power supply. The current pulse in the rails produces a highly magnetic field, which is combined with the current in the projectile and produces a very large propulsion force. The final muzzle velocity of railgun developed by the US Navy can achieve 2.5 km/s with a 20-kg projectile [1]. In the rail gun mode, the projectile must contact the rail to complete the pulse current loop. This creates mechanical friction so that this model is not suitable for application of the HTS Maglev system.

The coil gun mode is based on the magnetic forces between a coaxial coil and the projectile. The coaxial coil is usually designed as the barrel and can produce a highly magnetic field under a large current pulse. The projectile is made of ferromagnetic materials or permanent magnets, which can produce a large force with the high field of the coaxial coil. This force produces a high muzzle velocity. Because of the axisymmetric magnetic force configuration in the radial direction, the projectile in the coaxial coil could achieve stable suspension and have no mechanical friction. Because the large magnetic force derives from the highly magnetic field gradient, the length of coaxial coil should not be too long. The barrel of a coil gun usually consists of several segments of coaxial coil to improve the gradient of the magnetic field and thus of the force. However, the control of the coil gun is more complicated than that of the rail gun, because it is necessary to control the pulse current in each individual coil segment according to the location of projectile. Limited by the performance of electronic power devices under high frequency and large current, it is difficult to acquire a very high muzzle velocity for the coil gun. The coaxial coils could be fabricated with superconducting wires or tapes which have far lager current density than copper. A HTS bulk could acquire a high trapped flux density up to 17.24 T [2], which is far larger than the residual flux density of a PM. Thus, the magnetized HTS bulk is an ideal candidate for the projectile of the coil gun.

The linear motor mode is composed of two modules: one is the propulsion module, which is composed of segments of a linear motor, including an induction or a synchronous style. The other is the support module, which can be divided into three types: a mechanical slide type, the electromagnetic suspension (EMS) type, and the HTS Maglev type.

https://doi.org/10.1515/9783110538434-010

A mechanical slide type can be made of smooth metal or composite materials to reduce the friction coefficient, but mechanical friction cannot be eliminated absolutely, so it is not suitable for very-high-speed launch.

The EMS-type Maglev could acquire a stable suspension under the precise active control of the magnetic forces between the electromagnetic magnet and the ferromagnetic alloy suspended object. The response time of electronic power devices and the feedback control system is the main problem for the application of EMS type in the launch system with very high speed.

Because of the stable levitation ability without active control, the HTS Maglev type is an ideal candidate for a high-speed electromagnetic launch system. Another advantage of the HTS Maglev type is the self-recovery ability, which means the HTS Maglev system could redress the minor deviation caused by a disturbance or oscillation automatically. Compared with rail gun mode or coil gun mode, the linear motor mode is more suitable for the application of a HTS Maglev system.

## 10.2 Repeatable electromagnetic launch system

The most important application of HTS Maglev system combined with a linear motor is the repeatable electromagnetic launch (REL) system of aircrafts and spacecrafts. The report of the Air Force Office of Science Research (AFOSR) of the USA indicates that the launch cost of REL system is about 600 USD/kg, which is much cheaper than the 2000 USD/kg on the space shuttle [3]. The National Aeronautics and Space Administration (NASA) of the USA has funded several research programs of REL technology, and they believe the REL has the most potential for the near-earth orbit launch, over the next 20–30 years [4].

Under the sponsorship of NASA, Jayawant designed a REL system based on an induction linear motor and an EMS support system [5]. This REL system was designed to boost the launch in the initial 10–20 s, which could reduce the fuel of spacecraft dramatically. The research of Jayawant indicated that the EMS support system had drastic vibration, small force density, and stiffness; in other words, the EMS support system was not suitable for the REL launch of heavy-load spacecraft [5].

Schultz designed a new EMS support system for REL with support from NASA. This new EMS support system used LTS coils in place of copper wires, which could support a spacecraft weighing 590 t. Combined with a linear motor, this REL system could propel the spacecraft to 268 m/s, with an acceleration of 2$g$ [6].

Those REL systems were primarily designed as a launch auxiliary for the initial stage. Based on the design of the launch auxiliary system, a comprehensive REL launch scheme was proposed. To reduce the cost of linear motor, the REL launch scheme was usually designed to a circular or racetrack shape. The spacecraft accelerated circularly in the REL system to the required muzzle velocity, and then was released by the REL system with a specially designed switching device. Tidman completed the mechanical and dynamic analyses of a projectile in a closed-ring vacuum pipe REL system [7]. The final calculation results indicated the muzzle velocity could achieve 10 km/s. Hull proposed a REL launch system based on LTS magnets [8, 9]. This LTS REL system used Nb-Ti wires for the support apparatus and $Nb_3Sn$ to make the projectile. The large magnetic forces between the $Nb_3Sn$ projectile and the Nb-Ti support apparatus could maintain stable suspension and counteract the huge centrifugal force of the projectile at very high speed.

Launch Point Technologies, Inc., promoted a commercial superconductive REL system, which had the same basic principles as the LTS REL system proposed by Hull, called the "Launch Ring" [10]. The biggest problem of the Launch Ring system was also the huge centrifugal force of projectile under high speed. The Launch Ring system consisted of a rail ring of large diameter to reduce the centrifugal force. The larger the ring diameter, the smaller the required radial suspension forces to compensate for the centrifugal force. For example, a projectile of 200-kg mass needs 40 MN of radial suspension forces to compensate for the centrifugal force with a 1-km launch ring diameter, but only 5 MN when the launch ring diameter is 8 km. The final velocity was designed to be 10 km/s, which included an orbital velocity of 7.8 km/s with a margin velocity of 2.2 km/s to offset various launch losses. To achieve the final muzzle velocity, a segmented control linear synchronous motor was introduced, which included multiple motor sleds that could be accelerated simultaneously. According to the calculation, a 20-MW linear motor with 100 sleds, each sled 250 m long, could launch 50 tons per hour into orbit with the 8-km launch ring diameter. Compared with coil gun and rail gun systems, the maximal required power for the Launch Ring system is much smaller. According to the economic analyses by Launch Point Technologies, if the Launch Ring system can shoot 30,000 times per year, the cost of a 1-kg payload is 41 dollars, which is much lower than the cost of a rocket or space shuttle launch.

## 10.3 Preliminary research on HTS Maglev launch technology

After the successive development of first man-loading HTS vehicle [11]. (see Chapter 6), the application of the HTS Maglev system to electromagnetic launch was activated quickly in the Applied Superconductivity Laboratory (ASCLab), SWJTU. There are many parameters that needed to be investigated clearly and carefully for the practical development of a HTS Maglev launch system.

The loading capacity of the HTS Maglev launch system is the most important parameter, which decides the basic operation performance of a launch system. In the previous research on HTS Maglev vehicles, it was found that the levitation force relaxation increased with the increase in reloading. He investigated the levitation force relaxation of bulks YBCO under different loads [12]. From the experimental results, it was clear that the normalized relaxation rate of the levitation forces increased with the increase in load.

A numerical method to evaluate the levitation force relaxation under load based on the current vector potential (***T***) method was introduced (see Section 7.6.2), which could reveal visually the configuration and distribution of supercurrent within a HTS bulk. The simulation results showed nearly the same trend as the experimental results, which was helpful in the qualitative analyses of levitation characteristics. The calculation could display the influence of the load on the current density distribution clearly. From these results, it was found that the zone of current density $< 1 \times 10^7\,A/m^2$ decreased with the load increase; in other words, the shielding currents in the HTS bulk increased with the load increase.

The influence of the vertical inclination of the PMG on the levitation characteristics of HTS Maglev systems was investigated [13, 14]. To investigate the influence of the angle of vertical inclination, a HTS Maglev launch platform was designed and fabricated. The levitation forces, guidance, and driving forces were tested over the inclined angle range of 0° to 18°.

The experimental results indicated that the levitation gap increased with the increase in the incline angle. However, the levitation forces of HTS bulks and the driving forces of the linear motor decreased with the increase in vertical angle. When the vertical incline angle increased from 0° to 18°, the levitation gap increased 0.65 mm and the levitation forces decreased 10% [14]. The guidance forces were not influenced by the change in the levitation gap.

Furthermore, the influence of the transverse slope of the PMG on the HTS Maglev system was also investigated [15]. The experiments on transverse slope were tested by the SCML-1 system (see Section 5.3) with a specially designed measurement platform. The levitation and guidance forces of HTS bulks for transverse angles from 0° to 12° were tested. The results showed that the levitation and guidance forces increased and decreased with the increase in the transverse slope, respectively.

The dynamic response of HTS bulks above the PMG was very important to the practical application of the HTS Maglev launch system, especially under high-speed operation. The dynamic stiffness, the displacement and the levitation forces of the HTS bulks were measured under external vibration with excitation amplitudes of 1–4 mm and frequencies of 1–400 Hz [16]. The experimental results indicated that the levitation forces, the resonance frequency and the dynamic stiffness of the HTS bulks decreased with the increase of excitation amplitude, under the conditions of excitation frequency < 30 Hz and excitation amplitude > 3 mm. The influence of excitation frequency on the levitation performance of HTS bulks was weak, with exception of resonant frequencies. The energy loss due to vibration was the main reason that caused attenuation of levitation performance. Furthermore, the experimental results also showed that the serious reduction in levitation performance occurred mainly in the first several cycles of vibration, which meant that the HTS bulks could sustain external vibration in the vertical direction and recover to a new balance point, due to their strong flux pinning ability and levitation stiffness.

(a) Launch system including brake. (b) The PMG and linear motor of the launch system, besides the HTS Maglev vehicle "Century". (c) Launch system with adjustable launch angle.

**Fig. 10.1:** The photographs of the HTS Maglev launch system [17]. (a) The launch system including the brake. (b) PMG and linear motor, with the HTS Maglev vehicle "Century" shown on top. The top right corner is the linear motor control set. (c) Launch system with the adjustable launch angle of 0–30 degrees.

## 10.4 Prototypes of HTS Maglev launch system

Some prototypes of the HTS Maglev launch systems were constructed to test and verify their operation ability. A laboratory-scale HTS Maglev launch test system [17] was developed based on the previous vehicle development experience [18] in the ASCLab, SWJTU. Basic design considerations for the HTS Maglev launch system include two parallel PMGs of about 10 m, a levitation force of 500–600 N, a guidance force of 60 N, and an acceleration of 1–4 *g*.

First, for a HTS Maglev launch system a series of specialized technologies based on the research of the levitation and guidance between YBCO and the PMG must be considered. These include stability and reliability at high acceleration, the low-temperature cooling device, integration of the vehicle and the low-temperature device, tests of the HTS Maglev climbing up a large slope, techniques of HTS Maglev in an evacuated tube transport (ETT), etc. Therefore, the most important mission is to study the Maglev system efficiency of high acceleration with a new drive technology using HTS bulks and PMG.

This HTS Maglev launch system has no intrinsic difference compared to the HTS Maglev vehicle [11]. The launch system includes the vehicle, the NdFeB PMG, and the linear motor. The NdFeB PMG is 10.05 m long and 0.68 m wide. There is a linear motor drive set between the two parallel PMG. The frame of the PMG is 11.2 m long and 1.0 m wide. The PMG has a longitudinal slope of 0–30 degrees from the horizontal, and a transverse slope of 15 degrees.

The HTS Maglev launch test vehicle is 1.0 m long and 1.0 m wide. There are four on-board levitation devices with HTS YBCO bulks in the launch vehicle. The gap between the linear motor and its induction plane is 10 mm. Each linear motor is 695 mm long, 155 mm wide, and 80 mm thick. Total length of 15 single motors is about 10.4 m. The acceleration of the linear motor is 1–4 *g*. Figure 10.1 shows the photograph of the PMG of the HTS Maglev launch system. The linear motor is between the PMGs.

Tab. 10.1: Total levitation forces of four HTS Maglev devices on No. 3 PMG [17].

| Levitation gap (mm) | FCH 20 mm (N) | FCH 30 mm (N) | FCH 40 mm (N) | ZFC (N) |
|---|---|---|---|---|
| 20 | 73 | 213 | 3.8 | 506 |
| 15 | 230 | 422 | 534 | 743 |
| 10 | 506 | 743 | 866 | 1097 |
| 5 | 939 | 1191 | 1322 | 1540 |

To increase mechanical impact strength and decrease size and weight, both the launch vehicle and the low-temperature device should be formed as a whole. The low-temperature adiabatic device is a part of the entire launch vehicle. To increase dynamic strength, no liquid nitrogen cooling is used. There is no liquid nitrogen vessel, only a tube in the vehicle for external precooling with liquid nitrogen. The shape of the tube is determined by the application requirements. The liquid nitrogen flows into the adiabatic device through a tube before the experiment. The liquid nitrogen flow is stopped when the temperature of the YBCO bulk reaches 77 K.

The launch vehicle includes four HTS Maglev devices. Each HTS Maglev device is composed of 16 YBCO bulks of diameter 30 mm, insulation materials, and a cooling tube. In order to increase the mechanical impact and the dynamic strength, some special designs are used.

Levitation forces and guidance forces are measured by a self-developed HTS Maglev measurement system SCML-02 [19]. (see Section 5.3). The total levitation forces of four HTS Maglev devices on the No. 3 PMG (similar to the launch PMG) are listed in Tab. 10.1. The levitation gap is the distance between top surface of PMG and bottom surface of YBCO bulk. In the HTS Maglev launch test vehicle, the net levitation gap includes an insulation thickness of about 3–5 mm. The total levitation forces at the levitation gap of 15 mm are 422 N at the field cooling height (FCH) of 30 mm and 743 N at the zero field cooling height (ZFC). The total levitation forces at the levitation gap of 10 mm are 743 N at FCH of 30 mm and 1097 N at ZFC.

The total guidance forces, at FCH of 35 mm, are 66 N at a levitation gap of 15 mm and 86 N at a levitation gap of 10 mm. The above total levitation forces and guidance forces of the HTS Maglev launch test vehicle satisfy the design specification for a levitation force of 500–600 N and the guidance force of 60 N, which indicates the design validity of the HTS Maglev launch system. Research results of quasi-static HTS on PMG in the past several years provide strong technical support for the development of the HTS Maglev launch system.

The launch vehicle of liquid nitrogen-free or solid-state nitrogen is a good method. The method merit is to eliminate fluctuation of liquid nitrogen when the launch vehicle moves. This fluctuation will seriously affect the stability of launch vehicle. The solid-state nitrogen can be acquired by pumping air to reduce the pressure within the cryostat of launch body.

Another HTS Maglev assist launch system was demonstrated by Yang et al., at Beihang University (BUAA) [20]. The PMG of BUAA had a different configuration from that of ASCLab. It consisted of three PMs with opposite directions two iron plates. The maximum field values for these two iron plates were up to –1.5 T and 1.5 T respectively. The configuration of three PMs with opposite directions also brought a great lateral magnetic gradient which benefits the lateral stability of HTS Maglev system.

The entire PMG of BUAA was 10.05 m long and 0.68 m wide. The propulsion system was a linear induction motor (LIM), which is set between these two PMGs. The LIM consisted of 15 segments with the total length 10.4 m. The LIM was automatically controlled and could switch quickly between the acceleration and deceleration state. By the control of input current, the acceleration of the linear motor could be adjusted to be 1–4 *g*.

The vehicle of HTS Maglev launch system was 1.0 m long and 1.0 m wide. The launch vehicle contained four levitation cryostats with 36 YBCO bulks fixed inside. The YBCO bulk had a diameter of 30 mm and thickness of 18 mm. At a levitation gap of 15 mm, the total levitation force of the launch vehicle was 422 N and 743 N, at FCH of 30 mm and ZFC, respectively. At the levitation gap of 10 mm, the total levitation force of the launch vehicle was 743 N and 1097 N, at FCH of 30 mm and ZFC, respectively. At FCH of 35 mm, the total guidance force of the launch vehicle was 66 N and 86 N, at the levitation gap of 15 mm and 10 mm, respectively. The vehicle of the HTS Maglev launch system had a weight of 140 kg, which could reach a levitation gap of 6 mm at the 35 mm FCH. The maximal lateral force of the launch vehicle was 150 N at the levitation gap of 6 mm and the lateral displacement of 5 mm.

The group of Yang studied the effect of the FCH and the load weight on the dynamic force properties of a HTS Maglev system, including the resonant frequency and the stiffness in both horizontal and vertical vibrations [21]. Experimental results showed that the resonant frequency increased with decrease of the FCH or the load weight. At the 25-mm FCH, the resonant frequency approached a maximum value of 10 Hz. The analyses and discussions showed that the dynamic properties of the test vehicle were closely related to the motion process and the flux history in the superconductors. The effectiveness of lower FCH and preloading method to increase the dynamic stability of HTS Maglev system was tested and verified by this group.

Yang's group also investigated the damping properties of the HTS Maglev launch system at different operation conditions [22]. The experimental results showed a weak damping behavior of the HTS Maglev vehicle, and indicated that the damping of the Maglev vehicle was almost unaffected by the trapped flux of the bulks HTSC. The analyses indicated that the weak damping behavior of the HTS Maglev system derived from the strong flux pinning ability of the melt-textured bulk HTSC. In practical applications, it is better to introduce an additional damping device to increase the operation stability of the HTS Maglev system.

Although the muzzle velocities of the HTS Maglev launch test systems at the ASC Lab and BUAA are not very high, the basic application characteristics have

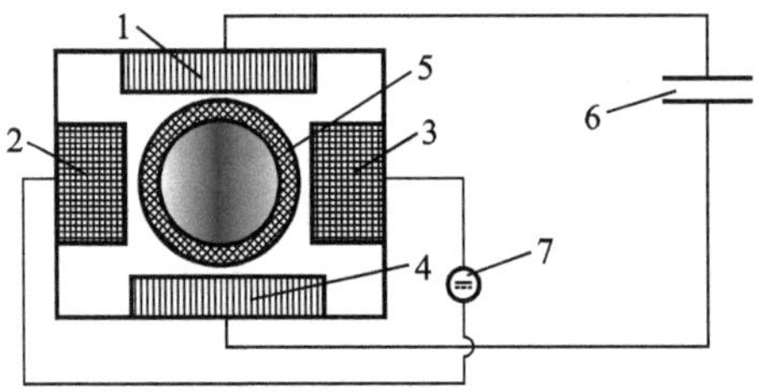


**Fig. 10.2:** The cross section of HTS REL system 1. Upper propulsion coil. 2. Left Maglev and balance coil. 3. Right Maglev and balance coil. 4. Bottom propulsion coil. 5. Projectile. 6. AC power supply. 7. DC power supply (W. Liu, J.S. Wang, G.T. Ma, et al., unpublished data).

been verified by those launch test systems. The experimental results and analyses indicate the HTS Maglev system is a good candidate for a REL launch system of very high speed.

## 10.5 Concepts of HTS REL system

Based on the LTS REL system, J. S. Wang et al. proposed a new REL system with the introduction of HTS technology (W. Liu, J.S. Wang, G.T. Ma, et al., unpublished data). As Fig. 10.2 shows, the HTS REL system consisted of seven parts: parts 1 and 4 were the Maglev and propulsion coils, respectively, which were used to produce the magnetic field for suspension and propulsion; parts 2 and 3 were the balance coils, which were used to produce a large magnetostatic field to counter the centrifugal force; part 5 was the projectile, which was made of bulk HTSC and with a hollow center to store liquid nitrogen or another cryogen. Through a special design of the configuration of the magnetic field, the HTS projectile could maintain stable suspension with parts 2 and 3 coils because of the flux pinning ability. Furthermore, when the projectile suffered a huge centrifugal force under high speed and produced a lateral displacement, a large supercurrent would be induced within the projectile because of the large field gradient from parts 2 and 3 coils. The supercurrent within the projectile combined with the magnetic field from parts 2 and 3 coils would produce a huge magnetic force to counter the centrifugal force.

To keep the stable suspension of projectile, the propulsion coils were designed symmetrically in the vertical direction, as shown in Fig. 10.2. An eddy current would be induced within the HTS projectile by the travelling wave field of parts 1 and 4 coils, which produced the propulsion forces. The propulsion coils require an AC power supply, and the Maglev & balance coils required a DC power supply, as Fig. 10.2 shows. The coils of parts 1-4 were fabricated with the HTS materials, which had a higher working temperature and were convenient to the design configuration.

Figure 10.3 shows the top view of entire structure of the circular HTS REL system. Part 1 is the balance guideway in the outlet barrel, which was used to keep stable suspension of the projectile. Parts 2 and 4 are the acceleration segments. Part 3 is the acceleration coil in the entrance barrel, which was used to accelerate the projectile to its original speed. Part 3 also includes the balance guideway to keep stable suspension of the projectile. Part 5 is the position detection device, which supplies the release

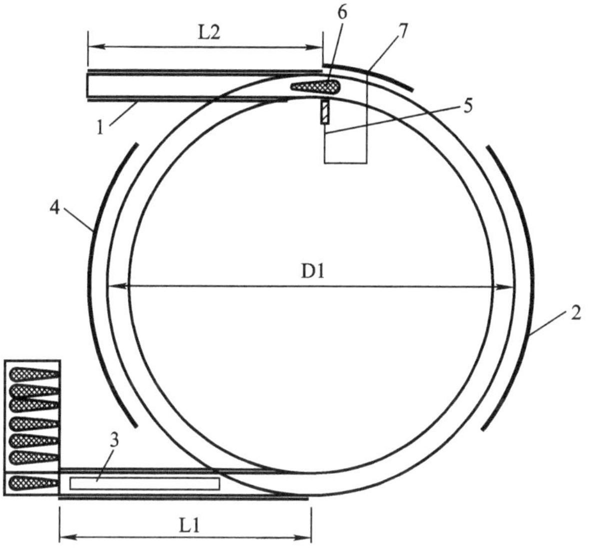


**Fig. 10.3:** Top view of the HTS REL system (W. Liu, J.S. Wang, G.T. Ma, et al., unpublished data).

signal of projectile to part 7, the electromechanical switching. Part 6 is the HTS projectile.

As Fig. 10.3 shows, only parts 2 and 4 segments were used to accelerate the projectile, which was significant in reducing the cost of the propulsion system. The difficult technology of this system was the AC power supply for the propulsion coils. For example, if the required muzzle velocity was 10 km/s, the pole pitch of the linear motor was 0.1 m, and then the synchronous frequency of linear motor was 50 kHz. For an inverter based on variable voltage, variable frequency technology, it was assumed that the carrier wave ratio was 20 and that the switching frequency of the power electronics component was 1 MHz, which was difficult to achieve especially under the high power level.

For this particular application, there are still some problems need to be solved. One is the huge electromagnetic stress produced by the large magnetostatic field of the balance coils and the supercurrent in the propulsion and balance coils. This huge electromagnetic stress may cause serious damage to the braced structures. Another problem is the large centrifugal force on the projectile, which can lead to serious overload and possible damage to the folding antennae, guided device, airfoil, inflatable structure etc. It is necessary to design the launch rings with a large enough diameter to reduce the damage from overload.

## 10.6 Studies on HTS linear synchronous motor

Linear synchronous motor (LSM) is the common drive system to a Maglev train. The on-board LSM motor may constitute nearly one third of the total weight of a Maglev

**Tab. 10.2:** Dimension parameters of three coils [23].

| Coil shape | Circular-shaped single pancake | Circular-shaped double pancake | Racetrack-shaped single pancake |
|---|---|---|---|
| Inner diameter (mm) | 38<br>– | 38<br>– | Long axis: 75<br>Short axis: 12 |
| Outer diameter (mm) | 41.5<br>– | 41.5<br>– | Long axis: 103.5<br>Short axis: 41.5 |
| Height of the coil (mm) | 4.1 | 9.2 | 4.1 |
| Turns | 12 | 24 | 100 |
| Length of the HTS type (mm) | 183 | 337 | 339 |
| Critical current (A) | 66.5 | 67.05 | 64.2 |

train. One way to reduce the size or weight of the motor is the usage of superconducting materials. A LSM prototype with a high-temperature superconducting excitation system has been developed to conceptually study operation features and to investigate the feasibility for its application in Maglev trains. This LSM prototype only makes use of superconducting components in the excitation system (i.e., the field windings), which is also the moving part of the system. In most LSM systems, the excitation system contains either magnets or electromagnets. When the stator generates a moving magnetic field, the magnetic field of the excitation system is forced to move, with a velocity equal to the frequency of the moving magnetic field multiplied by the pole pitch. The three-phase stator is made of conventional copper coils with a coil pitch of $w = 42$ mm and iron teeth to enhance the magnetic flux density along the $z$-axis, $B_z$.

### 10.6.1 The 2G HTS coil-type excitation system

Superconducting wire with conducts higher current replaces the copper wiring of the electromagnets of the moving part. The introduction of superconducting wire gives a stronger thrust force while reducing weight and increasing the power density. The superconducting coil is made of a Ru-doped YBCO-coated conductor since it has been found that ruthenium enhances the in-field performance of the critical current of 2G superconducting wires.

Single pancake and double pancake are two basic structures for the coil made of the superconducting tape. Three separate coils with different shapes were studied: circular-shaped single pancake, circular-shaped double pancake and racetrack-shaped single pancake [23]. The dimension parameters are shown in the Tab. 10.2. A power supply was employed to conduct DC current to the superconducting coil. The profiles of the magnetic field component along the $z$-direction and the $x$-direction generated by the racetrack-shaped superconducting coil, across its $x$-axis at $I_{DC} = 40$ A and a distance of 10 mm, are displayed in Fig. 10.4.

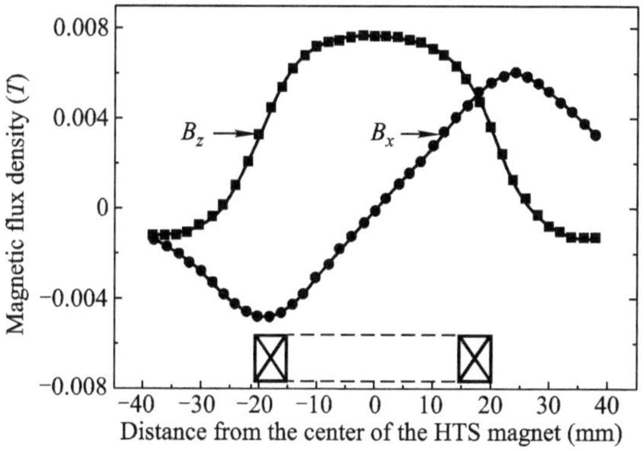


**Fig. 10.4:** Profile of the magnetic field generated by the racetrack-shaped single-pancake superconducting coil magnet when $I_{DC} = 40$ A measured at $z = 10$ mm [24].

Another superconducting coil was a single-pancake coil which consisted of 100 turns having the shape of a racetrack, 4.1 mm in height, with outer dimensions of 103.5 mm and 41.5 mm and inner dimensions of 75 and 12 mm [25]. The profiles of the magnetic field component along the $z$-direction and the $x$-direction generated by the superconducting coil magnet across its $x$-axis at $I_{DC} = 40$ A and a distance of 10 mm are displayed in Fig. 10.5. The maximum value of the magnetic flux density increased with addition of iron components to the superconducting coil magnet.

Four single-pancake racetrack coils were fabricated and connected in series. They consisted of a four-pole excitation system with a coil pitch of three times wider than the width of the coils [26–28]. The coil form was made of paramagnetic stainless steel with a hollow middle so that silicon steel laminations could be placed for further testing. The thickness of the coil form was 5 mm since the width of the coated conductor (CC) was 4.2 mm. The coils were flat and thin since each coil consisted only of $N = 10$ turns with one layer of Kapton insulation tape on each subsequent turn. The total thickness of the CC plus the insulation layer > 0.2 mm. The coils had the following dimensions: inner short radius $a_1 = 19$ mm, outer short radius $a_2 = 21$ mm, inner long radius $c_1 = 75.5$ mm, outer radius $c_2 = 77.5$ mm, and a height $2b = 4.2$ mm (Fig. 10.6a and b). A total of less than 16 m of CC wire was used for all four coils combined including all current leads and links between adjacent coils.

The critical current ($I_c$) of each coil was measured beforehand. Even though a higher standard was normally set on superconducting coils with a critical electric field of $E_c = 0.1$ V/cm, we set our $I_c$ to $E_c = 1$ V/cm since each coil used only consisted of 10 turns. The $n$-value was determined from the slope of the logarithm of current versus the logarithm of voltage near the critical current value. Even though direct current was applied to the superconducting coils, we plan to use alternating current in these same coils in the future for other experiments so their inductance was also measured and listed in Tab. 10.3. To determine the inductance, the voltage generated by each coil was

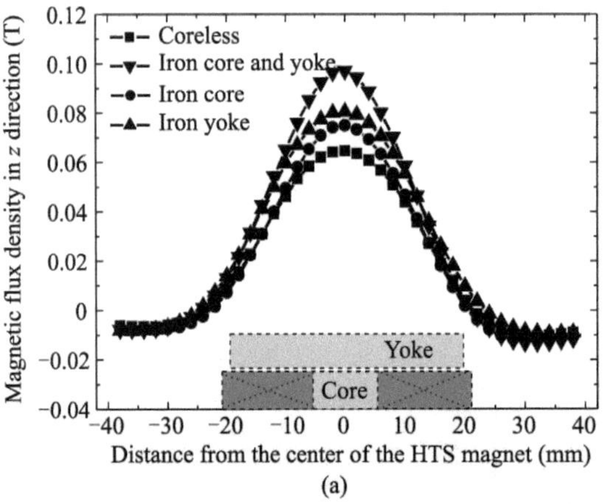


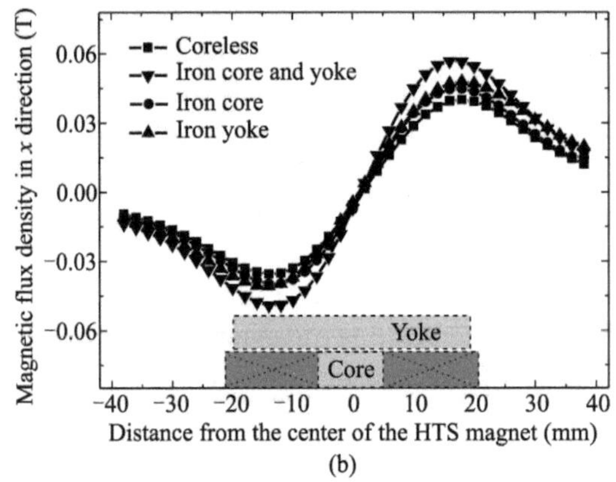


**Fig. 10.5:** Profile of the magnetic field generated by the superconducting coil magnet when $I_{DC} = 40$ A measured at $z = 10$ mm along the $z$-direction (a) and the $x$-direction [24] (b).

measured while the transport current varied at a constant rate. The ramping speeds were −1.0, −2.0, 1.0, and 2.0 A/s. The inductance value obtained for each speed was averaged. The center field along the $z$-axis of the coil was measured with a Hall probe at various current values. The field constant was determined by taking the slope of the center field versus the input current.

The coils were connected in series with subsequent opposite polarities making it a four pole excitation system with a pole pitch of 126 mm (Fig. 10.7). The excitation system was submerged in liquid nitrogen inside a homemade 'cryostat' made of Styrofoam reinforced with a silicon based low-temperature adhesive.

The three-phase stator was made of conventional copper coils with a coil pitch of $w = 42$ mm with iron teeth as its cores for the enhancement of the magnetic field density along the $z$-axis, $B_z$. At 10 mm above the surface of the stator, alternating current with $I_{AC} = 8$ A generated $B_z = 368$ Gauss. All experiments discussed in this